D1407349

HUGHES
ELECTRICAL TECHNOLOGY

HUGHES
ELECTRICAL TECHNOLOGY

EDWARD HUGHES

revised by Ian McKenzie Smith

B.Sc., Dip.A.Ed., C.Eng., M.I.E.E., M.I.E.R.E., F.I.Elec.I.E.
Depute Principal
Stow College
Glasgow

Sixth Edition

Copublished in the United States with
John Wiley & Sons, Inc., New York

Longman Scientific & Technical,
Longman Group UK Limited,
Longman House, Burnt Mill, Harlow,
Essex CM20 2JE, England
and Associated Companies throughout the world.

Copublished in the United States with
John Wiley & Sons, Inc., 605 Third Avenue, New York, NY 10158

© Longman Group Limited 1960
This edition © Longman Group UK Limited 1987

First Published 1960
Second Edition 1963
Third Edition 1967
Fourth Edition 1969
Fifth Edition 1977
Sixth Edition 1987
Second impression 1988

British Library Cataloguing in Publication Data
Hughes, Edward, *b. 1888—*
    Hughes electrical technology.—6th ed.
    1. Electrical engineering
    1. Title   II. McKenzie Smith, Ian
    III. Hughes, Edward *b. 1888*   Electrical
    technology
    621.3   TK 145
    ISBN 0-582-41372-9

Library of Congress Cataloging in Publication Data
Hughes, Edward, 1888—
    Hughes electrical technology.
    Includes index.
    1. Electric engineering.   2. Electronics.
I. Smith, Ian McKenzie.   II. Title.
TK 146.H9 1987   621.3   86-21483
ISBN 0-470-20733-7 (USA only)

Set in 10/11 pt Monophoto Times Roman
Produced by Longman Group (FE) Limited
Printed in Hong Kong

Contents

Preface to Sixth Edition

Electrical and electronic engineering continues to develop most rapidly and the sixth edition has taken this into account by encompassing a major review of the contents. In particular, a greatly increased emphasis has been placed on fundamental semiconductor electronic devices so that field effect transistors and operational amplifiers have been added to the junction transistor devices previously covered in the fifth edition. Chapters have also been added to cover digital systems and an introduction to microprocessors and their programming.

Electrical machines continue to hold an important place within the sixth edition but in recognition of the changes which have taken place in the control of machines, an extra chapter has been included to cover power electronic systems including the thyristor. All new material is supported by a substantial number of problems drawn from examination papers, and I am very grateful to SCOTVEC for their permission to use these questions.

The sixth edition comes in a completely new format with all the diagrams redrawn to comply with current practice. This volume therefore contains a complete introduction to the art of electrical and electronic engineering as practised in the late 1980's.

November 1986   IAN McKENZIE SMITH

Preface to First Edition

This volume covers the electrical engineering syllabuses of the Second and Third Year Courses for the Ordinary National Certificate in Electrical Engineering and of the First Year Course leading to a Degree of Engineering.

The rationalized M.K.S. system of units has been used throughout this book. The symbols, abbreviations and nomenclature are in accordance with the recommendations of the British Standards Institution; and, for the convenience of students, the symbols and abbreviations used in this book have been tabulated on pages xiii–xvii.

It is impossible to acquire a thorough understanding of electrical principles without working out a large number of numerical problems; and while doing this, students should make a habit of writing the solutions in an orderly manner, attaching the name of the unit wherever possible. When students tackle problems in examinations or in industry, it is important that they express their solutions in a way that is readily intelligible to others, and this facility can only be acquired by experience. Guidance in this respect is given by the 106 worked examples in the text, and the 670 problems afford ample opportunity for practice.

Most of the questions have been taken from examination papers; and for permission to reproduce these questions I am indebted to the University of London, the East Midland Educational Union, the Northern Counties Technical Examination Council, the Union of Educational Institutions and the Union of Lancashire and Cheshire Institutes.

I wish to express my thanks to Dr. F. T. Chapman, C.B.E., M.I.E.E., and Mr. E. F. Piper, A.M.I.E.E., for reading the manuscript and making valuable suggestions.

Hove,
April 1959

EDWARD HUGHES

Symbols, Abbreviations and Definitions

Based upon PD 5686: 1972 (BSI) and *The International System of Units* (1970), prepared jointly by the National Physical Laboratory, UK, and the National Bureau of Standards, USA, and approved by the International Bureau of Weights and Measures.

Notes on the use of symbols and abbreviations

1. A unit symbol is the same for the singular and the plural: for example, 10 kg, 5 V; and should be used only after a numerical value: for example, *m* kilograms, 5 kg.

2. Full point should be omitted after a unit symbol: for example, 5 mA, 10 μF, etc.

3. Full point should be used in a multi-word abbreviation: for example, e.m.f., p.d.

4. In a compound unit-symbol, the product of two units is preferably indicated by a dot, especially in manuscript. The dot may be dispensed with when there is no risk of confusion with another unit symbol: for example, N · m or N m, but not mN. A solidus (/) denotes division: for example, J/kg, r/min.

5. Only one multiplying prefix should be applied to a given unit: for example, 5 megagrams, not 5 kilokilograms.

6. A prefix should be applied to the numerator rather than the denominator: for example, MN/m^2 rather than N/mm^2.

7. The abbreviated forms a.c. and d.c. should be used only as adjectives: for example, d.c. motor, a.c. circuit.

8. A hyphen is inserted between the numerical value and the unit when the combination is used adjectivally: for example, a 240-volt (or 240-V) motor, a 2-ohm (or 2-Ω) resistor.

Units of length, volume, mass and time

Quantity	Unit	Symbol	Quantity	Unit	Symbol
Length	metre	m	Volume	cubic metre	m^3
	kilometre	km		litre (0.001 m^3)	l
Mass	kilogram	kg	Time	second	s
	megagram	Mg		minute	min
	or tonne (1 Mg)	t		hour	h

Electricity and Magnetism

Quantity	Quantity symbol	Unit	Unit symbol
Admittance	Y	siemens	S
Angular velocity	ω	radian per second	rad/s
Capacitance	C	farad	F
		microfarad	μF
		picofarad	pF
Charge or Quantity of electricity	Q	coulomb	C
Conductance	G	siemens	S
Conductivity	σ	siemens per metre	S/m
Current			
Steady or r.m.s. value	I	ampere	A
		milliampere	mA
		microampere	μA
Instantaneous value	i		
Maximum value	I_m		
Current density	J	ampere per square metre	A/m^2
Difference of potential			
Steady or r.m.s. value	V	volt	V
		millivolt	mV
		kilovolt	kV
Instantaneous value	v		
Maximum value	V_m		
Electric field strength	E	volt per metre	V/m
Electric flux	Ψ	coulomb	C
Electric flux density	D	coulomb per square metre	C/m^2
Electromotive force			
Steady or r.m.s. value	E	volt	V
Instantaneous value	e		
Maximum value	E_m		
Energy	W	joule	J
		kilojoule	kJ
		megajoule	MJ
		watt hour	W·h
		kilowatt hour	kW·h
		electronvolt	eV
Force	F	newton	N
Frequency	f	hertz	Hz
		kilohertz	kHz
		megahertz	MHz

Quantity	Quantity symbol	Unit	Unit symbol
Impedance	Z	ohm	Ω
Inductance, self	L	henry (plural, henrys)	H
Inductance, mutual	M	henry (plural, henrys)	H
Magnetic field strength	H	ampere per metre	A/m
Magnetic flux	Φ	weber	Wb
Magnetic flux density	B	tesla	T
Magnetomotive force	F	ampere	A
Permeability of free space or Magnetic constant	μ_0	henry per metre	H/m
Permeability, relative	μ_r		
Permeability, absolute	μ	henry per metre	H/m
Permittivity of free space or Electric constant	ε_0	farad per metre	F/m
Permittivity, relative	ε_r		
Permittivity, absolute	ε	farad per metre	F/m
Power	P	watt	W
		kilowatt	kW
		megawatt	MW
Reactance	X	ohm	Ω
Reactive voltampere	—	var	var
Reluctance	S	ampere per weber	A/Wb
Resistance	R	ohm	Ω
		microhm	$\mu\Omega$
		megohm	$M\Omega$
Resistivity	ρ	ohm metre	$\Omega \cdot m$
		microhm metre	$\mu\Omega \cdot m$
Susceptance	B	siemens	S
Torque	T	newton metre	$N \cdot m$
Voltampere	—	voltampere	VA
		kilovoltampere	kVA
Wavelength	λ	metre	m
		micrometre	μm

Light

Quantity	Quantity symbol	Unit	Unit symbol
Illuminance	E	lux	lx
Luminance (objective brightness)	L	candela per square metre	cd/m^2
Luminous flux	Φ	lumen	lm
Luminous intensity	I	candela	cd
Luminous efficacy	—	lumen per watt	lm/W

Selection of graphical symbols from BS 3939

Description	Symbol
Direct current or steady voltage	—
Alternating	~
Positive polarity	+
Negative polarity	—
Primary or secondary cell	
Battery of primary or secondary cells	
Fixed resistor	
Variable resistor	
Resistor with moving contact	
Filament lamp	
Crossing of conductor symbols on a diagram (no electrical connection)	
Junction of conductors	
Double junction of conductors	
Earth	
Capacitor: general symbol	
Polarized capacitor	
Winding	
Inductor and core	
Transformer	
Ammeter	Ⓐ
Voltmeter	Ⓥ
Wattmeter	Ⓦ

Description	Symbol
Galvanometer	
Motor	
Generator	
Make contact (normally open)	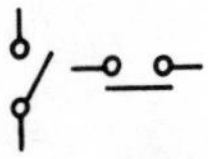
Break contact (normally closed)	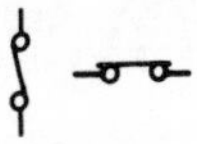
Rectifier	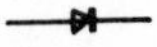
Zener diode	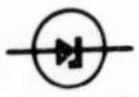
p–n–p transistor	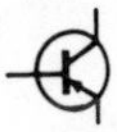
n–p–n transistor	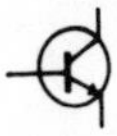
Amplifier	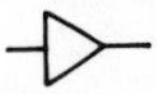

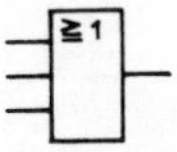

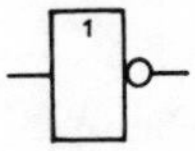

Binary logic elements
- AND
- OR
- NOT
- Exclusive OR
- Logic identity
- NAND
- NOR

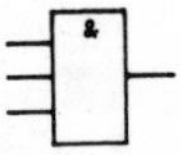

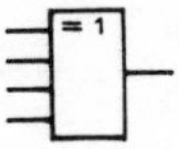

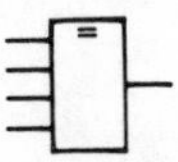

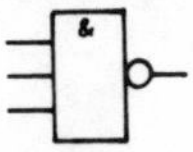

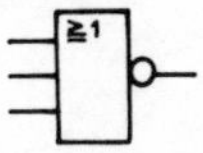

Electrical Machines

Term	Symbol
Number of armature conductors	Z
Number of commutator segments or bars	C
Number of pairs of poles	p
Number of parallel circuits	c
Number of phases	m
Number of turns	N

Abbreviations for Multiples and Sub-multiples

T	tera	10^{12}
G	giga	10^{9}
M	mega or	
	meg	10^{6}
k	kilo	10^{3}
d	deci	10^{-1}
c	centi	10^{-2}
m	milli	10^{-3}
μ	micro	10^{-6}
n	nano	10^{-9}
p	pico	10^{-12}

Greek letters used as symbols

Letter	Capital	Small
Alpha	—	α (angle, temperature coefficient of resistance, current amplification factor for common-base transistor)
Beta	—	β (current amplification factor for common-emitter transistor)
Delta	Δ (increment, mesh connection)	δ (small increment)
Epsilon	—	ϵ (permittivity)
Eta	—	η (efficiency)
Theta	—	θ (angle, temperature)
Lambda	—	λ (wavelength)
Mu	—	μ (micro, permeability, amplification factor)
Pi	—	π (circumference/diameter)
Rho	—	ρ (resistivity)
Sigma	Σ (sum of)	σ (conductivity)
Phi	Φ (magnetic flux)	ϕ (angle, phase difference)
Psi	Ψ (electric flux)	—
Omega	Ω (ohm)	ω (solid angle, angular velocity, angular frequency)

Miscellaneous

Term	Symbol	Term	Symbol
Approximately equal to	$\simeq$	Much less than	$\ll$
Proportional to	$\propto$	Base of natural logarithms	e
Infinity	∞	Common logarithm of x	$\log x$
Sum of	Σ	Natural logarithm of x	$\ln x$
Increment or finite	Δ, δ	Complex operator $\sqrt{-1}$	j
difference operator		Temperature	θ
Greater than	$>$	Time constant	T
Less than	$<$	Efficiency	η
Much greater than	$\gg$	Per unit	p.u.

Electronic semiconductor devices based on BS 3363

Current amplification factor for common-base transistor circuit $\qquad$ α

Current amplification factor for common-emitter transistor circuit $\qquad$ β

Subscripts for quantity symbols:

Anode	A, a
Cathode	K, k
Control terminal	G, g
Emitter	E, e
Base	B, b
Collector	C, c

D.C. and average values are indicated by upper-case subscripts, e.g.

I_C = direct collector current (no signal),

V_{CB} = direct collector-base p.d. in a transistor circuit (no signal).

Values of varying components are indicated by lower-case subscripts, e.g.

i_a and I_a = instantaneous and r.m.s. values respectively of the varying component of anode current,

v_{cb} and V_{cb} = instantaneous and r.m.s. values respectively of the varying component of the p.d. between collector and base in a transistor-circuit.

Sequence of subscripts: the first subscript denotes the terminal at which the current is measured, or where the terminal potential is measured with respect to the reference terminal; e.g. V_{CB} represents the direct potential of the collector with reference to the base in a transistor circuit.

Definitions of electric and magnetic SI units

The *ampere* (A) is that constant *current* which, if maintained in two straight parallel conductors of infinite length, of negligible circular cross-section, and placed 1 m apart in vacuum, would produce between these conductors a force equal to 2×10^{-7} N per metre of length.

The *coulomb* (C) is the *quantity of electricity* transported in 1 s by 1 A.

The *volt* (V) is the *difference of electrical potential* between two points of a conductor carrying a constant current of 1 A, when the power dissipated between these points is equal to 1 W.

The *ohm* (Ω) is the *resistance* between two points of a conductor when a constant difference of potential of 1 V, applied between these points, produces in this conductor a current of 1 A, the conductor not being a source of any electromotive force.

The *henry* (H) is the *inductance* of a closed circuit in which an e.m.f. of 1 V is produced when the electric current in the circuit varies uniformly at the rate of 1 A/s.

(*Note:* this also applies to the e.m.f. in one circuit produced by a varying current in a second circuit, i.e. mutual inductance.)

The *farad* (F) is the *capacitance* of a capacitor between the plates of which there appears a difference of potential of 1 V when it is charged by 1 C of electricity.

The *weber* (Wb) is the *magnetic flux* which, linking a circuit of one turn, produces in it an e.m.f. of 1 V when it is reduced to zero at a uniform rate in 1 s.

The *tesla* (T) is the *magnetic flux density* equal to 1 Wb/m^2.

Definitions of other derived SI units

The *newton* (N) is the *force* which, when applied to a mass of 1 kg, gives it an acceleration of 1 m/s^2.

The *pascal* (Pa) is the *stress* or *pressure* equal to 1 N/m^2.

The *joule* (J) is the *work done* when a force of 1 N is exerted through a distance of 1 m in the direction of the force.

The *watt* (W) is the *power* equal to 1 J/s.

The *hertz* (Hz) is the unit of *frequency*, namely the number of cycles per second.

The *lumen* (lm) is the *luminous flux* emitted within unit solid angle by a point source having a uniform intensity of 1 candela.

The *lux* (lx) is an *illuminance* of 1 lm/m^2.

1 The Electric Circuit

1.1 The International System of Units

The International System of Units, known as SI in every language, derives all the units in the various technologies from the following *seven* base units:

Quantity	Symbol	Unit	Abbreviation
length	l	metre	m
mass	m	kilogram	kg
time	t	second	s
electric current	I	ampere	A
temperature	T	kelvin	K
luminous intensity		candela	cd
amount of substance		mole	mol

The *metre* is the length equal to 1 650 763.73 wavelengths of the orange line in the spectrum of an internationally-specified krypton discharge lamp.

The *kilogram* is the mass of a platinum-iridium cylinder preserved at the International Bureau of Weights and Measures at Sèvres, near Paris.

The *second* is the interval occupied by 9 192 631 770 cycles of the radiation corresponding to the transition of the caesium-133 atom.

The *ampere* is defined on page 6.

The *kelvin* is 1/273.16 of the thermodynamic temperature of the triple point of water. On the Celsius scale, the temperature of the triple point of water is 0.01°C,

hence, $$0°C = 273.15 \text{ K}.$$

A temperature interval of 1°C = a temperature interval of 1 K.

The *candela* is defined on page 619.

The *mole* does not affect the units used in this book.

1.2 Unit of force

The SI unit of force is the *newton* (N), namely the *force which, when applied to a mass of 1 kg, gives it an acceleration of* 1 m/s^2. Hence the force F required to give a mass m an acceleration a is:

$$F \text{ [newtons]} = m \text{ [kilograms]} \times a \text{ [metres per second}^2] \quad (1.1)$$

Weight. The weight of a body is the gravitational force exerted by the earth on that body. Owing to the variation in the radius of the earth, the gravitational force on a given mass, at sea level, is different at different latitudes, as shown in fig. 1.1. It will be seen that the weight of a 1-kg mass at

Fig. 1.1 Variation of weight with latitude

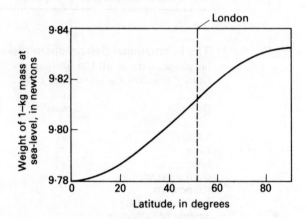

sea level in the London area is practically 9.81 N. For most purposes we can assume:

$$\text{the weight of a body} \simeq 9.81m \text{ newtons} \quad (1.2)$$

where m is the mass of the body in kilograms.

Example 1.1	*A force of* 50 N *is applied to a mass of* 200 kg. *Calculate the acceleration.*

Substituting in expression (1.1), we have:

$$50 \text{ [N]} = 200 \text{ [kg]} \times a$$

$$\therefore \quad a = 0.25 \text{ m/s}^2.$$

Example 1.2	*A steel block has a mass of* 80 kg. *Calculate the weight of the block at sea level in the vicinity of London.*

Since the weight of a 1-kg mass $\simeq 9.81$ N,

$$\therefore \quad \text{weight of the steel block} = 80 \text{ [kg]} \times 9.81 \text{ [N/kg]*}$$

$$= 784.8 \text{ N}.$$

---

* This expression can alternatively be stated thus:

$$\text{weight of block} \simeq 80 \text{ [kg]} \times 9.81 \text{ [m/s}^2] = 784.8 \text{ N}.$$

When the body under consideration is *stationary*, [N/kg] is better since it emphasizes the relationship between mass and weight, namely that a mass of 1 kg has a weight of approximately 9.81 N.

1.3 Unit of turning moment or torque

If a force F, in newtons, is acting at right angles to a radius r, in metres, from a point,

the turning moment or torque about that point

$$= Fr \text{ newton metres} \quad (1.3)$$

1.4 Unit of work or energy

The SI unit of energy is the *joule** (after the English physicist, James P. Joule, 1818–89). *The joule is the work done when a force of 1 N acts through a distance of 1 m in the direction of the force.* Hence, if a force F acts through distance d in its own direction,

$$\text{work done} = F \text{ [newtons]} \times d \text{ [metres]}$$

$$= Fd \text{ joules} \quad (1.4)$$

If a body having mass m, in kilograms, is moving with velocity v, in metres per second,

$$\text{kinetic energy} = \tfrac{1}{2}mv^2 \text{ joules} \quad (1.5)$$

If a body having mass m, in kilograms, is lifted vertically through height h, in metres, and if g is the gravitational acceleration, in metres per second^2, in that region,

the potential energy acquired by the body

$$= \text{work done in lifting the body}$$

$$= mgh \text{ joules}$$

$$\simeq 9.81mh. \quad (1.6)$$

Example 1.3   *A body having a mass of 30 kg is supported 50 m above the earth's surface. What is its potential energy relative to the ground?*
   If the body is allowed to fall freely, calculate its kinetic energy just before it touches the ground. Assume gravitational acceleration to be 9.81 m/s^2.

$$\text{Weight of body} = 30 \text{ [kg]} \times 9.81 \text{ [N/kg]} = 294.3 \text{ N},$$

$$\therefore \qquad \text{potential energy} = 294.3 \text{ [N]} \times 50 \text{ [m]} = 14\,715 \text{ J}.$$

   If v be the velocity of the body after it has fallen a distance s with an acceleration g,

$$v = \sqrt{(2gs)} = \sqrt{(2 \times 9.81 \times 50)} = 31.32 \text{ m/s},$$

and

$$\text{kinetic energy} = \tfrac{1}{2} \times 30 \text{ [kg]} \times (31.32)^2 \text{ [m/s]}^2 = 14\,715 \text{ J}.$$

Hence the whole of the initial potential energy has been converted into kinetic energy. When the body is finally brought to rest by

---

* *Joule* is pronounced 'jool', rhyming with 'tool'.

impact with the ground, practically the whole of this kinetic energy is converted into heat.

Specific heat capacity   Different substances absorb different amounts of heat to raise the temperature of a given mass of the substance by one degree. The heat required to raise the temperature of 1 kg of a substance by 1 degree is termed the *specific heat capacity* of that substance. Hence, if *c* represents the specific heat capacity of a substance in joules per kilogram kelvin, the heat required to raise the temperature of *m* kilograms of the substance by θ degrees

$$= mc\theta \text{ joules} \tag{1.7}$$

The following table gives the approximate values of the specific heat capacity of some well-known substances for a temperature range between 0°C and 100°C:

Substance	Specific heat capacity
Water	4190 J/kg · K
Copper	390 J/kg · K
Iron	500 J/kg · K
Aluminium	950 J/kg · K

Example 1.4   *Calculate the mass of copper that can be heated from 10°C to 80°C by the absorption of 100 kJ of thermal energy.*

Increase of temperature = 80 − 10 = 70°C = 70 K.

Quantity of heat = 100 kJ = 100 000 J.

Assuming the specific capacity of copper to be 390 J/kg · K,

$$100\,000\,[\text{J}] = m \times 390\,[\text{J/kg} \cdot \text{K}] \times 70\,[\text{K}]$$

∴   *m* = 3.66 kg.

1.5 Unit of power

Since power is the rate of doing work, it follows that the SI unit of power is the *joule/second* or *watt* (after the Scottish engineer, James Watt, 1736–1819). In practice, the watt is often found to be inconveniently small and so the *kilowatt* is frequently used. Similarly, when we are dealing with a large amount of energy, it is often convenient to express the latter in *kilowatt hours* rather than in joules.

$$1 \text{ kW} \cdot \text{h} = 1000 \text{ watt hours}$$

$$= 1000 \times 3600 \text{ watt seconds or joules}$$

$$= 3\,600\,000 \text{ J} = 3.6 \text{ MJ}.$$

If T is the torque, in newton metres, due to a force acting about an axis of rotation, and if n is the speed in revolutions per second,

$$\text{power} = \text{torque in newton metres}$$

$$\times \text{ speed in radians per second}$$

$$= T \times 2\pi n \text{ joules per second or watts}$$

$$= \omega T \text{ watts} \tag{1.8}$$

where ω = angular velocity in radians per second.

Example 1.5 *A stone block, having a mass of 120 kg, is hauled 100 m in 2 min along a horizontal floor. The coefficient of friction is 0.3. Calculate (a) the horizontal force required, (b) the work done, and (c) the power.*

(a) Weight of stone $\simeq 120\,[\text{kg}] \times 9.81\,[\text{N/kg}] = 1177.2\,\text{N}$,

∴ force required $= 0.3 \times 1177.2\,[\text{N}] = 353.16\,\text{N}$.

(b) Work done $= 353.16\,[\text{N}] \times 100\,[\text{m}] = 35\,316\,\text{J}$
$$= 35.3\,\text{kJ}.$$

(c) $$\text{Power} = \frac{35\,316\,[\text{J}]}{(2 \times 60)\,[\text{s}]} = 294.3\,\text{W}.$$

Example 1.6 *An electric motor is developing 10 kW at a speed of 900 r/min. Calculate the torque available at the shaft.*

$$\text{Speed} = \frac{900\,[\text{r/min}]}{60\,[\text{s/min}]} = 15\,\text{r/s}.$$

Substituting in expression (1.8), we have:

$$10\,000\,[\text{W}] = T \times 2\pi \times 15\,[\text{r/s}]$$

∴ $$T = 106\,\text{N} \cdot \text{m}.$$

Example 1.7 *An electric heater is required to heat 15 litres of water from 12°C to the boiling point (100°C). Calculate (a) the electrical energy consumed (i) in megajoules, (ii) in kilowatt hours, and (b) the cost of the energy consumed if the charge is 4.7 p/kW · h. Assume the specific heat capacity of water to be 4190 J/kg · K, 1 litre of water to have a mass of 1 kg and the efficiency of the heater to be 0.85.*

(a) Mass of water = 15 kg.

Rise of temperature $= 100 - 12 = 88°\text{C} = 88\,\text{K}$.

From expression (1.7),

$$\text{useful heat} = 15\,[\text{kg}] \times 4190\,[\text{J/kg} \cdot \text{K}] \times 88\,[\text{K}]$$

$$= 5\,530\,000\,\text{J} = 5.53\,\text{MJ}.$$

Since $$\text{efficiency} = \frac{\text{useful energy}}{\text{energy absorbed}}$$

i.e. $$0.85 = \frac{5.53\,[\text{MJ}]}{\text{energy absorbed}}$$

$\therefore$ energy absorbed $= 6.51\ \text{MJ}$.

Since $1\ \text{kW} \cdot \text{h} = 3.6\ \text{MJ}$

$\therefore$ energy absorbed $= \dfrac{6.51\ [\text{MJ}]}{3.6\ [\text{MJ/kW} \cdot \text{h}]} = 1.8\ \text{kW} \cdot \text{h}$.

(b)   Cost of energy $= 1.8\ [\text{kW} \cdot \text{h}] \times 4.7\ [\text{p/kW} \cdot \text{h}] = 8.5\ \text{p}$.

1.6 Electrical units

(a) Current   The unit of current is the *ampere* and is one of the SI base units mentioned in section 1.1.

     The *ampere* is defined as that *current which, if maintained in two straight parallel conductors of infinite length, of negligible circular cross-section, and placed* 1 *metre apart in a vacuum, would produce between these conductors a force of* 2×10^{-7} *newton per metre of length*. The conductors are attracted towards each other if the currents are in the same direction, whereas they repel each other if the currents are in opposite directions.

     The value of the current in terms of this definition can be determined by means of a very elaborately constructed balance in which the force between fixed and moving coils carrying the current is balanced by the force of gravity acting on a known mass.

(b) Quantity of electricity   The unit of electrical quantity is the *coulomb*, namely the *quantity of electricity passing a given point in a circuit when a current of* 1 *ampere is maintained for* 1 *second*. Hence,

$$Q\ [\text{coulombs}] = I\ [\text{amperes}] \times t\ [\text{seconds}].$$

1 ampere hour $= 3600$ coulombs.

     When a current is passed through an electrolyte, chemical decomposition takes place, and the mass of an element liberated by 1 coulomb is termed the *electrochemical equivalent* of that element.

If $z = $ electrochemical equivalent of an element in milligrams per coulomb,

and $I = $ current, in amperes, for time t, in seconds,

mass of element liberated $= zIt$ milligrams.

     The electrochemical equivalents of copper and silver are $0.3294\ \text{mg/C}$ and $1.1182\ \text{mg/C}$ respectively.

     The value of a direct current can be determined with considerable accuracy by passing the current through either a copper or a silver voltameter (i.e. two copper plates in a copper sulphate solution or two silver plates in a silver nitrate solution) for a given time and noting the increase in the mass of the negative plate (or cathode).

*(c) Potential difference** The unit of potential difference is the *volt*, namely *the difference of potential between two points of a conducting wire carrying a current of* 1 *ampere, when the power dissipated between these points is equal to* 1 *watt.*

(d) Resistance The unit of electric resistance is the *ohm*, namely *the resistance between two points of a conductor when a potential difference of* 1 *volt, applied between these points, produces in this conductor a current of* 1 *ampere, the conductor not being a source of any electromotive force.*

Alternatively, the *ohm* can be defined as *the resistance of a circuit in which a current of* 1 *ampere generates heat at the rate of* 1 *watt.*

If V represents the p.d., in volts, across a circuit having resistance R, in ohms, carrying a current I, in amperes, for time t, in seconds,

$$V = IR \text{ or } I = V/R \text{ [by Ohm's Law]}$$

$$\text{power} = IV = I^2R = V^2/R \text{ watts}$$

and heat generated $= I^2Rt = IVt$ joules.

(e) Electromotive force An electromotive force is that which tends to produce an electric current in a circuit, and the *unit of e.m.f. is the volt*. The principal sources of e.m.f. are:

(i) the electrodes of dissimilar materials immersed in an electrolyte, as in primary and secondary cells;

(ii) the relative movement of a conductor and a magnetic flux, as in electric generators and transformers; this source can, alternatively, be expressed as the variation of magnetic flux linked with a coil (sections 2.11 and 2.12);

(iii) the difference of temperature between junctions of dissimilar metals, as in thermo-junctions (section 26.7).

Consider a circuit of resistance R ohms to be connected across a cell having an internal resistance r ohms, as in fig. 1.2, and suppose I amperes to be the current, then:

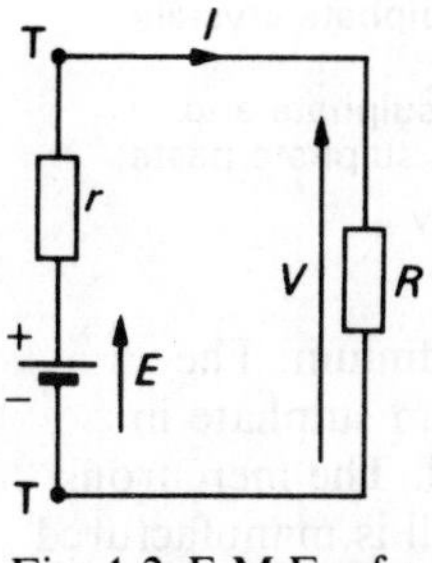

p.d. across $R = IR$ volts

and p.d. across $r = Ir$ volts.

Total power dissipated in R and $r = I^2(R + r)$ watts.

The e.m.f. of the cell is responsible for maintaining the electric current in the circuit; and the value of the e.m.f., E volts, must be such that the electrical power generated by chemical action in the cell is equal to that dissipated as heat

Fig. 1.2 E.M.F. of a cell

---

* The term *voltage* originally meant a difference of potential expressed in volts; but it is now used as a synonym for potential difference irrespective of the unit in which it is expressed. For instance, the voltage between the lines of a transmission system may be 400 kV, while in communication and electronic circuits, the voltage between two points may be 5 μV. The term *potential difference* is generally abbreviated to *p.d.*

in the resistance of the circuit,

i.e.
$$IE = I^2(R + r)$$

∴
$$E = I(R + r)$$

It follows that the difference of potential, V volts, across terminals TT (fig. 1.2) is given by:

$$V = IR = E - Ir$$

1.7 Electrical reference standards

It was mentioned in section 1.6(a) that the absolute value of the ampere is determined by means of a current balance. The fixed and moving coils are made to known dimensions and from these dimensions and the relative position of the coils, the value of the current to exert a certain force can be calculated. Also, the absolute value of the ohm can be determined by means of a copper disc rotated in a uniform magnetic field. Both of these methods require such elaborate and expensive equipment that they are only carried out at infrequent intervals in national laboratories. During intervals between the absolute determinations, use is made of standards of resistance and of e.m.f. in the form of manganin*-wire resistors and Weston cells respectively.

The Weston cadmium standard cell is shown in section in fig. 1.3. The positive electrode is mercury and the negative

Fig. 1.3 Weston cadmium standard cell

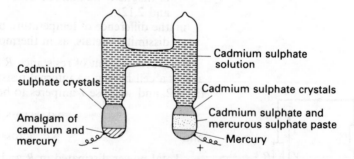

Cadmium sulphate crystals

Cadmium sulphate solution

Cadmium sulphate crystals

Amalgam of cadmium and mercury

Cadmium sulphate and mercurous sulphate paste

Mercury

electrode is an amalgam of mercury and cadmium. The electrolyte is a saturated solution of cadmium sulphate in water slightly acidulated with sulphuric acid. The mercurous sulphate acts as a depolarizer. When this cell is manufactured to a specification prescribed by the International Electrotechnical Commission, its e.m.f. is exactly 1.018 59 volts at 20°C. Consequently the volt can be taken as 1/1.018 59 of the e.m.f. of a cadmium cell at 20°C. The e.m.f.

---

* Manganin is an alloy of copper, manganese and nickel, with sometimes a trace of iron; its temperature coefficient of resistance can be as low as 0.000 003/°C. (See *Dictionary of Applied Physics*, Vol. II, p. 710.)

of the cell falls by about 40 microvolts per degree rise of temperature. Since the internal resistance is about $1000\,\Omega$, this cell is intended only as a standard of e.m.f. and not as a source of electrical energy.

1.8 Resistance

A resistor (i.e. a wire or other form of material used simply because of its resistance) is said to be *linear* if the current through the resistor is proportional to the p.d. across its terminals. If the resistance varies with the magnitude of the current or voltage, the resistor is said to be *non-linear*. A semiconductor junction rectifier (section 10.7) is an example of a non-linear resistor. In this book, resistors will normally be assumed to be linear, i.e. their resistance will be assumed to remain constant when the temperature is maintained constant.

If resistors of resistance R_1, R_2, R_3, etc., are connected in series, the total resistance R is given by:

$$R = R_1 + R_2 + R_3 + \cdots \tag{1.9}$$

If the resistors are connected in parallel, the total resistance is given by R, where:

$$\frac{1}{R} = \frac{1}{R_1} + \frac{1}{R_2} + \frac{1}{R_3} + \cdots \tag{1.10}$$

The reciprocal of the resistance is termed the conductance (symbol G), the unit of conductance being 1 *siemens** (abbreviation, S).

For conductances G_1, G_2, G_3, etc., in parallel, the total conductance G is given by:

$$G = G_1 + G_2 + G_3 + \cdots \tag{1.11}$$

If $\quad l$ = length of an electrical circuit, in metres
and $\quad A$ = cross-sectional area of the circuit, in square metres

$$\text{resistance of circuit} = R = \rho l/A \text{ ohms} \tag{1.12}$$

where $\qquad\qquad \rho$ = resistivity of the material.

$$= \frac{R\,[\text{ohms}] \times A\,[\text{metres}^2]}{l\,[\text{metres}]}$$

$$= RA/l \text{ ohm metres},$$

e.g. the resistivity of annealed copper at 20°C is $1.725 \times 10^{-8}\,\Omega\,\text{m}$.

Because it is a common one, an important case is that of

---

* The brothers Werner von Siemens (1816–92) and Sir William Siemens, F.R.S. (1823–83) were pioneers of electrical engineering, scientifically and industrially. William Siemens was, in 1872, the first President of the I.E.E.

two resistors in parallel:

$$\frac{1}{R} = \frac{1}{R_1} + \frac{1}{R_2} = \frac{R_1 + R_2}{R_1 R_2}$$

$$\therefore \qquad R = \frac{R_1 R_2}{R_1 + R_2} \qquad\qquad (1.13)$$

If the two resistors R_1 and R_2 are connected in parallel and the currents flowing in them are I_1 and I_2 respectively, the total current being I, then:

$$E = I_1 R_1 = I_2 R_2 = I \times \frac{R_1 R_2}{R_1 + R_2}$$

$$\therefore \qquad I_1 = \frac{R_2}{R_1 + R_2} I \qquad\qquad (1.14)$$

and

$$I_2 = \frac{R_1}{R_1 + R_2} I \qquad\qquad (1.15)$$

This expression for the division of current between two parallel paths avoids the necessity to calculate the corresponding voltage and is a useful shortcut in network analysis. It must be remembered that it may only be applied to two parallel resistors.

Example 1.8   *For the network shown in fig. 1.4, calculate the supply current I if R_1 dissipates energy at the rate of 20 W.*

$$P_1 = I_1^2 R_1$$

$$\therefore \qquad I_1 = \left(\frac{P_1}{R_1}\right)^{1/2} = \left(\frac{20}{5}\right)^{1/2} = 2\,\text{A}$$

$$I_1 = \frac{R_2}{R_1 + R_2} I$$

$$\therefore \qquad I = \frac{R_1 + R_2}{R_2} I_1 = \frac{5 + 10}{10} \times 2$$

$$= 3\,\text{A}$$

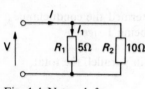

Fig. 1.4 Network for example 1.8

1.9 Temperature coefficient of resistance

The resistance of all pure metals increases with increase of temperature, whereas the resistance of carbon, electrolytes and insulating materials decreases with increase of temperature. Certain alloys, such as manganin (section 1.7) show practically no change of resistance for a considerable variation of temperature. For a moderate range of temperature, such as 100°C, the change of resistance is usually proportional to the change of temperature; and the ratio of the change of resistance per degree change of temperature to the resistance at some definite temperature,

Fig. 1.5 Variation of
resistance of copper with
temperature

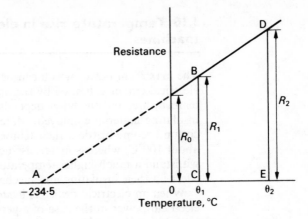

adopted as standard, is termed *the temperature coefficient of resistance* and is represented by the Greek letter α.

The variation of resistance of copper for a range over which copper conductors are usually operated is represented by the graph in fig. 1.5. If this graph is extended backwards, the point of intersection with the horizontal axis is found to be $-234.5°C$. Hence, for a copper conductor having a resistance of $1\,\Omega$ at $0°C$, the change of resistance for $1°C$ change of temperature is $(1/234.5)\,\Omega$, namely $0.004\,264\,\Omega$,

$$\therefore \quad \alpha_0 = \frac{0.004\,264\,[\Omega/°C]}{1\,[\Omega]} = 0.004\,264/°C.$$

In general, if a material having a resistance R_0 at $0°C$, taken as the standard temperature, has a resistance R_1 at θ_1 and R_2 at θ_2, and if α_0 is the temperature coefficient of resistance at $0°C$,

$$R_1 = R_0(1 + \alpha_0\theta_1) \quad \text{and} \quad R_2 = R_0(1 + \alpha_0\theta_2)$$

$$\therefore \qquad \frac{R_1}{R_2} = \frac{1 + \alpha_0\theta_1}{1 + \alpha_0\theta_2} \qquad (1.16)$$

In some countries, the standard temperature is taken to be $20°C$, which is roughly the average atmospheric temperature. This involves using a different value for the temperature coefficient of resistance, e.g. the temperature coefficient of resistance of copper at $20°C$ is $0.003\,92/°C$. Hence, for a material having a resistance R_{20} at $20°C$ and temperature coefficient of resistance α_{20} at $20°C$, the resistance R_t at temperature θ is given by:

$$R_t = R_{20}\{1 + \alpha_{20}(\theta - 20)\} \qquad (1.17)$$

Hence if the resistance of a coil, such as a field winding of an electrical machine, is measured at the beginning and at the end of a test, the mean temperature rise of the whole coil can be calculated.

1.10 Temperature rise in electrical apparatus and machines

The maximum power which can be dissipated as heat in an electrical circuit is limited by the maximum permissible temperature, and the latter depends upon the nature of the insulating material employed. Materials such as paper and cotton become brittle if their temperature is allowed to exceed about 100°C; whereas materials such as mica and glass can withstand a much higher temperature without any injurious effect on their insulating and mechanical properties.

When an electrical device is *loaded* (e.g. when supplying electrical power in the case of a generator, mechanical power in the case of a motor or acting as an amplifier in the case of a transistor), the temperature rise of the device is largely due to the I^2R loss in the conductors; and the greater the load, the greater is the loss and therefore the higher the temperature rise. The *full load* or *rated output* of the device is the maximum output obtainable under specified conditions, e.g. for a specified temperature rise after the device has been loaded continuously for a period of minutes or hours.

The temperature of a coil may be measured by (i) a thermometer, (ii) the increase of resistance of the coil and (iii) thermo-junctions embedded in the coil. The third method enables the distribution of temperature throughout the coil to be determined but is only possible if the thermo-junctions are inserted in the coil when the latter is being wound. Since the heat generated at the centre of the coil has to flow outwards, the temperature at the centre may be considerably higher than that at the surface.

The temperature of an electronic device, especially one incorporating a semiconductor junction, is of paramount importance since even a small rise of temperature above the maximum permissible level rapidly leads to a catastrophic breakdown.

1.11 Maximum power transfer

Let us consider a source, such as a battery or a d.c. generator, having an e.m.f. E and an internal resistance r, as shown enclosed by the dotted rectangle in fig. 1.6. A variable resistor is connected across terminals A and B of the source. If the value of the load resistance is R, then

$$I = E/(r + R)$$

and   power transferred to load $= I^2R$

$$= \frac{E^2R}{(r + R)^2} = \frac{E^2R}{r^2 + 2rR + R^2}$$

$$= \frac{E^2}{(r^2/R) + 2r + R} \qquad (1.18)$$

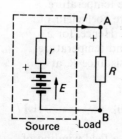

Fig. 1.6 Resistance matching

This power is a maximum when the denominator of (1.18) is a minimum,

i.e. when $$\frac{\mathrm{d}}{\mathrm{d}R}\{(r^2/R) + 2r + R\} = 0$$

$\therefore$ $$-(r^2/R^2) + 1 = 0$$

or $$R = r \qquad (1.19)$$

To check that this condition gives the minimum and not the maximum value of the denominator in expression (1.18), expression $\{-(r^2/R^2) + 1\}$ should be differentiated with respect to R, thus:

$$\frac{\mathrm{d}}{\mathrm{d}R}\{1 - (r^2/R^2)\} = 2r^2/R^3.$$

Since this quantity is positive, expression (1.19) is the condition for the denominator of (1.18) to be a minimum and therefore the output power to be a maximum. Hence the power transferred from the source to the load is a maximum when the resistance of the load is equal to the internal resistance of the source. This condition is referred to as *resistance matching*.

Resistance matching is of importance in communications and electronic circuits where the source usually has a relatively high resistance and where it is desired to transfer the largest possible amount of power from the source to the load. In the case of d.c. generators and secondary batteries, the internal resistance is so low that it is impossible to satisfy the above condition without overloading the source.

1.12 Network theorems

A network consists of a number of branches or circuit elements, considered as a unit, and is said to be *passive* if it contains no source of e.m.f. The equivalent resistance between any two terminals of a passive network is the ratio of the p.d. across the two terminals to the current flowing into (or out of) the network. When a network contains a source of e.m.f., it is said to be *active*.

We shall now consider the principal theorems that have been developed for solving problems on electrical networks.

1.13 Superposition theorem

In a linear network containing more than one source of e.m.f., the resultant current in any branch is the algebraic sum of the currents that would be produced by each e.m.f., acting alone, all the other sources of e.m.f. being replaced meanwhile by their respective internal resistances.

Let us consider the application of this theorem to the solution of the following problem.

Example 1.9   *Battery A in fig. 1.7 has an e.m.f. of 6.0 V and an internal resistance of 2.0 Ω. The corresponding values for battery B are 4.0 V and 3.0 Ω respectively. The two batteries are connected in parallel across a 10-Ω resistor R. Calculate the current in each branch of the network.*

Fig. 1.7 Circuit diagram for example 1.9

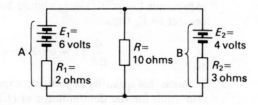

Fig. 1.8(*a*) represents the network with battery A only.

$$\text{Equivalent resistance of } R \text{ and } R_2 \text{ in parallel} = \frac{10 \times 3}{10 + 3} = 2.31 \ \Omega.$$

$$\therefore \qquad I_1 = \frac{6}{2 + 2.31} = 1.392 \ \text{A}.$$

Hence,

$$I_2 = 1.392 \times \frac{3}{10 + 3} = 0.321 \ \text{A}$$

and

$$I_3 = 1.392 - 0.321 = 1.071 \ \text{A}.$$

Fig. 1.8(*b*) represents the network with battery B only.

$$\text{Equivalent resistance of } R \text{ and } R_1 \text{ in parallel} = \frac{2 \times 10}{2 + 10} = 1.667 \ \Omega.$$

$$\therefore \qquad I_4 = \frac{4}{3 + 1.667} = 0.856 \ \text{A}.$$

Fig. 1.8 Diagrams for solution of example 1.9

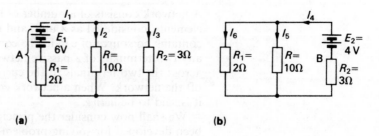

(a)   (b)

Hence,

$$I_5 = 0.856 \times \frac{2}{2 + 10} = 0.143 \ \text{A}$$

and

$$I_6 = 0.856 - 0.143 = 0.713 \ \text{A}.$$

Superimposing the results for fig. 1.8(*b*) on those for fig. 1.8(*a*), we have:

$$\text{resultant current through A} = I_1 - I_6$$

$$= 1.392 - 0.713 = 0.679 \ \text{A}$$

and resultant current through $B = I_4 - I_3$

$$= 0.856 - 1.071$$

$$= -0.215\,\text{A},$$

i.e. battery B is being charged at 0.215 A.

Resultant current through $R = I_2 + I_5$

$$= 0.321 + 0.143 = 0.464\,\text{A}.$$

1.14 Kirchhoff's Laws

First Law *The total current flowing towards a node* is equal to the total current flowing away from that node, i.e. the algebraic sum of the currents flowing towards a node is zero.* Thus at node C in fig. 1.9,

$$I_1 + I_2 = I_3 \quad \text{or} \quad I_1 + I_2 - I_3 = 0$$

In general, $$\sum I = 0 \qquad (1.20)$$

where $\sum$ represents the algebraic sum.

Fig. 1.9 Circuit to illustrate Kirchhoff's Laws

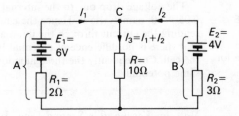

Second Law† *In a closed circuit, the algebraic sum of the products of the current and the resistance of each part of the circuit is equal to the resultant e.m.f. in the circuit.* Thus for the closed circuit involving E_1, E_2, R_1 and R_2 in fig. 1.9,

$$E_1 - E_2 = I_1 R_1 - I_2 R_2$$

and for the mesh involving E_2, R_2 and R,

$$E_2 = I_2 R_2 + I_3 R$$

In general, $$\sum E = \sum IR. \qquad (1.21)$$

Example 1.10 *Using Kirchhoff's Laws, calculate the current in each branch of the network shown in fig. 1.9, the e.m.f.s and resistances of which are the same as those of fig. 1.7.*

---

* A *node* of a network is defined as the junction poing of two or more branches of that network; e.g. C in fig. 1.9 is a node of three branches.

† See Note on p. 16 concerning Kirchhoff's Second Law.

Applying Kirchhoff's Laws to the circuit formed by A and R, we have:

$$6 = 2I_1 + 10(I_1 + I_2) = 12I_1 + 10I_2 \qquad (1.22)$$

Similarly, for the closed circuit formed by A and B,

$$6 - 4 = 2 = 2I_1 - 3I_2 \qquad (1.23)$$

Multiplying (1.23) by 6 and subtracting from (1.22), we have:

$$-6 = 28I_2$$

$$I_2 = -0.215\,\text{A}.$$

Substituting for I_2 in (1.23), we have:

$$I_1 = 1 - 1.5 \times 0.2143 = 0.679\,\text{A}$$

and

$$I_3 = 0.678 - 0.214 = 0.464\,\text{A}.$$

Example 1.11

Three similar primary cells are connected in series to form a closed circuit as shown in fig. 1.10. Each cell has an e.m.f. of 1.5 V and an internal resistance of 30 Ω. Calculate the current and show that points A, B and C are at the same potential.

In fig. 1.10, E and R represent the e.m.f. and internal resistance respectively of each cell.

$$\text{Total e.m.f.} = 1.5 \times 3 = 4.5\,\text{V},$$

$$\text{total resistance} = 30 \times 3 = 90\,\Omega$$

$$\therefore \qquad \text{current} = 4.5/90 = 0.05\,\text{A}.$$

The voltage drop due to the internal resistance of each cell is 0.05×30, namely 1.5 V. Hence the e.m.f. of each cell is absorbed in sending the current through the internal resistance of that cell, so that there is no difference of potential between the two terminals of the cell. Consequently the three junctions A, B and C are at the same potential.*

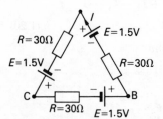

Fig. 1.10 Circuit diagram for example 1.11

Note on Kirchhoff's Second Law. In textbooks, this law is expressed in various ways of which the following is typical: *In a closed circuit, the algebraic sum of the e.m.f.s is equal to the algebraic sum of the voltage drops.* This suggests that there must be differences of potential between various points of the circuit. Actually, there may be *no* p.d. between any two points of the circuit. For instance, if a cylindrical magnet is moved along the axis of a homogeneous metal ring of uniform cross-sectional area, as in fig. 2.19, the e.m.f. induced in the ring circulates a current.

If e and i be the instantaneous values of the e.m.f. and current respectively, and if R and l be the resistance and

---

* This result has important practical applications, e.g. in connection with the non-existence of a third harmonic in the terminal voltage of a delta-connected 3-phase machine or transformer. The same principle can be applied to explain why there is no magnetic leakage in a toroid uniformly wound with a magnetizing winding — the m.m.f. per unit length is absorbed in sending the magnetic flux through the reluctance of that length, irrespective of how small that length may be. Hence, all points of the toroid are at the same magnetic potential.

mean periphery of the ring respectively,

$$e = iR, \quad \text{so that } e/l = i \times R/l.$$

This means that the e.m.f. induced in a length of, say, 1 mm of the ring is absorbed in sending current through the resistance of that millimetre. Consequently, there is no difference of potential across that millimetre length of the ring — in fact, *all points of the ring are at the same potential*.

Similarly, in example 1.11, p. 16, it is shown that when three exactly similar cells are connected in series, there is no p.d. between any two available points of the circuit.

It follows that the definition of Kirchhoff's Second Law given above is applicable *only when the sources of e.m.f. have no internal resistance*. The definition given on p. 15 is applicable to *all* closed circuits.

1.15 Thévenin's Theorem

The current through a resistor R connected across any two points A *and* B *of an active network* (i.e. a network containing one or more sources of e.m.f.) *is obtained by dividing the p.d. between* A *and* B, *with R disconnected, by* (R + r), *where r is the resistance of the network measured between points* A *and* B *with R disconnected and the sources of e.m.f. replaced by their internal resistances.*

An alternative way of stating Thévenin's Theorem is as follows: *An active network having two terminals* A *and* B *can be replaced by a constant-voltage source having an e.m.f. E and an internal resistance r. The value of E is equal to the open-circuit p.d. between* A *and* B, *and r is the resistance of the network measured between* A *and* B *with the load disconnected and the sources of e.m.f. replaced by their internal resistances.*

Suppose A and B in fig. 1.11(*a*) to be the two terminals of a network consisting of resistors having resistances R_2 and R_3 and a battery having an e.m.f. E_1 and an internal resistance R_1. It is required to determine the current through a load of resistance R connected across AB. With the load disconnected as in fig. 1.11(*b*),

$$\text{current through } R_3 = \frac{E_1}{R_1 + R_3}$$

and

$$\text{p.d. across } R_3 = \frac{E_1 R_3}{R_1 + R_3}.$$

Since there is no current through R_2,

$$\text{p.d. across AB} = V = \frac{E_1 R_3}{R_1 + R_3}.$$

Fig. 1.11(*c*) shows the network with the load disconnected

Fig. 1.11 Networks to
illustrate Thévenin's
Theorem

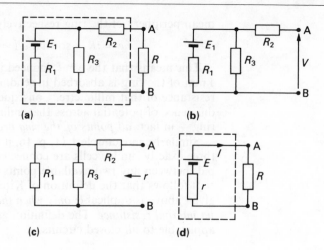

(a)   (b)

(c)   (d)

and the battery replaced by its internal resistance R_1.

Resistance of network between A and B $= r = R_2 + \dfrac{R_1 R_3}{R_1 + R_3}$.

Thévenin's Theorem merely states that the active network
enclosed by the dotted line in fig. 1.11(a) can be replaced by
the very simple circuit enclosed by the dotted line in
fig. 1.11(d) and consisting of a source having an e.m.f. E equal
to the open-circuit potential difference V between A and B,
and an internal resistance r, where V and r have the values
determined above. Hence,

$$\text{current through } R = I = \frac{E}{r + R}.$$

Thévenin's Theorem — sometimes referred to as Helmholtz's
Theorem — is an application of the Superposition Theorem.
Thus, if a source having an e.m.f. E equal to the open-circuit
p.d. between A and B in fig. 1.11(b) were inserted in the
circuit between R and terminal A in fig. 1.11(a), the positive
terminal of the source being connected to A, no current
would flow through R. Hence, this source could be regarded
as circulating through R a current superimposed upon but
opposite in direction to the current through R due to E_1
alone. Since the resultant current is zero, it follows that a
source of e.m.f. E connected in series with R and the
equivalent resistance r of the network, as in fig. 1.11(d), would
circulate a current I having the same value as that through R
in fig. 1.11(a); but in order that the direction of the current
through R may be from A towards B, the polarity of the
source must be as shown in fig. 1.11(d).

Example 1.12 *C and D in fig. 1.12(a) (which is similar to fig. 1.8) represent the two
terminals of an active network. Calculate the current through R.*

With R disconnected as in fig. 1.12(b),

$$I_1 = \frac{6 - 4}{2 + 3} = 0.4 \text{ A}$$

Fig. 1.12 Circuit diagrams
for example 1.12

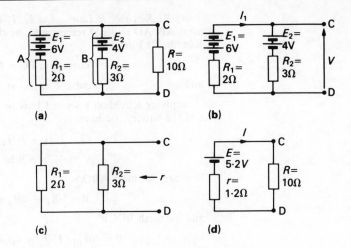

(a)   **(b)**

(c)   **(d)**

and   p.d. across $CD = E_1 - I_1 R_1$

i.e. $V = 6 - (0.4 \times 2) = 5.2 \, V.$

When the e.m.f.s are removed as in fig. 1.12(*c*),

$$\text{total resistance between C and D} = \frac{2 \times 3}{2 + 3}$$

i.e. $r = 1.2 \, \Omega.$

Hence the network AB in fig. 1.12(*a*) can be replaced by a single
source having an e.m.f. of 5.2 V and an internal resistance of 1.2 Ω,
as in fig. 1.12(*d*); consequently,

$$I = \frac{5.2}{1.2 + 10} = 0.464 \, A,$$

namely the value obtained in examples 1.9 and 1.10.

Example 1.13   *The resistances of the various arms of an unbalanced Wheatstone
bridge are given in fig. 1.13. The battery has an e.m.f. of 2.0 V and a
negligible internal resistance. Determine the value and direction of the
current in the galvanometer circuit BD, using (a) Kirchhoff's Laws
and (b) Thévenin's Theorem.*

Fig. 1.13 Network for
example 1.13

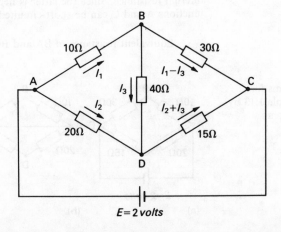

$E = 2 \, volts$

(a) *By Kirchhoff's Laws.* Let I_1, I_2 and I_3 be the currents in arms AB, AD and BD respectively, as shown in fig. 1.13. Then by Kirchhoff's First Law,

$$\text{current in BC} = I_1 - I_3$$

and
$$\text{current in DC} = I_2 + I_3.$$

Applying Kirchhoff's Second Law to the mesh formed by ABC and the battery, we have:

$$2 = 10I_1 + 30(I_1 - I_3)$$
$$= 40I_1 - 30I_3 \tag{1.24}$$

Similarly for mesh ABDA,

$$0 = 10I_1 + 40I_3 - 20I_2 \tag{1.25}$$

and for mesh BDCB,

$$0 = 40I_3 + 15(I_2 + I_3) - 30(I_1 - I_3)$$
$$= -30I_1 + 15I_2 + 85I_3 \tag{1.26}$$

Multiplying (1.25) by 3 and (1.26) by 4, and adding the two expressions thus obtained, we have:

$$0 = -90I_1 + 460I_3$$
$$\therefore \qquad I_1 = 5.111I_3.$$

Substituting for I_1 in (1.24), we have:

$$I_3 = 0.0115\,\text{A} = 11.5\,\text{mA}.$$

Since the value of I_3 is positive, the direction of I_3 is that assumed in fig. 1.13, namely from B to D.

(b) *By Thévenin's Theorem.* Since we require to find the current in the 40-Ω resistor between B and D, the first step is to remove this resistor, as in fig. 1.14(a). Then:

$$\text{p.d. between A and B} = 2 \times \frac{10}{10 + 30} = 0.5\,\text{V}$$

and
$$\text{p.d. between A and D} = 2 \times \frac{20}{20 + 15} = 1.143\,\text{V}$$

$$\therefore \qquad \text{p.d. between B and D} = 0.643\,\text{V},$$

B being positive relative to D. Consequently, current in the 40-Ω resistor, when connected between B and D, will flow from B to D.

The next step is to replace the battery by a resistance equal to its internal resistance. Since the latter is negligible in this problem, junctions A and C can be short-circuited as in fig. 1.14(b).

$$\text{Equivalent resistance of BA and BC} = \frac{10 \times 30}{10 + 30} = 7.5\,\Omega$$

Fig. 1.14 Diagrams for solution of example 1.13 by Thévenin's Theorem

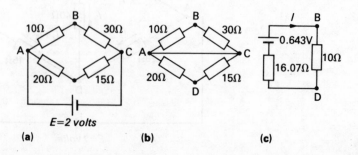

(a)   **(b)**   **(c)**

and   equivalent resistance of AD and CD $= \dfrac{20 \times 15}{20 + 15} = 8.57\,\Omega$,

∴   total resistance of network between B and D

$$= 16.07\,\Omega.$$

Hence the network of fig. 1.14(a) is equivalent to a source having an e.m.f. of 0.643 V and an internal resistance of 16.07 Ω as in fig. 1.14(c).

∴   current through BD $= \dfrac{0.643}{16.07 + 40} = 0.0115\,\text{A}$

$$= 11.5\,\text{mA from B to D}.$$

1.16 Delta–star transformation

Fig. 1.15(a) shows three resistors R_1, R_2 and R_3 connected in a closed mesh or *delta* to three terminals A, B and C, *their numerical subscripts 1, 2 and 3, being opposite to the terminals A, B and C respectively*. It is possible to replace these delta-connected resistors by three resistors R_a, R_b and R_c connected respectively between the same terminals A, B and C and a common point S, as in fig. 1.15(b). Such an arrangement is said to be *star-connected*. It will be noted that the letter

Fig. 1.15 Delta–star transformation

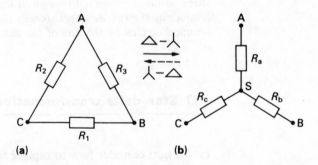

(a)   (b)

subscripts are now those of the terminals to which the respective resistors are connected. If the star-connected network is to be equivalent to the delta-connected network, the resistance between any two terminals in fig. 1.15(b) must be the same as that between the same two terminals in fig. 1.15(a). Thus, if we consider terminals A and B in fig. 1.15(a), we have a circuit having a resistance R_3 in parallel with a circuit having resistances R_1 and R_2 in series; hence

$$R_{AB} = \frac{R_3(R_1 + R_2)}{R_1 + R_2 + R_3} \tag{1.27}$$

For fig. 1.15(b), we have:

$$R_{AB} = R_a + R_b \qquad (1.28)$$

In order that the networks of fig. 1.15(a) and (b) may be equivalent to each other, the values of R_{AB} represented by expressions (1.27) and (1.28) must be equal,

$$\therefore \qquad R_a + R_b = \frac{R_1 R_3 + R_2 R_3}{R_1 + R_2 + R_3} \qquad (1.29)$$

Similarly,

$$R_b + R_c = \frac{R_1 R_2 + R_1 R_3}{R_1 + R_2 + R_3} \qquad (1.30)$$

and

$$R_a + R_c = \frac{R_1 R_2 + R_2 R_3}{R_1 + R_2 + R_3} \qquad (1.31)$$

Subtracting (1.30) from (1.29), we have:

$$R_a - R_c = \frac{R_2 R_3 - R_1 R_2}{R_1 + R_2 + R_3} \qquad (1.32)$$

Adding (1.31) and (1.32) and dividing by 2, we have:

$$R_a = \frac{R_2 R_3}{R_1 + R_2 + R_3} \qquad (1.33)$$

Similarly,

$$R_b = \frac{R_3 R_1}{R_1 + R_2 + R_3} \qquad (1.34)$$

and

$$R_c = \frac{R_1 R_2}{R_1 + R_2 + R_3} \qquad (1.35)$$

These relationships may be expressed thus: *the equivalent star resistance connected to a given terminal is equal to the product of the two delta resistances connected to the same terminal divided by the sum of the delta resistances.*

1.17 Star–delta transformation

Let us next consider how to replace the star-connected network of fig. 1.15(b) by the equivalent delta-connected network of fig. 1.15(a). Dividing equation (1.33) by equation (1.34), we have:

$$R_a/R_b = R_2/R_1$$
$$\therefore \qquad R_2 = R_1 R_a/R_b$$

Similarly, dividing (1.33) by (1.35), we have:

$$R_a/R_c = R_3/R_1$$
$$\therefore \qquad R_3 = R_1 R_a/R_c$$

Substituting for R_2 and R_3 in (1.33), we have:

$$R_1 = R_b + R_c + R_b R_c/R_a$$

Similarly, $$R_2 = R_c + R_a + R_c R_a / R_b$$
and $$R_3 = R_a + R_b + R_a R_b / R_c$$

These relationships may be expressed thus: *the equivalent delta resistance between two terminals is the sum of the two star resistances connected to those terminals plus the product of the same two star resistances divided by the third star resistance.*

1.18 Two-wire d.c. system of distribution

A d.c. system is usually supplied either from d.c. generators or from rectifiers at a voltage that is maintained approximately constant. In fact, one of the regulations governing the distribution of electrical energy stipulates that the voltage at a consumer's premises must not vary by more than ± 6 per cent; for instance, if the consumer is supplied at a nominal voltage of 240 V, the actual voltage should not exceed 254 V or fall below 226 V.

Fig. 1.16 gives the general arrangement of a distribution system. Two d.c. generators DD are shown connected in parallel to bus-bars BB. The bus-bars are two copper or

Fig. 1.16 A distribution system

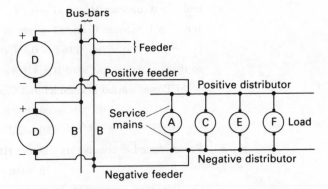

aluminium bars that extend the whole length of the switchboard, 'bus' being an abbreviation of 'omnibus', a Latin word meaning 'for all'. Thus, all the generators at the generating station and all the cables connecting the station to various 'feeding' points are connected to the bus-bars. The cables radiating from the station are called *feeders*, whereas *distributors* are cables to which the *service mains* supplying the individual consumers are connected. A distributor is connected at one or more points to feeders, but no service mains are connected to the latter.

In practice, the distributors are interconnected to form a network that is almost like that of a spider's web. This network is connected to feeders at suitable points. Such an arrangement has the advantage that if for some reason a feeder has to be disconnected, the current to the section normally supplied by that feeder can still be supplied through other feeders and distributors.

Example 1.14

Two loads, A and B (fig. 1.17), taking 50 A and 30 A respectively, are connected to a two-wire distributor at distances of 200 m and 300 m respectively from the feeding point, the p.d. at which is 120 V. The resistance of the distributor is 0.01 Ω per 100 m of single conductor. Find: (a) the p.d. across each load, (b) the cost of the energy wasted in the distributor if the above loads are maintained constant for 10 hours. Assume the cost of energy to be 4.7 p/kW · h.

(a) Current in CD and HG (fig. 1.17) = 30 A and current in EC and GF = 30 + 50 = 80 A

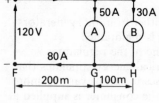

Fig. 1.17 Circuit diagram for example 1.14

Resistance of conductor EC = 0.01 × 200/100 = 0.02 Ω

and resistance of conductor CD = 0.01 Ω

Hence, p.d. between E and C = 80 × 0.02 = 1.6 V

Similarly, p.d. between G and F = 1.6 V

But p.d. between E and F = sum of the p.d.s in circuit ECGF

i.e. 120 = 1.6 + p.d. between C and G + 1.6.

If the voltage drop in the service main be neglected,

p.d. across load A = 120 − 3.2 = 116.8 V.

Also p.d. between C and D = 30 × 0.01 = 0.3 V

and p.d. between H and G = 0.3 V

But p.d. between C and G = sum of p.d.s in circuit CDHG

i.e. 116.8 = 0.3 + p.d. between D and H + 0.3

so that p.d. across load B = 116.8 − 0.6 = 116.2 V.

(b) Power wasted in conductors EC and FG

= current × voltage drops in EC and FG

= 80 × 3.2 = 256 W

Power wasted in conductors CD and HG

= 30 × 0.6 = 18 W

∴ total power wasted in distributor = 256 + 18 = 274 W

= 0.274 kW

and energy wasted in 10 hours = 0.274 × 10 = 2.74 kW · h

∴ cost of this energy = 2.74 × 4.7 = 12.9 p.

Example 1.15

A two-wire ring distributor (i.e. a distributor in which each conductor forms a complete circuit or loop, as in fig. 1.18) is 300 m long and is fed at 240 V at A. At a point B, 150 m from A, there is a load of 120 A and at C, 100 m in the opposite direction, there is a load of 80 A. The resistance per 100 m of single conductor is 0.03 Ω. Find: (a) the current in each section, (b) the p.d.s at B and C.

(a) Let *I* be the current, in amperes, from A to B in the positive conductor.

From Kirchhoff's First Law, it follows that the current from B to

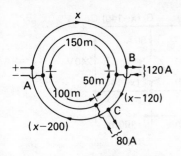

Fig. 1.18 Circuit diagram for example 1.15

C in positive conductor

$$= (I - 120)\,\text{A}$$

and current from C to A in positive conductor

$$= I - 120 - 80 = (I - 200)\,\text{A}.$$

Resistance of positive conductor between A and B

$$= 0.03 \times 150/100 = 0.045\,\Omega,$$

resistance of positive conductor between B and C

$$= 0.03 \times 50/100 = 0.015\,\Omega$$

and resistance of positive conductor between C and A

$$= 0.03\,\Omega.$$

Hence, voltage drop in positive conductor between A and B

$$= (I \times 0.045)\,\text{V},$$

voltage drop in positive conductor between B and C

$$= ((I - 120) \times 0.015)\,\text{V}$$

and voltage drop in positive conductor between C and A

$$= ((I - 200) \times 0.03)\,\text{V}.$$

Since there is no e.m.f. in the loop formed by the positive conductor, it follows from Kirchhoff's Second Law that:

$$I \times 0.045 + (I - 120) \times 0.015 + (I - 200) \times 0.03 = 0$$

$$\therefore \qquad\qquad I = 86.7\,\text{A}$$

$$= \text{current in section AB}.$$

$$\text{Current in section BC} = 86.7 - 120 = -33.3\,\text{A}$$

$$= 33.3\,\text{A from C to B in positive conductor}$$

and current in section CA $= 86.7 - 200 = -113.3\,\text{A}$

$$= 113.3\,\text{A from A to C in positive conductor}.$$

(*b*) Voltage drop in positive and negative conductors between A and B

$$= 86.7 \times 0.045 \times 2 = 7.8\,\text{V}$$

and voltage drop in positive and negative conductors between A and C

$$= 113.3 \times 0.03 \times 2 = 6.8\,\text{V}$$

$$\therefore \quad \text{p.d. across load at B} = 240 - 7.8 = 232.2\,\text{V}$$

and p.d. across load at C $= 240 - 6.8 = 233.2\,\text{V}$

The difference of 1 V between the p.d.s across the loads at B and C should agree with the voltage drop calculated from the current in section BC and the resistance of that section, namely $33.3 \times 0.015 \times 2 = 0.999$ V. The very slight discrepancy between the two values is due to the fact that the value of I was limited to three significant figures. This degree of accuracy is sufficient for most practical purposes.

Example 1.16   *A distributor, 450 m long, is loaded as shown in fig. 1.19. The p.d. at AB is 250 V and that at CD is 240 V. The resistance of the distributor is 0.02 Ω per 100 m of single conductor. Calculate the p.d. at each load point.*

Fig. 1.19 Circuit diagram for
example 1.16

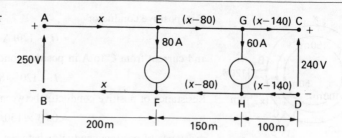

Let I be the current, in amperes, in AE and FB.
From Kirchhoff's First Law,

$$\text{current in EG and HF} = (I - 80)\,\text{A}$$

and $\qquad$ current in GC and DH $= (I - 140)\,\text{A}$.

$$\text{Resistance of AE and FB} = 0.02 \times 4 = 0.08\,\Omega,$$

$$\text{resistance of EG and HF} = 0.02 \times 3 = 0.06\,\Omega$$

and $\qquad$ resistance of GC and DH $= 0.02 \times 2 = 0.04\,\Omega$.

$\therefore \qquad$ voltage drop in AE and FB $= (0.08I)\,\text{V}$,

$$\text{voltage drop in EG and HF} = 0.06(I - 80)\,\text{V}$$

and $\qquad$ voltage drop in GC and DH $= 0.04(I - 140)\,\text{V}$.

But $\qquad$ voltage drop in AC and DB $= 250 - 240 = 10\,\text{V}$,

$\therefore \quad 0.08I + 0.06(I - 80) + 0.04(I - 140) = 10$

so that $\qquad\qquad\qquad\qquad\qquad I = 113.3\,\text{A}$.

Hence, $\qquad$ p.d. across load EF $= 250 - (113.3 \times 0.08)$

$$= 240.93\,\text{V}$$

and $\qquad$ p.d. across load GH $= 240.93 - (33.3 \times 0.06)$

$$= 238.93\,\text{V}.$$

Alternatively, $\qquad$ p.d. across GH $= 240 - (26.7 \times 0.04)$

$$= 238.93\,\text{V}.$$

1.19 Use of double subscripts

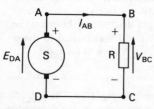

Fig. 1.20

Double-subscript notation is used to avoid ambiguity in the
direction of current, e.m.f. or p.d. In fig. 1.20, S represents a
d.c. source, the e.m.f. of which is acting from D towards A
and is therefore designated E_{DA}. The current in conductor AB
flows from A to B and is designated I_{AB}. For the simple
circuit of fig. 1.20, it is obvious that:

$$I_{AB} = I_{BC} = I_{CD} = I_{DA}.$$

The p.d. across the load is designated V_{BC} to symbolize
that the potential of B is positive with respect to that of C;

hence,

voltage drop across the load $= V_{BC} = R \times I_{BC}$.

If an arrow representing the direction of this voltage drop is drawn alongside the load, its head should point towards the end which is at the higher potential, i.e. towards B in fig. 1.20.

If the internal resistance of the source and the resistance of conductors AB and CD are negligible, then:

$$E_{DA} = V_{BC} = R \times I_{BC}$$

i.e. the voltage *rise* E_{DA} in the source is equal to voltage *drop* V_{BC} in the load.

When the double-subscript notation is applied to an a.c. circuit the sequence of the subscripts indicates the direction in which the current, the e.m.f. or the p.d. is assumed to be positive, e.g. in fig. 19.5, E_{CA} represents an alternating e.m.f. having its positive direction from C towards A in fig. 19.4; whereas in fig. 19.25, V_{RY} represents the voltage drop between lines R and Y, its instantaneous value being positive when R is positive with respect to Y.

Summary of important formulae

$$\theta\,[^\circ C] = T\,[K] - 273.15$$

$$\text{Force } F\,[\text{newtons}] = m\,[\text{kg}] \times a\,[\text{m/s}^2] \tag{1.1}$$

$$\text{Work done } W\,[\text{joules}] = F\,[\text{newtons}] \times d\,[\text{metres}] \tag{1.4}$$

$$\text{Kinetic energy } W\,[\text{joules}] = \tfrac{1}{2}m\,[\text{kg}] \times |v^2\,[\text{m/s}]^2 \tag{1.5}$$

$$\text{Potential energy } W\,[\text{joules}] = m\,[\text{kg}] \times g\,[\text{m/s}^2] \times h\,[\text{m}] \tag{1.6}$$

$$\simeq 9.81mg$$

$$\text{Power } P\,[\text{watts}] = 2\pi T\,[\text{N}\cdot\text{m}] \times n\,[\text{rad/s}] \tag{1.8}$$

$$1\,\text{kW}\cdot\text{h} = 3.6\,\text{MJ}$$

Heat required to raise the temperature of m kilograms of a substance having specific heat capacity c joules per kilogram kelvin through θ degrees

$$= mc\theta \text{ joules} \tag{1.7}$$

$$\text{Charge } Q\,[\text{coulombs}] = I\,[\text{amperes}] \times t\,[\text{seconds}]$$

$$\left.\begin{array}{l}\text{Mass of element liberated}\\\text{from electrolyte}\end{array}\right\} = zIt = zQ$$

Ohm's Law: $\qquad\qquad I = V/R,\ V = IR \text{ or } R = V/I$

$$\text{Electrical power } P = I^2R = IV = V^2/R \text{ watts}$$

$$\text{Electrical energy } W = IVt \text{ joules}$$

For resistors in series:

$$R = R_1 + R_2 + \cdots \tag{1.9}$$

For resistors in parallel:

$$1/R = 1/R_1 + 1/R_2 + \cdots \tag{1.10}$$

or $\quad G = G_1 + G_2 + \cdots \tag{1.11}$

$$R = \rho l / A \tag{1.12}$$

$$R_t = R_0(1 + \alpha_0 \theta)$$

$$= R_0(1 + 0.004\,26\theta) \text{ for annealed copper}$$

or $\quad R_t = R_{20}\{1 + \alpha_{20}(\theta - 20)\} \tag{1.15}$

$$= R_{20}\{1 + 0.003\,92(\theta - 20)\} \text{ for annealed copper}$$

For maximum power transfer,

$$R = r \tag{1.17}$$

EXERCISES I

1. A force of 80 N is applied to a mass of 200 kg. Calculate the acceleration in metres per second squared.
2. Calculate the force, in kilonewtons, required to give a mass of 500 kg an acceleration of 4 m/s².
3. A train having a mass of 300 Mg is hauled at a constant speed of 90 km/h along a straight horizontal track. The track resistance is 5 mN per newton of train weight. Calculate (a) the tractive effort in kilonewtons, (b) the energy in megajoules and in kilowatt hours expended in 10 minutes, (c) the power in kilowatts and (d) the kinetic energy of the train in kilowatt hours (neglecting rotational inertia).
4. The power required to drive a certain machine at 350 r/min is 600 kW. Calculate the driving torque in newton metres.
5. A d.c. motor connected to a 240-V supply is developing 20 kW at a speed of 900 r/min. Calculate the useful torque.
6. If the motor referred to in Q.5 has an efficiency of 0.88, calculate (a) the current and (b) the cost of the energy absorbed if the load is maintained constant for 6 h. Assume the cost of electrical energy to be 4.7 p/kW · h.
7. (a) An electric motor runs at 600 r/min when driving a load requiring a torque of 200 N · m. If the motor input is 15 kW, calculate the efficiency of the motor and the heat lost by the motor per minute, assuming its temperature to remain constant.
   (b) An electric kettle is required to heat 0.5 kg of water from 10°C to boiling point in 5 min, the supply voltage being 230 V. If the efficiency of the kettle is 0.80, calculate the resistance of the heating element. Assume the specific heat capacity of water to be 4.2 kJ/kg · K.   (S.A.N.C., O.1)
8. A pump driven by an electric motor lifts 1.5 m³ of water per minute to a height of 40 m. The pump has an efficiency of 90 per cent and the motor an efficiency of 85 per cent. Determine: (a) the power input to the motor; (b) the current taken from a 480-V supply; (c) the electrical energy consumed when the motor runs at this load for 8 h. Assume the mass of 1 m³ of water to be 1000 kg.

9. An electric kettle is required to heat 0.6 litre of water from 10°C to the boiling point in 5 min, the supply voltage being 240 V. The efficiency of the kettle is 78 per cent. Calculate: (*a*) the resistance of the heating element and (*b*) the cost of the energy consumed at 4.7 p/kW · h. Assume the specific heat capacity of water to be 4190 J/kg · K and 1 litre of water to have a mass of 1 kg.

10. An electric furnace is to melt 40 kg of aluminium per hour, the initial temperature of the aluminium being 12°C. Calculate: (*a*) the power required and (*b*) the cost of operating the furnace for 20 h, given that aluminium has the following thermal properties: specific heat capacity, 950 J/kg · K; melting point, 660°C; specific latent heat of fusion, 450 kJ/kg. Assume the efficiency of the furnace to be 85 per cent and the cost of electrical energy to be 4.7 p/kW · h.

11. A steady current of 5 A is passed through a copper coulometer for 20 min. Calculate the mass of copper deposited on the cathode, assuming the electrochemical equivalent of copper to be 0.33 mg/C.

12. A plate having a total surface area of 20 000 mm² is to be nickel-plated to a thickness of 0.15 mm. If the current available is 6 A, calculate the time required. Assume the electrochemical equivalent of nickel to be 0.304 mg/C and the density to be 8800 kg/m³.

13. Two coils having resistances of 5 Ω and 8 Ω respectively are connected across a battery having an e.m.f. of 6 V and an internal resistance of 1.5 Ω. Calculate: (*a*) the terminal voltage and (*b*) the energy in joules dissipated in the 5-Ω coil if the current remains constant for 4 min.

14. A coil of 12-Ω resistance is in parallel with a coil of 20-Ω resistance. This combination is connected in series with a third coil of 8-Ω resistance. If the whole circuit is connected across a battery having an e.m.f. of 30 V and an internal resistance of 2 Ω, calculate (*a*) the terminal voltage of the battery and (*b*) the power in the 12-Ω coil.

15. A coil of 20-Ω resistance is joined in parallel with a coil of *R* ohms resistance. This combination is then joined in series with a piece of apparatus A, and the whole circuit connected to 100-V mains. What must be the value of *R* so that A shall dissipate 600 W with 10 A passing through it? (U.L.C.I., O.1)

16. Two circuits, A and B, are connected in parallel to a 25-V battery, which has an internal resistance of 0.25 Ω. Circuit A consists of two resistors, 6 Ω and 4 Ω, connected in series. Circuit B consists of two resistors, 10 Ω and 5 Ω, connected in series. Determine the current flowing in and the potential difference across each of the four resistors. Also, find the power expended in the external circuit.   (N.C.T.E.C., O.1)

17. A load taking 200 A is supplied by copper and aluminium cables connected in parallel. The total length of conductor in each cable is 200 m, and each conductor has a cross-sectional area of 40 mm². Calculate: (i) the voltage drop in the combined cables; (ii) the current carried by each cable; (iii) the power wasted in each cable. Take the resistivity of copper and aluminium as 0.018 $\mu\Omega$ · m and 0.028 $\mu\Omega$ · m respectively.

18. A circuit, consisting of three resistances 12 Ω, 18 Ω and 36 Ω respectively joined in parallel, is connected in series with a fourth resistance. The whole is supplied at 60 V and it is found that the power dissipated in the 12-Ω resistance is 36 W. Determine the value of the fourth resistance and the total power dissipated in the group.   (U.E.I., O.1)

19. A coil consists of 2000 turns of copper wire having a cross-sectional area of 0.8 mm^2. The mean length per turn is 80 cm and the resistivity of copper is 0.02 $\mu\Omega \cdot$ m at normal working temperature. Calculate the resistance of the coil and the power dissipated when the coil is connected across a 110-V d.c. supply.

20. An aluminium wire 7.5 m long is connected in parallel with a copper wire 6 m long. When a current of 5 A is passed through the combination, it is found that the current in the aluminium wire is 3 A. The diameter of the aluminium wire is 1.0 mm. Determine the diameter of the copper wire. Resistivity of copper is 0.017 $\mu\Omega \cdot$ m; that of aluminium is 0.028 $\mu\Omega \cdot$ m.

21. The field winding of a d.c. motor is connected directly across a 440-V supply. When the winding is at the room temperature of 17°C, the current is 2.3 A. After the machine has been running for some hours, the current has fallen to 1.9 A, the voltage remaining unaltered. Calculate the average temperature throughout the winding, assuming the temperature coefficient of resistance of copper to be 0.004 26/°C at 0°C.

22. Define the term *resistance–temperature coefficient*.
    A conductor has a resistance of R_1 ohms at θ_1 °C, and consists of copper with a resistance–temperature coefficient α referred to 0°C. Find an expression for the resistance R_2 of the conductor at temperature θ_2°C.
    The field coil of a motor has a resistance of 250 Ω at 15°C. By how much will the resistance increase if the motor attains an average temperature of 45°C when running? Take $\alpha = 0.004\,28$/°C referred to 0°C.   (S.A.N.C., O.2)

23. Explain what is meant by the *temperature coefficient of resistance* of a material.
    A copper rod, 0.4 m long and 4.0 mm in diameter, has a resistance of 550 $\mu\Omega$ at 20°C. Calculate the resistivity of copper at that temperature.
    If the rod is drawn out into a wire having a uniform diameter of 0.8 mm, calculate the resistance of the wire when its temperature is 60°C. Assume the resistivity to be unchanged and the temperature coefficient of resistance of copper to be 0.004 26/°C at 0°C.   (App. El., L.U.)

24. A coil of insulated copper wire has a resistance of 150 Ω at 20°C. When the coil is connected across a 240-V supply, the current after several hours is 1.25 A. Calculate the average temperature throughout the coil, assuming the temperature coefficient of resistance of copper at 20°C to be 0.0039/°C.

25. An aluminium conductor has a resistance of 3.6 Ω at 20°C. What is its resistance at 50°C if the temperature coefficient of resistance of aluminium is 0.004 03/°C at 20°C?

26. Owing to a short circuit, a copper conductor having a cross-sectional area of 25 mm^2 carries a current of 20 000 A for 30 ms. Neglecting heat loss, calculate the temperature rise of the conductor. Assume the specific heat capacity of copper to be 390 J/kg $\cdot$ K, density to be 8900 kg/m^3 and resistivity to be 0.018 $\mu\Omega \cdot$ m.

27. A certain generator has an open-circuit voltage of 12 V and an internal resistance of 40 Ω. Calculate: (*a*) the load resistance for maximum power transfer; (*b*) the corresponding values of the terminal voltage and of the power supplied to the load.
    If the load resistance were increased to twice the value for maximum power transfer, what would be the power absorbed by the load?

28. A battery having an e.m.f. of 105 V and an internal resistance of 1 Ω is connected in parallel with a d.c. generator of e.m.f. 110 V

and internal resistance of 0.5 Ω to supply a load having a resistance of 8 Ω. Calculate: (i) the currents in the battery, the generator and the load; (ii) the potential difference across the load. (U.E.I., O.1)

29. State Kirchhoff's Laws and apply them to the solution of the following problem.

Two batteries, A and B, are connected in parallel, and an 80-Ω resistor is connected across the battery terminals. The e.m.f. and the internal resistance of battery A are 100 V and 5 Ω respectively, and the corresponding values for battery B are 95 V and 3 Ω respectively. Find (a) the value and direction of the current in each battery and (b) the terminal voltage. (U.E.I., O.1)

30. State Kirchhoff's Laws for an electric circuit, giving an algebraic expression for each law.

A network of resistors has a pair of input terminals AB connected to a d.c. supply and a pair of output terminals CD connected to a load resistor of 120 Ω. The resistances of the network are AC = BD = 180 Ω, and AD = BC = 80 Ω. Find the ratio of the current in the load resistor to that taken from the supply. (S.A.N.C., O.1)

31. State Kirchhoff's Laws as applied to an electrical circuit.

A secondary cell having an e.m.f. of 2 V and an internal resistance of 1 Ω is connected in series with a primary cell having an e.m.f. of 1.5 V and an internal resistance of 100 Ω, the negative terminal of each cell being connected to the positive terminal of the other cell. A voltmeter having a resistance of 50 Ω is connected to measure the terminal voltage of the cells. Calculate the voltmeter reading and the current in each cell. (App. El., L.U.)

32. State and explain Kirchhoff's Laws relating to electric circuits. Two storage batteries, A and B, are connected in parallel for charging from a d.c. source having an open-circuit voltage of 14 V and an internal resistance of 0.15 Ω. The open-circuit voltage of A is 11 V and that of B is 11.5 V; the internal resistances are 0.06 Ω and 0.05 Ω respectively. Calculate the initial charging currents.

What precautions are necessary when charging batteries in parallel? (App. El., L.U.)

33. State Kirchhoff's Laws as applied to an electrical circuit.

Two batteries A and B are joined in parallel. Connected across the battery terminals is a circuit consisting of a battery C in series with a 25-Ω resistor, the negative terminal of C being connected to the positive terminals of A and B. Battery A has an e.m.f. of 108 V and an internal resistance of 3 Ω, and the corresponding values for battery B are 120 V and 2 Ω. Battery C has an e.m.f. of 30 V and a negligible internal resistance. Determine (a) the value and direction of the current in each battery and (b) the terminal voltage of battery A. (App. El., L.U.)

34. A network is arranged as shown in fig. A. Calculate the value of the current in the 8-Ω resistor by (a) the Superposition Theorem, (b) Kirchhoff's Laws and (c) Thévenin's Theorem.

35. Calculate the voltage across AB in the network shown in fig. B and indicate the polarity of the voltage, using (a) Kirchhoff's Laws and (b) delta–star transformation.

36. A network is arranged as in fig. C. Calculate the equivalent resistance between (a) A and B, and (b) A and N.

37. A network is arranged as in fig. D, and a battery having an e.m.f. of 2 V and negligible internal resistance is connected

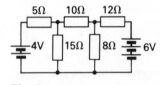

Fig. A

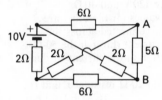

Fig. B

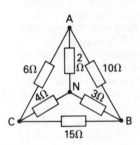

Fig. C

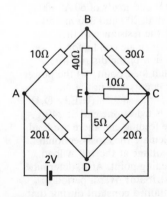

Fig. D

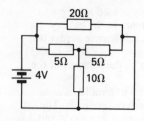

Fig. E

across AC. Determine the value and direction of the current in branch BE.

38. Calculate the value of the current through the 40-Ω resistor in fig. E.

39. Using Thévenin's Theorem, calculate the current through the 10-Ω resistor in fig. F.

The Electric Circuit

Fig. F

40. State your own interpretation of Thévenin's Theorem and use it to solve the following problem.

    Two batteries are connected in parallel. The e.m.f. and internal resistance of one battery are 120 V and 10 Ω respectively and the corresponding values for the other are 150 V and 20 Ω. A resistor of 50 Ω is connected across the battery terminals. Calculate (*a*) the current through the 50-Ω resistor, and (*b*) the value and direction of the current through each battery.

    If the 50-Ω resistor were reduced to 20-Ω resistance, find the new current through it.   (E.M.E.U., O.2)

41. Three resistors having resistances 50 Ω, 100 Ω and 150 Ω are star-connected to terminals A, B and C respectively. Calculate the resistances of equivalent delta-connected resistors.

42. Three resistors having resistances 20 Ω, 80 Ω and 30 Ω are delta-connected between terminals AB, BC and CA respectively. Calculate the resistances of equivalent star-connected resistors.

43. With the aid of delta and star connection diagrams, state the basic equations from which the delta–star and star–delta conversion equations can be derived.

    A star network, in which N is the star point, is made up as follows: A–N = 70 Ω, B–N = 100 Ω and C–N = 90 Ω. Find the equivalent delta network. If the above star and delta networks were superimposed, what would be the measured resistance between terminals A and C?   (E.M.E.U., O.2)

44. A two-wire distributor is fed at 250 V and loads of 60 A, 100 A and 120 A are taken at distances of 100, 200 and 250 metres respectively from the feeding end. If the resistance of the distributor is 0.1 Ω per 1000 metres of single conductor, calculate: (*a*) the current in each section of the distributor; (*b*) the potential difference across each load point; (*c*) the power dissipated in the distributor; (*d*) the energy lost in the distributor in 8 hours.   (U.E.I., O.1)

45. A two-wire distributor, 400 m long, is fed at one end at 240 V. At points 250 m and 400 m from the feeding end there are loads of 200 A and 160 A respectively. Calculate the cross-sectional area of each core in order that the voltage at the 160-A load may be 96 per cent of that at the feeding point. Also, determine the cost of the energy loss in the distributor over a period of 6 hours if the above load were maintained constant during that time. Assume the resistivity of the conductor at working

temperature to be $0.02\,\mu\Omega \cdot$ m, and the cost of electrical energy
to be $4.7\,\text{p/kW} \cdot \text{h}$. (App. El., L.U.)

46. A two-wire distributor, fed at one end, supplies a load A of 70 A
at a distance of 200 m and another load B of 50 A at a distance
of 300 m from the feeding end. The cross-sectional area of each
conductor is $75\,\text{mm}^2$ and the resistivity of copper is
$0.018\,\mu\Omega \cdot$ m. If the voltage across load A is 250 V, calculate:
(*a*) the voltages at the feeding point and across load B; (*b*) the
total power wasted in the cable.

47. A two-wire ring main, 4 km in length, is fed at 250 V at point A.
Respective loads of 80 A and 100 A are applied at point B,
1.5 km from point A, and point C, 2 km in the opposite
direction. If the resistance per 100 metres of single conductor is
$0.003\,\Omega$, calculate: (*a*) the current in each section; (*b*) the
voltages at points B and C.

48. A two-core cable is 400 m long and its resistance is $0.04\,\Omega$ per
100 m of single conductor. There are loads of 60 A, 80 A and
30 A at distances 100 m, 150 m and 250 m respectively from one
end. If the cable is fed at 240 V at each end, calculate (*a*) the
current in each section and (*b*) the voltage across each
load. (App. El., L.U.)

49. A two-wire distributor AB, 1200 m long, is fed at end A at
246 V and at end B at 242 V. There are concentrated loads of
70 A and 110 A at points 200 m and 800 m respectively from end
A. The resistance of the distributor (go and return) is
$0.1\,\Omega/1000$ m. Calculate the p.d. across each load.

2 Electromagnetism

2.1 Magnetic field

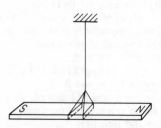

Fig. 2.1 A suspended permanent magnet

If a permanent magnet is suspended so that it is free to swing in a horizontal plane, as in fig. 2.1, it takes up a position such that a particular end points towards the earth's North Pole. That end is therefore said to be the *north-seeking* end of the magnet. Similarly, the other end is the *south-seeking* end. For short, these are referred to as the *north* (or N) and *south* (or S) poles respectively of the magnet.

Let us place a permanent magnet on a table, cover it over with a sheet of smooth cardboard and sprinkle steel filings uniformly over the sheet. Slight tapping of the latter causes the filings to set themselves in curved chains between the poles, as shown in fig. 2.2. The shape and density of these chains enable one to form a mental picture of the magnetic condition of the space or 'field' around a bar magnet and lead to the idea of *lines of magnetic flux*. It should be noted, however, that these lines of magnetic flux have no physical existence; they are purely imaginary and were introduced by Michael Faraday as a means of visualizing the distribution and density of a magnetic field. It is important to realize that the magnetic flux permeates the *whole* of the space occupied by that flux.

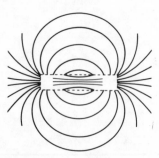

Fig. 2.2 Use of steel filings for determining distribution of magnetic field

2.2 Direction of magnetic field

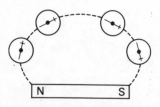

Fig. 2.3 Use of compass needles for determining direction of magnetic field

The direction of a magnetic field is taken as that in which the north-seeking pole of a magnet points when the latter is suspended in the field. Thus, if a bar magnet rests on a table and four compass needles are placed in positions indicated in fig. 2.3, it is found that the needles take up positions such that their axes coincide with the corresponding chain of filings (fig. 2.2) and their N poles are all pointing along the dotted line from the N pole of the magnet to its S pole. The lines of magnetic flux are assumed to pass through the magnet, emerge from the N pole and return to the S pole.

2.3 Characteristics of lines of magnetic flux

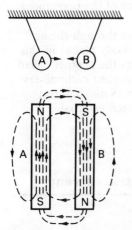

In spite of the fact that lines of magnetic flux have no physical existence, they do form a very convenient and useful basis for explaining various magnetic effects and for calculating their magnitudes. For this purpose, lines of magnetic flux are assumed to have the following properties:

1. *The direction of a line of magnetic flux at any point in a non-magnetic medium, such as air, is that of the north-seeking pole of a compass-needle placed at that point.*

2. *Each line of magnetic flux forms a closed loop*, as shown by the dotted lines in figs. 2.4 and 2.5. This means that a line of flux emerging from any point at the N-pole end of a magnet passes through the surrounding space back to the S-pole end and is then assumed to continue through the magnet to the point at which it emerged at the N-pole end.

3. *Lines of magnetic flux never intersect*. This follows from the fact that if a compass needle is placed in a magnetic field, its north-seeking pole will point in one direction only, namely in the direction of the magnetic flux at that point.

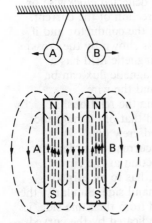

Fig. 2.4 Attraction between magnets

4. *Lines of magnetic flux are like stretched elastic cords, always trying to shorten themselves.* This effect can be demonstrated by suspending two permanent magnets, A and B, parallel to each other, with their poles arranged as in fig. 2.4. The distribution of the resultant magnetic field is indicated by the dotted lines. The lines of magnetic flux passing between A and B behave as if they were in tension, trying to shorten themselves and thereby causing the magnets to be attracted towards each other. In other words, unlike poles attract each other.

5. *Lines of magnetic flux which are parallel and in the same direction repel one another.* This effect can be demonstrated by suspending the two permanent magnets, A and B, with their N poles pointing in the same direction, as in fig. 2.5. It will be seen that in the space between A and B the lines of flux are practically parallel and are in the same direction. These flux lines behave as if they exerted a lateral pressure on one another, thereby causing magnets A and B to repel each other. Hence like poles repel each other.

Fig. 2.5 Repulsion between magnets

2.4 Magnetic induction and magnetic screening

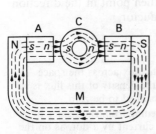

Fig. 2.6 Magnetic induction and screening

In fig. 2.6, N and S are the poles of a U-shaped permanent magnet M, A and B are soft-magnetic (e.g. steel) rectangular blocks attached to the magnet and C is a hollow cylinder of soft-magnetic material placed midway between A and B. The dotted lines in fig. 2.6 represent the paths of the magnetic flux due to the permanent magnet. It will be seen that this flux passes through A, B and C, making them into temporary magnets with the polarities indicated by *n* and *s*, i.e. A, B and C are magnetized by *magnetic induction*. Being magnetically soft, A, B and C lose almost the whole of their magnetism when they are removed from the influence of the permanent magnet M.

Fig. 2.6 also shows that no* flux passes through the air space inside cylinder C. Consequently, a body placed in this space would be found to be screened from the magnetic field around it. Magnetic screens are used to protect cathode-ray tubes (section 27.5) and instruments such as moving-iron ammeters and voltmeters (section 26.6) from external magnetic fields.

2.5 Magnetic field due to an electric current

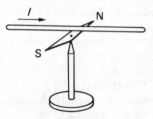

Fig. 2.7 Oersted's experiment

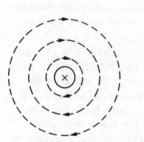

Fig. 2.8 Magnetic flux due to current in a straight conductor

When a conductor carries an electric current, a magnetic field is produced around that conductor — a phenomenon discovered by Oersted at Copenhagen in 1820. He found that when a wire carrying an electric current was placed above a magnetic needle (fig. 2.7) and in line with the normal direction of the latter, the needle was deflected clockwise or anticlockwise, depending upon the direction of the current. Thus it is found that if we look along the conductor and if the current is flowing away from us, as shown by the cross† inside the conductor in fig. 2.8, the magnetic field has a clockwise direction and the lines of magnetic flux can be represented by concentric circles around the wire.

A convenient method of representing the relationship between the direction of a current and that of its magnetic field is to place a corkscrew or a woodscrew (fig. 2.9) alongside the conductor carrying the current. In order that the screw may travel in the same direction as the current, namely towards the right in fig. 2.9, it has to be turned clockwise when viewed from the left hand side. Similarly, the direction of the magnetic field, viewed from the same side, is clockwise around the conductor, as indicated by the curved arrow F.

An alternative method of deriving this relationship is to grip the conductor with the *right* hand, with the thumb outstretched parallel to the conductor and pointing in the direction of the current; the fingers then point in the direction of the magnetic flux around the conductor.

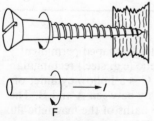

Fig. 2.9 Right-hand screw rule

---

* Actually there must be some magnetic flux across the space inside the soft-magnetic cylinder C, but the density of this flux is so low that, for most purposes, it can be assumed to be zero.

† It is usual to represent a current receding from the reader by a cross, as in fig. 2.8, and an approaching current by a dot, as on the left-hand conductor in fig. 2.12. The cross is the tail feathers of the departing arrow and the dot is the approaching point of the arrow.

2.6 Magnetic field of a solenoid

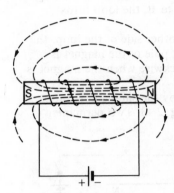

Fig. 2.10 Solenoid with a steel core

If a coil is wound on a steel rod, as in fig. 2.10, and connected to an accumulator, the steel becomes magnetized and behaves like a permanent magnet. The magnetic field of the electromagnet is represented by the dotted lines and its direction by the arrowheads.

The direction of the magnetic field produced by a current in a solenoid may be deduced by applying either the screw or the grip rule. Thus, if the axis of the screw is placed along that of the solenoid and if the screw is turned in the direction of the current, it travels in the direction of the magnetic field *inside* the solenoid, namely towards the right in fig. 2.10.

The grip rule can be expressed thus: if the solenoid is gripped with the *right* hand, with the fingers pointing in the direction of the current, then the thumb outstretched parallel to the axis of the solenoid points in the direction of the magnetic field *inside* the solenoid.

2.7 Force on a conductor carrying current across a magnetic field

In section 2.5 it was shown that a conductor carrying a current can produce a force on a magnet situated in the vicinity of the conductor. By Newton's Third Law of Motion, namely that to every force there must be an equal and opposite force, it follows that the magnet must exert an equal force on the conductor. One of the simplest methods of demonstrating this effect is to take a copper wire, about 2 mm in diameter, and bend it into a rectangular loop as represented by BC in fig. 2.11. The two tapered ends of the loop dip into mercury contained in cups, one directly above the other, the cups being attached to metal rods P and Q carried by a wooden upright rod D. A current of about 5 A is passed through the loop and the N pole of a permanent

Fig. 2.11 Force on conductor carrying current across a magnetic field

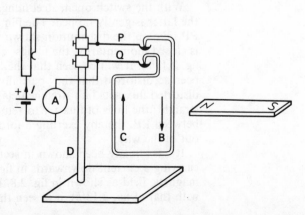

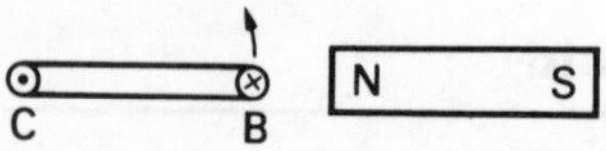

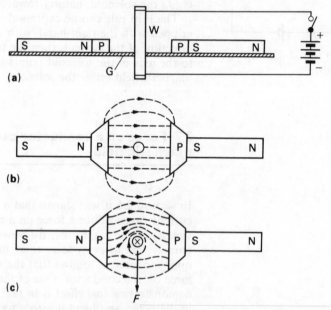

Fig. 2.12 Direction of force on conductor in fig. 2.11

magnet NS is moved towards B. If the current in this wire is flowing downwards, as indicated by the arrow in fig. 2.11, it is found that the loop, when viewed from above, turns counterclockwise, as shown in plan in fig. 2.12. If the magnet is reversed and again brought up to B, the loop turns clockwise.

If the magnet is placed on the other side of the loop, the latter turns clockwise when the N pole of the magnet is moved near to C, and counterclockwise when the magnet is reversed.

These effects can be explained by the simple apparatus shown in elevation and plan in fig. 2.13. Two permanent

Fig. 2.13 Flux distribution with and without current

magnets NS rest on a sheet of paper or glass G, and steel pole-pieces P are added to increase the area of the magnetic field in the gap between them. Midway between the pole-pieces is a wire W passing vertically downwards through G and connected through a switch to a 6-V battery capable of giving a very large current for a short time.

With the switch open, steel filings are sprinkled over G and the latter is gently tapped. The filings in the space between PP take up the distribution shown in fig. 2.13(*b*). If the switch is closed momentarily, the filings rearrange themselves as in fig. 2.13(*c*). It will be seen that the lines of magnetic flux have been so distorted that they partially surround the wire. This distorted flux acts like stretched elastic cords bent out of the straight; the lines of flux try to return to the shortest paths between PP, thereby exerting a force *F* urging the conductor out of the way.

It has already been shown in section 2.5 that a wire W carrying a current downwards in fig. 2.13(*a*) produces a magnetic field as shown in fig. 2.8. If this field is compared with that of fig. 2.13(*b*), it is seen that on the upper side the two fields are in the same direction, whereas on the lower side they are in opposition. Hence, the combined effect is to strengthen the magnetic field on the upper side and weaken it on the lower side, thus giving the distribution shown in fig. 2.13(*c*).

By combining diagrams similar to figs. 2.13(*b*) and 2.8, it is easy to understand that if either the current in W or the polarity of magnets NS is reversed, the field is strengthened on the lower side and weakened on the upper side of diagrams corresponding to fig. 2.13(*b*), so that the direction of the force acting on W is the reverse of that shown in fig. 2.13(*c*).

On the other hand, if both the current through W and the polarity of the magnets are reversed, the *distribution* of the resultant magnetic field and therefore the direction of the force on W remain unaltered.

By observation of the experiments, it can also be noted that the mechanical force exerted by the conductor always acts in a direction perpendicular to the plane of the conductor and the magnetic field direction. The direction is given by the left-hand rule illustrated in fig. 2.14.

The rule can be summarized as follows:

Fig. 2.14 Left-hand rule

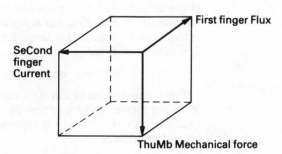

First finger Flux

SeCond finger Current

ThuMb Mechanical force

1. Hold the thumb, first finger and second finger of the left hand in the manner indicated by fig. 2.14, whereby they are mutually at right angles.
2. Point the First finger in the Field direction.
3. Point the seCond finger in the Current direction.
4. The thuMb then indicates the direction of the Mechanical force exerted by the conductor.

By trying this with your left hand, you can readily demonstrate that if either the current or the direction of the field is reversed then the direction of the force is also reversed. Also you can demonstrate that, if both current and field are reversed, the direction of the force remains unchanged.

2.8 Magnitude of the force on a conductor carrying current across a magnetic field

With the apparatus of fig. 2.11, it can be shown qualitatively that the force on a conductor carrying a current at right-angles to a magnetic field is increased (*a*) when the current in the conductor is increased and (*b*) when the magnetic field is made stronger by bringing the magnet nearer to the conductor. With the aid of more elaborate apparatus, the force on the conductor can be measured for various currents and various densities of the magnetic field, and it is found that:

force on conductor $\propto$ current $\times$ (flux density)

$\times$ (length of conductor)

If F = force on conductor in newtons,

I = current through conductor in amperes

and l = length, in metres, of conductor at right-angles to magnetic field,

F [newtons] $\propto$ flux density $\times l$ [metres] $\times I$ [amperes].

The *unit of flux density* is taken as *the density of a magnetic field such that a conductor carrying* 1 *ampere at right-angles to that field has a force of* 1 *newton per metre acting upon it*. This unit is termed a *tesla** (T). Hence, for a flux density of B teslas,

$$\text{force on conductor} = BlI \text{ newtons} \qquad (2.1)$$

For a magnetic field having a cross-sectional area of A square metres and a uniform flux density of B teslas, the *total flux* in *webers*† (Wb) is represented by the Greek capital letter Φ (phi), where

$$\Phi \text{ [webers]} = B \text{ [teslas]} \times A \text{ [metres}^2\text{]}$$

or B [teslas] = Φ [webers]$/A$ [metres^2]   (2.2)

i.e.   1 tesla = 1 weber per metre^2.

The *weber* may be defined either (i) as that magnetic flux which, when cut at a uniform rate by a conductor in 1 second, generates an e.m.f. of 1 volt (section 2.11), or (ii) as that magnetic flux which, linking a circuit of one turn, induces in it an e.m.f. of 1 volt when the flux is reduced to zero at a uniform rate in 1 second (section 2.12).

---

* Nikola Tesla (1857–1943), a Yugoslav who emigrated to U.S.A. in 1884, was a very famous electrical inventor. In 1888 he patented 2-phase and 3-phase synchronous generators and motors.

† Wilhelm Eduard Weber (1804–91), a German physicist, was the first to develop a system of absolute electrical and magnetic units.

Example 2.1   *A conductor carries a current of 800 A at right-angles to a magnetic field having a density of 0.5 T. Calculate the force on the conductor in newtons per metre length.*

From expression (2.1),

$$\text{force per metre length} = 0.5\,[\text{T}] \times 1\,[\text{m}] \times 800\,[\text{A}]$$
$$= 400\,\text{N}.$$

2.9 Electromagnetic induction

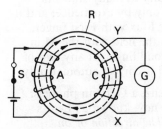

Fig. 2.15 Electromagnetic induction

On 29 August 1831, Michael Faraday (1791–1867) made the great discovery of *electromagnetic induction*, namely a method of obtaining an electric current with the aid of magnetic flux. He wound two coils, A and C, on a steel ring R, as in fig. 2.15, and found that when switch S was closed, a deflection was obtained on galvanometer G; and that when S was opened, G was deflected in the reverse direction. A few weeks later he found that when a permanent magnet NS was moved relative to a coil C (fig. 2.16), galvanometer G was deflected in one direction when the magnet was moved towards the coil and in the reverse direction when the magnet was withdrawn; and it was this experiment that finally convinced Faraday that an electric current could be produced by the movement of magnetic flux relative to a coil. Faraday also showed that the magnitude of the induced e.m.f. is proportional to the rate at which the magnetic flux passing through the coil is varied. Alternatively, we can say that when a conductor cuts or is cut by magnetic flux, an e.m.f. is generated in the conductor and the magnitude of the generated e.m.f. is proportional to the rate at which the conductor cuts or is cut by the magnetic flux.

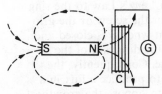

Fig. 2.16 Electromagnetic induction

2.10 Direction of induced e.m.f.

Two methods are available for deducing the direction of the induced or generated e.m.f., namely (*a*) Fleming's* Right-hand Rule and (*b*) Lenz's Law. The former is empirical, but the latter is fundamental in that it is based upon electrical principles.

(a) Fleming's Right-hand Rule *If the first finger of the right hand be pointed in the direction of the magnetic flux, as in fig. 2.17, and if the thumb be pointed in the direction of motion*

---

* John Ambrose Fleming (1849–1945) was Professor of Electrical Engineering at University College, London.

Fig. 2.17 Fleming's right-hand rule

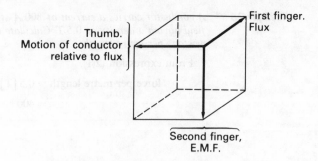

of the conductor **relative** *to the magnetic field, then the second finger, held at right-angles to both the thumb and the first finger, represents the direction of the e.m.f.* The manipulation of the thumb and fingers and their association with the correct quantity present some difficulty to many students. Easy manipulation can be acquired only by experience; and it may be helpful to associate **F**ield or **F**lux with **F**irst finger, **M**otion of the conductor relative to the field with the **M** in thu**M**b and e.m.f. with the **E** in s**E**cond finger. If any two of these are correctly applied, the third is correct automatically.

(b) Lenz's Law In 1834 Heinrich Lenz, a German physicist (1804–65), enunciated a simple rule, now known as Lenz's Law, which can be expressed thus: *The direction of an induced e.m.f. is always such that it tends to set up a current opposing the motion or the change of flux responsible for inducing that e.m.f.*

Let us consider the application of Lenz's Law to the ring shown in fig. 2.15. By applying either the screw or the grip rule given in section 2.6, we find that when S is closed and the battery has the polarity shown, the direction of the magnetic flux in the ring is clockwise. Consequently, the current in C must be such as to try to produce a flux in a counterclockwise direction, tending to oppose the growth of the flux due to A, namely the flux which is responsible for the e.m.f. induced in C. But a counterclockwise flux in the ring would require the current in C to be passing through the coil from X to Y (fig. 2.15). Hence, this must also be the direction of the e.m.f. induced in C.

2.11 Magnitude of the generated or induced e.m.f.

Fig. 2.18 represents the elevation and plan of a conductor AA situated in an airgap between poles NS. Suppose AA to be carrying a current, *I* amperes, in the direction shown. By applying either the screw or the grip rule of section 2.5, it is found that the effect of this current is to strengthen the field on the right and weaken that on the left of A, so that there is a force of *BlI* newtons (section 2.8) urging the conductor

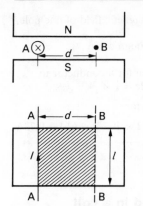

Fig. 2.18 Conductor moved
across magnetic field

towards the left, where B is the flux density in teslas and l is the length in metres of conductor in the magnetic field. Hence, a force of this magnitude has to be applied in the opposite direction to move A towards the right.

The work done in moving conductor AA through a distance d metres to position BB in fig. 2.18 is $(Bll \times d)$ joules. If this movement of AA takes place at a uniform velocity in t seconds, the e.m.f. induced in the conductor is constant at, say, E volts. Hence the electrical power generated in AA is IE watts and the electrical energy is IEt watt-seconds or joules. Since the mechanical energy expended in moving the conductor horizontally across the gap is all converted into electrical energy, then

$$IEt = Blld$$

$$\therefore \quad E = Bld/t = Blv \text{ volts,}$$

where v is the velocity in metres per second. But $Bld =$ the total flux, Φ webers, in the area shown shaded in fig. 2.18. This flux is cut by the conductor when the latter is moved from AA to BB. Hence

$$E \text{ [volts]} = \frac{\Phi \text{ [webers]}}{t \text{ [seconds]}} \tag{2.3}$$

i.e. the e.m.f., in volts, generated in a conductor is equal to the rate (in webers per second) at which the magnetic flux is cutting or being cut by the conductor; and the *weber* may therefore be defined as *that magnetic flux which, when cut at a uniform rate by a conductor in* 1 *second, generates an e.m.f. of* 1 *volt.*

In general, if a conductor cuts or is cut by a flux of $d\Phi$ webers in dt seconds,

$$\text{e.m.f. generated in conductor} = d\Phi/dt \text{ volts.} \tag{2.4}$$

Example 2.2   *Calculate the e.m.f. generated in the axle of a car travelling at* 80 *km/h, assuming the length of the axle to be* 2 *m and the vertical component of the earth's magnetic field to be* 40 *μT (microteslas).*

$$80 \text{ km/h} = \frac{(80 \times 1000) \text{ [m]}}{3600 \text{ [s]}}$$

$$= 22.2 \text{ m/s}$$

Vertical component of earth's field $= 40 \times 10^{-6}$ T,

$$\therefore \quad \text{flux cut by axle} = 40 \times 10^{-6} \text{ [T]} \times 2 \text{ [m]}$$

$$\times 22.2 \text{ [m/s]}$$

$$= 1776 \times 10^{-6} \text{ Wb/s}$$

and   e.m.f. generated in axle $= 1776 \times 10^{-6}$ V

$$= 1776 \text{ } \mu\text{V.}$$

Example 2.3   *A four-pole generator has a magnetic flux of* 12 *mWb per pole. Calculate the average value of the e.m.f. generated in one of the armature conductors while it is moving through the magnetic flux of one pole, if the armature is driven at* 900 *r/min.*

When a conductor moves through the magnetic field of one pole, it cuts a magnetic flux of 12 mWb.

Time taken for a conductor to move through one revolution $= \frac{60}{900} = \frac{1}{15}$ s.

Since the machine has 4 poles, time taken for a conductor to move through the magnetic field of one pole $= \frac{1}{4} \times \frac{1}{15} = \frac{1}{60}$ s,

$\therefore$ average e.m.f. generated in one conductor

$$= (12 \times 10^{-3}) \div (\tfrac{1}{60}) = 0.72 \text{ V}.$$

2.12 Magnitude of e.m.f. induced in a coil

Suppose the magnetic flux through a coil of N turns to be increased by Φ webers in t seconds due to, say, the relative movement of the coil and a magnet (fig. 2.16). Since each of the lines of magnetic flux cuts* each turn, one turn can be regarded as a conductor cut by Φ webers in t seconds; hence, from expression (2.3), the average e.m.f. induced in each turn is Φ/t volts. The current due to this e.m.f., by Lenz's Law, tries to prevent the increase of flux, i.e. tends to set up an opposing flux. Thus, if the magnet NS in fig. 2.16 is moved towards coil C, the flux passing from left to right through the latter is increased. The e.m.f. induced in the coil circulates a current in the direction represented by the dot and cross in fig. 2.19, where — for simplicity — coil C is represented as one turn. The effect of this current is to distort the magnetic field as shown by the dotted lines, thereby tending to push the coil away from the magnet. By Newton's Third Law of Motion, there must be an equal and opposite force tending to oppose the movement of the magnet.

Owing to the fact that the induced e.m.f. circulates a current tending to oppose the increase of flux through the coil, its direction is regarded as negative; hence

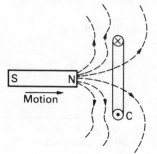

Fig. 2.19 Distortion of magnetic field by induced current

$$\left.\begin{array}{c}\text{average e.m.f. induced}\\ \text{in 1 turn}\end{array}\right\} = -\Phi/t \text{ volts}$$

$$= -\left\{\begin{array}{l}\text{average rate of } \textit{change } \text{of}\\ \text{flux in webers per second}\end{array}\right.$$

---

*It is immaterial whether we consider the e.m.f. as being due to change of flux linked with a coil or due to the coil cutting or being cut by lines of flux; the result is exactly the same. The fact of the matter is that we do not know what is really happening; but we can calculate the effect by imagining the magnetic field in the form of lines of flux, some of which expand from nothing when the field is increased or collapse to nothing when the field is reduced. In so doing they may be regarded as cutting the turns of the coil, or alternatively, the effect may be regarded as being due merely to a change in the number of flux-linkages.

and

$$\left.\begin{array}{l} \text{average e.m.f. induced} \\ \qquad \text{in coil} \end{array}\right\} = -N\Phi/t \text{ volts} \qquad (2.5)$$

$$= -\left\{\begin{array}{l} \text{average rate of } \textit{change} \text{ of} \\ \qquad \text{flux-linkages per second.} \end{array}\right.$$

The term 'flux-linkages' merely means the product of the flux in webers and the number of turns with which the flux is linked. Thus if a coil of 20 turns has a flux of 0.1 weber through it, the flux-linkages = $0.1 \times 20 = 2$ weber-turns.

From expression (2.5) it follows that:

instantaneous value of e.m.f., in volts, induced in a coil

= −rate of change of flux-linkages, in weber-turns per second

or
$$e = -\frac{\mathrm{d}}{\mathrm{d}t}(N\Phi) \text{ volts} \qquad (2.6)$$

This relationship is usually known as *Faraday's Law*, though it was not stated in this form by Faraday.

From expression (2.5) we can define* the *weber as that magnetic flux which, linking a circuit of one turn, induces in it an e.m.f. of 1 volt when the flux is reduced to zero at a uniform rate in 1 second.*

Next, let us consider the case of the two coils, A and C, shown in fig. 2.15. Suppose that when switch S is closed, the flux in the ring increases by Φ webers in t seconds. Then if coil A has N_1 turns,

average e.m.f. induced in A = $-N_1\Phi/t$ volts.

The minus sign signifies that this e.m.f., in accordance with Lenz's Law, is acting in opposition to the e.m.f. of the battery, thereby trying to prevent the growth of the current. This interpretation of the minus sign is a matter of convention, and is explained in section 4.3.

If coil C is wound with N_2 turns, and if all the flux produced by coil A passes through C,

average e.m.f. induced in C = $-N_2\Phi/t$ volts.

In this case the minus sign signifies that the e.m.f. circulates a current in such a direction as to tend to set up a flux in opposition to that produced by the current in coil A, thereby delaying the growth of flux in the ring.

In general, if the magnetic flux through a coil increases by $\mathrm{d}\Phi$ weber in $\mathrm{d}t$ second,

e.m.f. induced in coil = $-N \cdot \mathrm{d}\Phi/\mathrm{d}t$ volts $\qquad (2.7)$

Example 2.4   *A magnetic flux of 400 μWb passing through a coil of 1200 turns is reversed in 0.1 s. Calculate the average value of the e.m.f. induced in the coil.*

The magnetic flux has to decrease from 400 μWb to zero and then increase to 400 μWb in the reverse direction; hence the *increase* of flux in the original direction is $-800\ \mu$Wb.

* This is an alternative to the definition already given on p. 43.

Substituting in expression (2.5), we have:

$$\text{average e.m.f. induced in coil} = -\frac{1200 \times (-800 \times 10^{-6})}{0.1}$$

$$= 9.6\,\text{V}.$$

This e.m.f. is positive because its direction is the same as the original direction of the current, at first tending to prevent the current decreasing and then tending to prevent it increasing in the reverse direction.

Summary of important formulae

$$\text{Force on conductor} = BlI \text{ newtons} \qquad (2.1)$$

$$\text{E.M.F. generated in conductor} = Blv \text{ volts}$$

$$= d\Phi/dt \text{ volts} \qquad (2.4)$$

$$\text{E.M.F. induced in coil} = -\frac{d}{dt}(N\Phi) \text{ volts} \qquad (2.6)$$

EXERCISES 2

1. A current-carrying conductor is situated at right-angles to a uniform magnetic field having a density of 0.3 T. Calculate the force (in newtons per metre length) on the conductor when the current is 200 A.

2. Calculate the current in the conductor referred to in Q. 1 when the force per metre length of the conductor is 15 N.

3. A conductor, 150 mm long, is carrying a current of 60 A at right-angles to a magnetic field. The force on the conductor is 3 N. Calculate the density of the field.

4. The coil of a moving-coil loudspeaker has a mean diameter of 30 mm and is wound with 800 turns. It is situated in a radial magnetic field of 0.5 T. Calculate the force on the coil, in newtons, when the current is 12 mA.

5. The armature of a certain motor has 900 conductors and the current per conductor is 24 A. The flux density in the airgap under the poles is 0.6 T. The armature core is 160 mm long and has a diameter of 250 mm. Assume that the core is smooth (i.e. there are no slots and the winding is on the cylindrical surface of the core) and also assume that only two-thirds of the conductors are simultaneously in the magnetic field. Calculate (*a*) the torque in newton metres and (*b*) the mechanical power developed, in kilowatts, if the speed is 700 r/min.

   (*Note.* In the case of slotted cores, the flux density in the slots is very low, so that there is very little torque on the conductors; nearly all the torque is exerted on the teeth.)

6. Explain what happens when a long straight conductor is moved through a uniform magnetic field at constant velocity. Assume that the conductor moves perpendicularly to the field.

   If the ends of the conductor are connected together through an ammeter, what will happen?

   A conductor, 0.6 m long, is carrying a current of 75 A and is
placed at right-angles to a magnetic field of uniform flux
density. Calculate the value of the flux density if the mechanical
force on the conductor is 30 N.   (U.L.C.I., O.1)

7. State Lenz's Law.

   A conductor, 500 mm long, is moved at a uniform speed at
right-angles to its length and to a uniform magnetic field having
a density of 0.4 T. If the e.m.f. generated in the conductor is 2 V
and the conductor forms part of a closed circuit having a
resistance of 0.5 Ω, calculate: (i) the velocity of the conductor in
metres per second; (ii) the force acting on the conductor in
newtons; (iii) the work done in joules when the conductor has
moved 600 mm.   (U.E.I., O.1)

8. A wire, 100 mm long, is moved at a uniform speed of 4 m/s at
right-angles to its length and to a uniform magnetic field.
Calculate the density of the field if the e.m.f. generated in the
wire is 0.15 V.

   ، If the wire forms part of a closed circuit having a total
resistance of 0.04 Ω, calculate the force on the wire in newtons.

9. Give three practical applications of the mechanical force exerted
on a current-carrying conductor in a magnetic field.

   A conductor of active length 30 cm carries a current of 100 A
and lies at right-angles to a magnetic field of density 0.4 T.
Calculate the force in newtons exerted on it. If the force causes
the conductor to move at a velocity of 10 m/s, calculate (*a*) the
e.m.f. induced in it and (*b*) the power in watts developed
by it.   (E.M.E.U., O.1)

10. The axle of a certain motor car is 1.5 m long. Calculate the
e.m.f. generated in it when the car is travelling at 140 km/h.
Assume the vertical component of the earth's magnetic field to
be 40 μT.

11. An aeroplane having a wing span of 50 m is flying horizontally
at a speed of 600 km/h. Calculate the e.m.f. generated between
the wing tips, assuming the vertical component of the earth's
magnetic field to be 40 μT. Is it possible to measure this e.m.f.?

12. A copper disc, 250 mm in diameter, is rotated at 300 r/min
about a horizontal axis through its centre and perpendicular to
its plane. If the axis points magnetic north and south, calculate
the e.m.f. between the circumference of the disc and the axis.
Assume the horizontal component of the earth's field to be
18 μT.

13. A coil of 1500 turns gives rise to a magnetic flux of 2.5 mWb
when carrying a certain current. If this current is reversed in
0.2 s, what is the average value of the e.m.f. induced in the coil?

14. A short coil of 200 turns surrounds the middle of a bar magnet.
If the magnet sets up a flux of 80 μWb, calculate the average
value of the e.m.f. induced in the coil when the latter is removed
completely from the influence of the magnet in 0.05 s.

15. The flux through a 500-turn coil increases uniformly from zero
to 200 μWb in 3 ms. It remains constant for the fourth
millisecond and then decreases uniformly to zero during the fifth
millisecond. Draw to scale a graph representing the variation of
the e.m.f. induced in the coil.

16. State Lenz's Law. The field coils of a 6-pole d.c. generator, each
having 500 turns, are connected in series. When the field is
excited, there is a magnetic flux of 0.02 Wb/pole. If the field
circuit is opened in 0.02 s and the residual magnetism is
0.002 Wb/pole, calculate the average e.m.f. induced across the
field terminals. In which direction is this e.m.f. directed relative
to the direction of the field current?   (U.L.C.I., O.1)

17. Two coils, A and B, are wound on the same ferromagnetic core. There are 300 turns on A and 2800 turns on B. A current of 4 A through coil A produces a flux of 800 μWb in the core. If this current is reversed in 20 ms, calculate the average e.m.f.s induced in coils A and B.

18. A six-pole motor has a magnetic flux of 0.08 Wb per pole and the armature is rotating at 700 r/min. Calculate the average e.m.f. generated per conductor.

19. A four-pole armature is to generate an average e.m.f. of 1.4 V per conductor, the flux per pole being 15 mWb. Calculate the speed at which the armature must rotate.

3 Magnetic Circuit

3.1 Introductory

One of the characteristics of lines of magnetic flux is that each line is a closed loop (section 2.3); for instance, in fig. 2.10, the dotted lines represent the flux passing through the steel core and returning through the surrounding air space. The complete closed path followed by any group of lines of magnetic flux is referred to as a *magnetic circuit*. One of the simplest forms of magnetic circuit is shown in fig. 2.15, where the steel ring R provides the path for the magnetic flux.

3.2 Magnetomotive force; magnetic field strength

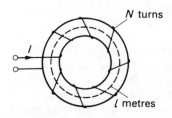

Fig. 3.1 A toroid

In an electric circuit, the current is due to the existence of an electromotive force. By analogy, we may say that in a magnetic circuit the magnetic flux is due to the existence of a *magnetomotive force* (m.m.f.) caused by a current flowing through one or more turns. The value of the m.m.f. is proportional to the current and to the number of turns, and is descriptively expressed in *ampere turns*; but for the purpose of dimensional analysis, it is expressed in *amperes*, since the number of turns is dimensionless. Hence the unit of magnetomotive force is the *ampere*.

If a current of I amperes flows through a coil of N turns, as shown in fig. 3.1, the magnetomotive force is the *total* current linked with the magnetic circuit, namely IN amperes. If the magnetic circuit is homogeneous and of uniform cross-sectional area, the magnetomotive force per metre length of the magnetic circuit is termed the *magnetic field strength* and is represented by the symbol H. Thus, if the mean length of the magnetic circuit of fig. 3.1 is l metres,

$$H = IN/l \text{ amperes per metre} \tag{3.1}$$

3.3 Permeability of free space or magnetic constant

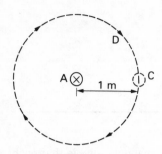

Fig. 3.2 Magnetic field at 1-metre radius due to current in a long straight conductor

Suppose A in fig. 3.2 to represent the cross-section of a long straight conductor, situated in a vacuum and carrying a current of one ampere towards the paper; and suppose the return path of this current to be some considerable distance away from A so that the effect of the return current on the magnetic field in the vicinity of A may be neglected. The lines of magnetic flux surrounding A will, by symmetry, be in the form of concentric circles, as already described in section 2.5, and the dotted circle D in fig. 3.2 represents the path of one of these lines of flux at a radius of 1 metre. Since conductor A and its return conductor form one turn, the magnetomotive force acting on path D is 1 ampere; and since the length of this line of flux is 2π metres, the magnetic field strength, H, at a radius of 1 m is $1/(2\pi)$ amperes per metre.

If the flux density in the region of line D is B teslas, it follows from expression (2.1) that the force per metre length on a conductor C (parallel to A) carrying 1 ampere at right-angles to this flux is given by:

Force per metre length $= B\,[\text{T}] \times 1\,[\text{m}] \times 1\,[\text{A}] = B$ newtons.

But from the definition of the ampere given in section 1.6(*a*), this force is 2×10^{-7} newton,

$\therefore$ flux density at 1-m radius from conductor carrying 1 A

$$= B = 2 \times 10^{-7} \text{ tesla}.$$

Hence,

$$\frac{\text{flux density at C}}{\text{magnetic field strength at C}} = \frac{B}{H} = \frac{2 \times 10^{-7}\,[\text{T}]}{1/2\pi\,[\text{A/m}]}$$

$$= 4\pi \times 10^{-7} \text{ H/m.*}$$

The ratio B/H for the above condition is termed the *permeability of free space* and is represented by the symbol μ_0. The value of this ratio is almost exactly the same whether the conductor A of fig. 3.2 is assumed to be situated in a vacuum (or free space) or in air or in any other non-magnetic material. Hence,

$$\mu_0 = \frac{B}{H} \text{ for a vacuum and non-magnetic materials}$$

$$= 4\pi \times 10^{-7} \text{ H/m} \tag{3.2}$$

and magnetic field strength for non-magnetic materials

$$= H = \frac{B}{\mu_0} = \frac{B}{4\pi \times 10^{-7}} \text{ amperes per metre} \tag{3.3}$$

---

* It is shown in the footnote on p. 97 that the units of absolute permeability are *henrys per metre*; e.g., $\mu_0 = 4\pi \times 10^{-7}$ H/m.

Example 3.1   *A coil of 200 turns is wound uniformly over a wooden ring having a mean circumference of 600 mm and a uniform cross-sectional area of 500 mm². If the current through the coil is 4 A, calculate (a) the magnetic field strength, (b) the flux density and (c) the total flux.*

(*a*) Mean circumference = 600 mm = 0.6 m,

$$\therefore \qquad H = 4 \times 200/0.6 = 1333 \text{ A/m}.$$

(*b*) From expression (3.2):

$$\text{flux density} = \mu_0 H = 4\pi \times 10^{-7} \times 1333$$

$$= 0.001\,675\text{ T} = 1675\,\mu\text{T}.$$

(*c*) Cross-sectional area = 500 mm² = 500 × 10⁻⁶ m²

$$\therefore \qquad \text{total flux} = 1675\,[\mu\text{T}] \times (500 \times 10^{-6})\,[\text{m}^2]$$

$$= 0.8375\,\mu\text{Wb}.$$

Example 3.2   *Calculate the magnetomotive force required to produce a flux of 0.015 Wb across an airgap 2.5 mm long, having an effective area of 200 cm².*

$$\text{Area of airgap} = 200 \times 10^{-4} = 0.02 \text{ m}^2$$

$$\therefore \qquad \text{flux density} = \frac{0.015\,[\text{Wb}]}{0.02\,[\text{m}^2]} = 0.75 \text{ T}.$$

From expression (3.3),

$$\text{magnetic field strength for gap} = \frac{0.75}{4\pi \times 10^{-7}} = 597\,000 \text{ A/m}.$$

$$\text{Length of gap} = 2.5 \text{ mm} = 0.0025 \text{ m},$$

∴   m.m.f. required to send flux across gap

$$= 597\,000\,[\text{A/m}] \times 0.0025\,[\text{m}] = 1492 \text{ A}.$$

3.4 Relative permeability

In section 2.6 it was shown that the magnetic flux inside a coil is intensified when a steel core is inserted. It follows that if the non-magnetic core of a toroid, such as that shown in fig. 3.1, is replaced by a steel core, the flux produced by a given m.m.f. is greatly increased; and *the ratio of the flux density produced in a material to the flux density produced in a vacuum* (or in a non-magnetic core) *by the same magnetic field strength* is termed the *relative permeability* and is denoted by the symbol μ_r. For air, $\mu_r = 1$; but for certain nickel-iron alloys, it may be as high as 100 000.

The value of the relative permeability of a ferromagnetic* material varies considerably for different values of the magnetic field strength, and it is usually convenient to

* A ferromagnetic material is one containing iron, cobalt, nickel or gadolinium. Steels contain iron and are the most common form of ferromagnetic materials.

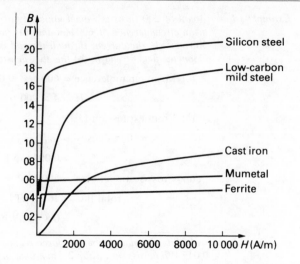

Fig. 3.3 Magnetization
characteristics of soft-
magnetic materials

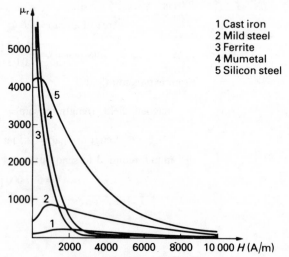

Fig. 3.4 μ_r/H characteristics
for soft-magnetic materials

represent the relationship between the flux density and the
magnetic field strength graphically as in fig. 3.3; and the
curves in figs. 3.4 and 3.5 represent the corresponding values
of the relative permeability plotted against the magnetic field
strength and the flux density respectively.

From expression (3.2), $B = \mu_0 H$ for a non-magnetic
material; hence, for a material having a relative
permeability μ_r,

$$B = \mu_r \mu_0 H$$

$\therefore$ *absolute permeability* $\mu = B/H = \mu_r \mu_0$ (3.4)

3.5 Reluctance

Let us consider a ferromagnetic ring having a cross-sectional
area of A square metres and a mean circumference of

Fig. 3.5 μ_r/B characteristics
for soft-magnetic materials

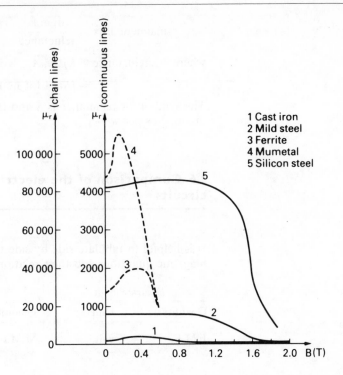

μ_r (chain lines)

μ_r (continuous lines)

1 Cast iron
2 Mild steel
3 Ferrite
4 Mumetal
5 Silicon steel

l metres (fig. 3.1), wound with N turns carrying a current
I amperes; then

$$\text{total flux} = \Phi = \text{flux density} \times \text{area} = BA \qquad (3.5)$$

and $\qquad$ m.m.f. $= F = $ magnetic field strength

$$\times \text{ length} = Hl \qquad (3.6)$$

Dividing (3.5) by (3.6), we have:

$$\frac{\Phi}{F} = \frac{BA}{Hl} = \mu_r \mu_0 \times \frac{A}{l}$$

$$\therefore \qquad \Phi = \frac{F}{l/\mu_r \mu_0 A} \qquad (3.7)$$

so that $\qquad \dfrac{F}{\Phi} = l/\mu_r \mu_0 A$

$$= \textit{reluctance} \text{ of magnetic circuit}.$$

This expression is similar in form to:

$$\frac{E}{I} = \rho l/A$$

for the electric circuit. The denominator, $l/\mu_r \mu_0 A$, in
expression (3.7) is similar in form to $\rho l/A$ for the resistance of
a conductor except that the absolute permeability, $\mu_r \mu_0$, for
the magnetic material corresponds to the reciprocal of the
resistivity, namely the conductivity of the electrical material.

$\qquad$ Since the m.m.f. is equal to the total number of amperes
($= IN$) acting on the magnetic circuit,

$$\therefore \quad \text{magnetic flux} = \frac{\text{m.m.f.}}{\text{reluctance}} \qquad (3.8)$$

where   reluctance $= l/\mu_r\mu_0 A$ $\qquad (3.9)$

$$= l/\mu_0 A \text{ for non-magnetic materials.}$$

The symbol for reluctance is S and the quantity is expressed in amperes per weber.

3.6 Comparison of the electric and magnetic circuits

It is helpful to tabulate side by side the various electric and magnetic quantities and their relationships, thus:

Electric circuit		Magnetic circuit	
Quantity	Unit	Quantity	Unit
E.M.F.	volt	M.M.F.	ampere
—	—	Magnetic field strength	ampere per metre
Current	ampere	Magnetic flux	weber
Current density	ampere per metre^2	Magnetic flux density	tesla
Resistance	ohm	Reluctance	ampere per weber
$\left(= \rho \cdot \dfrac{l}{A}\right)$		$\left(= \dfrac{1}{\mu_r\mu_0} \cdot \dfrac{l}{A}\right)$	
Current = e.m.f./resistance		Flux = m.m.f./reluctance	

One important difference between the electric and magnetic circuits is the fact that energy must be supplied to *maintain* the flow of electricity in a circuit, whereas the magnetic flux, once it is set up, does not require any further supply of energy. For instance, once the flux produced by a current in a solenoid has attained its maximum value, the energy subsequently absorbed by that solenoid is all dissipated as heat due to the resistance of the winding. However, the magnetic circuit stores energy in its field whereas the electric circuit immediately releases its energy as heat.

Example 3.3   *A mild-steel ring having a cross-sectional area of 500 mm^2 and a mean circumference of 400 mm has a coil of 200 turns wound uniformly around it. Calculate (a) the reluctance of the ring and (b) the current required to produce a flux of 800 μWb in the ring.*

(a)   Flux density in ring $= \dfrac{800 \times 10^{-6}\,[\text{Wb}]}{500 \times 10^{-6}\,[\text{m}^2]} = 1.6\,\text{T}.$

From fig. 3.5, the relative permeability of mild steel for a flux density of 1.6 T is about 380.

$$\therefore \quad \text{reluctance of ring} = \frac{0.4}{380 \times 4\pi \times 10^{-7} \times 5 \times 10^{-4}}$$

$$= 1.677 \times 10^6 \text{ A/Wb}.$$

(*b*) From expression (3.7),

$$800 \times 10^{-6} = \frac{\text{m.m.f.}}{1.677 \times 10^6}$$

$$\therefore \quad \text{m.m.f. } F = 1342 \text{ A}$$

and $\qquad$ magnetizing current $= F/N = 1342/200 = 6.7$ A.

Alternatively, from expression (3.4),

$$H = \frac{B}{\mu_r \mu_0} = \frac{1.6}{380 \times 4\pi \times 10^{-7}}$$

$$= 3350 \text{ A/m},$$

$$\therefore \quad \text{m.m.f.} = 3350 \times 0.4 = 1340 \text{ A}$$

and $\qquad$ magnetizing current $= 1340/200 = 6.7$ A.

3.7 Composite magnetic circuit

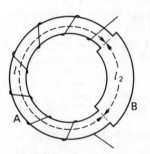

Fig. 3.6 Composite magnetic circuit

Suppose a magnetic circuit to consist of two specimens of steel, A and B, arranged as in fig. 3.6. If l_1 and l_2 be the mean lengths in metres of the magnetic circuits of A and B respectively, A_1 and A_2 their cross-sectional areas in square metres, and μ_1 and μ_2 their *absolute* permeabilities,

then $\qquad$ reluctance of A $= \dfrac{l_1}{\mu_1 A_1}$

and $\qquad$ reluctance of B $= \dfrac{l_2}{\mu_2 A_2}$.

If a coil is wound on core A as in fig. 3.6 and if the magnetic flux is assumed to be confined to the steel core, then

$$\text{total reluctance of magnetic circuit} = \frac{l_1}{\mu_1 A_1} + \frac{l_2}{\mu_2 A_2}$$

and $\quad$ total flux $= \Phi = \dfrac{\text{m.m.f. of coil}}{\text{total reluctance}} = \dfrac{NI}{\dfrac{l_1}{\mu_1 A_1} + \dfrac{l_2}{\mu_2 A_2}}$ $\quad$ (3.10)

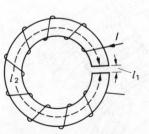

Fig. 3.7 Steel ring with an airgap

Fig. 3.7 shows a steel ring with an airgap. If the gap is very short, practically all the flux goes straight across from the one face to the other, so that the area of the gap may be assumed to be the same as that of the ring. Hence, if A is the cross-sectional area and l_1 and l_2 are the lengths of the steel core and gap respectively,

$$\text{total reluctance} = \frac{l_1}{\mu_1 A} + \frac{l_2}{\mu_0 A}$$

3.8 Magnetic leakage and fringing

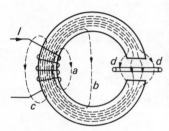

Fig. 3.8 Magnetic leakage and fringing

Suppose *dd* in fig. 3.8 to represent a metal ring symmetrically situated relative to the airgap in the steel ring, and suppose the magnetizing winding to be concentrated over a short length of the core. As far as ring *dd* is concerned, the flux passing through it can be regarded as the *useful* flux and that which returns by such paths as *a*, *b* and *c* is *leakage* flux.

The useful flux passing across the gap tends to bulge outwards as shown roughly in fig. 3.8, thereby increasing the effective area of the gap and reducing the flux density in the gap. This effect is referred to as *fringing*; and the longer the airgap, the greater is the fringing.

The distinction between useful and leakage fluxes may be more obvious if we consider an electrical machine. For instance, fig. 3.9 shows two poles of a six-pole machine. The armature slots have, for simplicity, been omitted. Some of the dotted lines do not enter the armature core and thus do not assist in generating an e.m.f. in the armature winding; consequently, they represent leakage flux. On the other hand, some of the flux passes between the pole tips and the armature core, as shown in fig. 3.9, and is referred to as fringing flux. Since this fringing flux is cut by the armature conductors, it forms part of the useful flux.

From figs. 3.8 and 3.9 it is seen that the effect of leakage flux is to increase the total flux through the exciting winding, and

$$\text{leakage factor} = \frac{\text{total flux through exciting winding}}{\text{useful flux}} \quad (3.11)$$

The value of the leakage factor for electrical machines is usually about 1.15–1.25.

Fig. 3.9 Magnetic leakage and fringing in a machine

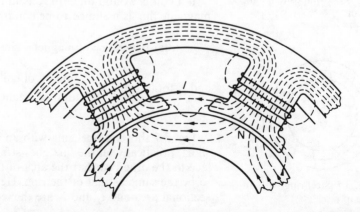

Example 3.4

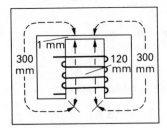

Fig. 3.10 Magnetic circuit for example 3.4

A magnetic circuit is made of mild steel arranged as in fig. 3.10. The centre limb is wound with 500 turns and has a cross-sectional area of 800 mm². Each of the outer limbs has a cross-sectional area of 500 mm². The airgap has a length of 1.0 mm. Calculate the current required to set up a flux of 1.3 mWb in the centre limb, assuming no magnetic leakage and fringing. The mean lengths of the various magnetic paths are shown on the diagram.

$$\text{Flux density in centre limb} = \frac{(1.3 \times 10^{-3})\,[\text{Wb}]}{(800 \times 10^{-6})\,[\text{m}^2]} = 1.625\,\text{T}.$$

From fig. 3.3, value of H for mild steel $\simeq 3800$ A/m,

$$\therefore \qquad \text{m.m.f. for centre limb} = 3800 \times 0.12 = 456\,\text{A}.$$

Since half the flux returns through one outer limb and half through the other, the two outer limbs are magnetically equivalent to a single limb having a cross-sectional area of 1000 mm² and a length of 300 mm,

$$\therefore \quad \text{flux density in outer limbs} = \frac{(1.3 \times 10^{-3})\,[\text{Wb}]}{(1000 \times 10^{-6})\,[\text{m}^2]} = 1.3\,\text{T}.$$

From fig. 3.3, value of H for mild steel $\simeq 850$ A/m

$$\text{m.m.f. for outer limbs} = 850 \times 0.3 = 255\,\text{A}.$$

$$\text{Flux density in airgap} = 1.625\,\text{T}$$

$$\therefore \qquad \text{value of } H \text{ for gap} = \frac{1.625}{4\pi \times 10^{-7}}$$

$$= 1.292 \times 10^6\,\text{A/m}$$

and $\qquad$ m.m.f. for gap $= 1.292 \times 10^6 \times 0.001$

$$= 1292\,\text{A}.$$

Hence, $\qquad$ total m.m.f. $= 456 + 255 + 1292$

$$= 2003\,\text{A}$$

and $\qquad$ magnetizing current $= 2003/500 = 4\,\text{A}.$

Example 3.5

A magnetic circuit is made up of steel laminations shaped as in fig. 3.11. The width of the core is 40 mm and the core is built up to a depth of 50 mm, of which 8 per cent is taken up by insulation between the laminations. The gap is 2.0 mm long and the effective area of the gap is 2500 mm². The coil is wound with 800 turns. If the leakage factor is 1.2, calculate the magnetizing current required to produce a flux of 0.0025 Wb across the airgap.

Fig. 3.11 Magnetic circuit for example 3.5

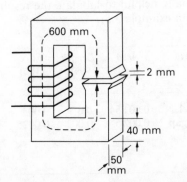

$$\text{Flux density in airgap} = \frac{(2.5 \times 10^{-3})\,[\text{Wb}]}{(2500 \times 10^{-6})\,[\text{m}^2]} = 1\,\text{T},$$

$$\therefore \qquad \text{value of } H \text{ for gap} = \frac{1\,[\text{T}]}{4\pi \times 10^{-7}\,[\text{H/m}]} = 796\,000\,\text{A/m}$$

and $\qquad$ m.m.f. for gap $= 796\,000 \times 0.002$

$$= 1592\,\text{A}.$$

Total flux through coil $=$ flux in gap $\times$ leakage factor

$$= 0.0025 \times 1.2 = 0.003\,\text{Wb}.$$

Since only 92 per cent of the cross-section of the core consists of steel,

$$\therefore \qquad \text{area of steel in core} = 40 \times 50 \times 0.92 = 1840\,\text{mm}^2$$

$$= 0.001\,84\,\text{m}^2$$

and $\quad$ flux density in core $= 0.003\,[\text{Wb}]/0.001\,84\,[\text{m}^2] = 1.63\,\text{T}.$

From fig. 3.3, value of H for laminations $\simeq 4000\,\text{A/m}$.

It will be evident from fig. 3.8 that when there is magnetic leakage, the flux density is not uniform over the whole length of the core. It is impossible, however, to allow for the variation of magnetic field strength due to this variation of flux density; and the usual practice is to assume that the magnetic field strength estimated for the region of maximum density applies to the whole of the core, thereby erring on the safe side.

Hence, $\qquad$ m.m.f. for core $= 4000 \times 0.6 = 2400\,\text{A}$

and $\qquad\qquad$ total m.m.f. $= 1592 + 2400 = 3992\,\text{A}$

$$\therefore \qquad \text{magnetizing current} = 3992/800 \simeq 5\,\text{A}.$$

It should be appreciated that owing to the difficulty of calculating the exact flux densities at various points of the magnetic circuit and to the uncertainty regarding the exact magnetic property of the ferromagnetic material, the results obtained in magnetic calculations are only approximately correct.

When we are dealing with composite magnetic circuits, it is usually helpful to tabulate the results thus (using the values from example 3.5):

Part	Area (m^2)	Length (m)	Flux (Wb)	Flux density (T)	Amperes/ metre (A/m)	M.M.F. (A)
Steel	0.001 84	0.6	0.003	1.63	4 000	2400
Airgap	0.002 5	0.002	0.0025	1.0	796 000	1592
					Total m.m.f. $=$	3992

3.9 Kirchhoff's Laws for the magnetic circuit

Kirchhoff's Laws for the electric circuit are given in section 1.14. These laws can also be applied to the magnetic circuit thus:

First Law

The total magnetic flux towards a junction is equal to the total magnetic flux away from that junction. This law follows from the fact that each line of flux forms a closed path; for instance, if a ferromagnetic core is arranged as shown in fig. 3.12 and if a coil C wound on limb L carries a current, the magnetic flux through C divides at P, some flux passing along limb M and the remainder along limb N, to join again at Q. There is no break or discontinuity in any of the lines of flux at P and Q; consequently the total flux from L towards P is exactly the same as the sum of the fluxes from P towards M and N, i.e.

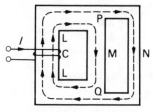

Fig. 3.12 Magnetic circuit to illustrate Kirchhoff's Laws

$$\Phi_L = \Phi_M + \Phi_N$$

or

$$\Phi_L - \Phi_M - \Phi_N = 0.$$

In general,

$$\sum \Phi = 0,$$

where $\sum$ represents the algebraic sum.

Second Law

In a closed magnetic circuit, the algebraic sum of the product of the magnetic field strength and the length of each part of the circuit is equal to the resultant magnetomotive force. For instance, if H_L is the magnetic field strength required for limb L and l_L is the length of the circuit from Q via L to P, and if H_M and l_M are the corresponding values for limb M and H_N and l_N are those for the limb extending from P via N to Q, then:

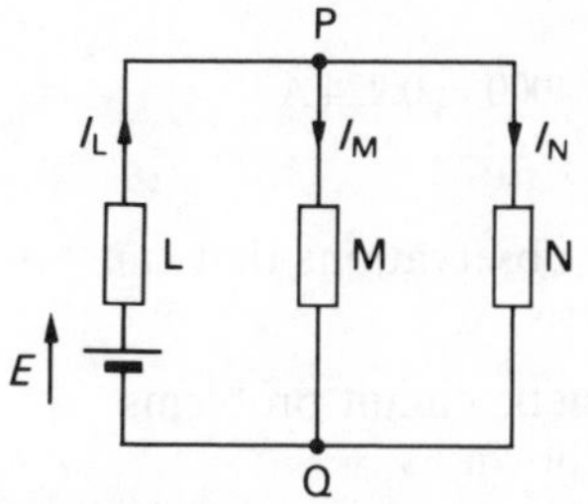

$$\text{total m.m.f. of coil C} = H_L l_L + H_M l_M$$
$$= H_L l_L + H_N l_N$$

and

$$0 = H_M l_M - H_N l_N.$$

In general,

$$\sum \text{m.m.f.} = \sum Hl.$$

Fig. 3.13 Electric circuit equivalent to fig. 3.12

It may be helpful to compare the magnetic circuit of fig. 3.12 with the corresponding electric circuit of fig. 3.13.

Example 3.6

A magnetic circuit is shown in fig. 3.14 and is of uniform cross-section throughout. The ferromagnetic core has a mean length of 220 mm and cross-sectional area 50 mm². The airgap has a length of 1.0 mm and the same effective cross-sectional area. A coil of 4000 turns is wound on the core and the magnetic characteristic of the material is also given in fig. 3.14. Estimate the current in the coil to produce a flux density of 0.9 T in the airgap, assuming that all flux passes through both parts of the magnetic circuit.

For the core,   flux density $B = 0.9\,\text{T}$.

The magnetization characteristic is given in the form of a nomogram which is a method of avoiding drawing the characteristic yet presents

Fig. 3.14 Magnetic circuit
and characteristic for
example 3.6

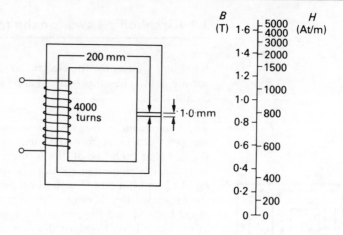

the same information in a graphic form. From the nomogram,

$$\text{value of } H = 820 \text{ A/m}$$

Using the magnetic interpretation of Kirchhoff's Second Law, we require to determine the m.m.f. both for the core and for the airgap in order to calculate the total m.m.f.

For the core,

$$\text{m.m.f. } F_c = H_c l_c = 820 \times 220 \times 10^{-3} = 180 \text{ A}$$

For the airgap,

$$H_g = \frac{B}{\mu_0 \mu_r} = \frac{0.9}{4\pi \times 10^{-7} \times 1} = 716\,000 \text{ A/m}$$

$$\text{m.m.f. } F_g = H_g l_g = 716\,000 \times 1 \times 10^{-3} = 716 \text{ A}$$

$$\text{Total m.m.f. } F = F_c + F_g = 180 + 716 = 896 \text{ A}$$

$$= IN.$$

$$\text{Magnetizing current } I = 896/4000 = 0.224 \text{ A}$$

This example leads to a number of observations that can be made about magnetic circuit analysis:

1. The form of solution in most magnetic circuit problems takes a standard form which can be shown as

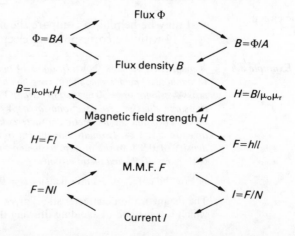

Most problems on magnetic circuits are solved by starting at or near one end of the above chain and moving along towards the other end. In example 3.6, we started at B and worked down the chain in order to obtain I.

2. The airgap m.m.f. calculation can be shortened by noting that a flux density of 1.0 T in air requires a magnetic field strength approximately 800 000 A/m. If this figure is remembered then the airgap m.m.f. in any problem can be quickly obtained by multiplying it by the required flux density. In example 3.6, the m.m.f. would have been given as $800\,000 \times 0.9 = 720\,000$ A/m which would have made a negligible difference to the final current estimated.

3. The magnetic circuit reluctances can be operated in a similar manner to electric circuit resistances; thus series reluctances can be added while parallel reluctances can be combined by the reciprocal method. In example 3.6, the solution could have been obtained as follows:

As before, $\quad B = 0.9\,\mathrm{T} \quad$ and $\quad H = 820\,\mathrm{A/m}$

For the core,

$$\text{relative permeability} = \frac{B}{\mu_0 H} = \frac{0.9}{4\pi \times 10^{-7} \times 820} = 873$$

hence

$$\text{reluctance } S_\mathrm{c} = \frac{l}{\mu_0 \mu_\mathrm{r} A} = \frac{1 \times 10^{-3}}{4\pi \times 10^{-7} \times 873 \times 50 \times 10^{-6}}$$

$$= 4\,010\,000\,\mathrm{A/Wb}$$

For the airgap,

$$\text{reluctance } S_\mathrm{g} = \frac{l}{\mu_0 \mu_\mathrm{r} A} = \frac{1 \times 10^{-3}}{4\pi \times 10^{-7} \times 1 \times 50 \times 10^{-6}}$$

$$= 15\,910\,000\,\mathrm{A/Wb}$$

The reluctances of the core and the airgap are effectively in series and therefore the total reluctance is given by

$$S = S_\mathrm{c} + S_\mathrm{g} = 4\,010\,000 + 15\,910\,000$$

$$= 19\,920\,000\,\mathrm{A/Wb}.$$

Total flux $\Phi = BA = 0.9 \times 50 \times 10^{-6} = 45 \times 10^{-6}\,\mathrm{Wb}$.

M.M.F. $F = \Phi S = 45 \times 10^{-6} \times 19\,920\,000 = 896\,\mathrm{A}$

hence current $I = 896/4000 = 0.224$ A as before.

4. In example 3.6, the relative importance of the airgap in the magnetic circuit is quite appreciable. The airgap is only 1.0 mm long yet its reluctance is approximately four times that of the steel core which has a length of 220 mm and also has a rather poor relative permeability for a steel. Had a better steel been used, the reluctance of the core would tend toward being negligible.

5. The airgap length was stated as 1.0 mm. This accuracy is required since even a length of 1.1 mm would increase the total reluctance by approximately 1600 000 A/Wb. As it is, the accuracy of the measurement of the airgap length is suspect and for this reason it is impossible to calculate the

reluctance of the airgap accurately. Hence to calculate the coil current is unrealistic and it is better to accept that such calculations are merely estimations.

6. Had there not been an airgap in the core of example 3.6, the current required to produce a flux density of 0.9 T would have been about 0.045 A instead of 0.224 A. This extra demand for current raises the question as to why an airgap should be introduced at all.

Generally, airgaps appear in magnetic circuits for two purposes. The first is to permit part of a magnetic circuit to move. A simple illustration of this is the magnetic relay shown in fig. 3.15, in which the closing of the airgap permits the movement of the crank which operates the switching contacts.

Fig. 3.15 Magnetic relay

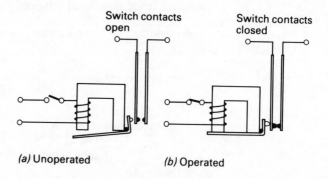

(a) Unoperated   *(b)* Operated

The second purpose of an airgap is to make the magnetization characteristic of the circuit more linear. In electrical machines such as transformers, a linear characteristic is most important in order to maintain a linear relationship between voltage and current. This importance will become apparent after you have read about electromagnetism in Chapter 4 and a.c. circuits in Chapters 6 and 7.

Fig. 3.16 shows the effect of the flux/m.m.f. characteristic on a steel-core magnetic circuit due to the introduction of a 0.2-mm airgap, and the introduction of a 1.0-mm airgap. For a rigid core, the cross-sectional area and the circuit length are constant, therefore the characteristic is similar to that of the magnetization (B/H) characteristic drawn to a different scale.

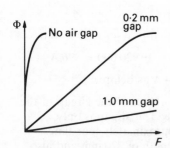

Fig. 3.16 Effect of airgaps on Φ/F characteristic

Because of the much larger m.m.f.s required for the airgaps, it might appear from fig. 13.16 that all of the characteristics shown are more or less linear and that no advantage is derived from the introduction of an airgap. However, the characteristics can be adjusted to the same horizontal range by plotting each to a base of the ratio of the m.m.f. to the corresponding saturation m.m.f., as shown in fig. 3.17. This diagram now clearly shows how the introduction of even a very small airgap makes the characteristic more linear and that it is not necessary to have a large airgap, as this would only make an unnecessary increase in the current required for magnetization.

Fig. 3.17 Airgap
linearization

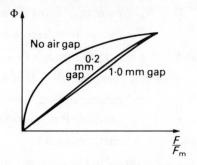

3.10 Determination of the magnetization curve for a ferromagnetic material

(a) By means of a fluxmeter

Fig. 3.18 shows a steel ring of uniform cross-section, uniformly wound with a coil P, thereby eliminating magnetic leakage. Coil P is connected to a battery through a reversing switch RS, an ammeter A and a variable resistor R_1. Another coil S, which need not be distributed around the ring, is connected through a two-way switch K to fluxmeter F which is a special type of permanent-magnet moving-coil instrument.

Fig. 3.18 Determination of
the magnetization curve for
a steel ring

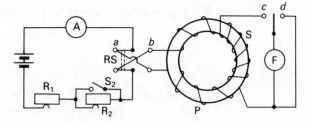

Current is led into and out of the moving coil of F by fine wires or ligaments so arranged as to exert negligible control over the position of the moving coil. When the flux in the ring is varied, the e.m.f. induced in S sends a current through the fluxmeter and produces a deflection that is proportional to the change of flux-linkages in coil S. Switch S_2 and resistor R_2 are not required for this test, but will be used when this circuit is employed for determining the hysteresis loop (section 3.12).

The current through coil P is adjusted to a desired value by means of R_1 and switch RS is then reversed several times to bring the steel into a 'cyclic' condition, i.e. into a condition such that the flux in the ring reverses from a certain value in one direction to the same value in the reverse direction. During this operation, switch K should be on d, thereby short-circuiting the fluxmeter. With switch RS on, say, a, switch K is moved over to c, the current through P is reversed by moving RS quickly over to b and the fluxmeter deflection is noted.

If N_P = number of turns on coil P,

l = mean circumference of the ring, in metres

and I = current through P, in amperes,

magnetic field strength = $H = IN_P/l$ amperes per metre.

If $\theta = \begin{cases} \text{fluxmeter deflection when current} \\ \quad \text{through P is reversed} \end{cases}$

and c = fluxmeter constant

= no. of weber-turns per unit of scale deflection,

change of flux-linkages with coil S = $c\theta$ (3.12)

If the flux in the ring changes from Φ to $-\Phi$ when the current through coil P is reversed, and if N_S is the number of turns on S,

$$\left.\begin{array}{c}\text{change of flux-linkages} \\ \text{with coil S}\end{array}\right\} = \text{change of flux} \times \text{no. of turns on S}$$

$$= 2\Phi N_S \qquad (3.13)$$

Equating (3.12) and (3.13), we have:

$$2\Phi N_S = c\theta$$

so that $$\Phi = \frac{c\theta}{2N_S} \text{ webers.}$$

If A = cross-sectional area of ring in square metres.

$$\text{flux density in ring} = B = \Phi/A$$

$$= \frac{c\theta}{2AN_S} \text{ teslas.} \qquad (3.14)$$

The test is performed with different values of the current; and from the data, a graph representing the variation of flux density with magnetic field strength can be plotted, as in fig. 3.3.

(b) By means of a ballistic galvanometer   A ballistic galvanometer has a moving coil suspended between the poles of a permanent magnet, but the coil is wound on a *non-metallic* former, so that there is very little damping when the coil has a resistor of high resistance in series. The first deflection or 'throw' is proportional to the number of coulombs discharged through the galvanometer if the duration of the discharge is short compared with the time of one oscillation.

If $\theta = \begin{cases} \text{first deflection or 'throw' of the ballistic galvanometer} \\ \quad \text{when the current through coil P is reversed} \end{cases}$

and k = ballistic constant of the galvanometer

= quantity of electricity in coulombs per unit deflection,

$$\left.\begin{array}{c}\text{quantity of electricity through} \\ \text{galvanometer}\end{array}\right\} = k\theta \text{ coulombs} \quad (3.15)$$

If Φ = flux produced in ring by I amperes through P and t = time, in seconds, of reversal of flux,

average e.m.f. induced in S = $N_S \times 2\Phi/t$ volts.

If R = total resistance of the secondary circuit,

quantity of electricity through ballistic galvanometer

= average current × time

$$= \frac{2\Phi N_S}{tR} \times t = 2\Phi N_S/R \text{ coulombs} \qquad (3.16)$$

Equating (3.15) and (3.16), we have:

$$k\theta = 2\Phi N_S/R$$

$$\therefore \qquad \Phi = \frac{k\theta R}{2N_S} \text{ webers}.$$

The values of the flux density, etc., can then be calculated as already described for method (*a*).

3.11 Hysteresis

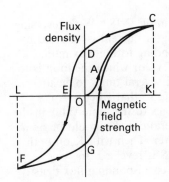

Fig. 3.19 Hysteresis loop

If we take a closed steel ring which has been completely demagnetized* and measure the flux density with increasing values of the magnetic field strength, the relationship between the two quantities is represented by curve OAC in fig. 3.19. If the value of H is then reduced, it is found that the flux density follows curve CD, and that when H has been reduced to zero, the flux density remaining in the steel is OD and is referred to as the *remanent flux density*.

If H is increased in the reverse direction, the flux density decreases, until at some value OE, the flux has been reduced to zero. The magnetic field strength OE required to wipe out the residual magnetism is termed the *coercive force*. Further increase of H causes the flux density to grow in the reverse direction as represented by curve EF. If the reversed magnetic field strength OL is adjusted to the same value as the maximum value OK in the initial direction, the final flux density LF is the same as KC.

If the magnetic field strength is varied backwards from OL to OK, the flux density follows a curve FGC similar to curve CDEF, and the closed figure CDEFGC is termed the *hysteresis loop*.

If hysteresis loops for a given steel ring are determined for different maximum values of the magnetic field strength, they are found to lie within one another, as shown in fig. 3.20. The apexes A, C, D and E of the respective loops lie on the B/H curve determined with increasing values of H. It will be seen

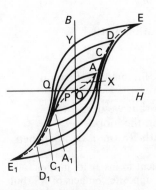

Fig. 3.20 A family of hysteresis loops

* The simplest method of demagnetizing is to reverse the magnetizing current a large number of times while, at the same time, gradually reducing the current to zero.

that the value of the remanent flux density depends upon the value of the peak magnetization; thus, for loop A, the remanent flux density is OX, whereas for loop E, corresponding to a maximum magnetization that is approaching saturation, the remanent flux density is OY. The value of the remanent flux density obtained when the maximum magnetization reaches the saturation value of the material is termed the *remanence* of that material. Thus for the material having the hysteresis loops of fig. 3.20, the remanence is approximately OY.

The value of the coercive force in fig. 3.20 varies from OP for loop AA_1 to OQ for loop EE_1; and the value of the coercive force when the maximum magnetization reaches the saturation value of the material is termed the *coercivity* of that material. Thus, for fig. 3.20, the coercivity is approximately OQ. The value of the coercivity varies enormously for different materials, being about 40 000 A/m for Alnico (an alloy of iron, aluminium, nickel, cobalt and copper, used for permanent magnets) and about 3 A/m for Mumetal (an alloy of nickel, iron, copper and molybdenum).

3.12 Determination of the hysteresis loop

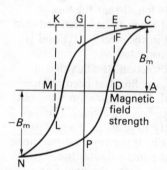

Fig. 3.21 Plotting a hysteresis loop

The circuit arrangement is shown in fig. 3.18.* With reversing switch RS on side *a* and S_2 closed, the current is adjusted by means of R_1 to the value corresponding to the maximum magnetic field strength OA (fig. 3.21) for which the loop is to be determined. RS is then reversed several times to ensure that the steel ring is in a cyclic state so that when OA is reversed, the flux density AC changes to the same value in the reverse direction. During these operations, K should be on *d*, thereby short-circuiting the fluxmeter. Then with K on *c*, the deflection θ_m of F is noted when RS is reversed. In section 3.10 it was shown that the corresponding flux density B_m is given by:

$$B_m = \frac{c}{2AN_S} \times \theta_m$$

Hence points C and N of fig. 3.21 can be plotted.
From expression (3.14), it follows that

$$\textit{change of flux density} = \frac{c}{AN_S} \times \text{fluxmeter deflection}$$

and it is with the aid of this formula that we derive the remaining points required for plotting the loop.
To derive portion CFJ of loop. With RS on *a*, S_2 is opened

---

* In actual practice, it is more convenient to insert R_2 in the connection between a pair of diagonally opposite contacts of RS; but the arrangement shown in fig. 3.18 makes it easier to follow the procedure.

and R_2 adjusted to give a current corresponding to a magnetic field strength OD (fig. 3.21). S_2 is then closed to bring the value of H back to OA, and RS is reversed several times to restore the steel to the cyclic condition. Fluxmeter deflection θ_1 is then noted when S_2 is opened.

$$\text{Corresponding change of flux density} = \frac{c}{AN_S} \times \theta_1.$$

This change of flux density is represented by EF in fig. 3.21, so that the actual flux density corresponding to magnetic field strength OD = B_m − EF = DF.

This test is repeated with different values of OD down to zero, the latter value being obtained by opening the circuit of R_2 so that when S_2 is opened, the current is reduced to zero and the corresponding change of flux density is represented by GJ.

To derive portion JLN of loop. To obtain point L, S_2 is opened and R_2 adjusted to give a current corresponding to magnetic field strength OM. S_2 is closed and RS reversed several times to bring the steel back to the cyclic state. Then S_2 is opened and RS simultaneously moved over from a to b, and the deflection θ_2 of the fluxmeter is noted. (Actually, S_2 should be opened the slightest fraction of a second before RS is reversed so as to ensure that the magnetic field strength on the negative side does not exceed OM.) Then:

$$\textit{change} \text{ of flux density} = \text{KL} = \frac{c}{AN_S} \times \theta_2.$$

When the change of flux density exceeds B_m, as shown for KL in fig. 3.21, then:

$$\left.\begin{array}{l}\textit{actual} \text{ flux density corresponding to} \\ \text{magnetic field strength OM}\end{array}\right\} = \text{KL} - B_m = \text{ML}.$$

This test is repeated for sufficient number of negative values of the magnetic field strength to enable curve JLN to be drawn. The return portion NPC of the loop is exactly the reverse of CJN.

3.13 Current-ring theory of magnetism

It may be relevant at this point to consider why the presence of a ferromagnetic material in a current-carrying coil increases the value of the magnetic flux and why magnetic hysteresis occurs in ferromagnets. As long ago as 1823, André-Marie Ampère — after whom the unit of current was named — suggested that the increase in the magnetic flux might be due to electric currents circulating within the molecules of the ferromagnet. Subsequent discoveries have confirmed this suggestion, and the following brief explanation may assist in giving some idea of the current-ring theory of magnetism.

An atom consists of a nucleus of positive electricity surrounded, at distances large compared with their diameter, by electrons, which are charges of negative electricity. The electrons revolve in orbits around the nucleus, and each electron also spins around its own axis — somewhat like a gyroscope — and the magnetic characteristics of ferromagnetic materials appear to be due mainly to this electron spin. The movement of an electron around a circular path is equivalent to a minute current flowing in a circular ring. In a ferromagnetic atom, e.g. iron, four more electrons spin round in one direction than in the reverse direction, and the axes of spin of these electrons are parallel with one another; consequently, the effect is equivalent to four current rings producing magnetic flux in a certain direction.

The ferromagnetic atoms are grouped together in *domains*, each about 0.1 mm in width; and in any one domain the magnetic axes of all the atoms are parallel with one another. In an unmagnetized bar of ferromagnetic material, the magnetic axes of different domains are in various directions so that their magnetizing effects cancel one another. Between adjacent domains there is a region or 'wall', about 10^{-4} mm thick, within which the direction of the magnetic axes of the atoms changes gradually from that of the axes in one domain to that of the axes in the adjacent domain.

When an unmagnetized bar of ferromagnetic material, e.g. steel, is moved into a current-carrying solenoid, there are sudden tiny increments of the magnetic flux as the magnetic axes of the various domains are orientated so that they coincide with the direction of the m.m.f. due to the solenoid, thereby increasing the magnitude of the flux. This phenomenon is known as the *Barkhausen effect* and can be demonstrated by winding a search coil on the steel bar and connecting it through an amplifier to a loudspeaker. The sudden increments of flux due to successive orientation of the various domains, while the steel bar is being moved into the solenoid, induce e.m.f. impulses in the search coil and the effect can be heard as a rustling noise.

It follows that when a current-carrying solenoid has a ferromagnetic core the magnetic flux can be regarded as consisting of two components:

(a) the flux produced by the solenoid without a ferromagnetic core;

(b) the flux due to ampere-turns equivalent to the current rings formed by the spinning electrons in the orientated domains. This component reaches its maximum value when all the domains have been orientated so that their magnetic axes are in the direction of the magnetic flux. The core is then said to be saturated.

These effects are illustrated in fig. 3.22, where graph P represents the variation of the actual flux density with magnetic field strength for a ferromagnetic material. The straight line Q represents the variation of flux density with magnetic field strength if there were no ferromagnetic material present; thus for magnetic field strength OA,

Fig. 3.22 Actual and intrinsic flux densities

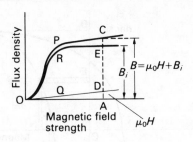

$$AD = \mu_0 \times OA.$$

For the same magnetic field strength OA, the actual flux density B is represented by AC. Hence the flux density due to the presence of the ferromagnetic core = AC − AD = AE. This flux density is referred to as the *intrinsic magnetic flux density* or *magnetic polarization* and is represented by symbol B_i,

i.e. $$B_i = B - \mu_0 H$$

The variation of the intrinsic flux density with magnetic field strength is represented by graph R in fig. 3.22, from which it will be seen that when the magnetic field strength exceeds a certain value, the intrinsic flux density remains constant, i.e. the ferromagnetic material is magnetically saturated, or in other words, all the domains have been orientated so that their axes are in the direction of the magnetic field.

This alignment of the domains has a certain amount of stability, depending upon the quality of the ferromagnetic material and the treatment it has received during manufacture. Consequently, the core may retain much of its magnetism after the external magnetomotive force has been removed, as already discussed in section 3.11, the remanent flux being maintained by the m.m.f. due to the electronic current-rings in the iron. A permanent magnet can therefore be regarded as an electromagnet, the relatively large flux being due to the high value of the *inherent m.m.f.* (i.e. coercive force × length of magnet) retained by the steel after it has been magnetized.

A disturbance in the alignment of the domains necessitates the expenditure of energy in taking a specimen of a ferromagnetic material through a cycle of magnetization. This energy appears as heat in the specimen and is referred to as *hysteresis loss*.

3.14 Hysteresis loss

Suppose fig. 3.23 to represent the hysteresis loop obtained on a steel ring of mean circumference l metres and cross-sectional area A square metres. Let N be the number of turns on the magnetizing coil.

Fig. 3.23 Hysteresis loss

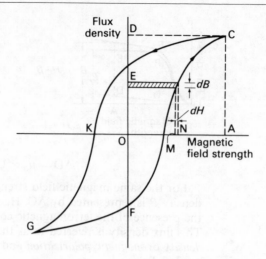

Let dB = increase of flux density when the magnetic field strength is increased by a very small amount MN in dt seconds, and i = current in amperes corresponding to OM.

i.e. $$OM = Ni/l \qquad (3.17)$$

Instantaneous e.m.f. induced in winding

$$= - A \times dB \times N/dt \text{ volts}$$

and component of applied voltage to neutralize this e.m.f.

$$= AN \times dB/dt \text{ volts}$$

$\therefore$ instantaneous power supplied to magnetic field

$$= i \times AN \times dB/dt \text{ watts}$$

and energy supplied to magnetic field in time dt second

$$= iAN \times dB \text{ joules.}$$

From (3.17) $\quad i = l \times OM/N$

$\therefore$ energy supplied to magnetic field in time dt

$$= (l \times OM/N) \times AN \times dB \text{ joules}$$

$$= OM \times dB \times lA \text{ joules}$$

$$= \text{area of shaded strip, joules per metre}^3$$

$\therefore$ energy supplied to magnetic field when H is increased

from zero to OA= area FJCDF joules per metre^3

Similarly, energy returned from magnetic field when H is reduc

from OA to zero= area CDEC joules per metre^3

$\therefore$ net energy absorbed by magnetic field

$$= \text{area FJCEF joules per metre}^3$$

Hence, hysteresis loss for a complete cycle

$$= \text{area of loop GFCEG joules per metre}^3$$

$$(3.18)$$

If the hysteresis loop is plotted to scales of 1 cm to x amperes per metre along the horizontal axis and 1 cm to y teslas along the vertical axis, and if a represents the area of the loop in square centimetres, then:

hysteresis loss per cycle $= axy$ joules per metre^3

$$= a \times (NI/l) \times \Phi/A \text{ joules per metre}^3$$

where $\qquad NI/l = \begin{cases} \text{magnetic field strength per} \\ \quad \text{centimetre of scale} \end{cases}$

and $\qquad\qquad \Phi/A =$ flux density per centimetre of scale

Hence, total hysteresis loss per cycle, in joules

$$= lA \times a \times (NI/l) \times \Phi/A$$

$$= a \times NI \times \Phi$$

$$= (\text{area of loop in centimetres}^2)$$

$$\times (\text{amperes per centimetre of scale})$$

$$\times (\text{webers per centimetre of scale}) \qquad (3.19)$$

The total hysteresis loss/cycle in a specimen can therefore be determined from the area of a loop drawn by plotting the total flux, in webers, against the total m.m.f. in amperes.

From the areas of a number of loops similar to those shown in fig. 3.20, the hysteresis loss per cycle was found by Steinmetz* to be proportional to $(B_{max})^{1.6}$ for a certain quality of steel, where B_{max} represents the maximum value of the flux density. It is important to realize that this index of 1.6 is purely empirical and has no theoretical basis. In fact, indices as high as 3.5 have been obtained, and in general we may say that

hysteresis loss per cubic metre per cycle $\propto (B_{max})^x$

where $x = 1.5$–2.5, depending upon the quality of the steel and the range of flux density over which the measurement has been made.

From the above proof, it is evident that the hysteresis loss is proportional to the volume and to the number of cycles through which the magnetization is taken. Hence,

if $\quad f =$ frequency of alternating magnetization in hertz

and $\quad v =$ volume of steel in cubic metres,

$$\text{hysteresis loss} = kvf(B_{max})^x \text{ watts} \qquad (3.20)$$

where k is a constant for a given specimen and a given range of flux density.

The magnitude of the hysteresis loss depends upon: (*a*) the composition of the specimen, e.g. nickel–iron alloys such as Mumetal and Permalloy have very narrow hysteresis loops due to their low coercive force, and consequently have very low hysteresis loss; and (*b*) the heat treatment and the

---

* Charles Proteus Steinmetz, U.S.A. electrical engineer, 1865–1923.

mechanical handling to which the specimen has been subjected. For instance, if a material has been bent or cold-worked, the hysteresis loss is increased, but the specimen can usually be restored to its original magnetic condition by a strain-relieving heat treatment.

Example 3.7

The area of the hysteresis loop obtained with a certain specimen of steel was 9.3 cm². The co-ordinates were such that 1 cm = 1000 A/m and 1 cm = 0.2 T. Calculate (a) the hysteresis loss per cubic metre per cycle and (b) the hysteresis loss per cubic metre at a frequency of 50 Hz. (c) If the maximum flux density was 1.5 T, calculate the hysteresis loss per cubic metre for a maximum density of 1.2 T and a frequency of 30 Hz, assuming the loss to be proportional to $(B_{max})^{1.8}$.

(a) $\left.\begin{array}{l} \text{Area of hysteresis loop in} \\ \quad BH \text{ units} \end{array}\right\} = 9.3 \times 1000 \times 0.2 = 1860$

∴ $\left.\begin{array}{l} \text{hysteresis loss per cubic} \\ \quad \text{metre per cycle} \end{array}\right\} = 1860 \text{ joules.}$

(b) $\left.\begin{array}{l} \text{Hysteresis loss per cubic} \\ \quad \text{metre at 50 Hz} \end{array}\right\} = 1860 \times 50 = 93\,000 \text{ W.}$

(c) From expression (3.20),

$$93\,000 = k \times 1 \times 50 \times (1.5)^{1.8}$$

$$\log(1.5)^{1.8} = 1.8 \times 0.176 = 0.3168 = \log 2.074$$

∴ $$k = \frac{93\,000}{50 \times 2.074} = 896$$

For $B_{max} = 1.2$ T and $f = 30$ Hz,

$\left.\begin{array}{l} \text{hysteresis loss per} \\ \quad \text{cubic metre} \end{array}\right\} = 896 \times 30 \times (1.2)^{1.8} = 37\,350 \text{ W.}$

Or,

$\left.\begin{array}{l} \text{hysteresis loss at} \\ \text{1.2 T and 30 Hz} \end{array}\right\} = 93\,000 \times \frac{30}{50} \times \left(\frac{1.2}{1.5}\right)^{1.8} = 37\,350 \text{ W.}$

3.15 Condition for minimum volume of a permanent magnet

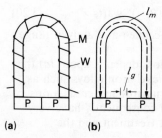

(a)   (b)

Fig. 3.24 Permanent-magnet steel core with soft-iron pole-shoes

Let M in fig. 3.24 be a steel core to be magnetized by passing a current through winding W, and let PP be soft-iron pole-shoes of negligible reluctance. Suppose curve CDE in fig. 3.25 to represent the demagnetizing portion of the hysteresis loop for a maximum magnetic field strength OA obtained with the closed magnetic circuit of fig. 3.24(a). It was pointed out in section 3.13 that the remanent flux density OD may be regarded as being maintained by the *inherent magnetomotive force* due to electronic currents in the magnet. The value of the inherent m.m.f. is given by the product of the coercive force OE and the length l_m of the magnet.

With the pole-shoes PP in contact, as in fig. 3.24(a), let us

Fig. 3.25 Demagnetizing
portion of hysteresis loop

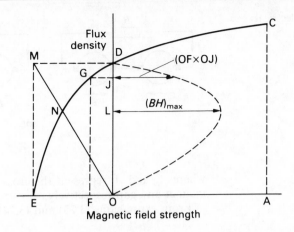

reduce the magnetic field strength from OA to zero and then
increase it to OF in the *reverse* direction, so that the flux
density in the magnet is reduced to FG (or OJ). The
corresponding value of the demagnetizing magnetomotive
force is (OF × l_m).

Alternatively, after the magnetic field strength has been
reduced from OA to zero, with the pole-shoes in contact as in
fig. 3.24(a), let us pull the pole-shoes apart. The effect is to
increase the reluctance of the magnetic circuit and therefore
to reduce the value of the flux density. Suppose l_g to be the
length of the gap required to reduce the flux density in the
magnet to FG. If B_g be the corresponding flux density in the
gap, then

m.m.f. required for gap = $B_g l_g / \mu_0$ amperes

= magnetic potential drop in gap.

As far as the reduction of the flux density in the *magnet* is
concerned, the magnetic potential drop in the gap produces
the same effect* as the demagnetizing m.m.f., OF × l_m, and
the two quantities can therefore be equated thus:

$$\text{OF} \times l_m = B_g l_g / \mu_0 \tag{3.23}$$

Neglecting magnetic leakage, we have:

total flux in magnet = total flux in gap

i.e.
$$\text{OJ} \times A_m = B_g A_g \tag{3.24}$$

---

* This effect is analogous to that shown for the electrical circuit of
fig. 3.26. In fig. 3.26(a), a counter e.m.f., e, is introduced to reduce
the current to I,

∴
$$E - e = IR \tag{3.21}$$

In fig. 3.26(b), the resistance is increased by the introduction of r
to give the same current I,

∴
$$E = I(R + r) \quad \text{or} \quad E - Ir = IR \tag{3.22}$$

Comparison of (3.21) and (3.22) shows that the electric potential
drop Ir across r has the same effect as the counter e.m.f. e,

i.e.
$$e = Ir.$$

Fig. 3.26 An electrical
analogue

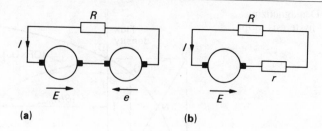

(a)   **(b)**

where A_m and A_g represent the cross-sectional areas of the
magnet and gap respectively.

From expressions (3.23) and (3.24), we have:

$$\text{volume of magnet} = l_m A_m = \frac{B_g l_g}{\mu_0 \times \text{OF}} \times \frac{B_g A_g}{\text{OJ}}$$

$$= \frac{B_g^2 l_g A_g}{\mu_0 (\text{OF} \times \text{OJ})} \tag{3.25}$$

From expression (3.25), it is evident that for a given flux
density in and given dimensions of the gap, the volume of the
magnet is a minimum when the product (OF × OJ) is a
maximum. The variation of this product for different flux
densities in the *magnet core* is represented by curve K in
fig. 3.25, from which it follows that the volume of magnet is a
minimum when it is operated at flux density OL. The value of
the product $(BH)_{max}$ for a steel specimen is the best criterion
of its suitability for use as a permanent magnet. The following
table gives the magnetic data for some of the principal
magnet materials:

Material	Remanence in teslas	Coercivity in A/m	Value of B for $(BH)_{max}$	Value of H for $(BH)_{max}$	$(BH)_{max}$ in J/m^3
1 per cent carbon steel	0.9	4 000	0.6	−2 600	1 560
6 per cent tungsten steel	1.05	5 200	0.7	−3 750	2 620
35 per cent cobalt steel	0.9	20 000	0.6	−13 000	7 800
Alnico (Fe–Al–Ni–Co–Cu)	0.8	40 000	0.52	−26 000	13 500
Alcomax III	1.25	52 000	1.0	−42 000	42 000
Ticonal G (Fe–Al–Ni–Co–Cu)	1.35	47 000	1.1	−42 000	46 000

An alternative construction which gives approximate
optimum working conditions is to draw a vertical line at E in
fig. 3.25 and a horizontal line at D. The lines intersect at M.
The straight line OM cuts the demagnetization curve at point
N. The coordinates of N correspond very closely to those
giving $(BH)_{max}$; in fact, if the demagnetization curve were a
rectangular hyperbola, the values would be identical.

Example 3.8 *Estimate the minimum volume of (a) a tungsten steel magnet and (b) an Alcomax III magnet to maintain a flux density of 0.5 T across an airgap, 4.0 mm long, having an effective cross-sectional area of 6 cm^2. Assume the data given in the above table.*

(a) If H_1 and B_1 represent the values of the magnetic field strength and the flux density in the magnet, respectively, for $(BH)_{max}$, then for tungsten steel, $H_1 = -3750$ A/m and $B_1 = 0.7$ T.
Substituting in expression (3.23), we have:

$$3750 \times l_m = 0.5 \times 0.004/(4\pi \times 10^{-7})$$

$$\therefore \qquad l_m = 0.425 \text{ m} = 42.5 \text{ cm}.$$

Substituting in expression (3.24), we have:

$$0.7 \times A_m = 0.5 \times 6 \times 10^{-4}$$

$$\therefore \qquad A_m = 4.29 \times 10^{-4} \text{ m}^2 = 4.29 \text{ cm}^2$$

$$\therefore \qquad \text{minimum volume of tungsten magnet} = 42.5 \times 4.29$$

$$= 182.5 \text{ cm}^3.$$

(b) For Alcomax III, $H_1 = -42\,000$ A/m and $B_1 = 1.0$ T.
From expression (3.23),

$$42\,000 \times l_m = 0.5 \times 0.004/(4\pi \times 10^{-7})$$

$$\therefore \qquad l_m = 0.038 \text{ m} = 3.8 \text{ cm}.$$

From expression (3.24),

$$1.0 \times A_m = 0.5 \times 6 \times 10^{-4}$$

$$\therefore \qquad A_m = 3 \times 10^{-4} \text{ m}^2 = 3 \text{ cm}^2.$$

$$\therefore \qquad \text{minimum volume of Alcomax III magnet}$$

$$= 3.8 \times 3 = 11.4 \text{ cm}^3$$

Hence

$$\frac{\text{minimum volume of tungsten magnet}}{\text{minimum volume of Alcomax III magnet}} = \frac{182.5}{11.4} = 16$$

i.e. the volume of an Alcomax III magnet is only a sixteenth of that of a tungsten magnet for the same duty.

3.16 Load line of a permanent magnet

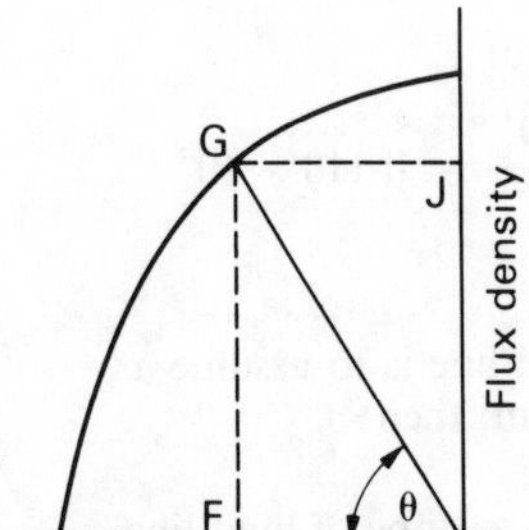

Fig. 3.27 Load line of a permanent magnet

Suppose the demagnetization curve of a magnet to be as shown in fig. 3.27, and suppose the gap between the pole-shoes to be such as to reduce the flux density in the magnet to OJ, as already discussed in section 3.15. Then, using the same symbols as before, we have from expression (3.23):

$$OF = B_g l_g/(\mu_0 l_m) \qquad (3.26)$$

and from expression (3.24):

$$OJ = B_g A_g/A_m \qquad (3.27)$$

Dividing (3.27) by (3.26), we have:

$$\frac{\text{OJ}}{\text{OF}} = \frac{\mu_0 l_m A_g}{l_g A_m} = \tan\theta \qquad (3.28)$$

$$= \text{slope of } load\ line\ \text{OG}.$$

It is evident from expression (3.28) that the slope of the load line is determined by the geometrical form of the magnet and the airgap, and is independent of the shape of the demagnetization curve.

Expression (3.28) also shows that for a given cross-sectional area of the magnet and given dimensions of the gap, the slope of the load line is directly proportional to the length, l_m, of the magnet. This means that the shorter the magnet, the smaller is the value of the flux, an effect that is attributed to *self-demagnetization* in some textbooks. This is an unfortunate term, since it is based upon the assumption of a hypothetical free pole at each end of the magnet. There is no evidence for the existence of such poles.

Example 3.9   *A Ticonal X magnet is 30 mm long and has a cross-sectional area of 200 mm². It is fitted with pole-shoes, of negligible reluctance, having an airgap, 8.0 mm long, of cross-sectional area 500 mm². The demagnetization curve for Ticonal X is given in fig. 3.28. Calculate the flux density in the airgap, assuming the magnetic leakage to be negligible.*

Fig. 3.28 Demagnetization curve for example 3.9

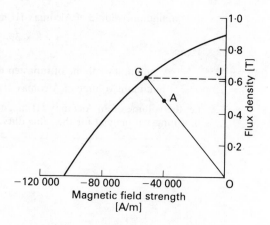

Substituting in expression (3.28), we have:

$$\frac{\text{OJ}}{\text{OF}} = \frac{4\pi \times 10^{-7} \times 0.03 \times 500 \times 10^{-6}}{0.008 \times 200 \times 10^{-6}} = 0.118 \times 10^{-4}$$

$$= \text{slope of load line.}$$

The simplest method of locating the load line is to assume a magnetic field strength of, say, $-40\,000$ A/m; then:

$$\left.\begin{array}{c}\text{corresponding flux density}\\ \text{in magnet}\end{array}\right\} = 40\,000 \times \text{slope of load line}$$

$$= 40\,000 \times 0.118 \times 10^{-4}$$

$$= 0.472\ \text{T}.$$

With these values of $-40\,000$ A/m and 0.472 T, point A is plotted, and load line OA drawn and produced to meet the demagnetization curve at G. From fig. 3.28,

actual value of flux density in magnet $= OJ = 0.615$ T

$\therefore$ total flux $= 0.615 \times 200 \times 10^{-6} = 123 \times 10^{-6}$ Wb

and flux density in gap $= \dfrac{123 \times 10^{-6}}{500 \times 10^{-6}} = 0.246$ T.

3.17 Barium-ferrite magnets

The original non-metallic permanent magnet is the natural lodestone composed of ferrous ferrite ($FeOFe_2O_3$). Modern permanent magnets of this class are made by mixing powdered ferrous oxide with powdered barium carbonate together with a suitable bonding resin. The mixture is then pressed to the desired form and subjected to heat and magnetic treatments.

A barium-ferrite magnet may possess a coercivity as high as $240\,000$ A/m, but the remanence is relatively low, being of the order of 0.2–0.4 T. The demagnetization curve of a barium-ferrite magnet bearing the trade name Magnadur 2 is given in Q. 27, p. 89.

These ceramic magnets are electrical insulators and have a resistivity of about $10^6\,\Omega \cdot$ m.

3.18 Magnetic field of a long solenoid

Fig. 3.29(*a*) shows a uniformly wound solenoid, the axial length, *l*, of which is very large compared with the diameter. The magnetic flux density at various points inside the solenoid can be determined by means of a small search-coil S connected to a ballistic galvanometer. The deflections of the galvanometer are noted for the reversal of a given current through the solenoid winding P with the search-coil in different positions along the axis of the solenoid. The curve in fig. 3.29(*b*) represents the result obtained with a solenoid having a length equal to 10 times the diameter. It is found that between two planes, A and B, situated 0.1*l* on each side of the centre C of such a solenoid, the variation of the flux density inside the solenoid is only about 0.1 per cent. This length of the solenoid may therefore be regarded as a part of a uniformly-wound ring (section 3.2); hence the magnetic field strength inside the central portion of a long solenoid is

Fig. 3.29 The magnetic field
of a long solenoid

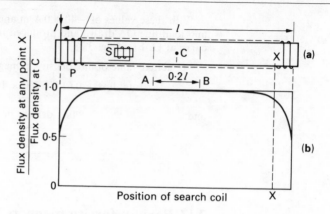

$I \times$ number of turns per metre length,

i.e. $H = NI/l$ amperes per metre   (3.29)

where I = the magnetizing current in amperes,

N = total number of turns on the solenoid,

and l = length of solenoid in metres.

Example 3.10   *A solenoid, 1.2 m long, is uniformly wound with 800 turns. A short coil,
wound with 50 turns having a mean diameter of 30 mm, is placed at
the centre of the solenoid and connected to a ballistic galvanometer.
The total resistance of the search-coil circuit is 2000 Ω. When a
current of 5 A through the solenoid is reversed, the deflection on the
galvanometer scale is 85 divisions. Calculate the ballistic constant
(coulombs per unit deflection), assuming the damping of the
galvanometer to be negligible and the coils to be coaxial.*

$$\left.\begin{array}{c}\text{Magnetic field strength at}\\ \text{centre of solenoid}\end{array}\right\} = 5 \times 800/1.2 = 3333 \text{ A/m}$$

$$\therefore \quad \left.\begin{array}{c}\text{flux density at centre}\\ \text{of solenoid}\end{array}\right\} = 4\pi \times 10^{-7} \times 3333$$

$$= 0.004\ 185 \text{ T}.$$

$$\text{Area of search-coil} = \frac{\pi}{4} \times (0.03)^2 = 0.000\ 706\ 5 \text{ m}^2.$$

$$\text{Flux through search-coil} = 0.004\ 185 \times 0.000\ 706\ 5$$

$$= 2.96 \times 10^{-6} \text{ Wb}.$$

If the current is reversed in t seconds,

average e.m.f. induced in search-coil

$$= \frac{2 \times 2.96 \times 10^{-6} \times 50}{t} = \frac{296}{t} \times 10^{-6} \text{ volt}$$

and average current through galvanometer

$$= \frac{296 \times 10^{-6}}{t \times 2000} = \frac{0.148}{t} \times 10^{-6} \text{ ampere}$$

$$\therefore \quad \text{quantity of electricity through galvanometer}$$

$$= 0.148 \times 10^{-6} \text{ coulomb}$$

and ballistic constant of galvanometer

$$= 0.148 \times 10^{-6}/85$$

$$= 1.74 \times 10^{-9} \text{ coulomb per division}.$$

This is one of the standard methods used for determining the ballistic constant of a galvanometer.

3.19 Magnetic energy in a non-magnetic medium

For a non-magnetic medium, the relative permeability is unity, so that the magnetic field strength and the flux density are proportional to each other and the relationship between them is represented by a straight line OD in fig. 3.30.

It was shown in section 3.14 that when the flux density is increased by dB due to an increase MN of the magnetic field strength,

$$\left.\begin{array}{r}\text{energy supplied to} \\ \text{magnetic circuit}\end{array}\right\} = \left\{\begin{array}{l}\text{area of shaded strip,} \\ \text{joules per metre}^3.\end{array}\right.$$

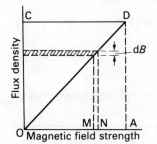

Fig. 3.30 *B/H* relationship for a non-magnetic medium

Hence, for a maximum flux density OC in fig. 3.30,

$$\left.\begin{array}{r}\text{total energy stored} \\ \text{in magnetic field}\end{array}\right\} = \left\{\begin{array}{l}\text{area of triangle OCD in} \\ \text{joules per metre}^3\end{array}\right.$$

$$= \tfrac{1}{2}\text{OA} \times \text{OC} = \frac{1}{2} \cdot \frac{\text{OC}}{\mu_0} \cdot \text{OC}$$

$$= \frac{\text{OC}^2}{2\mu_0} \text{ joules per metre}^3$$

Consequently, for a flux density of B teslas,

$$\left.\begin{array}{r}\text{energy stored in a non-} \\ \text{magnetic medium}\end{array}\right\} = \tfrac{1}{2}HB$$

$$= \frac{B^2}{2\mu_0} \text{ joules per metre}^3 \qquad (3.30)$$

3.20 Magnetic pull between two ferromagnetic surfaces

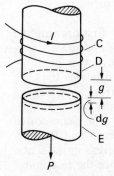

Fig. 3.31 Magnetic pull between two steel surfaces

Consider a fixed steel core D, wound with coil C, and a movable steel core E, as shown in fig. 3.31. Suppose D and E to have the same cross-sectional area of A square metres. Also, suppose the opposite surfaces of D and E to be g metres apart and that a current I amperes through coil C produces a flux density of B teslas in the airgap. From expression (3.30),

energy stored per cubic metre of airgap $= B^2/2\mu_0$ joules.

Let P be the force of attraction in newtons between the two surfaces of D and E. Suppose core E to be moved a short distance dg metres away from D and that the current through coil C be increased at the same time so as to keep the flux density unaltered. No e.m.f. will therefore have been induced in coil C, so that no energy will have been supplied either from the elctrical circuit to the magnetic circuit or vice versa. Hence all the energy stored in the *additional* volume of airgap must have been derived from the work done when force P newtons acted through a distance dg metres, namely $P \times dg$ joules,

i.e.
$$P \times dg = (B^2/2\mu_0) \times A \times dg$$
$$\therefore \qquad P = B^2 A/2\mu_0 \text{ newtons} \qquad (3.31)$$

3.21 Force between two long parallel conductors carrying electric current

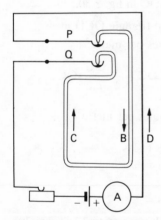

Fig. 3.32 Force between two parallel current-carrying conductors

When two current-carrying conductors are parallel to each other, there is a force acting on each of the conductors, an effect that can be easily demonstrated by means of the apparatus referred to in section 2.7. Thus, in fig. 3.32, the rectangular loop BC has its tapered ends dipping into mercury in cups supported one directly above the other by rods P and Q, as already shown in fig. 2.11, and a current of about 10–15 amperes is passed through the loop. Part of the electrical circuit consists of a long straight rod D which can be placed alongside B, as shown in plan in fig. 3.33. When the currents in D and B are in opposite directions, as in fig. 3.33(a), the two conductors repel each other and the loop (viewed from above) is deflected clockwise. On the other hand, if rod D is turned through 180° so that the currents in B and D are in the same direction, as in fig. 3.33(b), the conductors attract each other and the loop turns counterclockwise.

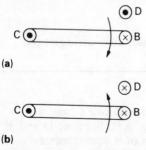

Fig. 3.33 Force between two parallel current-carrying conductors

These effects are most easily explained by first drawing the magnetic fields produced by each conductor and then combining these fields. Thus, fig. 3.34(a) shows two conductors, A and B, each carrying current towards the paper. The lines of magnetic flux due to current in A alone are represented by the uniformly-dotted circles in fig. 3.34(a), and those due to B alone are represented by the chain-dotted circles. It is evident that in the space between A and B the two fields tend to neutralize each other, whereas in the space outside A and B they assist each other. Hence the resultant distribution is somewhat as shown in fig. 3.34(b). Since magnetic flux behaves like a stretched elastic cord, the effect

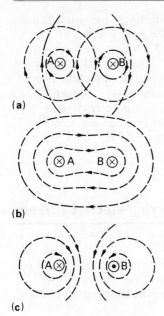

(a)

(b)

(c)

Fig. 3.34 Magnetic fields due to parallel current-carrying conductors

is to try to move conductors A and B towards each other; in other words, there is a force of attraction between A and B.

If the current in B is reversed, the magnetic fields due to A and B assist each other in the space between the conductors and the resultant distribution of the flux will be as shown in fig. 3.34(*c*). The lateral pressure between the lines of flux exerts a force on the conductors tending to push them apart (section 2.3(5)).

3.22 Magnitude of force between two parallel current-carrying conductors

Let C in fig. 3.35 represent the cross-section of a long straight conductor carrying a current I amperes towards the paper. The return conductor is assumed to be some distance away, so that its magnetic field may be neglected.

The magnetic flux set up by the current can be represented as concentric circles around the conductor. Let us therefore consider the flux in the cylinder of radius x metres and thickness dx metres, shown dotted in fig. 3.35. Since the outward and return conductors form one turn, it follows that the magnetic field strength acting on this path is $I/(2\pi x)$ amperes per metre; and since the relative permeability of air and other non-magnetic mediums is unity,

$$\text{flux density at radius } x = I\mu_0/2\pi x \text{ teslas} \qquad (3.32)$$

Hence the flux density around a conductor is inversely proportional to the distance from the centre of the conductor, as represented by the hyperbola in fig. 3.35(*b*). The density is a maximum at the surface of the conductor and is equal to $I\mu_0/2\pi r$ teslas, where r is the radius of the conductor in metres.

Let us next consider two parallel conductors A and B (fig. 3.36) carrying I_1 and I_2 amperes respectively and let the distance between their centres be d metres. From expression (3.32), it follows that the flux density due to I_1 alone at a distance of d metres from the centre of A is $I_1\mu_0/2\pi d$ teslas; and from fig. 3.36 it is seen that conductor B is carrying I_2 amperes at right-angles to the field set up by I_1 amperes in conductor A. Hence from expression (2.1),

$$\left.\begin{array}{l}\text{force per metre length} \\ \text{of conductor B}\end{array}\right\} = \frac{I_1\mu_0}{2\pi d} \cdot I_2$$

$$= 2 \times 10^{-7} \times I_1 I_2/d \text{ newtons} \qquad (3.33)$$

Similarly,

$$\left.\begin{array}{l}\text{force per metre length} \\ \text{of A}\end{array}\right\} = 2 \times 10^{-7} \times I_1 I_2/d \text{ newtons.}$$

If the currents I_1 and I_2 are in the same direction, as in fig. 3.36, there is a force of attraction between the conductors;

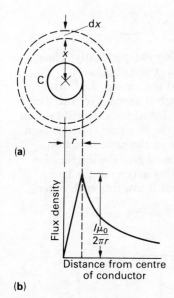

(a)

Flux density

$\dfrac{I\mu_0}{2\pi r}$

Distance from centre of conductor

(b)

Fig. 3.35 Magnetic field around a long straight conductor situated in a non-magnetic medium

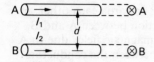

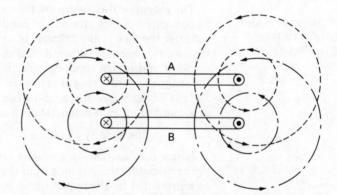

Fig. 3.36 Two parallel current-carrying conductors situated in a non-magnetic medium

but if the currents are in opposite directions, the conductors repel each other (section 3.21). If $I_1 = I_2 = 1$ A and if $d = 1$ m, it follows from expression (3.33) that the force per metre length of conductor is 2×10^{-7} N. This value forms the basis for the definition of the ampere, given in section 1.6, namely *that current which, if maintained in two straight parallel conductors of infinite length, of negligible circular cross-section, and placed 1 metre apart in a vacuum, would produce between these conductors a force of 2×10^{-7} newtons per metre of length.*

3.23 Force between coils carrying electric current

Suppose A and B in fig. 3.37 to represent the cross-section of two coils carrying currents in the directions shown by the dots and crosses. Let us first consider the distribution of the

Fig. 3.37 Magnetic fields due to currents in A and B separately

magnetic fields due to the coils acting independently. Thus, current through A alone gives the flux distribution represented by the uniformly-dotted lines in fig. 3.37, while current in the same direction through B alone gives the distribution indicated by the chain-dotted lines. It will be seen that in the space between the coils the two fields oppose each other, while on the outside they are in the same direction. Consequently, the combined effect is to give the distribution shown in fig. 3.38. Since magnetic flux acts as if it is in tension, it tends to move coils A and B towards each other.

Fig. 3.38 Resultant magnetic field when currents in A and B are in the same direction

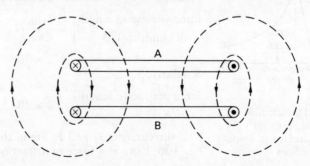

On the other hand, if the current through B is reversed, the direction of the arrowheads on the chain-dotted lines in fig. 3.37 is reversed. Consequently, the magnetic fields of A and B are in the same direction in the space between the coils and in opposition outside the coils, so that the resultant distribution becomes that shown in fig. 3.39. The magnetic flux between the coils exerts a lateral pressure, thereby producing a force of repulsion between coils A and B (see section 2.3(5)).

Fig. 3.39 Resultant magnetic field when currents in A and B are in opposite directions

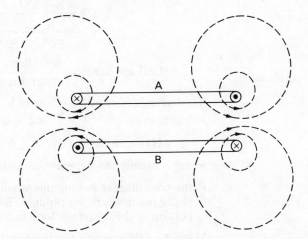

This attraction and repulsion between coils carrying an electric current has been applied in the current balance used at the National Physical Laboratory for determining the absolute value of an electric current (section 1.6).

Summary of important formulae

$$\text{Magnetomotive force} = NI \text{ amperes}$$

$$\text{Magnetic field strength} = H = NI/l \text{ amperes per metre} \tag{3.1}$$

$$\text{Magnetic flux} = \Phi = \frac{\text{m.m.f.}}{\text{reluctance}} \text{ webers}$$

$$\text{Flux density} = B = \Phi/A \text{ teslas}$$

$$B/H = \mu = \text{absolute permeability}$$

$$= \mu_0\mu_r \tag{3.4}$$

$$\mu_0 = 4\pi \times 10^{-7} \text{ henry per metre} \tag{3.2}$$

$$\mu_r = \frac{B \text{ in a material}}{B \text{ in vacuum due to same magnetic field strength}}.$$

$$H = \frac{B}{\mu} = \frac{B}{\mu_0\mu_r}$$

For a homogeneous magnetic-circuit of uniform cross-section,

$$\text{reluctance} = \frac{I}{\mu A} \tag{3.7}$$

and

$$\text{m.m.f.} = \Phi \cdot \frac{l}{\mu A} = \frac{Bl}{\mu}$$

For a composite magnetic circuit,

$$\Phi = \frac{NI}{\dfrac{l_1}{\mu_1 A_1} + \dfrac{l_2}{\mu_2 A_2}} \tag{3.10}$$

$$\text{Leakage factor} = \frac{\text{total flux}}{\text{useful flux}} \tag{3.11}$$

$$\left.\begin{array}{r}\text{Hysteresis loss/cubic} \\ \text{metre/cycle}\end{array}\right\} = \left\{\begin{array}{l}\text{area of hysteresis} \\ \text{loop, in joules}\end{array}\right. \tag{3.18}$$

$$\text{Hysteresis loss} = kvf(B_{\max})^x \tag{3.20}$$

where x usually lies between 1.5 and 2.5.

Permanent magnet has minimum volume when it is operated at the point where the product (BH) for the demagnetizing portion of the hysteresis loop is a maximum (3.25)

Magnetic field strength inside central portion of long solenoid

$$= I \times \text{turns per metre} \tag{3.29}$$

Magnetic energy stored in non-magnetic medium

$$= \frac{B^2}{2\mu_0} \text{ joules per metre}^3 \tag{3.30}$$

Magnetic pull between two ferromagnetic surfaces

$$= \frac{B^2 A}{2\mu_0} \text{ newtons} \tag{3.31}$$

Flux density around wire carrying current

$$= \frac{I\mu_0}{2\pi x} \text{ teslas} \tag{3.32}$$

Force between two conductors carrying current

$$= \frac{2 \times 10^{-7} I_1 I_2}{d} \text{ newtons per metre} \tag{3.33}$$

EXERCISES 3

Data of B/H, when not given in question, should be taken from fig. 3.3.

1. A mild steel ring has a mean diameter of 160 mm and a cross-sectional area of 300 mm². Calculate: (a) the m.m.f. to produce

a flux of 400 μWb, and (b) the corresponding values of the reluctance of the ring and of the relative permeability.

2. A steel magnetic circuit has a uniform cross-sectional area of 5 cm^2 and a length of 25 cm. A coil of 120 turns is wound uniformly over the magnetic circuit. When the current in the coil is 1.5 A, the total flux is 0.3 mWb; when the current is 5 A, the total flux is 0.6 mWb. For each value of current, calculate: (a) the magnetic field strength; (b) the relative permeability of the steel. (U.E.I., O.1)

3. A mild steel ring has a mean circumference of 500 mm and a uniform cross-sectional area of 300 mm^2. Calculate the m.m.f. required to produce a flux of 500 μWb.

   An airgap, 1.0 mm in length, is now cut in the ring. Determine the flux produced if the m.m.f. remains constant. Assume the relative permeability of the mild steel to remain constant at 1200. (S.A.N.C., O.1)

4. A steel ring has a mean diameter of 15 cm, a cross-section of 20 cm^2 and a radial airgap of 0.5 mm cut in it. The ring is uniformly wound with 1500 turns of insulated wire and a magnetizing current of 1 A produces a flux of 1 mWb in the airgap. Neglecting the effect of magnetic leakage and fringing, calculate: (a) the reluctance of the magnetic circuit; (b) the relative permeability of the steel. (U.E.I., O.1)

5. (a) A steel ring, having a mean circumference of 750 mm and a cross-sectional area of 500 mm^2, is wound with a magnetizing coil of 120 turns. Using the following data, calculate the current required to set up a magnetic flux of 630 μWb in the ring.

Flux density (T)	0.9	1.1	1.2	1.3
Magnetic field strength (A/m)	260	450	600	820

   (b) The airgap in a magnetic circuit is 1.1 mm long and 2000 mm^2 in cross-section. Calculate: (i) the reluctance of the airgap, and (ii) the m.m.f. to send a flux of 700 microwebers across the airgap. (U.L.C.I., O.1)

6. A magnetic circuit consists of a cast steel yoke which has a cross-sectional area of 200 mm^2 and a mean length of 120 mm. There are two airgaps, each 0.2 mm long. Calculate: (a) the m.m.f. required to produce a flux of 0.05 mWb in the airgaps; (b) the value of the relative permeability of cast steel at this flux density.

   The magnetization curve for cast steel is given by the following:

B (T)	0.1	0.2	0.3	0.4
H (A/m)	170	300	380	460

   (W.J.E.C., O.1)

7. An electromagnet has a magnetic circuit that can be regarded as comprising three parts in series, viz.: (a) length of 80 mm and cross-sectional area 60 mm^2; (b) a length of 70 mm and cross-sectional area 80 mm^2; (c) an airgap of length 0.5 mm and cross-sectional area 60 mm^2.

   Parts (a) and (b) are of a material having magnetic characteristics given by the following table:

H (A/m)	100	210	340	500	800	1500
B (T)	0.2	0.4	0.6	0.8	1.0	1.2

   Determine the current necessary in a coil of 4000 turns wound on part (b) to produce in the airgap a flux density of 0.7 T. Magnetic leakage may be neglected. (E.M.E.U., O.1)

8. A certain magnetic circuit may be regarded as consisting of

three parts, A, B and C in series, each one of which has a uniform cross-sectional area. Part A has a length of 300 mm and a cross-sectional area of 450 mm². Part B has a length of 120 mm and a cross-sectional area of 300 mm². Part C is an airgap 1.0 mm in length and of cross-sectional area 350 mm². Neglecting magnetic leakage and fringing, determine the m.m.f. necessary to produce a flux of 0.35 mWb in the airgap.

The magnetic characteristic for parts A and B is given by:

H (A/m)	400	560	800	1280	1800
B (T)	0.7	0.85	1.0	1.15	1.25

(N.C.T.E.C., O.2)

9. A magnetic circuit made of silicon steel is arranged as in fig. A. The centre limb has a cross-sectional area of 800 mm² and each of the side limbs has a cross-sectional area of 500 mm². Calculate the m.m.f. required to produce a flux of 1 mWb in the centre limb, assuming the magnetic leakage to be negligible.

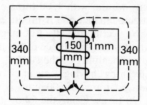

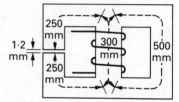

Fig. A   Fig. B

10. A magnetic core made of mild steel has the dimensions shown in fig. B. There is an airgap 1.2 mm long in one side limb and a coil of 400 turns is wound on the centre limb. The cross-sectional area of the centre limb is 1600 mm² and that of each side limb is 1000 mm². Calculate the exciting current required to produce a flux of 1000 µWb in the airgap. Neglect any magnetic leakage and fringing.

11. An electromagnet with its armature has a core length of 400 mm and a cross-sectional area of 500 mm². There is a total airgap of 1.8 mm. Assuming a leakage factor of 1.2, calculate the m.m.f. required to produce a flux of 400 µWb in the armature. Points on the B/H curve are as follows:

Flux density (T)	0.8	1.0	1.2
Magnetic field strength (A/m)	800	1000	1600

12. Define the *ampere* in terms of S.I. units.

A coil P of 300 turns is uniformly wound on a ferromagnetic ring having a cross-sectional area of 500 mm² and a mean diameter of 400 mm. Another coil S of 30 turns wound on the ring is connected to a fluxmeter having a constant of 200 microweber-turns per division. When a current of 1.5 A through P is reversed, the fluxmeter deflection is 96 divisions. Calculate: (a) the flux density in the ring, and (b) the magnetic field strength. (E.M.E.U., O.1)

13. The magnetization curve of a ring specimen of steel is determined by means of a fluxmeter. The specimen, which is uniformly wound with a magnetizing winding of 1000 turns, has a mean length of 65 cm and a cross-sectional area of 7.5 cm². A search coil of 2 turns is connected to the fluxmeter, which has a constant of 0.1 mWb-turn per scale division. When the magnetizing current is reversed, the fluxmeter deflection is noted, the following table being the readings which are

obtained:

Magnetizing current (amperes)	0.2	0.4	0.6	0.8	1.0
Fluxmeter deflection (divisions)	17.8	31.2	38.0	42.3	44.7

Explain the basis of the method and calculate the values of the magnetic field strength and flux density for each set of readings. (W.J.E.C., O.2)

14. A ring specimen of mild steel has a cross-sectional area of 6 cm^2 and a mean circumference of 30 cm. It is uniformly wound with two coils, A and B, having 90 turns and 300 turns respectively. Coil B is connected to a ballistic galvanometer having a constant of 1.1×10^{-8} coulomb per division. The total resistance of this secondary circuit is 200 000 Ω. When a current of 2.0 A through coil A is reversed, the galvanometer gives a maximum deflection of 200 divisions. Neglecting the damping of the galvanometer, calculate the flux density in the steel and its relative permeability. (E.M.E.U., O.1)

15. Two coils of 2000 and 100 turns respectively are wound uniformly on a non-magnetic toroid having a mean circumference of 1 m and a cross-sectional area of 500 mm^2. The 100-turn coil is connected to a ballistic galvanometer, the total resistance of this circuit being 5.1 kΩ. When a current of 2.5 A in the 2000-turn coil is reversed, the galvanometer is deflected through 100 divisions. Neglecting damping, calculate the ballistic constant for the instrument. (W.J.E.C., O.2)

16. Explain with the aid of diagrams and brief notes how a hysteresis loop may be obtained for a ferromagnetic specimen using a fluxmeter.

A hysteresis loop is plotted against a horizontal axis which scales 1000 A/m = 1 cm and a vertical axis which scales 1 T = 5 cm. If the area of the loop is 9 cm^2 and the overall height is 14 cm, calculate: (*a*) the hysteresis loss in joules per cubic metre per cycle; (*b*) the maximum flux density; (*c*) the hysteresis loss in watts per kilogram, assuming the density of the material to be 7800 kg/m^3. (E.M.E.U., O.2)

17. (*a*) Explain briefly the occurrence of hysteresis loss and eddy-current loss in a transformer core and indicate how these losses may be reduced.

(*b*) The area of a hysteresis loop for a certain magnetic specimen is 160 cm^2 when drawn to the following scales: 1 cm represents 25 A/m and 1 cm represents 0.1 T. Calculate the hysteresis loss in watts per kilogram when the specimen is carrying an alternating flux at 50 Hz, the maximum value of which corresponds to the maximum value attained on the hysteresis loop. The density of the material is 7600 kg/m^3. (W.J.E.C., O.2)

18. Explain how the energy losses in a sample of ferromagnetic material subjected to an alternating magnetic field depend on the frequency and the flux density. What particular property of the material can be used as a measure of the magnitude of each type of loss?

The area of a hysteresis loop plotted for a sample of steel is 67.1 cm^2, the maximum flux density being 1.06 T. The scales of *B* and *H* are such that 1 cm = 0.12 T and 1 cm = 7.07 A/m. Find the loss due to hysteresis if 750 g of this steel were subjected to an alternating magnetic field of maximum flux density 1.06 T at a frequency of 60 Hz. The density of the steel is 7700 kg/m^3. (U.E.I., O.2)

19. The hysteresis loop for a certain ferromagnetic ring is drawn to the following scales: 1 cm to 300 A and 1 cm to 100 μWb. The

area of the loop is 37 cm². Calculate the hysteresis loss per cycle.

20. In a certain transformer, the hysteresis loss is 300 W when the maximum flux density is 0.9 T and the frequency 50 Hz. What would be the hysteresis loss if the maximum flux density were increased to 1.1 T and the frequency reduced to 40 Hz? Assume the hysteresis loss over this range to be proportional to $(B_{max})^{1.7}$.

21. The demagnetization curve for a certain magnet steel is represented by the following data:

H (A/cm)	0	−144	−240	−304	−360	−400	−430
B (T)	0.6	0.5	0.4	0.3	0.2	0.1	0

Derive a graph showing the variation of (BH) with the flux density and estimate the values of B and H at which the product (BH) is a maximum.

22. Name *three* materials which are suitable for permanent magnets. Explain the characteristics of these materials which render them suitable for this purpose.

The demagnetization curve for a certain permanent-magnet steel is as follows:

B (T)	0	0.1	0.2	0.3	0.4	0.5	0.6	0.64
H (A/cm)	−488	−464	−432	−382	−336	−240	−96	0

Determine the minimum dimensions of a magnet made from this material to maintain a flux density of 0.5 T across an airgap the dimensions of which are: length 1 cm, effective cross-sectional area 4 cm².   (U.L.C.I., O.2)

23. (a) Derive the condition which must be observed to ensure that a permanent magnet producing a given airgap field shall have minimum volume.

(b) An airgap of 300 mm² cross-section and 2.5 mm length is to be supplied with a flux density of 0.12 T. The demagnetization curve for the magnetic material used is as follows:

B (T)	0	0.25	0.5	0.75	1.0	1.25	1.40
H (A/m)	−4630	−4370	−4000	−3500	−2800	−1550	0

Calculate the dimensions of the magnet using the minimum volume of material. Neglect magnetic leakage. (W.J.E.C., O.2)

24. Show that provided reasonable simplifying assumptions are made, the minimum volume of a given permanent-magnet material needed to establish a specified flux in a specified airgap is obtained when the value of $(B) \times (-H)$ for the material is a maximum.

A magnet of minimum volume is to be made from a material having a demagnetization curve given by:

H (A/mm)	0	−12	−20	−25	−28	−30	−32
B (T)	0.7	0.6	0.5	0.4	0.3	0.2	0

It is to produce a flux density of 0.4 T in an airgap of length 3 mm and cross-sectional area 500 mm². Magnetic leakage can be neglected, and pole pieces of negligible reluctance assumed. Determine the approximate length and cross-section of magnet material.   (App. El., L.U.)

25. The demagnetization curve of the steel of a permanent magnet is given by:

H (A/m)	−1000	−2000	−3000	−4000
B (T)	0.65	0.56	0.37	0.12

Find the length and cross-section of the magnet of minimum volume that would provide a flux density of 0.1 T in an airgap 3 mm long and of cross-section 10 cm². Assume a leakage flux between the poles of the magnet of 50 per cent of the useful flux. (S.A.N.C., O.2)

Note. If $\quad k = \text{leakage factor} = \dfrac{\text{total flux in magnet}}{\text{total flux in gap}}$,

$$\text{total flux in magnet} = kB_g A_g$$

$$\therefore \qquad A_m = kB_g A_g / \text{OJ}$$

where OJ = flux density in magnet corresponding to $(BH)_{max}$.

26. A certain magnet is 50 mm long and has a cross-sectional area of 300 mm². It is fitted with pole-shoes of negligible reluctance, arranged to give an airgap 6.0 mm in length and 600 mm² in cross-sectional area. If the demagnetization curve is represented by the data given in Q. 21, determine the flux density in the airgap. Neglect magnetic leakage and fringing.

27. The demagnetization curve for the barium-ferrite magnet, Magnadur 2, is as follows:

H (A/mm)	0	-40	-80	-120	-160	-170	-175
B (T)	0.39	0.345	0.295	0.24	0.18	0.15	0

The magnet has a length of 35 mm and a cross-sectional area of 600 mm², and is fitted with pole pieces of negligible reluctance. Between the pole pieces there is an airgap 12 mm in length and 800 mm² in cross-sectional area. Neglecting magnetic leakage and fringing, determine the flux density in the gap.

28. A solenoid M, 1.2 m long and 80 mm diameter, is uniformly wound with 750 turns. A search coil N, 25 mm in diameter and wound with 30 turns, is mounted co-axially midway along the solenoid. Calculate the average e.m.f. induced in N when a current of 5 A through M is reversed in 0.2 s.

29. Each of the two airgaps of a moving-coil instrument is 2.5 mm long and has a cross-sectional area of 600 mm². If the flux density is 0.08 T, calculate the total energy stored in the magnetic field of the airgaps.

30. Given that the energy stored per unit volume of a magnetic field is $B^2/(2\mu_0)$, derive an expression for the pull between two magnetized surfaces.

A lifting magnet of inverted U-shape is formed out of a steel bar 60 cm long and 10 cm² in cross-sectional area. Exciting coils of 750 turns each are wound on the two side limbs and are connected in series. There is an effective airgap of 0.1 mm between the load and the magnet at each pole. Assuming that the reluctance of the load is negligible and ignoring fringing at the airgaps and magnetic leakage, calculate the current required in the winding to lift a mass of 200 kg. Use the following magnetic data for the steel:

B (T)	1.56	1.57	1.58	
H (A/m)	2800	3000	3200	(U.L.C.I., O.2)

31. A steel ring having a mean diameter of 350 mm and a cross-sectional area of 240 mm² is broken by a parallel-side airgap of length 12 mm. Short pole pieces of negligible reluctance extend the effective cross-sectional area of the airgap to 1200 mm². Taking the relative permeability of the steel as 700 and neglecting leakage, determine the current necessary in 300 turns of wire wound on the ring to produce a flux density of 0.25 T in the gap.

Evaluate also the tractive force between the poles.

(App. El., L.U.)

32. Each of the two pole faces of a lifting magnet has an area of 150 cm², and this may also be taken as the cross-sectional area of the 40-cm long flux path in the magnet. Determine the m.m.f. needed on the magnet if it is to lift a 900-kg steel block separated by 0.5 mm from the pole faces. Assume the magnetic leakage factor to be 1.2, neglect fringing of the gap flux and the reluctance of the flux path in the steel block, and take the magnetization curve of the magnet material to be given by:

H (A/m)	400	600	800	1200	1600
B (T)	0.81	0.98	1.10	1.24	1.35

(App. El., L.U.)

33. A long straight conductor, situated in air, is carrying a current of 500 A, the return conductor being far removed. Calculate the magnetic field strength and the flux density at a radius of 80 mm.

34. (a) The flux density in air at a point 40 mm from the centre of a long straight conductor A is 0.03 T. Assuming that the return conductor is a considerable distance away, calculate the current in A.

   (b) In a certain magnetic circuit, having a length of 500 mm and a cross-sectional area of 300 mm², an m.m.f. of 200 A produces a flux of 400 μWb. Calculate (i) the reluctance of the magnetic circuit and (ii) the relative permeability of the core.

35. Two long parallel conductors, spaced 40 mm between centres, are each carrying a current of 5000 A. Calculate the force in newtons per metre length of conductor.

36. Two long parallel conductors P and Q, situated in air and spaced 8 cm between centres, carry currents of 600 A in opposite directions. Calculate the values of the magnetic field strength and of the flux density at points A, B and C in fig. C, where the dimensions are given in centimetres.

   Calculate also the values of the same quantities at the same points if P and Q are each carrying 600 A in the same direction.

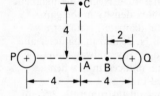

Fig. C

4 Inductance in a D.C. Circuit

4.1 Inductive and non-inductive circuits

Let us consider what happens when a coil L (fig. 4.1) and a resistor R, connected in parallel, are switched across a battery B. L consists of a large number of turns wound on a steel core D (or it might be the field winding of a generator or motor), and R is connected in series with a *centre-zero* ammeter A_2.

Fig. 4.1 Inductive and non-inductive circuits

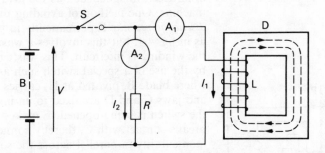

When switch S is closed, it is found that the current I_2 through R increases almost instantly to its final value, whereas the current i_1* through L takes an appreciable time to grow — as indicated in fig. 4.2. The final value, I_1, is equal to $\dfrac{\text{battery voltage V}}{\text{resistance of coil L}}$. In fig. 4.2, I_2 has been shown a little larger than I_1, but this is of no importance.

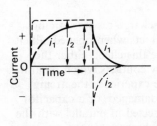

Fig. 4.2 Variation of switch-on and switch-off currents

When S is opened, current through L decreases comparatively slowly, but the current through R instantly reverses its direction and becomes the same current as i_1; in other words, the current of L is circulating round R.

Let us now consider the reason for the difference in the behaviour of the currents in L and R.

The growth of current in L is accompanied by an increase of flux — shown dotted — in the steel core D. But it has been pointed out in section 2.10 that any change in the flux linked with a coil is accompanied by an e.m.f. induced in that coil, the direction of which — described by Lenz's Law — is always such as to oppose the change responsible for inducing the

---

* A lower case letter is used to represent the instantaneous value of a varying quantity.

e.m.f., namely the growth of current in L. In other words, the induced e.m.f. is acting in opposition to the current and, therefore, to the applied voltage. In circuit R, the flux is so small that its induced e.m.f. is negligible.

When switch S is opened, the currents in both L and R tend to decrease; but any decrease of i_1 is accompanied by a decrease of flux in D and therefore by an e.m.f. induced in L in such a direction as to oppose the decrease of i_1. Consequently the induced e.m.f. is now acting in the same direction as the current. But it is evident from fig. 4.1 that after S has been opened, the only return path for L's current is that via R; hence the reason why i_1 and i_2 are now one and the same current.

If the experiment is repeated without R, it is found that the growth of i_1 is unaffected; but when S is opened there is considerable arcing at the switch due to the maintenance of the current across the gap by the e.m.f. induced in L. The more quickly S is opened, the more rapidly does the flux in D collapse and the greater is the e.m.f. induced in L. This is the reason why it is dangerous to break quickly the full excitation of an electromagnet such as the field winding of a d.c. machine. One method of avoiding this risk is to connect a *discharge resistor* R permanently in parallel with the winding, as in fig. 4.1; but this involves a waste of energy in R while the winding is in circuit. This waste of energy can be avoided by the use of a special switch such as that shown in fig. 4.3, where blade B, pivoted at H, carries an extension E. Hinge H and jaws C and D are fixed to an insulating panel P. When the switch is being opened, E makes contact with D before B breaks contact with C, thereby connecting R and F momentarily in parallel across the source. After B has broken contact with C, resistor R and winding F form a closed circuit and the current decreases relatively slowly so that there is no risk of a dangerously high e.m.f. being induced in F. The resistance of the discharge resistor is usually about the same as that of the field winding.

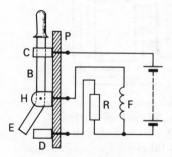

Fig. 4.3 Field-discharge switch

Even with the assistance of the discharge resistor, arcing will still take place at the switch contacts when opened, albeit on a reduced scale. To further reduce this effect, either the switch can be provided with auxiliary contacts which separate after the main contacts and therefore experience the arcing and thus leave the main contacts undamaged, or a capacitor (described in Chapter 5) can be connected in parallel with the switch thus diverting the current which produces the arc away from the switch.

Any circuit in which a change of current is accompanied by a change of flux, and therefore by an induced e.m.f., is said to be *inductive* or to possess *self inductance* or merely *inductance*. It is impossible to have a perfectly *non-inductive* circuit, i.e. a circuit in which no flux is set up by a current; but for most purposes a circuit which is not in the form of a coil may be regarded as being practically non-inductive — even the open helix of an electric fire is almost non-inductive. In cases where the inductance has to be reduced to the smallest possible value — for instance, in resistance boxes — the wire is bent

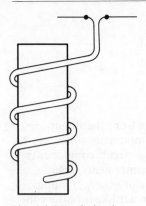

Fig. 4.4 Non-inductive resistor

back on itself, as shown in fig. 4.4, so that the magnetizing effect of one conductor is neutralized by that of the adjacent conductor. The wire can then be coiled round an insulator S without increasing the inductance.

4.2 Unit of inductance

The unit of inductance is termed the *henry*, in commemoration of a famous American physicist, Joseph Henry (1797–1878), who, quite independently, discovered electromagnetic induction within a year after it had been discovered in Britain by Michael Faraday in 1831. *A circuit has an inductance of 1 henry (or 1 H) if an e.m.f. of 1 volt is induced in the circuit when the current varies uniformly at the rate of 1 ampere per second.* If either the inductance or the rate of change of current be doubled, the induced e.m.f. is doubled. Hence if a circuit has an inductance of L henrys and if the current *increases* from i_1 to i_2 amperes in t seconds:

$$\left.\begin{array}{c}\text{average rate of change}\\ \text{of current}\end{array}\right\} = \frac{i_2 - i_1}{t} \text{ amperes per second,}$$

and

$$\left.\begin{array}{c}\text{average induced}\\ \text{e.m.f.}\end{array}\right\} = -L \times \text{rate of change of current}$$

$$= -L \times \frac{i_2 - i_1}{t} \text{ volts} \qquad (4.1)$$

Considering instantaneous values, if di = increase of current, in amperes, in time dt seconds,

$$\text{rate of change of current} = di/dt \text{ amperes per second}$$

and e.m.f. induced in circuit $= -L \cdot di/dt$ volts $\qquad (4.2)$

The minus sign in expressions (4.1) and (4.2) reminds us that the direction of the induced e.m.f. is opposite to that of the current increase or decrease as the case might be. The basis of this reaction is one more familiar to mechanical systems.

When a force is applied to a mechanical system, the system reacts by deforming, or by mass-accelerating, or by dissipating or absorbing energy. A comparable state exists when a force (voltage) is applied to an electrical system which accelerates (accepts magnetic energy in an inductor) or dissipates energy in heat (in a resistor). The comparable state to deformation is the acceptance of potential energy in a capacitor, which is dealt with in Chapter 5.

Consider a series circuit as shown in fig. 4.5. Applying Kirchhoff's Voltage Law, we observe that the sum of the e.m.f.s is equal to the sum of the volt drops. Here we have the applied e.m.f. E and the e.m.f. of the inductor e_L. There is only the volt drop across the resistor v_R. It follows that

$$E + e_L = v_R$$

Fig. 4.5 Polarity of e.m.f. in a circuit diagram

hence
$$E = v_R - e_L$$
$$= iR + L \cdot \frac{di}{dt}$$

$$\therefore \qquad e_L = -L \cdot \frac{di}{dt}$$

This interpretation of the diagram arises from the intention that we clearly identify active circuit components which are sources of e.m.f. as distinct from inactive circuit components which provide only volt drops. Active components include batteries, generators and (because they store energy) inductors and capacitors. A charged capacitor can act like a battery for a short period of time and a battery possesses an e.m.f. which, when the battery is part of a circuit fed from a source of voltage, acts against the passage of current through it from the positive terminal to the negative terminal, i.e. the process whereby a battery is charged. An inductor opposes the increase of current in it by acting against the applied voltage by means of an opposing e.m.f.

Returning to the negative sign in the expression (4.2), we therefore observe that the polarity results from our determination to treat the potential difference across the inductor as being an e.m.f. and therefore our conventions require that we show the p.d. acting in the direction of the circuit current. The negative sign therefore is a result of convention, but it serves to remind us that although the p.d. arrow suggests aid to the current, in fact the e.m.f. opposes.

When the e.m.f. opposes the increase of current in the inductor, energy is taken into the inductor. Similarly, when the e.m.f. opposes the decrease of current in the inductor, energy is supplied from the inductor back into the electric circuit.

Example 4.1

If the current through a coil having an inductance of 0.5 H is reduced from 5 A to 2 A in 0.05 s, calculate the mean value of the e.m.f. induced in the coil.

Average rate of change of current
$$= (2 - 5)/0.05 = -60 \text{ amperes per second}$$
From (4.1) average e.m.f. induced in coil
$$= -0.5 \times (-60) = 30 \text{ V}.$$

The direction of the induced e.m.f. is the same as that of the current, opposing its decrease.

4.3 Inductance in terms of flux-linkages per ampere

Suppose a current of I amperes through a coil of N turns to produce a flux of Φ webers, and suppose the reluctance of the magnetic circuit to remain constant so that the flux is proportional to the current. Also, suppose the inductance of the coil to be L henrys.

If the current is increased from zero to I amperes in t seconds,

average rate of change of current $= I/t$ amperes per second,

$\therefore$ average e.m.f. induced in coil $= -LI/t$ volts $\qquad$ (4.3)

In section 2.12, it was explained that the value of the e.m.f., in volts, induced in a coil is equal to the rate of change of flux-linkages per second. Hence, when the flux increases from zero to Φ webers in t seconds,

average rate of change of flux $= \Phi/t$ webers per second

and average e.m.f. induced in coil $= -N\Phi/t$ volts $\qquad$ (4.4)

Equating expressions (4.3) and (4.4), we have:

$$-LI/t = -N\Phi/t$$

$\therefore \qquad L = N\Phi/I \text{ henrys} \qquad (4.5)$

$$= \text{flux-linkages per ampere}.$$

Considering instantaneous values,

if $\qquad d\phi = $ increase of flux, in webers,

due to an increase di amperes in dt seconds,

rate of change of flux $= d\phi/dt$ webers per second

and $\qquad$ induced e.m.f. $= -N \cdot d\phi/dt$ volts $\qquad$ (4.6)

Equating (4.2) and (4.6), we have:

$$-L \cdot di/dt = -N \cdot d\phi/dt$$

$\therefore \qquad L = N \cdot d\phi/di \qquad (4.7)$

$$= \frac{\text{change of flux-linkages}}{\text{change of current}}$$

For a coil having a magnetic circuit of constant reluctance, the flux is proportional to the current; consequently, $d\phi/di$ is equal to the flux per ampere, so that

$$L = \text{flux-linkages per ampere}$$

$$= N\Phi/I \text{ henrys} \qquad (4.8)$$

This expression gives us an alternative method of defining the unit of inductance, namely: *a coil possesses an inductance of 1 henry if a current of 1 ampere through the coil produces a flux-linkage of 1 weber-turn.*

Example 4.2 *A coil of 300 turns, wound on a core of non-magnetic material, has an inductance of 10 mH. Calculate (a) the flux produced by a current of 5 A and (b) the average value of the e.m.f. induced when a current of 5 A is reversed in 8 ms (milliseconds).*

(a) From expression (4.5),

$$10 \times 10^{-3} = 300 \times \Phi/5$$

$\therefore \qquad \Phi = 0.167 \times 10^{-3} \text{ Wb} = 167 \,\mu\text{Wb}.$

(b) When a current of 5 A is reversed, the flux decreases from $167\,\mu\text{Wb}$ to zero and then increases to $167\,\mu\text{Wb}$ in the reverse direction,

$$\therefore \qquad \text{change of flux} = -334\,\mu\text{Wb}$$

$$\text{and} \qquad \text{average rate of change of flux} = \frac{-334 \times 10^{-6}}{8 \times 10^{-3}}$$

$$= -0.041\,75\,\text{Wb/s},$$

$$\therefore \qquad \text{average e.m.f. induced in coil} = -(-0.041\,75) \times 300$$

$$= 12.5\,\text{V}.$$

Alternatively, since the current changes from 5 to $-5\,\text{A}$,

$$\text{average rate of change of current} = -\frac{5 \times 2}{8 \times 10^{-3}}$$

$$= -1250\,\text{A/s}.$$

Hence, from expression (4.2),

$$\text{average e.m.f. induced in coil} = -0.01 \times (-1250)$$

$$= 12.5\,\text{V}.$$

The sign is positive because the e.m.f. is acting in the direction of the original current, at first trying to prevent the current decreasing to zero and then opposing its growth in the reverse direction.

4.4 Factors determining the inductance of a coil

Let us first consider a coil uniformly wound on a *non-magnetic* ring of uniform section — similar to that of fig. 3.1. From expression (3.2), it follows that the flux density, in teslas, in such a ring is $4\pi \times 10^{-7} \times$ the amperes per metre. Consequently, if l be the length of the magnetic circuit in metres and A its cross-sectional area in square metres, then for a coil of N turns with a current I amperes:

$$\text{magnetic field strength} = IN/l$$

$$\text{and} \qquad \text{total flux} = \Phi = BA = \mu_0 HA$$

$$= 4\pi \times 10^{-7} \times (IN/l)A$$

Substituting for Φ in expression (4.5) we have:

$$\text{inductance} = L = 4\pi \times 10^{-7} \times AN^2/l \text{ henrys} \qquad (4.8)$$

Hence the inductance is proportional to the square of the number of turns and to the cross-sectional area, and is inversely proportional to the length of the magnetic circuit.

If the coil is wound on a closed ferromagnetic core, such as a ring, the problem of defining the inductance of such a coil becomes involved due to the fact that the variation of flux is no longer proportional to the variation of current. Suppose the relationship between the magnetic flux and the magnetizing current to be as shown in fig. 4.6; then if the

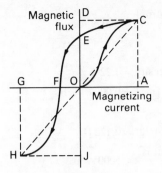

Fig. 4.6 Variation of
magnetic flux with
magnetizing current for a
closed ferromagnetic circuit

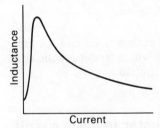

Fig. 4.7 Inductance of a
ferromagnetic-cored coil

core has initially no residual magnetism, an increase of
current from zero to OA causes the flux to increase from zero
to AC, but when the current is subsequently reduced to zero,
the decrease of flux is only DE. If the current is then
increased to OC in the reverse direction, the change of flux is
EJ. Consequently, we can have an infinite number of
inductance values, depending upon the particular variation of
current that we happen to consider.

Since we are usually concerned with the effect of
inductance in an a.c. circuit, where the current varies from a
maximum in one direction to the same maximum in the
reverse direction, it is convenient to consider the value of the
inductance as being the ratio of the change of flux-linkages to
the change of current when the latter is reversed. Thus, for
the case shown in fig. 4.6:

$$\text{inductance of coil} = \frac{DJ}{AG} \times \text{number of turns}.$$

This value of inductance is the same as if the flux varied
linearly along the dotted line COH in fig. 4.6.

If μ_r represents the value of the relative permeability
corresponding to the maximum value AC of the flux, then the
inductance of the steel-cored coil, as defined above, is μ_r
times that of the same coil with a non-magnetic core. Hence,
from expression (4.8), we have:

inductance of a ferromagnetic-cored coil (for reversal of flux)

$$= 4\pi \times 10^{-7} \times \frac{AN^2}{l} \times \mu_r \text{ henrys.*} \quad (4.9)$$

The variations of relative permeability with magnetic field
strength for various qualities of steel are shown in fig. 3.4;
hence it follows from expression (4.9) that as the value of an
alternating current through a coil having a closed steel circuit
is increased, the value of the inductance increases to a
maximum and then decreases, as shown in fig. 4.7. It will
now be evident that when the value of the inductance of such
a coil is stated, it is also necessary to specify the current
variation for which that value has been determined.

Example 4.3 *A ring of 'Stalloy' stampings having a mean circumference of 400 mm
and a cross-sectional area of 500 mm² is wound with 200 turns.
Calculate the inductance of the coil corresponding to a reversal of a
magnetizing current of (a) 1 A, (b) 10 A.*

---

* From expression (4.9), L (henrys) $= \mu_0\mu_r \times A$ (metres²) $\times N^2/l$
(metres)

$$\therefore \quad \text{absolute permeability} = \mu_0\mu_r = \frac{L \text{ (henrys)} \times l \text{ (metres)}}{N^2 \times A \text{ (metres}^2)}$$

$$= \frac{Ll}{N^2 A} \text{ henrys per metre},$$

hence the units of absolute permeability are *henrys per metre* (or
H/m); e.g. $\mu_0 = 4\pi \times 10^{-7}$ H/m.

(a)
$$H = \frac{1\,[\text{A}] \times 200\,[\text{turns}]}{0.4\,[\text{m}]} = 500\,\text{A/m}$$

From fig. 3.3:

corresponding flux density = 1.22 T

∴ total flux = 1.22 [T] × 0.0005 [m²] = 0.000 61 Wb.

From expression (4.5),

inductance = (0.000 61 × 200)/1.0 = 0.122 H.

(b)
$$H = \frac{10\,[\text{A}] \times 200\,[\text{turns}]}{0.4\,[\text{m}]} = 5000\,\text{A/m}.$$

From fig. 3.3:

corresponding flux density = 1.58 T

∴ total flux = 1.58 × 0.0005 = 0.000 79 Wb

and inductance = (0.000 79 × 200)/10 = 0.0158 H.

Example 4.4 *If a coil of 200 turns be wound on a non-magnetic core having the same dimensions as the 'Stalloy' ring for example 4.3, calculate its inductance.*

From expression (4.8) we have:

$$\text{inductance} = \frac{(4\pi \times 10^{-7})\,[\text{H/m}] \times 0.0005\,[\text{m}^2] \times (200)^2\,[\text{turns}^2]}{0.4\,[\text{m}]}$$

$$= 0.000\,062\,8\,\text{H}$$

$$= 62.8\,\mu\text{H}.$$

A comparison of the results from examples 4.3 and 4.4 for a coil of the same dimensions shows why a ferromagnetic core is used when a large inductance is required.

4.5 Ferromagnetic-cored inductor in a d.c. circuit

An inductor (i.e. a piece of apparatus used primarily because it possesses inductance) is frequently used in the output circuit of a rectifier to smooth out any variation (or ripple) in the direct current. If the inductor were made with a closed ferromagnetic circuit, the relationship between the flux and the magnetizing current would be represented by curve OBD in fig. 4.8. It will be seen that if the current increases from OA to OC, the flux increases from AB to CD. If this increase takes place in *t* seconds, then:

average induced e.m.f. = −number of turns

× rate of change of flux

$$= -N \times (\text{CD} - \text{AB})/t \text{ volts} \quad (4.10)$$

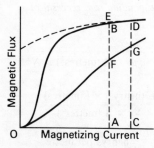

Fig. 4.8 Effect of inserting an airgap in a ferromagnetic core

Let L_i be the *incremental* inductance of the coil over this range of flux variation, i.e. the effective value of the inductance when the flux is not proportional to the magnetizing current and varies over a relatively small range, then

$$\text{average induced e.m.f.} = -L_i \times (OC - OA)/t \text{ volts.} \quad (4.11)$$

Equating (4.10) and (4.11), we have:

$$L_i \times (OC - OA)/t = N \times (CD - AB)/t$$

$$\therefore \qquad\qquad L_i = N \times \frac{CD - AB}{OC - OA} \qquad\qquad (4.12)$$

$$= N \times \text{average slope of curve BD}.$$

From fig. 4.8 it is evident that the slope is very small when the core is saturated. This effect is accentuated by hysteresis; thus if the current is reduced from OC to OA, the flux decreases from CD only to AE, so that the effective inductance is still further reduced.

If a short radial airgap were made in the ferromagnetic ring, the flux produced by current OA would be reduced to some value AF. For the reduced flux density in the core, the total m.m.f. required for the ferromagnet and the gap is approximately proportional to the flux; and for the same increase of current, AC, the increase of flux = CG − AF. As (CG − AF) may be much greater than (CD − AB), we have the curious result that the effective inductance of a ferromagnetic-cored coil in a d.c. circuit may be increased by the introduction of an airgap.

An alternative method of increasing the flux-linkages per ampere and maintaining this ratio practically constant is to make the core of compressed magnetic dust, such as small particles of ferrite or nickel–iron alloy, bound by shellac. This type of coil is used for 'loading' telephone lines, i.e. for inserting additional inductance at intervals along a telephone line to improve its transmission characteristics.

Expression (4.12) indicates that the inductance L of a magnetic system need not be a constant. It follows that if L can vary with time then expression (4.2) requires to be stated as

$$e = -\frac{d(L_i)}{dt}$$

It can be shown that this expands to give

$$e = -L \cdot \frac{di}{dt} - i \cdot \frac{dL}{dt}$$

It is for this reason that inductance is no longer defined in terms of the rate of change of current since this presumes that the inductance is constant, which need not be the case in practice.

It is interesting to note that the energy conversion associated with $-L \cdot di/dt$ is stored in the magnetic field yet the energy associated with the other term is partially stored in the magnetic field, while the remainder is converted to mechanical energy which is the basis of motor or generator action.

Example 4.5   *A laminated steel ring is wound with 200 turns. When the magnetizing current varies between 5 and 7 A, the magnetic flux varies between 760 and 800 μWb. Calculate the incremental inductance of the coil over this range of current variation.*

From expression (4.12) we have:

$$L_i = 200 \times \frac{(800 - 760) \times 10^{-6}}{(7 - 5)} = 0.004 \, \text{H}.$$

4.6 Graphical derivation of curve of current growth in an inductive circuit

In section 4.1 the growth of current in an inductive circuit was discussed qualitatively; we shall now consider how to derive the curve showing the growth of current in a circuit of known resistance and inductance (assumed constant).

When dealing with an inductive circuit it is convenient to separate the effects of inductance and resistance by representing the inductance L as that of an inductor or coil having no resistance and the resistance R as that of a resistor having no inductance, as shown in fig. 4.9. It is evident from the latter that the current ultimately reaches a steady value I (fig. 4.10), where $I = V/R$.

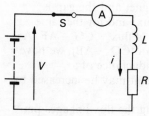

Fig. 4.9 Inductive circuit

Fig. 4.10 Growth of current in an inductive circuit

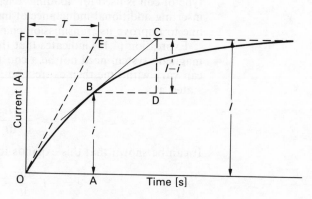

Let us consider any instant A during the growth of the current. Suppose the current at that instant to be i amperes, represented by AB in fig. 4.10. The corresponding p.d. across R is Ri volts. Also at that instant the rate of change of the current is given by the slope of the curve at B, namely the slope of the tangent to the curve at B.

But the slope of BC $= \dfrac{\text{CD}}{\text{BD}} = \dfrac{I - i}{\text{BD}}$ amperes per second.

Hence e.m.f. induced in L at instant A

$$= -L \times \text{rate of change of current}$$
$$= -L \times (I - i)/\text{BD volts}.$$

The total applied voltage V is absorbed partly in providing the voltage drop across R and partly in neutralizing the e.m.f. induced in L:

i.e.
$$V = Ri + L \times (I - i)/\text{BD}.$$

Substituting RI for V, we have:

$$RI = Ri + L \times (I - i)/\text{BD}$$
$$\therefore \qquad R(I - i) = L \times (I - i)/\text{BD}$$

hence
$$\text{BD} = L/R.$$

In words, this expression means that the rate of growth of current at any instant is such that if the current continued increasing at that rate, it would reach its maximum value of I amperes in L/R seconds. Hence this period is termed the *time constant* of the circuit and is usually represented by the symbol T:

i.e.
$$\text{time constant} = T = L/R \text{ seconds.} \qquad (4.13)$$

Immediately after switch S is closed, the rate of growth of the current is given by the slope of the tangent OE drawn to the curve at the origin; and if the current continued growing at this rate, it would attain its final value in time
$\text{FE} = T$ seconds.

From expression (4.13) it follows that the greater the inductance and the smaller the resistance, the larger is the time constant and the longer it takes for the current to reach its final value. Also this relationship can be used to derive the curve representing the growth of current in an inductive circuit, as illustrated by the following example.

Example 4.6

A coil having a resistance of $4\,\Omega$ and a constant inductance of $2\,H$ is switched across a 20-V d.c. supply. Derive the curve representing the growth of the current.

From (4.13), time constant $= T = 2/4 = 0.5\,\text{s}$.

Final value of current $= I = 20/4 = 5\,\text{A}$.

With the horizontal and vertical axes suitably scaled, as in fig. 4.11, draw a horizontal dotted line at the level of 5 A. Along this line mark off a period $\text{MN} = T = 0.5\,\text{s}$, and join ON.

Take any point P relatively near the origin and draw a horizontal dotted line $\text{PQ} = T = 0.5\,\text{s}$, and at Q draw a vertical dotted line QS. Join PS.

Repeat the operation from a point X on PS, Z on XY, etc.

A curve touching OP, PX, XZ, etc., represents the growth of the current. The greater the number of points used in the construction, the more accurate is the curve.

Fig. 4.11 Graph for
example 4.6

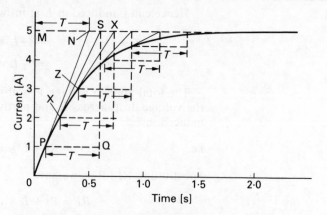

4.7 Mathematical derivation of curve of current growth in an inductive circuit

Let us again consider the circuit shown in fig. 4.9, and suppose i amperes to be the current t seconds after the switch is closed, and di amperes to be the increase of current in dt seconds, as in fig. 4.12. Then:

rate of change of current = $\mathrm{d}i/\mathrm{d}t$ amperes per second

and induced e.m.f. = $-L \cdot \mathrm{d}i/\mathrm{d}t$ volts.

Since $\left.\begin{matrix} \text{total applied} \\ \text{voltage} \end{matrix}\right\} = \left\{\begin{matrix} \text{p.d. across } R + \text{voltage to} \\ \text{neutralize induced e.m.f.} \end{matrix}\right.$

$\therefore \qquad\qquad V = Ri + L \cdot \mathrm{d}i/\mathrm{d}t \qquad\qquad (4.14)$

so that $\qquad V - Ri = L \cdot \mathrm{d}i/\mathrm{d}t$

and $\qquad\qquad \dfrac{V}{R} - i = \dfrac{L}{R} \cdot \dfrac{\mathrm{d}i}{\mathrm{d}t}.$

Fig. 4.12 Growth of current
in an inductive circuit

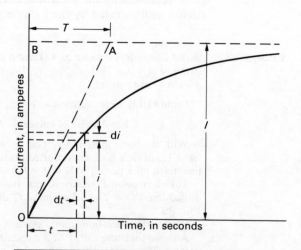

*British Standards Institution recommends the use of ln and log in preference to $\log_e$ and $\log_{10}$ respectively.

But V/R = final value of current = (say) I,

$$\therefore \qquad \frac{R}{L}\,dt = \frac{di}{I - i}$$

Integrating both sides, we have:

$$\frac{Rt}{L} = -\ln(I - i)^* + A$$

where A = the constant of integration.

At the instant of closing the switch, $t = 0$ and $i = 0$, so that $A = \ln I$,

$$\therefore \qquad \frac{Rt}{L} = -\ln(I - i) + \ln I$$

$$= \ln\frac{I}{I - i} \qquad (4.15)$$

Hence, $\qquad (I - i)/I = e^{-Rt/L}$

$$\therefore \qquad i = I(1 - e^{-Rt/L}) \qquad (4.16)$$

This exponential relationship is often referred to as the Helmholtz Equation.

Immediately after the switch is closed, the rate of change of the current is given by the slope of tangent OA drawn to the curve at the origin. If the current continued growing at this initial rate, it would attain its final value, I amperes, in T seconds, the *time constant* of the circuit (section 4.6). From fig. 4.12, it is seen that:

initial rate of growth of current = I/T amperes per second.

At the instant of closing the switch, $i = 0$; hence from expression (4.14),

$$V = L \times \text{initial rate of change of current},$$

$$\therefore \qquad \text{initial rate of change of current} = V/L$$

Hence, $\qquad I/T = V/L$

so that $\qquad T = LI/V = L/R$ seconds $\qquad (4.17)$

Substituting for R/L in (4.16), we have:

$$i = I(1 - e^{-t/T}) \qquad (4.18)$$

For $t = T$, $\qquad i = I(1 - 0.368) = 0.632I$,

hence the time constant is the time required for the current to attain 63.2 per cent of its final value.

Example 4.7

A coil having a resistance of 4 Ω and a constant inductance of 2 H is switched across a 20-V d.c. supply. Calculate (a) the time constant, (b) the final value of the current and (c) the value of the current 1 s after the switch is closed.

(a) Time constant = $L/R = 2/4 = 0.5$ s.

(b) Final value of current = $V/R = 20/4 = 5$ A.

(c) Substituting $t = 1$, $T = 0.5$ s and $I = 5$ A in (4.18),

$$i = 5(1 - e^{-1/0.5}) = 5(1 - e^{-2})$$

From mathematical tables, $e^{-2} = 0.1353$.

$$\therefore \qquad i = 5(1 - 0.1353) = 4.323 \text{ A}.$$

If such mathematical tables are not available, this section of the question can be solved by using expression (4.15), thus:

$$Rt/L = \ln I/(I - i)$$

$$= 2.303 \log I/(I - i)$$

$$\therefore \qquad \log I/(I - i) = \frac{4 \times 1}{2 \times 2.303} = 0.8684$$

and

$$I/(I - i) = 7.386$$

so that

$$i = 4.323 \text{ A}.$$

4.8 Mathematical derivation of curve of current decay in an inductive circuit

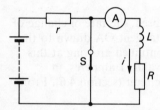

Fig. 4.13 Short-circuited inductive circuit

Fig. 4.13 represents a coil, of inductance L henrys and resistance R ohms, in series with a resistor r across a battery. The function of r is to prevent the battery current becoming excessive when switch S is closed. Suppose the steady current through the coil to be I amperes when S is open. Also, suppose i amperes in fig. 4.14 to be the current through the coil t seconds after S is closed. Since the external applied voltage V of equation (4.14) is zero as far as the coil is concerned, we have:

$$0 = Ri + L \cdot di/dt$$

$$\therefore \qquad Ri = -L \cdot di/dt \qquad (4.19)$$

Fig. 4.14 Decay of current in an inductive circuit

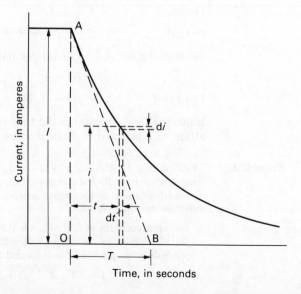

Time, in seconds

In this expression, $\mathrm{d}i$ is numerically negative since it represents a decrease of current.

Hence, $$(R/L)\,\mathrm{d}t = -\mathrm{d}i/i$$

Integrating both sides, we have:

$$(R/L)t = -\ln i + A$$

where A = the constant of integration.

At the instant of closing switch S, $t = 0$ and $i = I$, so that

$$0 = -\ln I + A$$

$\therefore$ $$(R/L)t = \ln I - \ln i = \ln I/i$$

Hence, $$I/i = \mathrm{e}^{Rt/L}$$

and $$i = I\,\mathrm{e}^{-Rt/L} \tag{4.20}$$

Immediately after S is closed, the rate of decay of the current is given by the slope of the tangent AB in fig. 4.14, and

initial rate of change of current $= -I/T$ amperes per second.

Also, from equation (4.19), since initial value of i is I,

initial rate of change of current $= -RI/L$ amperes per second.

Hence, $RI/L = I/T$

so that $T = L/R = time\ constant$ of circuit,

namely, the value already deduced in section 4.6.

The curve representing the decay of the current can be derived graphically by a procedure similar to that used in section 4.6 for constructing the curve representing the current growth in an inductive circuit.

4.9 Energy stored in an inductor

If the current in a coil having a *constant* inductance of L henrys grows at a uniform rate from zero to I amperes in t seconds, the average value of the current is $\frac{1}{2}I$ and the e.m.f. induced in the coil is $-(L \times I/t)$ volts. The value of the applied voltage required to neutralize this induced e.m.f. is therefore LI/t volts. The product of the current and the component of the applied voltage to neutralize the induced e.m.f. represents the power absorbed by the magnetic field associated with the coil.

Hence average power absorbed by the magnetic field

$$= \tfrac{1}{2}I \times LI/t \text{ watts}$$

and total energy absorbed by the magnetic field

$$= \text{average power} \times \text{time}$$

$$= \tfrac{1}{2}I \times (LI/t) \times t$$

$$= \tfrac{1}{2}LI^2 \text{ joules} \tag{4.21}$$

Let us now consider the general case of a current increasing at a uniform or a non-uniform rate in a coil having a *constant* inductance L henrys. If the current increases by di ampere in dt second,

$$\text{induced e.m.f.} = -L \cdot di/dt \text{ volts},$$

and if i is the value of the current at that instant,

$$\left.\begin{array}{l}\text{energy absorbed by the magnetic}\\ \text{field during time } dt \text{ second}\end{array}\right\} = iL \cdot (di/dt) \cdot dt$$

$$= Li \cdot di \text{ joules}.$$

Hence total energy absorbed by the magnetic field when the current increases from 0 to I amperes

$$= L \int_0^I i \cdot di = L \times \frac{1}{2}\left[i^2\right]_0^I$$

$$= \tfrac{1}{2}LI^2 \text{ joules} \tag{4.21}$$

From expression (4.9), $L = N^2 \mu A/l$ for a homogeneous magnetic circuit of uniform cross-sectional area.

$$\therefore \quad \text{energy per cubic metre } \omega_f = \tfrac{1}{2}I^2N^2\mu/l^2 = \tfrac{1}{2}\mu H^2$$

$$= \tfrac{1}{2}HB = \tfrac{1}{2}B^2/(\mu_r\mu_0) \text{ joules}. \tag{4.22}$$

This expression has been derived on the assumption that μ_r remains constant. When the coil is wound on a closed ferromagnetic core, the variation of μ_r renders this expression inapplicable and the energy has to be determined graphically as already explained in section 3.14. For non-magnetic materials, $\mu_r = 1$ and the energy stored per cubic metre is $\tfrac{1}{2}B^2/\mu_0$ joules, namely the expression already derived in section 3.19.

When an inductive circuit is opened, the current has to die away and the magnetic energy has to be dissipated. If there is no resistor in parallel with the circuit, the energy is largely dissipated as heat in the arc at the switch. With a parallel resistor, as described in section 4.1, the energy is dissipated as heat generated by the decreasing current in the total resistance of the circuit in which that current is flowing.

Example 4.8   *A coil has a resistance of 5 Ω and an inductance of 1.2 H. The current through the coil is increased uniformly from zero to 10 A in 0.2 s, maintained constant for 0.1 s and then reduced uniformly to zero in 0.3 s. Plot graphs representing the variation with time of (a) the current, (b) the induced e.m.f., (c) the p.d.s across the resistance and the inductance, (d) the resultant applied voltage and (e) the power to and from the magnetic field. Assume the coil to be wound on a non-metallic core.*

The variation of current is represented by graph A in fig. 4.15; and since the p.d. across the resistance is proportional to the current, this p.d. increases from zero to (10 A × 5 Ω), namely 50 V, in 0.2 s, remains constant at 50 V for 0.1 s and then decreases to zero in 0.3 s, as represented by graph B.

During the first 0.2 s, the current is increasing at the rate of

10/0.2, namely 50 A/s,

$\therefore$ corresponding induced e.m.f. $= -50 \times 1.2 = -60$ V.

During the following 0.1 s, the induced e.m.f. is zero, and during the last 0.3 s, the current is decreasing at the rate of $-10/0.3$, namely -33.3 A/s,

$\therefore$ corresponding induced e.m.f. $= -(-33.3 \times 1.2) = 40$ V.

The variation of the induced e.m.f. is represented by the uniformly-dotted graph C in fig. 4.15. Since the p.d. applied across the inductance has to neutralize the induced e.m.f., its variation is represented by the chain-dotted graph D which is exactly similar to graph C except that the signs are reversed.

The resultant voltage applied to the coil is obtained by adding graphs B and D; thus the resultant voltage increases uniformly from 60 to 110 V during the first 0.2 s, remains constant at 50 V for the next 0.1 s and then changes uniformly from 10 to -40 V during the last 0.3 s, as shown by graph E in fig. 4.14.

The power supplied to the magnetic field increases uniformly from zero to (10 A $\times$ 60 V), namely 600 W, during the first 0.2 s. It is zero during the next 0.1 s. Immediately the current begins to decrease, energy is being returned from the magnetic field to the electric circuit, and the power decreases uniformly from (-40 V $\times$ 10 A), namely -400 W, to zero as represented by graph F.

The positive shaded area enclosed by graph F represents the energy ($= \frac{1}{2} \times 600 \times 0.2 = 60$ J) absorbed by the magnetic field

Fig. 4.15 Graphs for example 4.8

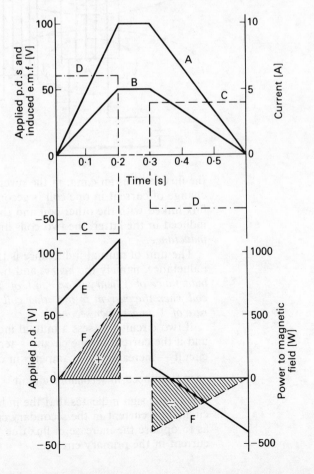

during the first 0.2 s; and the negative shaded area represents the energy ($= \frac{1}{2} \times 400 \times 0.3 = 60$ J) returned from the magnetic field to the electric circuit during the last 0.3 s. The two areas are obviously equal in magnitude, i.e. all the energy supplied to the magnetic field is returned to the electric circuit.

4.10 Mutual inductance

If two coils A and C are placed relative to each other as in fig. 4.16, then when S is closed, some of the flux produced by the current in A becomes linked with C and the e.m.f. induced in C circulates a momentary current through galvanometer G. Similarly when S is opened the collapse of

Fig. 4.16 Mutual inductance

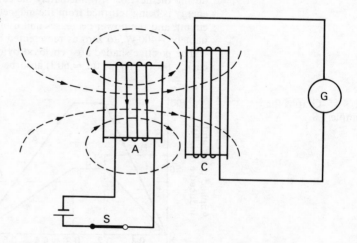

the flux induces an e.m.f. in the reverse direction in C. Since a change of current in one coil is accompanied by a change of flux linked with the other coil and therefore by an e.m.f. induced in the latter, the two coils are said to have *mutual inductance*.

The unit of mutual inductance is the same as for self inductance, namely the *henry*; and *two coils have a mutual inductance of 1 henry if an e.m.f. of 1 volt is induced in one coil when the current in the other coil varies uniformly at the rate of 1 ampere per second*.

If two circuits possess a mutual inductance of M henrys and if the current in one circuit — termed the *primary* circuit — increases by di amperes in dt seconds,

e.m.f. induced in *secondary* circuit $= - M \cdot \mathrm{d}i/\mathrm{d}t$ volts   (4.23)

The minus sign indicates that the induced e.m.f. tends to circulate a current in the secondary circuit in such a direction as to oppose the increase of flux due to the increase of current in the primary circuit.

If $\mathrm{d}\phi$ webers is the increase of flux linked with the secondary circuit due to the increase of $\mathrm{d}i$ amperes in the primary,

$$\left.\begin{array}{c} \text{e.m.f. induced in} \\ \text{secondary circuit} \end{array}\right\} = -N_2 \cdot \mathrm{d}\phi/\mathrm{d}t \text{ volts} \qquad (4.24)$$

where $N_2 =$ number of secondary turns. From expressions (4.23) and (4.24),

$$M \cdot \mathrm{d}i/\mathrm{d}t = N_2 \cdot \mathrm{d}\phi/\mathrm{d}t$$

$$\therefore \qquad M = N_2 \cdot \mathrm{d}\phi/\mathrm{d}i$$

$$= \frac{\text{change of flux-linkages with secondary}}{\text{change of current in primary}} \cdot \quad (4.25)$$

If the relative permeability of the magnetic circuit remains constant, the ratio $\mathrm{d}\phi/\mathrm{d}i$ must also remain constant and is equal to the flux per ampere, so that

$$M = \frac{\text{flux-linkages with secondary}}{\text{current in primary}} = N_2\Phi_2/I_1 \quad (4.26)$$

where Φ_2 is the flux linked with the secondary circuit due to a current I_1 in the primary circuit.

The mutual inductance between two circuits, A and B, is precisely the same whether we assume A to be the primary and B the secondary, or vice versa; for instance, if the two coils are wound on a non-metallic cylinder, as in fig. 4.17, then, from expression (4.21),

$$\left.\begin{array}{c} \text{energy in the magnetic field due} \\ \text{to current } I_A \text{ in coil A alone} \end{array}\right\} = \tfrac{1}{2}L_A I_A^2 \text{ joules}$$

$$\text{and } \left.\begin{array}{c} \text{energy in the magnetic field due} \\ \text{to current } I_B \text{ in coil B alone} \end{array}\right\} = \tfrac{1}{2}L_B I_B^2 \text{ joules.}$$

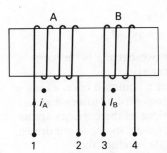

Fig. 4.17 Mutual inductance

Suppose the current in B to be maintained steady at I_B amperes in the direction shown in fig. 4.17, and the current in A to be increased by $\mathrm{d}i$ amperes in $\mathrm{d}t$ seconds, then:

$$\text{e.m.f. induced in B} = -M_{12} \cdot \mathrm{d}i/\mathrm{d}t \text{ volts}$$

where $M_{12} =$ mutual inductance when A is primary.

If the direction of i_A is that indicated by the arrowhead in fig. 4.17, then, by Lenz's Law, the direction of the e.m.f. induced in B is anticlockwise when the coil is viewed from the right-hand end; i.e. the induced e.m.f. is in opposition to I_B and the p.d. across terminals 3 and 4 has to be increased by $M_{12} \cdot \mathrm{d}i/\mathrm{d}t$ volts to maintain I_B constant. Hence the *additional* electrical energy absorbed by coil B in time $\mathrm{d}t$

$$= I_B M_{12}(\mathrm{d}i/\mathrm{d}t) \times \mathrm{d}t = I_B M_{12} \cdot \mathrm{d}i \text{ joules.}$$

Since I_B remains constant, the $I^2 R$ loss in B is unaffected, and there is no e.m.f. induced in coil A apart from that due to the increase of i_A; therefore this additional energy supplied to coil B is absorbed by the magnetic field. Hence, when the current in A has increased to I_A,

$$\left.\begin{array}{c}\text{total energy in}\\ \text{magnetic field}\end{array}\right\} = \tfrac{1}{2}L_A I_A^2 + \tfrac{1}{2}L_B I_B^2 + \int_0^{I_A} I_B M_{12} \cdot \mathrm{d}i$$

$$= \tfrac{1}{2}L_A I_A^2 + \tfrac{1}{2}L_B I_B^2 + M_{12} I_A I_B \text{ joules}$$

If the direction of either I_A or I_B were reversed, the direction of the e.m.f. induced in B, while the current in A was increasing, would be the same as that of I_B, and coil B would then be acting as a generator. By the time the current in A would have reached its steady value I_A, the energy withdrawn from the magnetic field and generated in coil B would be $M_{12} I_A I_B$ joules, and final energy in magnetic field would be:

$$\tfrac{1}{2}L_A I_A^2 + \tfrac{1}{2}L_B I_B^2 - M_{12} I_A I_B \text{ joules}.$$

Hence, in general, total energy in magnetic field

$$= \tfrac{1}{2}L_A I_A^2 + \tfrac{1}{2}L_B I_B^2 \pm M_{12} I_A I_B \text{ joules} \qquad (4.27)$$

the sign being positive when the ampere-turns due to I_A and I_B are additive, and negative when they are in opposition.

If M_{21} were the mutual inductance with coil B as primary, it could be shown by a similar procedure that the total energy in the magnetic field

$$= \tfrac{1}{2}L_A I_A^2 + \tfrac{1}{2}L_B I_B^2 \pm M_{21} I_A I_B \text{ joules}.$$

Since the final conditions are identical in the two cases, the energies must be the same,

$$\therefore \qquad\qquad M_{12} I_A I_B = M_{21} I_A I_B$$

or $$\qquad\qquad M_{12} = M_{21} = (\text{say})\ M,$$

i.e. the mutual inductance between two circuits is the same whichever circuit is taken as the primary.

When the two coils are shown on a common core, as in fig. 4.17, it is obvious that the magnetomotive forces due to I_A and I_B are additive when the directions of the currents are as indicated by the arrowheads. If, however, the coils are drawn as in fig. 4.18, it is impossible to state whether the magnetomotive forces due to currents I_A and I_B are additive or in opposition; and it is to remove this ambiguity that the dot notation has been adopted. Thus, in figs. 4.17 and 4.18, dots are inserted at ends 1 and 3 of the coils to indicate that when currents *enter both* coils (or *leave both* coils) at these ends, as in fig. 4.18(a), the magnetomotive forces of the coils are additive, and the mutual inductance is then said to be *positive*. But if I_A enters coil A at the dotted end and I_B *leaves* coil B at the dotted end, as in fig. 4.18(b), the m.m.f.s of the

Fig. 4.18 Application of the dot notation

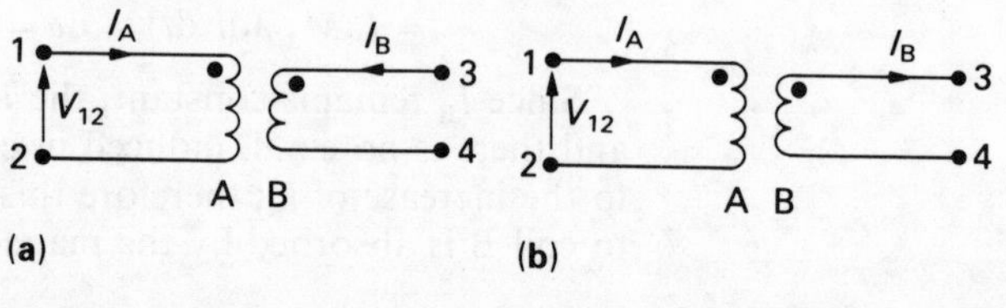

(a)   (b)

coils are in opposition and the mutual inductance is then said to be *negative*.

The following illustrates the application of the dot notation but anticipates the explanation of complex notation given in Chapter 8.

If I_A and I_B represent the complex currents in coils A and B respectively,

V_{12} represents the complex voltage applied across terminals 1–2, the voltage being assumed positive when terminal 1 is positive with respect to terminal 2,

R_A and L_A represent the resistance and self inductance respectively of coil A,

M represents the mutual inductance of coils A and B, and ω represents the angular frequency ($= 2\pi f$),

then if the currents are assumed to have their positive directions as indicated by the arrowheads in fig. 4.18(*a*),

$$V_{12} = I_A(R_A + j\omega L_A) + j\omega M I_B.$$

But if the currents have their positive directions as indicated in fig. 4.18(*b*),

$$V_{12} = I_A(R_A + j\omega L_A) - j\omega M I_B.$$

A further application of the dot notation is given at the end of section 4.12.

4.11 Relationship between mutual inductance and self inductances of two coils: coupling coefficient

Suppose a ring of non-magnetic material to be wound *uniformly* with two coils, A and B, the turns of one coil being as close as possible to those of the other coil, so that the whole of the flux produced by current in one coil is linked with all the turns of the other coil. If coil A has N_1 turns and B has N_2 turns, and if the reluctance of the magnetic circuit is S amperes per weber, then, from expression (4.5), the self inductances of A and B are

$$L_1 = N_1\Phi_1/I_1 = N_1^2\Phi_1/(I_1 N_1) = N_1^2/S \qquad (4.28)$$

and $\qquad L_2 = N_2\Phi_2/I_2 = N_2^2/S \qquad (4.29)$

where Φ_1 and Φ_2 are the magnetic fluxes due to I_1 in coil A and I_2 in coil B respectively, and

$$S = I_1 N_1/\Phi_1 = I_2 N_2/\Phi_2.$$

Since the whole of flux Φ_1 due to I_1 is linked with coil B, it follows from expression (4.26) that

$$M = N_2\Phi_1/I_1 = N_1 N_2\Phi_1/(I_1 N_1)$$

$$= N_1 N_2/S \qquad (4.30)$$

Hence, from (4.28), (4.29) and (4.30),

$$L_1 L_2 = N_1^2 N_2^2 / S^2 = M^2$$

so that

$$M = \sqrt{(L_1 L_2)} \qquad\qquad (4.31)$$

We have assumed that

(a) the reluctance remains constant, and
(b) the magnetic leakage is zero, i.e. that all the flux produced by one coil is linked with the other coil.

The first assumption means that expression (4.31) is strictly correct only when the magnetic circuit is of non-magnetic material. It is, however, approximately correct for a ferromagnetic core if the latter has one or more gaps of air or other non-magnetic material, since the reluctance of such a magnetic circuit is approximately constant.

When there is magnetic leakage, i.e. when all the flux due to current in one coil is not linked with the other coil,

$$M = k\sqrt{(L_1 L_2)} \qquad\qquad (4.32)$$

where k is termed the *coupling coefficient*. 'Coupling coefficient' is a term much used in radio work to denote the degree of coupling between two coils; thus, if the two coils are close together, most of the flux produced by current in one coil passes through the other and the coils are said to be *tightly* coupled. If the coils are well apart, only a small fraction of the flux is linked with the secondary, and the coils are said to be *loosely* coupled.

Example 4.9   *Calculate the mutual inductance between the solenoid and the search coil of example 3.10. If the self inductance of the solenoid is 2 mH and that of the search coil is 25 μH, find the coupling coefficient between the two coils.*

For a long solenoid (section 3.18),

$$\text{flux density at centre} = \frac{I N \mu_0}{l} = \frac{800 I \times 4\pi \times 10^{-7}}{1.2}$$

$$= 0.000\,838 I \text{ tesla}$$

$$\therefore \quad \text{flux through search coil} = 0.000\,838 I \times \pi \times (0.015)^2$$

$$= 0.592 I \times 10^{-6} \text{ Wb}.$$

Since there is no ferromagnetic material in the solenoid, the change of flux is proportional to the change of current; hence, from (4.26).

$$\text{mutual inductance} = \frac{50 \times 0.592 I \times 10^{-6}}{I} = 29.6 \times 10^{-6} \text{ H}$$

$$= 29.6\,\mu\text{H}.$$

In expression (4.32), the self and mutual inductances must be expressed in the same units,

$$\therefore \quad \text{coupling coefficient} = \frac{29.6}{\sqrt{(2000 \times 25)}} = 0.132.$$

4.12 Inductance of inductively coupled coils connected in series

Fig. 4.19(a) shows two coils A and B wound coaxially on an insulating cylinder, with terminals 2 and 3 joined together. It will be evident that the fluxes produced by a current i through the two coils are in the same direction, and the coils are said to be *cumulatively coupled*. Suppose A and B to

Fig. 4.19 Cumulative and differential coupling of two coils connected in series

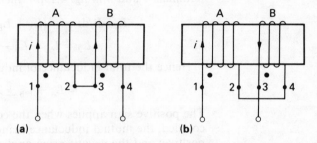

(a)   (b)

have self inductances L_A and L_B henrys respectively and a mutual inductance M henrys, and suppose the arrowheads to represent the positive direction of the current. If the current increases by di amperes in dt seconds,

$$\left.\begin{array}{l}\text{e.m.f. induced in A due to}\\ \text{its self inductance}\end{array}\right\} = -L_A \cdot di/dt \text{ volts}$$

and
$$\left.\begin{array}{l}\text{e.m.f. induced in B due to}\\ \text{its self inductance}\end{array}\right\} = -L_B \cdot di/dt \text{ volts}.$$

Also,
$$\left.\begin{array}{l}\text{e.m.f. induced in A due to}\\ \text{increase of current in B}\end{array}\right\} = -M \cdot di/dt \text{ volts}$$

and
$$\left.\begin{array}{l}\text{e.m.f. induced in B due to}\\ \text{increase of current in A}\end{array}\right\} = -M \cdot di/dt \text{ volts}.$$

The minus signs signify that the direction of all the e.m.f.s is opposite to that of the current. Hence:

total e.m.f. induced in A and B $= -(L_A + L_B + 2M) \cdot di/dt$.

If the windings between terminals 1 and 4 be regarded as a single circuit having a self inductance L_1 henrys, then for the same increase di amperes in dt seconds,

e.m.f. induced in the whole circuit $= -L_1 \cdot di/dt$ volts.

But the e.m.f. induced in the whole circuit is obviously the same as the sum of the e.m.f.s induced in A and B, i.e.

$$-L_1 \cdot di/dt = -(L_A + L_B + 2M) \cdot di/dt$$

$$\therefore \qquad L_1 = L_A + L_B + 2M \qquad\qquad (4.33)$$

Let us next reverse the direction of the current in B relative to that in A by joining together terminals 2 and 4, as in fig. 4.19(b). With this differential coupling, the e.m.f.,

$M \cdot di/dt$, induced in coil A due to an increase di amperes in dt seconds in coil B, is in the same direction as the current and is therefore in opposition to the e.m.f. induced in coil A due to its self inductance. Similarly, the e.m.f. induced in B by mutual inductance is in opposition to that induced by the self inductance of B. Hence,

total e.m.f. induced in A and B

$$= -L_A \cdot di/dt - L_B \cdot di/dt + 2M \cdot di/dt$$

If L_2 be the self inductance of the whole circuit between terminals 1 and 3 in fig. 4.19(*b*), then

$$-L_2 \cdot di/dt = -(L_A + L_B - 2M) \cdot di/dt$$

$$\therefore \qquad L_2 = L_A + L_B - 2M \qquad (4.34)$$

Hence the total inductance of inductively-coupled circuits

$$= L_A + L_B \pm 2M \qquad (4.35)$$

The positive sign applies when the coils are cumulatively coupled, the mutual inductance being then regarded as positive; and the negative sign applies when they are differentially coupled.

From expressions (4.33) and (4.34), we have

$$M = (L_1 - L_2)/4 \qquad (4.36)$$

i.e. the mutual inductance between two inductively-coupled coils is a quarter of the difference between the total self inductance of the circuit when the coils are cumulatively coupled and that when they are differentially coupled.

The above results can be derived by using complex notation (Chapter 8) together with the dot notation described in section 4.10. For simplicity, let us assume the resistance of the coils to be negligible.

If I_a represents the complex current in fig. 4.19(*a*), where the current is assumed positive when it enters *each* coil at the dotted end, and M is therefore positive,

$$\text{p.d. across coil A} = I_a(j\omega L_A + j\omega M)$$

and $$\text{p.d. across coil B} = I_a(j\omega L_B + j\omega M)$$

$\therefore$ p.d. across terminals 1–4, fig. 4.19(*a*)

$$= V_{14} = I_a(j\omega L_A + j\omega L_B + 2j\omega M) = j\omega L_1 I_a$$

$$\therefore \qquad L_1 = L_A + L_B + 2M.$$

In fig. 4.19(*b*), the current is assumed positive when it *enters* coil A at the dotted end and *leaves* coil B at the dotted end; consequently M is negative.

Hence, if I_b represents the complex current in this case,

$$\text{p.d. across coil A} = I_b(j\omega L_A - j\omega M)$$

and $$\text{p.d. across coil B} = I_b(j\omega L_B - j\omega M)$$

$\therefore$ p.d. across terminals 1–3, fig. 4.19(*b*)

$$= V_{13} = I_b(j\omega L_A + j\omega L_B - 2j\omega M) = j\omega L_2 I_b$$

$$\therefore \qquad L_2 = L_A + L_B - 2M.$$

Hence $\qquad M = (L_1 - L_2)/4$

Summary of important formulae

$$\text{Induced e.m.f.} = -L \cdot di/dt \text{ volts} \qquad (4.2)$$

For a magnetic circuit having constant relative permeability,

$$L = \text{flux-linkages per ampere} \qquad (4.5)$$

For a homogeneous magnetic circuit of uniform section and constant relative permeability,

$$L = N^2 \cdot \mu_0 \mu_r A/l \text{ henrys} \qquad (4.9)$$

For R and L, in series, connected across a d.c. supply,

$$\text{instantaneous current} = i = I(1 - e^{-Rt/L}) \qquad (4.16)$$

$$= I(1 - e^{-t/T}) \qquad (4.18)$$

and $\qquad \text{time constant} = T = L/R \text{ seconds} \qquad (4.17)$

For decay of current in inductive circuit,

$$\text{instantaneous current} = i = I e^{-Rt/L} \qquad (4.20)$$

$$\text{Magnetic energy stored in inductor} = \tfrac{1}{2}LI^2 \text{ joules} \qquad (4.21)$$

For a magnetic circuit having constant relative permeability,

$$M = \frac{\text{flux-linkages with secondary}}{\text{current in primary}} \qquad (4.26)$$

Energy stored in magnetic field of coils A and B having mutual inductance

$$= \tfrac{1}{2}L_A I_A^2 + \tfrac{1}{2}L_B I_B^2 \pm M I_A I_B \qquad (4.27)$$

$$\text{Coupling coefficient} = \frac{M}{\sqrt{(L_1 L_2)}} \qquad (4.32)$$

For inductively coupled coils connected in series,

$$\text{total inductance} = L_A + L_B \pm 2M \qquad (4.35)$$

EXERCISES 4

1. A 1500-turn coil surrounds a magnetic circuit which has a reluctance of 6×10^6 A/Wb. What is the inductance of the coil?
2. Calculate the inductance of a circuit in which 30 V are induced when the current varies at the rate of 200 A/s.
3. At what rate is the current varying in a circuit having an inductance of 50 mH when the induced e.m.f. is 8 V?
4. What is the value of the e.m.f. induced in a circuit having an inductance of 700 μH when the current varies at a rate of 5000 A/s?

5. A certain coil is wound with 50 turns and a current of 8 A produces a flux of 200 μWb. Calculate (a) the inductance of the coil corresponding to a reversal of the current, (b) the average e.m.f. induced when the current is reversed in 0.2 s.

6. A toroidal coil of 100 turns is wound uniformly on a non-magnetic ring of mean diameter 150 mm. The cross-sectional area of the ring is 706 mm². Estimate: (a) the magnetic field strength at the inner and outer edges of the ring when the current is 2 A; (b) the current required to produce a flux of 0.5 μWb; (c) the self inductance of the coil.

   If the ring had a small radial airgap cut in it, state, giving reasons, what alterations there would be in the answers to (a), (b) and (c).   (S.A.N.C., O.1)

7. A coil consists of two similar sections wound on a common core. Each section has an inductance of 0.06 H. Calculate the inductance of the coil when the sections are connected (a) in series, (b) in parallel.

8. A steel rod, 1 cm diameter and 50 cm long, is formed into a closed ring and uniformly wound with 400 turns of wire. A direct current of 0.5 A is passed through the winding and produces a flux density of 0.75 T. If all the flux links with every turn of the winding, calculate: (a) the relative permeability of the steel; (b) the inductance of the coil; (c) the average value of the e.m.f. induced when the interruption of the current causes the flux in the steel to decay to 20 per cent of its original value in 0.01 s.   (U.L.C.I., O.1)

9. Explain, with the aid of diagrams, the terms *self inductance* and *mutual inductance*. In what unit are they measured? Define this unit.

   Calculate the inductance of a ring-shaped coil having a mean diameter of 200 mm wound on a wooden core of diameter 20 mm. The winding is evenly wound and contains 500 turns.

   If the wooden core is replaced by a ferromagnetic core which has a relative permeability of 600 when the current is 5 A, calculate the new value of inductance.   (U.L.C.I., O.1)

10. Name and define the unit of *self inductance*.

    A large electromagnet is wound with 1000 turns. A current of 2 A in this winding produces a flux through the coil of 0.03 Wb. Calculate the inductance of the electromagnet.

    If the current in the coil is reduced from 2 A to zero in 0.1 s, what average e.m.f. will be induced in the coil?
    (N.C.T.E.C., O.1)

11. State Faraday's and Lenz's laws.

    The field winding of a 4-pole separately-excited d.c. generator consists of 4 coils connected in series, each coil being wound with 1200 turns. If a current of 2 A produces a magnetic flux of 400 μWb, calculate: (a) the inductance of the field circuit; (b) the average value and direction of the induced e.m.f. if the field switch is opened at such a speed that the flux falls to the residual value of 20 μWb in 0.01 s.   (U.E.I., O.1)

12. Explain what is meant by the self inductance of a coil and define the practical unit in which it is expressed.

    A flux of 0.5 mWb is produced in a coil of 900 turns wound on a wooden ring by a current of 3 A. Calculate: (a) the inductance of the coil; (b) the average e.m.f. induced in the coil when a current of 5 A is switched off, assuming the current to fall to zero in 1 ms; (c) the mutual inductance between the coils, if a second coil of 600 turns was uniformly wound over the first coil.   (U.E.I., O.1)

13. Define the *ampere* in terms of SI units.

    A steel ring, having a mean circumference of 250 mm and a cross-sectional area of 400 mm², is wound with a coil of 70 turns. From the following data calculate the current required to set up a magnetic flux of 510 μWb.

B (teslas)	1.0	1.2	1.4
H (amperes/metre)	350	600	1250

    Calculate also: (*a*) the inductance of the coil at this current; (*b*) the self-induced e.m.f. if this current is switched off in 0.005 s. (E.M.E.U., O.1)

14. Explain the meaning of *self inductance* and define the unit in which it is measured.

    A coil consists of 750 turns and a current of 10 A in the coil gives rise to a magnetic flux of 1200 μWb. Calculate the inductance of the coil, and determine the average e.m.f. induced in the coil when this current is reversed in 0.01 s.

    (W.J.E.C., O.1)

15. Explain what is meant by the self inductance of an electric circuit and define the unit of self inductance.

    A non-magnetic ring having a mean diameter of 300 mm and a cross-sectional area of 500 mm² is uniformly wound with a coil of 200 turns. Calculate from first principles the inductance of the winding. (App. El., L.U.)

16. Two coils, A and B, have self inductances of 120 μH and 300 μH respectively. When a current of 3 A through coil Q is reversed, the deflection on a fluxmeter connected across B is 600 μWb-turns. Calculate: (*a*) the mutual inductance between the coils, (*b*) the average e.m.f. induced in coil B if the flux is reversed in 0.1 s and (*c*) the coupling coefficient.

17. A steel ring having a mean diameter of 20 cm and cross-section of 10 cm² has a winding of 500 turns upon it. The ring is sawn through at one point, so as to provide an airgap in the magnetic circuit. How long should this gap be, if it is desired that a current of 4 A in the winding should produce a flux density of 1.0 T in the gap? State the assumptions made in your calculation.

    What is the inductance of the winding when a current of 4 A is flowing through it?

    The permeability of free space is $4\pi \times 10^{-7}$ H/m and the data for the *B/H* curve of the steel are given below:

H (A/m)	190	254	360	525	1020	1530	2230
B (T)	0.6	0.8	1.0	1.2	1.4	1.5	1.6

    (App. El., L.U.)

18. A certain circuit has a resistance of 10 Ω and a constant inductance of 3 H. The current through this circuit is increased uniformly from 0 to 5 A in 0.6 s, maintained constant at 4 A for 0.1 s and then reduced uniformly to zero in 0.3 s. Draw to scale graphs representing the variation of (*a*) the current, (*b*) the induced e.m.f. and (*c*) the resultant applied voltage.

19. A coil having a resistance of 2 Ω and an inductance of 0.5 H has a current passed through it which varies in the following manner: (*a*) a uniform change from zero to 50 A in 1 s; (*b*) constant at 50 A for 1 s; (*c*) a uniform change from 50 A to zero in 2 s. Plot the current graph to a time base. Tabulate the potential difference applied to the coil during each of the above periods and plot the graph of potential difference to a time base. (U.L.C.I., O.2)

20. A coil wound with 500 turns has a resistance of $2\,\Omega$. It is found that a current of 3 A produces a flux of $500\,\mu\text{Wb}$. Calculate: (a) the inductance and the time constant of the coil; (b) the average e.m.f. induced in the coil when the flux is reversed in 0.3 s.

   If the coil is switched across a 10-V d.c. supply, derive graphically a curve showing the growth of the current, assuming the inductance to remain constant.

21. Explain the term *time constant* in connection with an inductive circuit.

   A coil having a resistance of $25\,\Omega$ and an inductance of 2.5 H is connected across a 50-V d.c. supply. Determine graphically: (a) the initial rate of growth of the current; (b) the value of the current after 0.15 s; and (c) the time required for the current to grow to 1.8 A. (E.M.E.U., O.1)

22. The field winding of a d.c. machine has an inductance of 10 H and takes a final current of 2 A when connected to a 200-V d.c. supply. Calculate: (i) the initial rate of growth of current; (ii) the time constant; and (iii) the current when the rate of growth is 5 A/s. (W.J.E.C., O.2)

23. A 200-V d.c. supply is suddenly switched across a relay coil which has a time constant of 3 ms. If the current in the coil reaches 0.2 A after 3 ms, determine the final steady value of the current and the resistance and inductance of the coil.

   Calculate the energy stored in the magnetic field when the current has reached its final steady value.

24. A coil of inductance 4 H and resistance $80\,\Omega$ is in parallel with a $200\text{-}\Omega$ resistor of negligible inductance across a 200-V d.c. supply. The switch connecting these to the supply is then opened, the coil and resistor remaining connected together. State, in each case for an instant immediately before and for one immediately after the opening of the switch: (a) the current through the resistor; (b) the current through the coil; (c) the e.m.f. induced in the coil; and (d) the voltage across the coil.

   Give rough sketch graphs, with explanatory notes, to show how these four quantities vary with time. Include intervals both before and after the opening of the switch, and mark on the graphs an approximate time scale. (App. El., L.U.)

25. A circuit consists of a $200\text{-}\Omega$ non-reactive resistor in parallel with a coil of 4 H inductance and $100\,\Omega$ resistance. If this circuit is switched across a 100-V d.c. supply for a period of 0.06 s and then switched off, calculate the current in the coil 0.012 s after the instant of switching off. What is the maximum p.d. across the coil?

26. Define the units of: (a) magnetic flux, and (b) inductance.

   Obtain an expression for the induced e.m.f. and for the stored energy of a circuit, in terms of its inductance, assuming a steady rise of current from zero to its final value and ignoring saturation.

   A coil, of inductance 5 H and resistance $100\,\Omega$, carries a steady current of 2 A. Calculate the initial rate of fall of current in the coil after a short-circuiting switch connected across its terminals has been suddenly closed. What was the energy stored in the coil, and in what form was it dissipated?
   (S.A.N.C., O.2)

27. If two coils have a mutual inductance of $400\,\mu\text{H}$, calculate the e.m.f. induced in one coil when the current in the other coil varies at a rate of 30 000 A/s.

28. If an e.m.f. of 5 V is induced in a coil when the current in an adjacent coil varies at a rate of 80 A/s, what is the value of the mutual inductance of the two coils?

29. If the mutual inductance between two coils is 0.2 H, calculate the e.m.f. induced in one coil when the current in the other coil is increased at a uniform rate from 0.5 to 3 A in 0.05 s.

30. If the toroid of Q. 6 has a second winding of 80 turns wound over the first winding of 100 turns, calculate the mutual inductance.

31. When a current of 2 A through a coil P is reversed, a deflection of 36 divisions is obtained on a fluxmeter connected to a coil Q. If the fluxmeter constant is 150 μWb · turns/div, what is the value of the mutual inductance of coils P and Q?

32. Explain the meaning of the terms *self inductance* and *mutual inductance* and define the unit by which each is measured.

    A long solenoid, wound with 1000 turns, has an inductance of 120 mH and carries a current of 5 A. A search coil of 25 turns is arranged so that it is linked by the whole of the magnetic flux. A ballistic galvanometer is connected to the search coil and the combined resistance of the search coil and galvanometer is 200 Ω. Calculate, from first principles, the quantity of electricity which flows through the galvanometer when the current in the solenoid is reversed. (U.L.C.I., O.2)

33. Define the unit of mutual inductance. A cylinder, 50 mm in diameter and 1 m long, is uniformly wound with 3000 turns in a single layer. A second layer of 100 turns of much finer wire is wound over the first one, near its centre. Calculate the mutual inductance between the two coils. Derive any formula used.
                                                    (App. El., L.U.)

34. A solenoid P, 1 m long and 100 mm in diameter, is uniformly wound with 600 turns. A search-coil Q, 30 mm in diameter and wound with 20 turns, is mounted co-axially midway along the solenoid. If Q is connected to a ballistic galvanometer, calculate the quantity of electricity through the galvanometer when a current of 6 A through the solenoid is reversed. The resistance of the secondary circuit is 0.1 MΩ. Find, also, the mutual inductance between the two coils.

35. When a current of 2 A through a coil P is reversed, a deflection of 43 divisions is obtained on a fluxmeter connected to a coil Q. If the fluxmeter constant is 150 μWb · turns/div, find the mutual inductance of coils P and Q. If the self inductances of P and Q are 5 mH and 3 mH respectively, calculate the coupling coefficient.

36. Two coils, A and B, have self inductances of 20 mH and 10 mH respectively and a mutual inductance of 5 mH. If the currents through A and B are 0.5 A and 2 A respectively, calculate:
    (*a*) the two possible values of the energy stored in the magnetic field; and (*b*) the coupling coefficient.

37. Two similar coils have a coupling coefficient of 0.25. When they are connected in series cumulatively, the total inductance is 80 mH. Calculate: (*a*) the self inductance of each coil; (*b*) the total inductance when the coils are connected in series differentially; and (*c*) the total magnetic energy due to a current of 2 A when the coils are connected in series (i) cumulatively and (ii) differentially.

38. Two coils, with terminals AB and CD respectively, are inductively coupled. The inductance measured between terminals AB is 380 μH and that between terminals CD is 640 μH. With B joined to C, the inductance measured between terminals AD is 1600 μH. Calculate: (*a*) the mutual inductance of the coils; and (*b*) the inductance between terminals AC when B is connected to D.

39. For two coils, A and B, in proximity to each other, the flux-linkages with B per unit current in A are equal to the linkages

with A produced by unit current flowing in B. Show that the mutual inductance between the coils is given by:

$$M = k\sqrt{L_a L_b} \text{ henrys},$$

where $k \not> 1$ and L_a and L_b are the respective self inductances.

When two identical coupled coils are connected in series, the inductance of the combination is found to be 80 mH. When the connections to one of the coils are reversed, a similar measurement indicates 20 mH. Find the coupling coefficient between the two coils. (S.A.N.C., O.2)

5 Electrostatics

5.1 Electrification by friction

Two glass rods rubbed with silk repel each other; similarly, two ebonite rods rubbed with fur repel each other. But the electrified glass and ebonite rods attract each other. The glass and ebonite therefore appear to be charged with different kinds of electricity; and these experiments show that bodies charged with the same kind of electricity repel, while bodies charged with opposite kinds of electricity attract one another.

It was about 1750 that Benjamin Franklin — an American — suggested that electricity was some form of fluid which passed from one body to the other when they were rubbed together, and that in the case of glass rubbed with silk, the electric fluid passed from the silk into the glass so that the glass contained a 'plus' or 'positive' amount of electricity. On the other hand, when ebonite was rubbed with fur, the electric fluid passed from the ebonite to the fur, leaving the ebonite with a 'minus' or 'negative' amount of electricity. Franklin's one-fluid theory has long since been discarded, but his convention still remains: thus, the glass is said to be positively charged and the ebonite negatively charged.

5.2 Structure of the atom

Every material is made up of one or more elements, an element being a substance composed entirely of atoms of the same kind; for instance, water is a combination of the elements hydrogen and oxygen, whereas common salt is a combination of the elements sodium and chlorine. The atoms of different elements differ in their structure, and this accounts for different elements possessing different characteristics.

Every atom consists of a relatively massive core or nucleus carrying a positive charge, around which *electrons* move in orbits at distances that are great compared with the size of the nucleus. Each *electron* has a mass of 9.11×10^{-31} kg and a *negative* charge, $-e$, equal to 1.602×10^{-19} C. The nucleus of every atom except that of hydrogen consists of *protons* and *neutrons*. Each *proton* carries a *positive* charge, e, equal in magnitude to that of an electron and its mass is 1.673×10^{-27} kg, namely 1836 times that of an electron. A *neutron*, on the other hand, carries *no* resultant charge and its mass is approximately the same as that of a proton. Under normal conditions, an atom is neutral, i.e. the total negative charge on its electrons is equal to the total positive charge on the protons.

The atom possessing the simplest structure is that of hydrogen — it consists merely of a nucleus of one proton together with a single electron which may be thought of as revolving in an orbit, of about 10^{-10} m diameter, around the proton, as in fig. 5.1(*a*).

Fig. 5.1(*b*) shows the arrangement of a helium atom. In this case, the nucleus consists of two protons and two neutrons, with two electrons orbiting in what is termed the K *shell*. The nucleus of a carbon atom has six protons and six neutrons and therefore carries a positive charge 6e. This nucleus is surrounded by six orbital electrons, each carrying a negative charge of $-e$, two electrons being in the K shell and four in the L shell, and their relative positions may be imagined to be as shown in fig. 5.1(*c*).

Fig. 5.1 Hydrogen, helium and carbon atoms

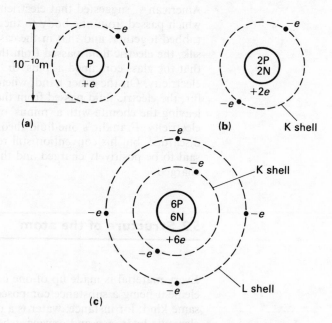

The farther away an electron is from the nucleus, the smaller is the force of attraction between that electron and the positive charge on the nucleus; consequently the easier it is to detach such an electron from the atom. When atoms are packed tightly together, as in a metal, each outer electron experiences a small force of attraction towards neighbouring nuclei, with the result that such an electron is no longer bound to any individual atom, but can move at random within the metal. These electrons are termed *free* or *conduction* electrons and only a slight external influence is required to cause them to drift in a desired direction.

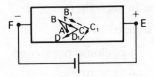

Fig. 5.2 Movement of a free electron

The full lines AB, BC, CD, etc., in fig. 5.2 represent paths of the random movement of one of the free electrons in a *metal* rod when there is *no* p.d. across terminals EF; i.e. the electron is accelerated in direction AB until it collides with an atom, with the result that it may rebound in direction BC, etc. Different free electrons move in different directions so as to maintain the electron density constant throughout the metal; in other words, there is no resultant drift of electrons towards either E or F.

If a cell is connected across terminals E and F so that E is positive relative to F, the effect is to modify the random movement of the electron as shown by the dotted lines AB_1, B_1C_1, and C_1D_1 in fig. 5.2, i.e. there is superimposed on the random movement a drift of the electron towards the positive terminal E; and the number of electrons reaching terminal E from the rod is the same as that entering the rod from terminal F. It is this drift of the electrons that constitutes the electric current in the circuit.

An atom which has lost or gained one or more free electrons is referred to as an *ion*; thus, for an atom which has lost one or more electrons, its negative charge is less than its positive charge, and such an atom is therefore termed a *positive ion*.

5.3 Movement of electrons in a conductor

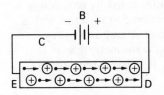

Fig. 5.3 Movement of electrons in a conductor

In section 5.2 it was explained how the drift of electrons in a desired direction can be produced by introducing a source of electromotive force into the circuit. In fig. 5.3, DE represents an enlarged view of a metal rod forming part of a closed circuit which includes a battery B. The circles with crosses represent positive ions, namely atoms which have lost one or more of their outermost electrons. These ions are locked in the structure of the metal and are therefore unable to move. The small black circles represent the free electrons moving from left to right. This procession or drift of the electrons takes place round the whole circuit, including battery B; i.e. the number of electrons emerging per second from the negative terminal of B is exactly the same as that entering the positive terminal per second. It follows that an electric current in a metal conductor consists of a movement of electrons from a point at the *lower* potential to a point at the *higher* potential, namely in the opposite direction to that taken as the conventional direction of the current. The latter was universally adopted long before the discovery of the electron, and so we continue to say that an electric current flows from a point at the *higher* potential to that at the *lower* potential.

Since each electron carries a negative charge of 1.602×10^{-19} coulomb, it follows that when the current in a circuit is 1 ampere (or 1 coulomb per second), the number of electrons passing any given point must be such that:

$1.602 \times 10^{-19} \times$ no. of electrons per second

$= 1$ coulomb per second

$\therefore$ no. of electrons per second $= 6.24 \times 10^{18}$

i.e. when the current in a circuit is 1 ampere, electrons are passing any given point of the circuit at the rate of 6.24×10^{18} per second.

5.4 Capacitor

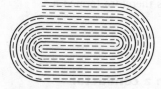

Fig. 5.4 Paper-insulated capacitor

Two metal plates, separated by an insulator, constitute a *capacitor*, namely an arrangement which has the capacity of storing electricity as an excess of electrons on one plate and a deficiency on the other.

A common type of capacitor used in practice consists of two strips of metal foil, represented by full lines in fig. 5.4, separated by strips of waxed paper, shown dotted, these strips being wound spirally, forming — in effect — two very large surfaces near to each other. The whole assembly is thoroughly soaked in hot paraffin wax. In radio receivers, some of the capacitors consist of two sets of metal vanes, one of which is fixed and the other set is so arranged that the vanes can be moved into and out of the space between the fixed vanes without touching the latter.

A charged capacitor may be regarded as a reservoir of electricity and its action can be demonstrated by connecting a capacitor of, say, 20 microfarads (section 5.8) in series with a resistor R, a centre-zero microammeter A and a two-way switch S, as in fig. 5.5. An electrostatic voltmeter ES (section 26.9) is connected across C. If R has a resistance of,

Fig. 5.5 Capacitor charged and discharged through a resistor

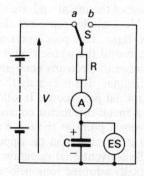

say, 1 megohm, it is found that when S is closed on *a*, the deflection on A rises immediately to its maximum value and then falls off to zero, as indicated by curve A in fig. 5.6. At the same time, the p.d. across C grows in the manner shown by curve M. When S is moved over to *b*, the current again

Fig. 5.6 Charging and
discharging currents and
p.d.s

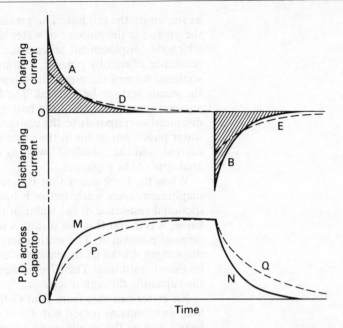

rises immediately to the same maximum but in the reverse
direction, and then falls off as shown by curve B. Curve N
shows the corresponding variation of p.d. across C.

If the experiment is repeated with a resistance of, say,
2 MΩ, it is found that the initial current, both on charging
and on discharging, is halved, but it takes about twice as long
to fall off, as shown by the dotted curves D and E. Curves P
and O represent the corresponding variation of the p.d.
across C during charge and discharge respectively.

The shaded area between curve A and the horizontal axis
in fig. 5.6 represents the product of the average charging
current (in amperes) and the time (in seconds), namely the
quantity of electricity (in coulombs) required to charge the
capacitor to a p.d. of V volts. Similarly the shaded area
enclosed by curve B represents the same quantity of electricity
obtainable during discharge.

5.5 Hydraulic analogy

The operation of charging and discharging a capacitor may
be more easily understood if we consider the hydraulic
analogy given in fig. 5.7, where P represents a piston operated
by a rod R, and D is a rubber diaphragm stretched across a
cylindrical chamber C. The cylinders are connected by pipes
E, E and are filled with water.

When no force is being exerted on P, the diaphragm is flat,
as shown dotted, and the piston is in position A. If P is
pushed towards the left, water is withdrawn from G and
forced into F and the diaphragm is in consequence distended,

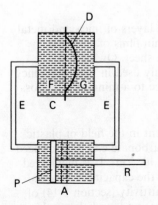

Fig. 5.7 Hydraulic analogy
of a capacitor

as shown by the full line. The greater the force applied to P, the greater is the amount of water displaced. But the rate at which this displacement takes place depends upon the resistance offered by pipes E, E; thus the smaller the cross-sectional area of the pipes, the longer is the time required for the steady state to be reached. The force applied to P is analogous to the e.m.f. of the battery, the quantity of water displaced corresponds to the charge, the rate at which the water passes any point in the pipes corresponds to the current, and the cylinder C with its elastic diaphragm is the analogue of the capacitor.

When the force exerted on P is removed, the distended diaphragm forces water out of F back into G; and if the frictional resistance of the water in the pipes exceeds a certain value, it is found that the piston is merely pushed back to its original position A. The strain energy stored in the diaphragm due to its distension is converted into heat by the frictional resistance. The effect is similar to the discharge of the capacitor through a resistor.

No water can pass from F to G through the diaphragm so long as it remains intact; but if it is strained excessively it bursts, just as the insulation in a capacitor is punctured when the p.d. across it becomes excessive.

5.6 Types of capacitors

Capacitors may be divided into the following main groups according to the nature of the dielectric:

(a) Air capacitors

This type usually consists of one set of fixed plates and another set of movable plates, and is mainly used for radio work where it is required to vary the capacitance.

(b) Paper capacitors

The electrodes consist of metal foils interleaved with paper impregnated with wax or oil and rolled into a compact form (fig. 5.4).

(c) Mica capacitors

This type consists either of alternate layers of mica and metal foil clamped tightly together or of thin films of silver sputtered on the two sides of a mica sheet. Owing to its relatively high cost, this type is mainly used in high-frequency circuits when it is necessary to reduce to a minimum the loss in the dielectric.

(d) Polycarbonate capacitors

Polycarbonate is a recent development in the field of plastic insulating materials. A film of polycarbonate can be produced in thicknesses down to $2 \mu m$ ($= 2 \times 10^{-6}$ m). It is metallized with aluminium and wound to form the capacitor elements.

Polycarbonate has a relative permittivity (section 5.14) of about 2.8; it possesses high resistivity and low dielectric loss. The power factor of a polycarbonate capacitor is about 0.001

at 1 kHz and about 0.005 at 1 MHz, and the maximum operating temperature is about 125–150°C. The significance of these figures will become apparent after reading Chapter 7.

(e) Electrolytic capacitors

The type most commonly used consists of two aluminium foils, one with an oxide film and one without, the foils being interleaved with a material such as paper saturated with a suitable electrolyte; for example, ammonium borate. The aluminium oxide film is formed on the one foil by passing it through an electrolytic bath of which the foil forms the positive electrode. The finished unit is assembled in a container — usually of aluminium — and hermetically sealed. The oxide film acts as the dielectric; and as its thickness in a capacitor suitable for a working voltage of 100 V is only about 0.15 μm, a very large capacitance is obtainable in a relatively small volume.

The main disadvantages of this type of capacitor are: (*a*) the insulation resistance is comparatively low and (*b*) it is only suitable for circuits where the voltage applied to the capacitor never reverses its direction. Electrolytic capacitors are mainly used where very large capacitances are required, e.g. for reducing the ripple in the voltage wave obtained from a rectifier (section 10.12).

Solid types of electrolytic capacitors have been developed to avoid some of the disadvantages of the wet electrolytic type. In one arrangement, the wet electrolyte is replaced by manganese dioxide. In another arrangement, the anode is a cylinder of pressed sintered tantalum powder coated with an oxide layer which forms the dielectric. This oxide has a conducting coat of manganese dioxide which acts as an electron conductor and replaces the ionic conduction of the liquid electrolyte in the wet type. A layer of graphite forms the connection with a silver or copper cathode and the whole is enclosed in a hermetically-sealed steel can.

(f) Tantalum electrolytic capacitors

These capacitors are much smaller than the corresponding aluminium electrolytic capacitors. The construction may take the form indicated in fig. 5.8, in which one plate consists of pressed, sintered tantalum powder coated with an oxide layer

Fig. 5.8 Sintered tantalum capacitor

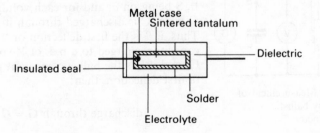

which is the dielectric. The case of brass, copper or even silver forms the other plate. Layers of manganese dioxide and graphite form the electrolyte.

(g) Other capacitors

There are further ranges of capacitors which include the metallized-paper capacitor, which is similar to the paper capacitor except that the paper is coated with a thin layer of metal on one side. Two such layers are rolled together to form a capacitor, being mounted in a sealed container. The advantage of this form is that it is self-healing, i.e. should a localized breakdown of the dielectric occur, the heat vaporizes the metallic coating and the conductor around the problem area is thereby removed. Such a capacitor may be used for spark suppression in a car ignition system.

Ceramic capacitors are formed from metallic coatings on the opposite faces of a thin slab of ceramic material. The capacitance values available are small but are useful in high-frequency applications, especially in cases where high temperatures may be experienced.

Various titanates have very high values of relative permittivities and lend themselves to the manufacture of extremely small capacitors. At a time of extreme miniaturization, titanate capacitors have a wide range of applications.

5.7 Relationship between the charge and the applied voltage

The method described in section 5.4 for determining the charge on each plate of a capacitor is instructive, but is unsuitable for accurate measurement. A much better method is to discharge the capacitor through a ballistic galvanometer (section 3.10), since the deflection of the latter is proportional to the charge.

Let us charge a capacitor C (fig. 5.9) to various voltages by means of a slider on a resistor R connected across a battery B, S being on *a*; and for each voltage, note the deflection of G when C is discharged through it by moving S over to *b*. Thus, if θ is the first deflection or 'throw' observed when the capacitor, charged to a p.d. of V volts, is discharged through G, and if k is the ballistic constant of G in coulombs per unit of first deflection, then:

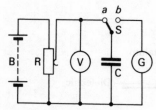

Fig. 5.9 Measurement of charge by ballistic galvanometer

$$\text{discharge through G} = Q = k\theta \text{ coulombs.}$$

It is found that for a given capacitor,

$$\frac{\text{charge on C [coulombs]}}{\text{p.d. across C [volts]}} = \text{a constant} \qquad (5.1)$$

5.8 Capacitance

The property of a capacitor to store an electric charge when its plates are at different potentials is referred to as its *capacitance*.

The unit of capacitance is termed the *farad* (symbol F) — a curtailment of 'Faraday' — and may be defined as *the capacitance of a capacitor between the plates of which there appears a difference of potential of 1 volt when it is charged by 1 coulomb of electricity.*

It follows from expression (5.1) and from the definition of the farad that:

$$\frac{\text{charge [coulombs]}}{\text{applied p.d. [volts]}} = \text{capacitance [farads]}$$

or in symbols, $Q/V = C$

$$\therefore \qquad Q = CV \qquad (5.2)$$

In practice, the farad is found to be inconveniently large and the capacitance is usually expressed in *microfarads* (μF) or in *picofarads* (pF),

where 1 microfarad $= 10^{-6}$ farad

and 1 picofarad $= 10^{-12}$ farad.

Example 5.1 *A capacitor having a capacitance of 80 μF is connected across a 500-V d.c. supply. Calculate the charge.*

From (5.2), charge $= (80 \times 10^{-6})$ [farad] $\times 500$ [volts]

$= 0.04$ coulombs.

5.9 Capacitors in parallel

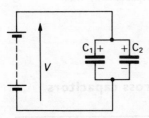

Fig. 5.10 Capacitors in parallel

Suppose two capacitors, having capacitances C_1 and C_2 farads respectively, to be connected in parallel (fig. 5.10) across a p.d. of V volts. The charge on C_1 is Q_1 coulombs and that on C_2 is Q_2 coulombs, where:

$$Q_1 = C_1 V \quad \text{and} \quad Q_2 = C_2 V.$$

If we were to replace C_1 and C_2 by a single capacitor of such capacitance C farads that the same total charge of $(Q_1 + Q_2)$ coulombs would be produced by the same p.d., then $Q_1 + Q_2 = CV.$

Substituting for Q_1 and Q_2, we have:

$$C_1 V + C_2 V = CV$$

$$\therefore \qquad C = C_1 + C_2 \qquad (5.3)$$

Hence *the resultant capacitance of capacitors in parallel is the arithmetic sum of their respective capacitances.*

5.10 Capacitors in series

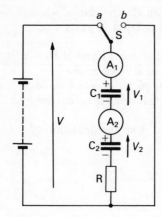

Fig. 5.11 Capacitors in series

Suppose C_1 and C_2 in fig. 5.11 to be two capacitors connected in series with suitable centre-zero ammeters A_1 and A_2, a resistor R and a two-way switch S. When S is put over to a, A_1 and A_2 are found to indicate exactly the same charging current, each reading decreasing simultaneously from a maximum to zero, as already shown in fig. 5.6. Similarly, when S is put over to b, A_1 and A_2 indicate similar discharges. It follows that during charge the displacement of electrons from the positive plate of C_1 to the negative plate of C_2 is exactly the same as that from the upper plate (fig. 5.11) of C_2 to the lower plate of C_1. In other words, the displacement of Q coulombs of electricity is the same in every part of the circuit, and the charge on each capacitor is therefore Q coulombs.

If V_1 and V_2 be the corresponding p.d.s across C_1 and C_2 respectively, then from (5.2):

$$Q = C_1 V_1 = C_2 V_2 \qquad (5.4)$$

so that
$$V_1 = \frac{Q}{C_1}, \quad \text{and} \quad V_2 = \frac{Q}{C_2}.$$

If we were to replace C_1 and C_2 by a single capacitor of capacitance C farads such that it would have the same charge Q coulombs with the same p.d. of V volts, then:

$$Q = CV, \quad \text{or} \quad V = Q/C.$$

But it is evident from fig. 5.11 that $V = V_1 + V_2$. Substituting for V, V_1 and V_2, we have:

$$\frac{Q}{C} = \frac{Q}{C_1} + \frac{Q}{C_2}$$

$$\therefore \quad \frac{1}{C} = \frac{1}{C_1} + \frac{1}{C_2} \qquad (5.5)$$

Hence *the reciprocal of the resultant capacitance of capacitors connected in series is the sum of the reciprocals of their respective capacitances.*

5.11 Distribution of voltage across capacitors in series

From expression (5.4),

$$V_2/V_1 = C_1/C_2 \qquad (5.6)$$

But
$$V_1 + V_2 = V$$

$$\therefore \qquad V_2 = V - V_1.$$

Substituting for V_2 in (5.6), we have:

$$(V - V_1)/V_1 = C_1/C_2$$

$$\therefore \qquad V_1 = V \times \frac{C_2}{C_1 + C_2} \qquad (5.7)$$

and

$$V_2 = V \times \frac{C_1}{C_1 + C_2} \qquad (5.8)$$

Example 5.2   *Three capacitors have capacitances of $2\,\mu F$, $4\,\mu F$ and $8\,\mu F$ respectively. Find the total capacitance when they are connected* (a) *in parallel,* (b) *in series.*

(*a*) From (5.3):

$$\text{total capacitance} = 2 + 4 + 8 = 14\,\mu F.$$

(*b*) If C be the resultant capacitance in microfarads when the capacitors are in series, then from (5.5):

$$\frac{1}{C} = \frac{1}{2} + \frac{1}{4} + \frac{1}{8} = 0.5 + 0.25 + 0.125 = 0.875$$

$$\therefore \qquad C = 1.143\,\mu F.$$

Example 5.3   *If two capacitors having capacitances of $6\,\mu F$ and $10\,\mu F$ respectively are connected in series across a 200-V supply, find* (a) *the p.d. across each capacitor,* (b) *the charge on each capacitor.*

(*a*) Let V_1 and V_2 be the p.d.s across the 6-μF and 10-μF capacitors respectively; then, from expression (5.7),

$$V_1 = 200 \times \frac{10}{6 + 10} = 125\ V$$

and

$$V_2 = 200 - 125 = 75\ V.$$

(*b*) Charge on each capacitor

$$= \text{charge on } C_1$$

$$= 6 \times 10^{-6} \times 125 = 0.000\,75\ C.$$

5.12 Relationship between the capacitance and the dimensions of a capacitor

It follows from expression (5.3) that if two similar capacitors are connected in parallel, the capacitance is double that of one capacitor. But the effect of connecting two similar capacitors in parallel is merely to double the area of each plate. In general, we may therefore say that the capacitance of a capacitor is proportional to the area of the plates.

On the other hand, if two similar capacitors are connected in series, it follows from expression (5.5) that the capacitance is halved. We have, however, doubled the thickness of the insulation between the plates that are connected to the supply. Hence we may say in general that the capacitance of a capacitor is inversely proportional to the distance between the plates; and the above relationships may be summarized thus:

$$\text{capacitance} \propto \frac{\text{area of plates}}{\text{distance between plates}}$$

5.13 Electric field strength and electric flux density

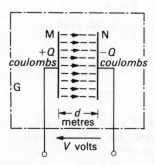

Fig. 5.12 A parallel-plate capacitor

Let us consider a capacitor consisting of metal plates, M and N, in a glass enclosure G, shown chain-dotted in fig. 5.12, from which all the air has been removed. Let A be the area in square metres of one side of each plate and let d be the distance in metres between the plates. Let Q be the charge in coulombs due to a p.d. of V volts between the plates.

Suppose the dimensions of the plates to be so large compared with the distance between them that we may assume negligible fringing of the electric flux, i.e. all the electric flux may be assumed to pass straight across from M to N, as shown by the dotted lines in fig. 5.12.

The *electric field strength* in the region between the two plates M and N is the *potential drop per unit length* or *potential gradient*, namely V/d volts per metre; and the *direction* of the electric field strength at any point is the direction of the mechanical force on a positive charge situated at that point, namely from the positively-charged plate M towards the negatively-charged plate N in fig. 5.12. The symbol for electric field strength is $\boldsymbol{E}$*,

i.e.
$$\boldsymbol{E} = V/d \text{ volts per metre} \tag{5.9}$$

In SI, *one* unit of electric flux is assumed to emanate from a positive charge of 1 coulomb and to enter a negative charge of 1 coulomb. Hence, if the charge on plates M and N is Q coulombs, electric flux between M and N $= \Psi = Q$ coulombs and

$$\text{electric flux density} = \boldsymbol{D} = Q/A \text{ coulombs per metre}^2 \tag{5.10}$$

where $A = $ area of dielectric, in square metres, at right-angles to direction of electric flux.

From expressions (5.9) and (5.10),

$$\frac{\text{electric flux density}}{\text{electric field strength}} = \frac{\boldsymbol{D}}{\boldsymbol{E}} = \frac{Q}{A} \div \frac{V}{d} = \frac{Q}{V} \times \frac{d}{A} = \frac{Cd}{A}.$$

In electromagnetism, the ratio of the magnetic flux density in a vacuum to the magnetic field strength is termed the

---

* Bold type is used to enable the symbols $\boldsymbol{E}$ and $\boldsymbol{D}$ for electric field strength and electric flux density respectively to be easily distinguished from the letters E and D used to represent other quantities.

permeability of free space and is represented by μ_0. Similarly, in electrostatics, the ratio of the electric flux density in a vacuum to the electric field strength is termed the *permittivity of free space* and is represented by ϵ_0.

Hence, $\qquad\qquad\qquad\qquad \epsilon_0 = Cd/A$

or $\qquad\qquad\qquad\qquad\quad C = \epsilon_0 A/d \qquad\qquad\qquad (5.11)$

The effect of filling the space between M and N (fig. 5.12) with air at atmospheric pressure is to increase the capacitance by 0.06 per cent compared with the value when the space is completely evacuated; hence for all practical purposes, expression (5.11) can be applied to capacitors having air dielectric.

The value of ϵ_0 can be determined experimentally by charging a capacitor, of known dimensions and with vacuum dielectric, to a p.d. of V volts and then discharging it through a ballistic galvanometer having a known ballistic constant k coulombs per unit deflection. If the deflection is θ divisions,

$$Q = CV = k\theta$$

$$\therefore \qquad \epsilon_0 = C \cdot \frac{d}{A} = \frac{k\theta}{V} \cdot \frac{d}{A}.$$

From carefully conducted tests it has been found that the value of ϵ_0 is 8.85×10^{-12} F/m (see footnote on p. 134).

Hence the capacitance of a parallel-plate capacitor with vacuum or air dielectric is given by:

$$C = \frac{(8.85 \times 10^{-12})\,[\text{F/m}] \times A\,[\text{m}^2]}{d\,[\text{m}]} \text{ farads} \qquad (5.12)$$

It may be mentioned at this point that there is a definite relationship between μ_0, ϵ_0 and the velocity of light and other electromagnetic waves; thus,

$$\frac{1}{\mu_0 \epsilon_0} = \frac{1}{4\pi \times 10^{-7} \times 8.85 \times 10^{-12}} \simeq 8.99 \times 10^{16}$$

$$\simeq (2.998 \times 10^8)^2$$

But the velocity of light $= 2.998 \times 10^8$ metres per second

$$\therefore \qquad \left.\begin{array}{r} \text{velocity of light} \\ \text{in metres per second} \end{array}\right\} = \frac{1}{\sqrt{(\mu_0 \epsilon_0)}}. \qquad (5.13)$$

This relationship was discovered by Prof. Clerk Maxwell in 1865 and enabled him to predict the existence of radio waves about twenty years before their effect was demonstrated experimentally by Prof. H. Hertz.

5.14 Relative permittivity

If the experiment described in section 5.13 is performed with a sheet of glass filling the space between plates M and N, it is found that the value of the capacitance is greatly increased; and the ratio of the capacitance of a capacitor having a certain material as dielectric to the capacitance of that capacitor with vacuum (or air) dielectric is termed the *relative permittivity* of that material and is represented by the symbol ϵ_r. (The term *dielectric constant* is obsolete.) Values of the relative permittivity of some of the most important insulating materials are given in the following table:

Material	Relative permittivity
Air	1.0006
Paper (dry)	2–2.5
Bakelite	4.5–5.5
Glass	5–10
Rubber	2–3.5
Mica	3–7
Porcelain	6–7
Barium titanate	6000
Insulating oil	3
Polythene	2–2.5

From expression (5.11), it follows that if the space between the metal plates of the capacitor in fig. 5.12 is filled with a dielectric having a relative permittivity ϵ_r,

$$\text{capacitance} = C = \frac{\epsilon_0 \epsilon_r A}{d} \text{ farads} \qquad (5.14)$$

$$= \frac{(8.85 \times 10^{-12})\,[\text{F/m}] \times \epsilon_r \times A\,[\text{m}^2]}{d\,[\text{m}]} \text{ farads}$$

and   charge due to a p.d. of V volts $= Q = CV$

$$= \frac{\epsilon_0 \epsilon_r A V}{d} \text{ coulombs},$$

$$\therefore \quad \frac{\text{electric flux density}}{\text{electric field strength}} = \frac{\boldsymbol{D}}{\boldsymbol{E}} = \frac{Q}{A} \div \frac{V}{d} = \frac{Qd}{VA} = \epsilon_0 \epsilon_r$$

$$= \epsilon = \text{absolute permittivity* (5.15)}$$

This expression is similar in form to expression (3.4) deduced for the magnetic circuit, namely:

$$\frac{\text{magnetic flux density}}{\text{magnetic field strength}} = \frac{B}{H} = \mu_0 \mu_r.$$

---

* From expression (5.15),

$$\text{absolute permittivity} = \epsilon_0 \epsilon_r = \frac{C\,[\text{farads}] \times d\,[\text{metres}]}{A\,[\text{metres}^2]}$$

$$= Cd/A \text{ farads per metre,}$$

hence the units of absolute permittivity are *farads per metre* (or F/m); e.g. $\epsilon_0 = 8.85 \times 10^{-12}\,\text{F/m}$.

5.15 Capacitance of a multi-plate capacitor

Fig. 5.13 Multi-plate
capacitor

Suppose a capacitor to be made up of n parallel plates, alternate plates being connected together as in fig. 5.13.

Let $\quad A$ = area of *one* side of each plate in square metres,

$\qquad d$ = thickness of dielectric in metres

and $\quad \epsilon_r$ = relative permittivity of the dielectric.

Fig. 5.13 shows a capacitor with seven plates, four being connected to A and three to B. It will be seen that each side of the three plates connected to B is in contact with the dielectric, whereas only one side of each of the outer plates is in contact with it. Consequently, the useful surface area of each set of plates is $6A$ square metres. For n plates, the useful area of each set is $(n-1)A$ square metres:

$$\therefore \qquad \text{capacitance} = \frac{\epsilon_0 \epsilon_r (n-1)A}{d} \text{ farads}$$

$$= \frac{8.85 \times 10^{-12} \epsilon_r (n-1)A}{d} \text{ farads} \quad (5.16)$$

Example 5.4

A capacitor is made with 7 metal plates connected as in fig. 5.13 and separated by sheets of mica having a thickness of 0.3 mm and a relative permittivity of 6. The area of one side of each plate is 500 cm². Calculate the capacitance in microfarads.

Using expression (5.16), we have $n = 7$, $A = 0.05 \text{ m}^2$, $d = 0.0003 \text{ m}$ and $\epsilon_r = 6$.

$$\therefore \qquad C = \frac{8.85 \times 10^{-12} \times 6 \times 6 \times 0.05}{0.0003} = 0.0531 \times 10^{-6} \text{ F}$$

$$= 0.0531 \,\mu\text{F}.$$

Example 5.5

A p.d. of 400 V is maintained across the terminals of the capacitor of example 5.4. Calculate (a) the charge, (b) the electric field strength or potential gradient and (c) the electric flux density in the dielectric.

(a) Charge $= Q = CV = 0.0531 \,[\mu\text{F}] \times 400 \,[\text{V}]$

$$= 21.24 \,\mu\text{C}.$$

(b) Electric field strength or potential gradient

$$= V/d = 400 \,[\text{V}]/0.0003 \,[\text{m}] = 1\,333\,000 \text{ V/m}$$

$$= 1333 \text{ kV/m}.$$

(c) Electric flux density

$$= Q/A = 21.24 \,[\mu\text{C}]/(0.05 \times 6) \,[\text{m}^2]$$

$$= 70.8 \,\mu\text{C/m}^2.$$

5.16 Capacitance of and electric field strength in a parallel-plate capacitor with composite dielectric

Suppose the space between metal plates M and N to be filled by dielectrics 1 and 2 of thickness d_1 and d_2 metres respectively, as shown in fig. 5.14(a).

Let Q = charge in coulombs due to p.d. of V volts

and A = area of each dielectric in square metres,

then $D = Q/A = \begin{cases} \text{electric flux density, in coulombs} \\ \text{per metre}^2, \text{ in A and B.} \end{cases}$

Fig. 5.14 Parallel-plate capacitor with two dielectrics

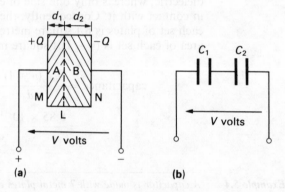

(a)   (b)

Let E_1 and E_2 = electric field strengths in 1 and 2 respectively; then if the relative permittivities of 1 and 2 are ϵ_1 and ϵ_2 respectively,

$$\text{electric field strength in A} = E_1 = \frac{D}{\epsilon_1\epsilon_0} = \frac{Q}{\epsilon_1\epsilon_0 A}$$

and $$\text{electric field strength in B} = E_2 = \frac{D}{\epsilon_2\epsilon_0} = \frac{Q}{\epsilon_2\epsilon_0 A}.$$

Hence, $$\frac{E_1}{E_2} = \frac{\epsilon_2}{\epsilon_1} \qquad (5.17)$$

i.e. for dielectrics having the same cross-sectional area in series, the electric field strengths (or potential gradients) are inversely proportional to their relative permittivities.

$$\left.\begin{array}{c}\text{Potential drop in} \\ \text{a dielectric}\end{array}\right\} = \text{electric field strength} \times \text{thickness}$$

$$\therefore \quad \left.\begin{array}{c}\text{p.d. between plate M and the boundary} \\ \text{surface L between 1 and 2}\end{array}\right\} = E_1 d_1$$

Hence all points on surface L are at the same potential, i.e. L is an *equipotential surface* and is at right-angles to the direction of the electric field strength. It follows that if a very thin metal foil were inserted between 1 and 2, it would not alter the electric field in the dielectrics. Hence the latter may be regarded as equivalent to two capacitances, C_1 and C_2, connected in series as in fig. 5.14(b),

where $$C_1 = \frac{\epsilon_1 \epsilon_0 A}{d_1} \quad \text{and} \quad C_2 = \frac{\epsilon_2 \epsilon_0 A}{d_2}$$

and total capacitance between plates M and N $= \dfrac{C_1 C_2}{C_1 + C_2}$.

Example 5.6
A capacitor consists of two metal plates, each 400 mm × 400 mm, spaced 6 mm apart. The space between the metal plates is filled with a glass plate 5 mm thick and a layer of paper 1 mm thick. The relative permittivities of the glass and paper are 8 and 2 respectively. Calculate (a) the capacitance, neglecting any fringing flux, and (b) the electric field strength in each dielectric in kilovolts per millimetre due to a p.d. of 10 kV between the metal plates.

(a) Fig. 5.15(a) shows a cross-section (not to scale) of the capacitor; and in fig. 5.15(b), C_p represents the capacitance of the paper layer between M and the equipotential surface L and C_g represents that of the glass between L and N. From expression (5.14) we have:

$$C_p = \frac{8.85 \times 10^{-12} \times 2 \times 0.4 \times 0.4}{0.001} = 2.83 \times 10^{-9}\,\text{F}$$

and $$C_g = \frac{8.85 \times 10^{-12} \times 8 \times 0.4 \times 0.4}{0.005} = 2.265 \times 10^{-9}\,\text{F}.$$

If C is the resultant capacitance between M and N,

$$\frac{1}{C} = \frac{10^9}{2.83} + \frac{10^9}{2.265} = 0.7955 \times 10^9$$

$$\therefore \qquad C = 1.257 \times 10^{-9}\,\text{F} = 0.001\,257\,\mu\text{F}.$$

Fig. 5.15 Diagrams for example 5.6

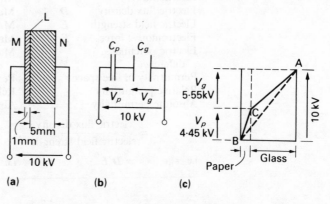

(b) Since C_p and C_g are in series across 10 kV, it follows from expression (5.7) that the p.d., V_p, across the paper is given by:

$$V_p = \frac{10 \times 2.265}{2.83 + 2.265} = 4.45\,kV$$

and $$V_g = 10 - 4.45 = 5.55\,\text{kV}.$$

These voltages are represented graphically in fig. 5.15(c).

$$\left.\begin{array}{l}\text{electric field strength in the}\\\text{paper dielectric}\end{array}\right\} = 4.45/1 = 4.45\,\text{kV/mm}$$

and $\left.\begin{array}{l}\text{electric field strength in the}\\ \text{glass dielectric}\end{array}\right\} = 5.55/5 = 1.11\,\text{kV/mm}.$

These electric field strengths are represented by the slopes of AC and CB for the glass and paper respectively in fig. 5.15(c). Had the dielectric between plates M and N been homogeneous, the electric field strength would have been $10/6 = 1.67\,\text{kV/mm}$, as represented by the slope of the dotted line AB in fig. 5.15(c).

From the result of example 5.6 it can be seen that the effect of using a composite dielectric of two materials having different relative permittivities is to increase the electric field strength in the material having the lower relative permittivity. This effect has very important applications in high-voltage work.

5.17 Comparison of electrostatic and electromagnetic terms

It may be helpful to compare the terms and symbols used in electrostatics with the corresponding terms and symbols used in electromagnetism:

Electrostatics		*Electromagnetism*	
Term	Symbol	Term	Symbol
Electric flux	Ψ	Magnetic flux	Φ
Electric flux density	D	Magnetic flux density	B
Electric field strength	E	Magnetic field strength	H
Electromotive force	E	Magnetomotive force	F
Electric potential difference	V	Magnetic potential difference	—
Permittivity of free space	ϵ_0	Permeability of free space	μ_0
Relative permittivity	ϵ_r	Relative permeability	μ_r
Absolute permittivity		Absolute permeability	

$$= \frac{\text{electric flux density}}{\text{electric field strength}} \qquad = \frac{\text{magnetic flux density}}{\text{magnetic field strength}}$$

i.e. $\epsilon_0\epsilon_r = \epsilon = D/E$ \qquad i.e. $\mu_0\mu_r = \mu = B/H$

5.18 Force on an isolated charge in an electric field

Suppose L in fig. 5.16 to be a very small metal sphere carrying a positive charge of q coulombs situated in the electric field between plates M and N separated by an air

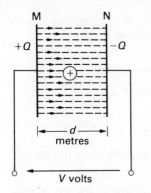

Fig. 5.16 An isolated charge in an electric field

dielectric, M being positive relative to N. Since like charges repel and unlike charges attract each other, there is a force acting on L urging it towards N.

The movement of a positive charge from M to N is equivalent to a momentary current flowing from a point M to a point N of a wire, where the potential of M is V volts above that of N. In such a case, energy absorbed, in joules

$$= \text{p.d. in volts} \times \text{current in amperes} \times \text{time in seconds}$$

$$= \text{p.d. in volts} \times \text{charge in coulombs}.$$

Hence, when a positive charge of q coulombs moves from M to N in fig. 5.16,

$$\text{energy absorbed} = Vq \text{ joules}.$$

Since this energy is due to the force on the charge acting through a distance d metres,

$$\begin{pmatrix} \text{force, in newtons,} \\ \text{on charge } q \end{pmatrix} \times \begin{pmatrix} \text{distance, in metres,} \\ \text{between M and N} \end{pmatrix} = Vq \text{ joules}$$

$$\therefore \qquad \text{force on charge} = qV/d = q\boldsymbol{E} \text{ newtons}$$

$$= \boldsymbol{E} \text{ newtons per coulomb} \qquad (5.18)$$

where $\boldsymbol{E}$ is the electric field strength in the dielectric.

If L in fig. 5.16 were carrying a negative charge, the force on L would urge it towards the positive plate M.

5.19 Deflection of an electron moving through a uniform electric field

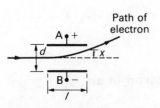

Fig. 5.17 Deflection of an electron by an electric field

Suppose an electron to have a velocity of v metres per second at right-angles to an electric field between two parallel plates A and B, as shown in fig. 5.17.

If $\qquad V = \text{p.d., in volts, between plates A and B}$

and $\qquad d = \text{distance, in metres, between the plates,}$

electric field strength between plates $= V/d$ volts per metre.

If $\boldsymbol{e} = \text{the negative charge, in coulombs, on an electron,}$ then from expression (5.18),

$$\text{force on electron} = \boldsymbol{e}V/d \text{ newtons}.$$

If plate A is positive relative to plate B, this force deflects the electron towards plate A, as shown in fig. 5.17.

If $\boldsymbol{m} = \text{mass of electron in kilograms,}$

$$\text{transverse force on electron} = \text{mass} \times \text{transverse acceleration}$$

i.e. $\qquad \boldsymbol{e}V/d = \boldsymbol{m} \times \text{transverse acceleration,}$

$$\therefore \qquad \text{transverse acceleration} = \frac{\boldsymbol{e}}{\boldsymbol{m}} \times \frac{V}{d} \text{ metres per second}^2.$$

If the axial length of the electric field is l metres, the time taken by an electron to traverse the electric field is l/v seconds,

and $\left.\begin{array}{c}\text{final transverse}\\ \text{velocity of electron}\end{array}\right\} = \left(\begin{array}{c}\text{transverse}\\ \text{acceleration}\end{array}\right) \times \text{time}$

$$= \frac{\mathbf{e}}{\mathbf{m}} \times \frac{V}{d} \times \frac{l}{v} \text{ metres per second}$$

$\therefore$ deflection of electron during its movement in the electric field

$$= x = \text{average transverse velocity} \times \text{time}$$

$$= \frac{1}{2} \times \frac{\mathbf{e}}{\mathbf{m}} \times \frac{V}{d} \times \left(\frac{l}{v}\right)^2 \text{ metres} \qquad (5.19)$$

Since $\mathbf{e} = 1.6 \times 10^{-19}$ coulomb and $\mathbf{m} = 9.1 \times 10^{-31}$ kilogram,

$\therefore$ $\mathbf{e/m} = 1.76 \times 10^{11}$ C/kg.

Substituting for $\mathbf{e/m}$ in expression (5.19), we have:

$$\left.\begin{array}{c}\text{deflection of}\\ \text{electron}\end{array}\right\} = 0.88 \times 10^{11} \times (V/d) \times (l/v)^2 \text{ metres} \quad (5.20)$$

Example 5.7   *An electron has a velocity of 10^7 m/s at right-angles to the electric field between the deflecting plates of a cathode-ray tube (section 27.6). The plates are 8 mm apart and 20 mm long, and the p.d. between the plates is 50 V. Calculate the distance through which the electron is deflected during its movement through the electric field.*

From the data given in the question, $V = 50$ V, $d = 0.008$ m, $l = 0.02$ m and $v = 10^7$ m/s. Substituting in expression (5.20), we have:

$$\text{deflection} = 0.88 \times 10^{11} \times (50/0.008) \times (0.02 \times 10^{-7})^2$$

$$= 0.0022 \text{ m} = 2.2 \text{ mm}.$$

5.20 Movement of a free electron in an electric field

Suppose M and N in fig. 5.18 to be two metal plates, d metres apart, in an evacuated glass vessel G, and suppose a p.d. of V volts to be maintained between the plates. If an electron were released from the negative plane N, it follows from section 5.18 that the work done in moving the electron from N to M is $\mathbf{e}V$ joules, where $\mathbf{e}$ is the negative charge, in coulombs, on the electron.

Suppose the electron to have zero initial velocity and a final velocity of v metres per second; then the final kinetic energy of the electron, immediately before its impact with

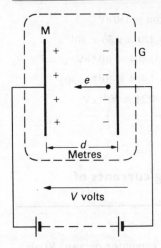

Fig. 5.18 Movement of a free electron in an electric field

plate M, is $\frac{1}{2}mv^2$ joules, where **m** is the mass of the electron in kilograms. Since there are no gas molecules in the space between M and N, there can be no loss of energy by collision of the electron with such molecules and all the work done on the electron is converted into kinetic energy,

$$\therefore \qquad \frac{1}{2}\mathbf{m}v^2 = \mathbf{e}V$$

and $\qquad v = \sqrt{(2V\mathbf{e}/\mathbf{m})}$ metres per second $\qquad$ (5.21)

This kinetic energy is converted into heat at the moment of impact of the electron with plate M.

Since **e** is 1.6×10^{-19} coulombs, an electron that is accelerated by a p.d. of 100 volts acquires kinetic energy of 1.6×10^{-17} joule. A more convenient unit for such small amounts of energy is the *electronvolt* (symbol, eV), namely the work done when an electron is moved through a p.d. of 1 volt; hence,

$$1 \text{ electronvolt} = 1.6 \times 10^{-19} \text{ joule} \qquad (5.22)$$

It follows that when an electron is moved through a p.d. of 100 V,

$$\text{work done} = 100 \text{ eV}.$$

Example 5.8   *The p.d. between the anode and cathode of a vacuum diode (section 27.1) is 150 V and the current through the diode is 15 mA. Calculate (a) the maximum velocity acquired by the electrons, assuming their initial velocity to be zero, (b) the number of electrons passing per second from cathode to anode and (c) the energy absorbed in 5 min in (i) joules, (ii) electronvolts. Assume electron charge = 1.6×10^{-19} C and electron mass = 9.11×10^{-31} kg.*

(*a*) From expression (5.21),

$$v = \sqrt{(2 \times 150 \times 1.6 \times 10^{-19})/(9.11 \times 10^{-31})}$$
$$= 7.27 \times 10^6 \text{ m/s}.$$

(*b*) Since a current of 1 A in a circuit corresponds to 6.24×10^{18} electrons passing a given point per second (see section 5.3),

$$\therefore \qquad \text{no. of electrons per second} = 6.24 \times 10^{18} \times 0.015$$
$$= 9.36 \times 10^{16}$$

(*c*) (i) Energy absorbed $= VIt = 150 \times 0.015 \times 5 \times 60$
$$= 675 \text{ J}.$$

Alternatively, energy absorbed per electron

$$= \frac{1}{2}\mathbf{m}v^2 = \frac{1}{2} \times 9.11 \times 10^{-31} \times (7.27)^2 \times 10^{12}$$
$$= 2.4 \times 10^{-17} \text{ J}.$$

$\therefore$ total energy absorbed in 5 minutes

$$= (\text{energy per electron}) \times \text{no. of electrons}$$
$$= 2.4 \times 10^{-17} \times 9.36 \times 10^{16} \times 300$$
$$= 675 \text{ J}.$$

(ii) Since $\qquad 1 \text{ electronvolt} = 1.6 \times 10^{-19}$ joule,

$$\therefore \qquad \text{energy absorbed} = 675/(1.6 \times 10^{-19})$$
$$= 4.21 \times 10^{21} \text{ eV}.$$

Alternatively, energy absorbed per electron = 150 eV

$$\therefore \quad \text{energy absorbed in 1 second} = 150 \times 9.36 \times 10^{16}$$
$$= 1.404 \times 10^{19} \text{ eV}$$

and   energy absorbed in 5 minutes = $1.404 \times 10^{19} \times 300$
$$= 4.21 \times 10^{21} \text{ eV}.$$

5.21 Charging and discharging currents of a capacitor

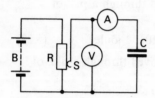

Fig. 5.19 Charging and discharging of a capacitor

Suppose C in fig. 5.19 to represent a capacitor of, say, 30 μF connected in series with a centre-zero microammeter A across a slider S and one end of a resistor R. A battery B is connected across R. If S is moved at a uniform speed along R, the p.d. applied to C, indicated by voltmeter V, increases uniformly from 0 to V volts, as shown by line OD in fig. 5.20.

If C is the capacitance in farads and if the p.d. across C increases uniformly from 0 to V volts in t_1 seconds,

$$\text{charging current} = i_1 = \frac{Q \text{ [coulombs or ampere seconds]}}{t_1 \text{ [seconds]}}$$

$$= CV/t_1 \text{ amperes,}$$

i.e. $\left.\begin{array}{l}\text{charging current} \\ \text{in amperes}\end{array}\right\} = \left\{\begin{array}{l}\text{rate of change of charge in} \\ \text{coulombs per second}\end{array}\right.$

$$= C \text{ [farads]} \times \text{rate of change of} $$
$$\text{p.d. in volts per second}.$$

Fig. 5.20 Voltage and current during charging and discharging of a capacitor

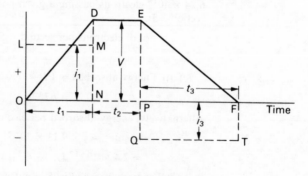

Since the p.d. across C increases at a uniform rate, the charging current, i_1, remains constant and is represented by the dotted line LM in fig. 5.20.

Suppose the p.d. across C to be maintained constant at V volts during the next t_2 seconds. Since the rate of change of p.d. is now zero, the current (apart from a slight leakage current) is zero and is represented by the dotted line NP. If the p.d. across C is then reduced to zero at a uniform rate by moving slider S backwards, the microammeter indicates a current i_3 flowing in the reverse direction, represented by the dotted line QT in fig. 5.20. If t_3 is the time in seconds for the p.d. to be reduced from V volts to zero,

then $$Q = -i_3 t_3 \text{ coulombs}$$

$$\therefore \qquad i_3 = -Q/t_3 = -C \times V/t_3 \text{ amperes}$$

i.e. $$\left. \begin{matrix} \text{discharge current} \\ \text{in amperes} \end{matrix} \right\} = \left\{ \begin{matrix} \text{rate of change of charge in} \\ \text{coulombs per second} \end{matrix} \right.$$

$$= C \,[\text{farads}] \times \text{rate of change of} \\ \text{p.d. in volts per second}.$$

Since $Q = i_1 t_1 = -i_3 t_3$ (assuming negligible leakage current through C),

$\therefore$ areas of rectangles OLMN and PQTF are equal.

In practice, it is seldom possible to vary the p.d. across a capacitor at a constant rate, so let us consider the general case of the p.d. across a capacitor of C farads being increased by dv volts in dt seconds. If the corresponding increase of charge is dq coulombs,

$$dq = C \cdot dv.$$

If the charging current at that instant is i amperes,

$$dq = i \cdot dt$$

$$\therefore \qquad i \cdot dt = C \cdot dv$$

and $$i = C \cdot dv/dt$$

$$= C \times \text{rate of change of p.d.} \qquad (5.23)$$

If the capacitor is being discharged and if the p.d. falls by dv volts in dt seconds, the discharge current is given by:

$$i = C \cdot dv/dt \qquad (5.24)$$

Since dv is now negative, the current is also negative.

5.22 Graphical derivation of curve of voltage across a capacitor connected in series with a resistor across a d.c. supply

In section 5.4 we derived the curves of the voltage across a capacitor during charging and discharging from the readings on an electrostatic voltmeter connected across the capacitor. We will now consider how these curves can be derived graphically from the values of the capacitance, the resistance and the applied voltage. At the instant when S is closed on *a* (fig. 5.5), there is no p.d. across C; consequently the whole of the voltage is applied across R and the initial value of the charging current $= I = V/R$.

Fig. 5.21 Growth of p.d. across a capacitor in series with a resistor

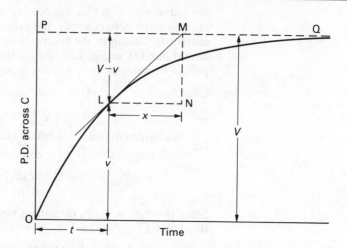

The growth of the p.d. across C is represented by the curve in fig. 5.21. Suppose v to be the p.d. across C and i to be the charging current t seconds after S is put over to a. The corresponding p.d. across R $= V - v$, where V is the terminal voltage of the battery.

Hence

$$iR = V - v$$

and

$$i = (V - v)/R \qquad (5.25)$$

If this current remained *constant* until the capacitor was fully charged, and if the time taken was x seconds:

$$\left.\begin{array}{c}\text{corresponding quantity}\\ \text{of electricity}\end{array}\right\} = ix = \frac{V - v}{R} \times x \text{ coulombs.}$$

With a constant charging current, the p.d. across C would have increased uniformly up to V volts, as represented by the tangent LM drawn to the curve at L.

But the charge added to the capacitor also

$$= \text{increase of p.d.} \times C$$

$$= (V - v) \times C$$

Hence

$$\frac{V - v}{R} \times x = C(V - v)$$

and $x = CR =$ the time constant, T, of the circuit   (5.26)

The construction of the curve representing the growth of the p.d. across a capacitor is therefore similar to that described in section 4.6 for the growth of current in an inductive circuit. Thus, OA in fig. 5.22 represents the battery voltage V, and AB the time constant T. Join OB, and from a point D fairly near the origin draw DE $= T$ seconds and draw EF perpendicularly. Join DF, etc. Draw a curve such that OB, DF, etc., are tangents to it.

From expression (5.25) it is evident that the instantaneous value of the charging current is proportional to $(V - v)$, namely the vertical distance between the curve and the horizontal line PQ in fig. 5.21. Hence the shape of the curve

Fig. 5.22 Growth of p.d. across a capacitor in series with a resistor

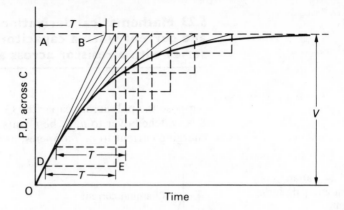

representing the charging current is the inverse of that of the p.d. across the capacitor and is the same for both charging and discharging currents (assuming the resistance to be the same), and its construction is illustrated by the following example.

Example 5.9

A 20-µF capacitor is charged to a p.d. of 400 V and then discharged through a 100 000-Ω resistor. Derive a curve representing the discharge current.

From (5.26):

$$\text{time constant} = 100\,000\,[\Omega] \times \frac{20}{1\,000\,000}\,[\text{F}] = 2\,\text{s}.$$

$$\left.\begin{array}{l}\text{Initial value of}\\\text{discharge current}\end{array}\right\} = \frac{V}{R} = \frac{400}{100\,000} = 0.004\,\text{A} = 4\,\text{mA}.$$

Hence draw OA in fig. 5.23 to represent 4 mA and OB to represent 2 seconds. Join AB. From a point C corresponding to, say, 3.5 mA, draw CD equal to 2 seconds and DE vertically. Join CE.

Fig. 5.23 Discharge current, example 5.9

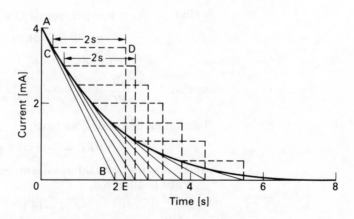

Repeat the construction at intervals of, say, 0.5 mA and draw a curve to which AB, CE, etc., are tangents. This curve represents the variation of discharge current with time.

5.23 Mathematical derivation of curves of voltage and current when a capacitor is connected in series with a resistor across a d.c. supply

Suppose the p.d. across capacitor C in fig. 5.5, t seconds after S is switched over to a, to be v volts, and the corresponding charging current to be i amperes, as indicated in fig. 5.24.

Fig. 5.24 Variation of current and p.d. during charging

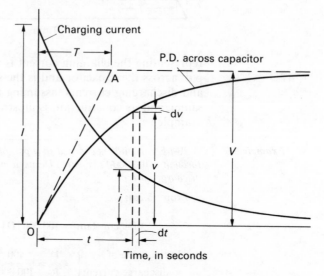

Also, suppose the p.d. to increase from v to $(v + \mathrm{d}v)$ volts in $\mathrm{d}t$ seconds, then, from expression (5.23),

$$i = C \cdot \mathrm{d}v/\mathrm{d}t$$

and corresponding p.d. across R $= Ri = RC \cdot \mathrm{d}v/\mathrm{d}t$.

But $V = \text{p.d. across C} + \text{p.d. across R}$

$$= v + RC \cdot \mathrm{d}v/\mathrm{d}t \qquad (5.27)$$

$\therefore$ $V - v = RC \cdot \mathrm{d}v/\mathrm{d}t$

so that $\dfrac{\mathrm{d}t}{RC} = \dfrac{\mathrm{d}v}{V - v}.$

Integrating both sides, we have:

$$t/RC = -\ln(V - v) + A$$

where $A =$ the constant of integration.
When $t = 0$, $v = 0$,

$\therefore$ $A = \ln V$

so that $\dfrac{t}{RC} = \ln \dfrac{V}{V - v}$

$\therefore$ $\dfrac{V}{V - v} = e^{t/RC}$

and $$v = V(1 - e^{-t/RC}) \qquad (5.28)$$

Also, $$i = C \cdot dv/dt = CV \cdot \frac{d}{dt}(1 - e^{-t/RC})$$

$$= (V/R)\, e^{-t/RC} \qquad (5.29)$$

At the instant of switching on, $t = 0$ and $e^{-0} = 1$,

$$\therefore \qquad \text{initial value of current} = V/R = \text{(say) } I.$$

This result is really obvious from the fact that at the instant of switching on there is no charge on C and therefore no p.d. across it. Consequently the whole of the applied voltage must momentarily be absorbed by R.

Substituting for V/R in expression (5.29), we have:

$$\text{instantaneous charging current} = i = I\, e^{-t/RC} \qquad (5.30)$$

If the p.d. across the capacitor continued increasing at the initial rate, it would be represented by OA, the tangent drawn to the initial part of the curve. If T be the *time constant* in seconds, namely the time required for the p.d. across C to increase from zero to its final value if it continued increasing at its initial rate, then:

$$\text{initial rate of increase of p.d.} = V/T \text{ volts per second} \qquad (5.31)$$

But it follows from (5.27) that at the instant of closing the switch on a, $v = 0$, then:

$$V = RC \cdot dv/dt$$

$$\therefore \qquad \text{initial rate of change of p.d.} = dv/dt = V/RC \qquad (5.32)$$

Equating (5.31) and (5.32), we have:

$$V/T = V/RC$$

$$\therefore \qquad T = RC \text{ seconds} \qquad (5.33)$$

Hence we can rewrite (5.28) and (5.30) thus:

$$v = V(1 - e^{-t/T}) \qquad (5.34)$$

and $$i = I\, e^{-t/T} \qquad (5.35)$$

Comparison of expressions (4.18) and (5.34) shows that the shape of the voltage growth across a capacitor is similar to that of the current growth in an inductive circuit.

5.24 Discharge of a capacitor through a resistor

Having charged capacitor C in fig. 5.5 to a p.d. of V volts, let us now move switch S over to b and thereby discharge the capacitor through R. The pointer of microammeter A is immediately deflected to a maximum value in the negative direction, and then the readings on both the microammeter and the voltmeter ES (fig. 5.5) decrease to zero as indicated in fig. 5.25.

Fig. 5.25 Variation of
current and p.d. during
discharge

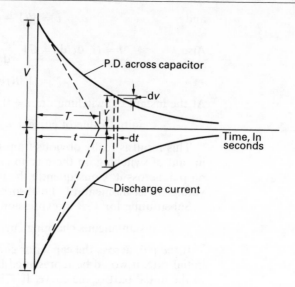

Suppose the p.d. across C to be v volts t seconds after S
has been moved to b, and the corresponding current to be
i amperes, as in fig. 5.25, then:

$$i = -v/R \qquad (5.36)$$

The negative sign indicates that the direction of the discharge
current is the reverse of that of the charging current.

Suppose the p.d. across C to change by dv volts in
dt seconds,

$$\therefore \qquad i = C \cdot \mathrm{d}v/\mathrm{d}t \qquad (5.37)$$

Since dv is now negative, i must also be negative, as already
noted. Equating (5.36) and (5.37), we have:

$$-v/R = C \cdot \mathrm{d}v/\mathrm{d}t$$

so that
$$\frac{\mathrm{d}t}{RC} = -\frac{\mathrm{d}v}{v}.$$

Integrating both sides, we have:

$$t/RC = -\ln v + A.$$

When $t = 0$, $v = V$, so that $A = \ln V$.

Hence
$$t/RC = \ln V/v$$

so that
$$V/v = \mathrm{e}^{t/RC}$$

and
$$v = V\,\mathrm{e}^{-t/RC} = V\,\mathrm{e}^{-t/T} \qquad (5.38)$$

Also,
$$i = -v/R = -\frac{V}{R}\,\mathrm{e}^{-t/RC} = -I\,\mathrm{e}^{-t/T} \qquad (5.39)$$

where $I =$ initial value of the discharge current $= V/R$.

Example 5.10   *An 8-μF capacitor is connected in series with a 0.5-MΩ resistor across
a 200-V d.c. supply. Calculate: (a) the time constant; (b) the initial
charging current; (c) the time taken for the p.d. across the capacitor to*

grow to 160 *V; and* (d) *the current and the p.d. across the capacitor* 4 s *after it is connected to the supply.*

(a) From (5.33),   time constant $= 0.5 \times 10^6 \times 8 \times 10^{-6}$

$$= 4 \, \text{s}.$$

(b)   Initial charging current $= \dfrac{V}{R} = \dfrac{200}{0.5 \times 10^6} \, \text{A}$

$$= 400 \, \mu\text{A}.$$

(c) From (5.34), $160 = 200(1 - e^{-t/4})$

$$\therefore \qquad e^{-t/4} = 0.2.$$

From mathematical tables, $t/4 = 1.61$

$$\therefore \qquad t = 6.44 \, \text{s}.$$

Or alternatively, $e^{t/4} = 1/0.2 = 5.$

$$\therefore \qquad (t/4)\log e = \log 5.$$

But $e = 2.718,$

$$\therefore \qquad t = \frac{4 \times 0.699}{0.4343} = 6.44 \, \text{s}.$$

(d) From (5.34) $v = 200(1 - e^{-4/4}) = 200(1 - 0.368)$

$$= 200 \times 0.632 = 126.4 \, \text{V}.$$

It will be seen that the time constant can be defined as the time required for the p.d. across the capacitor to grow from zero to 63.2 per cent of its final value.

From (5.35),

$$\text{corresponding current} = i = 400 \cdot e^{-1} = 400 \times 0.368$$

$$= 147 \, \mu\text{A}.$$

5.25 Displacement current in a dielectric

In the preceding sections we have considered the charging current as being the movement of electrons in the conductors connecting the source to the plates of the capacitor; e.g. when switch S is put over to *a* in fig. 5.11, electrons flow from the positive plate of C_1 via the battery to the negative plate of C_2, and the same number of electrons flow from the upper plate of C_2 to the lower plate of C_1. This current is referred to as *conduction* current.

Let us consider the capacitor of fig. 5.12, in which the metal plates M and N are in an evacuated glass enclosure G. There are no electrons in the space between the plates and therefore there cannot be any movement of electrons in this space when the capacitor is being charged. We know, however, that an electric field is being set up and that energy is being stored in the space between the plates; in other words, the space between the plates of a charged capacitor is in a state of electrostatic strain.

We do not know the exact nature of this strain (any more than we know the nature of the strain in a magnetic field), but Prof. Clerk Maxwell, in 1865, introduced the concept that any *change* in the electric flux in any region is equivalent to an electric current in that region, and he called this electric current a *displacement* current, to distinguish it from the *conduction* current referred to above.

This displacement current produces a magnetic field exactly as if it had been a conduction current. For instance, when a capacitor having circular parallel metal plates M and N (fig. 5.26) is being charged by a current *i* flowing in the direction shown, a magnetic field is created in the space between the plates, as indicated by the concentric dotted lines. The plane of these concentric circles is parallel to the plates.

This magnetic field disappears as soon as the displacement current ceases, i.e. as soon as the charge on the capacitor ceases to increase. When the capacitor is discharged, the magnetic field reappears in the reverse direction and again disappears when the discharge ceases. In other words, *the magnetic field is set* up only when the electric field is undergoing a change of intensity.* Hence, when a capacitor is being charged or discharged, we can say that the current is *continuous* around the *whole* circuit, being in the form of *conduction current* in the wires and *displacement current* in the dielectric of the capacitor. This means that we can apply Kirchhoff's First Law to plate M of the capacitor of fig. 5.26 by saying that the conduction current entering the plate is equal to the displacement current leaving that plate.

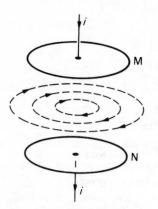

Fig. 5.26 Magnetic field due to displacement current

5.26 Energy stored in a charged capacitor

Suppose the p.d. across a capacitor of capacitance C farads to be increased from v to $(v + dv)$ volts in dt seconds. From (5.23), the charging current, i amperes, is given by:

$$i = C \cdot dv/dt.$$

Instantaneous value of power to capacitor $= iv$ watts

$$= vC \cdot dv/dt \text{ watts},$$

and $\left. \begin{array}{l} \text{energy supplied to capacitor during} \\ \text{interval } dt \end{array} \right\} = vC \cdot \dfrac{dv}{dt} \cdot dt$

$$= Cv \cdot dv \text{ joules}.$$

Hence total energy supplied to capacitor when p.d. is

---

* Conversely, an electric field appears whenever a magnetic field changes in intensity, and it is this reciprocity between electric and magnetic fields that forms the basis of electromagnetic theory dealing with the radiation of electromagnetic waves.

increased from 0 to V volts

$$= \int_0^V Cv \cdot dv = \tfrac{1}{2} C \left[v^2 \right]_0^V$$

$$= \tfrac{1}{2} C V^2 \text{ joules} \qquad (5.40)$$

For a capacitor with dielectric of thickness d metres and area A square metres,

$$\text{energy per cubic metre} = \frac{1}{2} \cdot \frac{CV^2}{Ad} = \frac{1}{2} \cdot \frac{\epsilon A}{d} \cdot \frac{V^2}{Ad}$$

$$= \tfrac{1}{2} \epsilon (V/d)^2 = \tfrac{1}{2} \epsilon \mathbf{E}^2$$

$$= \tfrac{1}{2} \mathbf{DE} = \tfrac{1}{2} \mathbf{D}^2 / \epsilon \text{ joules} \qquad (5.41)$$

These expressions are similar to expressions (4.22) for the energy stored per cubic metre of a magnetic field.

Example 5.11 *A 50-μF capacitor is charged from a 200-V supply. After being disconnected it is immediately connected in parallel with a 30-μF capacitor. Find: (a) the p.d. across the combination, and (b) the electrostatic energies before and after the capacitors are connected in parallel. The 30-μF capacitor is initially uncharged.*

(a) From (5.2),   charge $= (50 \times 10^{-6}) \, [\text{F}] \times 200 \, [\text{V}]$

$$= 0.01 \, \text{C}.$$

When the capacitors are connected in parallel, the total capacitance is $80 \, \mu$F, and the charge of 0.01 coulomb is divided between the two capacitors:

$\therefore$ $\qquad\qquad\qquad 0.01 \, [\text{C}] = (80 \times 10^{-6}) \, [\text{F}] \times \text{p.d.}$

$\therefore$ $\qquad\qquad$ p.d. across capacitors $= 125 \, \text{V}.$

(b) From (5.40) it follows that when the 50-μF capacitor is charged to a p.d. of 200 V:

$$\text{electrostatic energy} = \tfrac{1}{2} \times (50 \times 10^{-6}) \, [\text{F}] \times (200)^2 \, [\text{V}^2]$$

$$= 1 \, \text{J}.$$

With the capacitors in parallel:

$$\text{total electrostatic energy} = \tfrac{1}{2} \times 80 \times 10^{-6} \times (125)^2 = 0.625 \, \text{J}.$$

It is of interest to note that there is a reduction in the energy stored in the capacitors. This loss appears as heat in the resistance of the circuit by the current responsible for equalizing the p.d.s, in the spark that may occur when the capacitors are connected in parallel, and in electromagnetic radiation if the discharge is oscillatory.

5.27 Force of attraction between oppositely-charged plates

Let us consider two parallel plates M and N (fig. 5.27) immersed in a homogeneous fluid, such as air or oil, having an absolute permittivity ϵ. Suppose the area of the dielectric

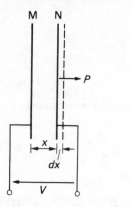

Fig. 5.27 Attraction between charged parallel plates

to be A square metres and the distance between M and N to be x metres. If the p.d. between the plates is V volts, then from (5.41),

$$\text{energy per cubic metre of dielectric} = \tfrac{1}{2}\epsilon(V/x)^2 \text{ joules}.$$

Suppose plate M to be fixed and N to be movable, and let F be the force of attraction, in newtons, between the plates. Let us next disconnect the charged capacitor from the supply and then pull plate N outwards through a distance dx metre. If the insulation of the capacitor is perfect, the charge on the plates remains constant. This means that the electric flux density and therefore the potential gradient in the dielectric must remain unaltered, the constancy of the potential gradient being due to the p.d. between plates M and N increasing in proportion to the distance between them. It follows from expression (5.41) that the energy per cubic metre of the dielectric remains constant. Consequently, all the energy in the *additional* volume of the dielectric must be derived from the work done when the force F newtons acts through distance dx metres, namely $F \cdot dx$ joules,

i.e. $F \cdot dx = \tfrac{1}{2}\epsilon(V/x)^2 \cdot A \cdot dx$

$\therefore \qquad F = \tfrac{1}{2}\epsilon A(V/x)^2 \text{ newtons}$ \hfill (5.42)

$\qquad\qquad = \tfrac{1}{2}\epsilon A \times (\text{potential gradient in volts per metre})^2.$

Example 5.12   *Two parallel metal discs, each 100 mm in diameter, are spaced 1 mm apart, the dielectric being air. Calculate the force, in newtons, on each disc when the p.d. between them is 1 kV.*

$$\text{Area of one side of each plate} = 0.7854 \times (0.1)^2$$
$$= 0.007\,854 \text{ m}^2.$$
$$\text{Potential gradient} = 1000\,[V]/0.001\,[m]$$
$$= 10^6 \text{ V/m}.$$

From expression (5.42),

$$\text{force} = \tfrac{1}{2} \times (8.85 \times 10^{-12})\,[\text{F/m}] \times 0.007\,854\,[\text{m}^2] \times (10^6)^2\,[\text{V/m}]^2$$
$$= 0.0348 \text{ N}.$$

5.28 Dielectric strength

If the p.d. between the opposite sides of a sheet of solid insulating material is increased beyond a certain value, the material breaks down. Usually this results in a tiny hole or puncture through the dielectric so that the latter is then useless as an insulator.

The potential gradient necessary to cause breakdown of an insulating medium is termed its *dielectric strength* and is usually expressed in megavolts per metre. The value of the dielectric strength of a given material decreases with increase of thickness, and the following table gives the approximate dielectric strengths of some of the most important materials:

Material	Thickness (mm)	Dielectric strength (MV/m)
Air (at normal pressure and temperature)	0.2	5.75
	0.6	4.92
	1	4.46
	6	3.27
	10	2.98
Mica	0.01	200
	0.1	176
	1.0	61
Glass (density 2.5)	1	28.5
	5	18.3
Ebonite	1	50
Paraffin-waxed paper	0.1	40–60

Summary of important formulae

$$Q \text{ [coulombs]} = C \text{ [farads]} \times V \text{ [volts]} \qquad (5.2)$$

$$1 \text{ microfarad} = 10^{-6} \text{ farad}$$

$$1 \text{ picofarad} = 10^{-12} \text{ farad}$$

For capacitors in parallel,

$$C = C_1 + C_2 + \cdots \qquad (5.3)$$

For capacitors in series,

$$\frac{1}{C} = \frac{1}{C_1} + \frac{1}{C_2} + \cdots \qquad (5.5)$$

For C_1 and C_2 in series,

$$V_1 = V \cdot \frac{C_2}{C_1 + C_2} \qquad (5.7)$$

$$\text{Electric field strength in dielectric} = E = V/d \qquad (5.9)$$

$$\text{Electric flux density} = D = Q/A \qquad (5.10)$$

$$\text{Absolute permittivity} = D/E = \epsilon = \epsilon_0 \epsilon_r \qquad (5.15)$$

$$\text{Permittivity of free space} = \epsilon_0 = 8.85 \times 10^{-12} \text{ F/m}$$

$$\frac{1}{\sqrt{(\mu_0 \epsilon_0)}} = 2.998 \times 10^8 \text{ m/s}$$

= velocity of electromagnetic waves in free space.

Relative permittivity of a material

$$= \frac{\text{capacitance of capacitor with that material as dielectric}}{\text{capacitance of same capacitor with vacuum dielectric}}$$

$$\left.\begin{array}{l}\text{Capacitance of parallel-plate capacitor}\\ \text{with } n \text{ plates}\end{array}\right\} = \frac{\epsilon(n-1)A}{d}$$

(5.16)

For two dielectrics, A and B, of same area, in series,

$$\frac{\text{electric field strength or potential gradient in A}}{\text{electric field strength or potential gradient in B}}$$

$$= \frac{\text{relative permittivity of B}}{\text{relative permittivity of A}} \quad (5.17)$$

$$\left.\begin{array}{l}\text{Force on isolated charge}\\ \text{in electric field}\end{array}\right\} = E \text{ newtons per coulomb} \quad (5.18)$$

Deflection of electron moving across electric field

$$= \frac{1}{2} \times \frac{e}{m} \times \frac{V}{d} \times \left(\frac{l}{v}\right)^2 \text{ metres} \quad (5.19)$$

Final velocity of free electron in electric field

$$= \sqrt{(2Ve/m)} \quad (5.21)$$

$$1 \text{ electronvolt} = 1.6 \times 10^{-19} \text{ joule} \quad (5.22)$$

$$\text{Charging current of capacitor} = dq/dt = C \cdot dv/dt \quad (5.23)$$

For R and C in series across d.c. supply,

$$v = V(1 - e^{-t/RC}) \quad (5.28)$$

and
$$i = I e^{-t/RC} \quad (5.30)$$

$$\text{Time constant} = T = RC \quad (5.33)$$

For C discharged through R,

$$v = V e^{-t/RC} \quad (5.38)$$

and
$$i = -I e^{-t/RC} \quad (5.39)$$

$$\text{Energy stored in capacitor} = \tfrac{1}{2}CV^2 \text{ joules} \quad (5.40)$$

$$\left.\begin{array}{l}\text{Energy per cubic}\\ \text{metre of dielectric}\end{array}\right\} = \tfrac{1}{2}\epsilon E^2 = \tfrac{1}{2}DE$$

$$= \tfrac{1}{2}D^2/\epsilon \text{ joules} \quad (5.41)$$

$$\left.\begin{array}{l}\text{Electrostatic attraction}\\ \text{between parallel plates}\end{array}\right\} = \tfrac{1}{2}\epsilon A(V/x)^2 \text{ newtons} \quad (5.42)$$

EXERCISES 5

1. A 20-μF capacitor is charged at a constant current of 5 μA for 10 min. Calculate the final p.d. across the capacitor and the corresponding charge in coulombs.
2. Three capacitors have capacitances of 10 μF, 15 μF and 20 μF respectively. Calculate the total capacitance when they are connected (a) in parallel, (b) in series.
3. A 9-μF capacitor is connected in series with two capacitors,

4 μF and 2 μF respectively, which are connected in parallel. Determine the capacitance of the combination.

If a p.d. of 20 V is maintained across the combination, determine the charge on the 9-μF capacitor and the energy stored in the 4-μF capacitor. (U.E.I., O.2)

4. Two capacitors, having capacitances of 10 μF and 15 μF respectively, are connected in series across a 200-V d.c. supply. Calculate: (*a*) the charge on each capacitor; (*b*) the p.d. across each capacitor. Also find the capacitance of a single capacitor that would be equivalent to these two capacitors in series.

5. Three capacitors of 2, 3 and 6 μF respectively are connected in series across a 500-V d.c. supply. Calculate: (*a*) the charge on each capacitor; (*b*) the p.d. across each capacitor; and (*c*) the energy stored in the 6-μF capacitor.

6. A certain capacitor has a capacitance of 3 μF. A capacitance of 2.5 μF is required by combining this capacitance with another. Calculate the capacitance of the second capacitor and state how it must be connected to the first.

7. A capacitor A is connected in series with two capacitors B and C connected in parallel. If the capacitances of A, B and C are 4, 3 and 6 μF respectively, calculate the equivalent capacitance of the combination.

If a p.d. of 20 V is maintained across the whole circuit, calculate the charge on the 3-μF capacitor.

8. Three capacitors, A, B and C, are connected in series across a 200-V d.c. supply. The p.d.s across the capacitors are 40, 70 and 90 V respectively. If the capacitance of A is 8 μF, what are the capacitances of B and C?

9. Two capacitors, A and B, are connected in series across a 200-V d.c. supply. The p.d. across A is 120 V. This p.d. is increased to 140 V when a 3-μF capacitor is connected in parallel with B. Calculate the capacitances of A and B.

10. Show from first principles that the total capacitance of two capacitors having capacitances C_1 and C_2 respectively, connected in parallel, is $C_1 + C_2$.

A circuit consists of two capacitors A and B in parallel connected in series with another capacitor C. The capacitances of A, B and C are 6 μF, 10 μF and 16 μF respectively. When the circuit is connected across a 400-V d.c. supply, calculate: (i) the potential difference across each capacitor; (ii) the charge on each capacitor. (U.E.I., O.1)

11. On what factors does the capacitance of a parallel-plate capacitor depend?

Derive an expression for the resultant capacitance when two capacitors are connected in series.

Two capacitors, A and B, having capacitances of 20 μF and 30 μF respectively, are connected in series to a 600-V d.c. supply. Determine the p.d. across each capacitor.

If a third capacitor C is connected in parallel with A and it is then found that the p.d. across B is 400 V, calculate the capacitance of C and the energy stored in it.

(N.C.T.E.C., O.2)

12. Derive an expression for the energy stored in a capacitor of C farads when charged to a potential difference of V volts.

A capacitor of 4 μF capacitance is charged to a p.d. of 400 V and then connected in parallel with an uncharged capacitor of 2 μF capacitance. Calculate the p.d. across the parallel capacitors and the energy stored in the capacitors before and after being connected in parallel. Explain the difference.

(E.M.E.U., O.1)

13. Derive expressions for the equivalent capacitance of a number of capacitors: (*a*) in series; (*b*) in parallel.

   Two capacitors of $4\,\mu F$ and $6\,\mu F$ capacitance respectively are connected in series across a p.d. of 250 V. Calculate the p.d. across each capacitor and the charge on each.

   The capacitors are disconnected from the supply p.d. and reconnected in parallel with each other, with terminals of similar polarity being joined together. Calculate the new p.d. and charge for each capacitor.

   What would have happened if, in making the parallel connection, the connections of one of the capacitors had been reversed? (U.L.C.I., O.2)

14. Show that the total capacitance of two capacitors having capacitances C_1 and C_2 connected in series is $C_1 C_2/(C_1 + C_2)$.

   A $5\text{-}\mu F$ capacitor is charged to a potential difference of 100 V and then connected in parallel with an uncharged $3\text{-}\mu F$ capacitor. Calculate the potential difference across the parallel capacitors. (U.E.I., O.1)

15. Find an expression for the energy stored in a capacitor of capacitance C farads charged to a p.d. of V volts.

   A $3\text{-}\mu F$ capacitor is charged to a p.d. of 200 V and then connected in parallel with an uncharged $2\text{-}\mu F$ capacitor. Calculate the p.d. across the parallel capacitors and the energy stored in the capacitors before and after being connected in parallel. Account for the difference. (App. El., L.U.)

16. Explain the terms *electric field strength and permittivity*.

   Two square metal plates, each of size 200 mm, are completely immersed in insulating oil of relative permittivity 5 and spaced 3 mm apart. A p.d. of 600 V is maintained between the plates. Calculate: (*a*) the capacitance of the capacitor; (*b*) the charge stored on the plates; (*c*) the electric field strength in the dielectric; (*d*) the electric flux density. (U.L.C.I., O.1)

17. A capacitor consists of two metal plates, each having an area of $900\,\text{cm}^2$, spaced 3.0 mm apart. The whole of the space between the plates is filled with a dielectric having a relative permittivity of 6. A p.d. of 500 V is maintained between the two plates. Calculate: (*a*) the capacitance; (*b*) the charge; (*c*) the electric field strength; (*d*) the electric flux density. (E.M.E.U., O.1)

18. Describe with the aid of a diagram what happens when a battery is connected across a simple capacitor comprising two metal plates separated by a dielectric.

   A capacitor consists of two metal plates, each having an area of $600\,\text{cm}^2$, separated by a dielectric 4 mm thick which has a relative permittivity of 5. When the capacitor is connected to a 400-V d.c. supply, calculate: (i) the capacitance; (ii) the charge; (iii) the electric field strength; (iv) the electric flux density. (U.E.I., O.1)

19. Define: (*a*) the farad; (*b*) the relative permittivity.

   A capacitor consists of two square metal plates of side 200 mm, separated by an air space 2.0 mm wide. The capacitor is charged to a p.d. of 200 V and a sheet of glass having a relative permittivity of 6 is placed between the metal plates immediately they are disconnected from the supply. Calculate: (*a*) the capacitance with air dielectric; (*b*) the capacitance with glass dielectric; (*c*) the p.d. across the capacitor after the glass plate has been inserted; (*d*) the charge on the capacitor. (U.E.I., O.1)

20. What factors affect the capacitance that exists between two parallel metal plates insulated from each other?

   A capacitor consists of two similar, square, aluminium plates, each 100 mm × 100 mm, mounted parallel and opposite each other. Calculate the capacitance when the distance between the plates is 1.0 mm and the dielectric is mica of relative permittivity 7.0.

If the plates are connected to a circuit which provides a constant current of 2 μA, how long will it take the potential difference of the plates to change by 100 V, and what will be the increase in the charge? (S.A.N.C., O.1)

21. What are the factors which determine the capacitance of a parallel-plate capacitor? Mention how a variation in each of these factors will influence the value of capacitance.

Calculate the capacitance in microfarads of a capacitor having 11 parallel plates separated by mica sheets 0.2 mm thick. The area of one side of each plate is 1000 mm^2 and the relative permittivity of mica is 5. (W.J.E.C., O.1)

22. A parallel-plate capacitor has a capacitance of 300 pF. It has 9 plates, each 40 mm × 30 mm, separated by mica having a relative permittivity of 5. Calculate the thickness of the mica.

23. A capacitor consists of two parallel metal plates, each of area 2000 cm^2 and 5.0 mm apart. The space between the plates is filled with a layer of paper 2.0 mm thick and a sheet of glass 3.0 mm thick. The relative permittivities of the paper and glass are 2 and 8 respectively. A potential difference of 5 kV is applied between the plates. Calculate: (a) the capacitance of the capacitor; (b) the potential gradient in each dielectric; (c) the total energy stored in the capacitor. (N.C.T.E.C., O.2)

24. Obtain from first principles an expression for the capacitance of a single-dielectric, parallel-plate capacitor in terms of the plate area, the distance between plates and the permittivity of the dielectric.

A sheet of mica, 1.0 mm thick and of relative permittivity 6, is interposed between two parallel brass plates 3.0 mm apart. The remainder of the space between the plates is occupied by air. Calculate the area of each plate if the capacitance between them is 0.001 μF. Assuming that air can withstand a potential gradient of 3 MV/m, show that a p.d. of 5 kV between the plates will not cause a flashover. (S.A.N.C., O.2)

25. Explain what is meant by electric field strength in a dielectric and state the factors upon which it depends.

Two parallel metal plates of large area are spaced at a distance of 10 mm from each other in air, and a p.d. of 5000 V is maintained between them. If a sheet of glass, 5.0 mm thick and having a relative permittivity of 6, is introduced between the plates, what will be the maximum electric field strength and where will it occur? (App. El., L.U.)

26. Two capacitors of capacitance 0.2 μF and 0.05 μF are charged to voltages of 100 V and 300 V respectively. The capacitors are then connected in parallel by joining terminals of corresponding polarity together. Calculate:

(a) the charge on each capacitor before being connected in parallel,

(b) the energy stored on each capacitor before being connected in parallel,

(c) the charge on the combined capacitors,

(d) the p.d. between the terminals of the combination,

(e) the energy stored in the combination.

27. A capacitor consists of two metal plates, each 200 mm × 200 mm, spaced 1.0 mm apart, the dielectric being air. The capacitor is charged to a p.d. of 100 V and then

discharged through a ballistic galvanometer having a ballistic constant of 0.0011 microcoulomb per scale division. The amplitude of the first deflection is 32 divisions. Calculate the value of the absolute permittivity of air.

Calculate also the electric field strength and the electric flux density in the air dielectric when the terminal p.d. is 100 V.

28. When the capacitor of Q. 27 is immersed in oil, charged to a p.d. of 30 V and then discharged through the same galvanometer, the first deflection is 27 divisions. Calculate: (a) the relative permittivity of the oil, (b) the electric field strength and the electric flux density in the oil when the terminal p.d. is 30 V and (c) the energy stored in the capacitor.

29. A capacitor is charged to a p.d. of 50 V and then discharged through a ballistic galvanometer having a constant of 1.6×10^{-10} coulomb per division. If the first deflection of the galvanometer is 124 divisions, what is the value of the capacitance? If the capacitor consists of two metal plates, each 300 mm × 300 mm, spaced 2.0 mm apart in air, calculate the absolute permittivity of air.

30. A 0.1-μF capacitor is charged to a p.d. of 5 V. Calculate: (a) the charge in microcoulombs; (b) the number of electrons displaced.

31. When the current in a wire is 600 A, what is the number of electrons per second passing a given point of the wire?

32. If the number of electrons passing per second through an ammeter is 7×10^{16}, what should be the ammeter reading?

33. If there are 3×10^{15} electrons passing per second between two metal surfaces and if the p.d. between the surfaces is 200 V, calculate: (a) the energy absorbed in 20 minutes (i) in joules (ii) in electronvolts, and (b) the final velocity of the electrons assuming that their initial velocity is zero.

34. In a thermionic valve, the current between the filament and the anode is 3 mA and the p.d. is 120 V. Calculate: (a) the number of electrons per second passing from the filament to the anode; (b) the energy absorbed in 5 minutes in (i) electronvolts and (ii) joules; and (c) the final velocity of the electrons, assuming their initial velocity to be zero.

35. An electron is situated between two parallel plates and just outside the negative plate. The plates are spaced 5.0 mm apart in an evacuated bulb and a p.d. of 400 V is maintained between them. Calculate the force on the electron.

If the electron, starting from rest, travels to the positive plate, calculate: (a) the kinetic energy of the electron immediately before impact with that plate, and (b) the time taken.

36. Explain what is meant by the terms, *electric field strength*, *relative permittivity* and *equipotential surface*.

An electron has a velocity of 1.5×10^7 m/s at right-angles to the uniform electric field between two parallel deflecting plates of a cathode-ray tube. If the plates are 25 mm long and spaced 9 mm apart and the p.d. between the plates is 75 V, calculate how far the electron is deflected sideways during its movement through the electric field. Assume electronic charge to be 1.6×10^{-19} C and electronic mass to be 9.1×10^{-31} kg.

(App. El., L.U.)

37. Define the *electronvolt*.

An electron, travelling at a velocity of 10^7 m/s, enters a transverse electric field at right-angles. If this field is 20 mm wide, determine: (a) the field strength necessary to deflect the path of the electron through 30°; (b) the resultant velocity of the electron on leaving the transverse field.

What would the angular deflection be if the initial velocity was doubled?

Derive any expression used. (W.J.E.C., O.2)

38. Derive an expression for the velocity acquired by a particle of mass **m** and charge **e**, which starts from rest and passes through a potential difference V in an electric field.

    Two parallel plates situated 20 mm apart in vacuum have a potential difference of 100 V between them. If the electron leaves the cathode with negligible velocity, determine: (*a*) the velocity of the electron when it strikes the anode; (*b*) the transit time.

    If the anode dissipation is 0.16 W, how many electrons per second pass between cathode and anode? (W.J.E.C., O.2)

39. Define the units of *electric field strength*, *electric flux density* and *permittivity*.

    An electron of charge 1.6×10^{-19} C is at rest in a vacuum between two parallel plates spaced 20 mm apart. If the potential difference between these plates is 500 V, calculate the force on the electron. (N.C.T.E.C., O.2)

40. A 20-μF capacitor is charged and discharged thus:

Steady charging current of 0.02 A	from 0 to 0.5 s
Steady charging current of 0.01 A	from 0.5 to 1.0 s
Zero current	from 1.0 to 1.5 s
Steady discharging current of 0.01 A	from 1.5 to 2.0 s
Steady discharging current of 0.005 A	from 2.0 to 4.0 s

    Draw graphs to scale showing how the current and the capacitor voltage vary with time.

41. Define the *time constant* of a circuit that includes a resistor and capacitor connected in series.

    A 100-μF capacitor is connected in series with an 8000-Ω resistor. Determine the time constant of the circuit. If the combination is connected suddenly to a 100-V d.c. supply, find: (*a*) the initial rate of rise of p.d. across the capacitor; (*b*) the initial charging current; (*c*) the ultimate charge in the capacitor; and (*d*) the ultimate energy stored in the capacitor. (W.J.E.C., O.2)

42. A 10-μF capacitor connected in series with a 50-kΩ resistor is switched across a 50-V d.c. supply. Derive graphically curves showing how the charging current and the p.d. across the capacitor vary with time.

43. A 2-μF capacitor is joined in series with a 2-MΩ resistor to a d.c. supply of 100 V. Draw a current-time graph and explain what happens in the period after the circuit is made, if the capacitor is initially uncharged.

    Calculate the current flowing and the energy stored in the capacitor at the end of an interval of 4 s from the start. (N.C.T.E.C., O.2)

44. Derive an expression for the current flowing at any instant after the application of a constant voltage V to a circuit having a capacitance C in series with a resistance R.

    Determine, for the case in which $C = 0.01\,\mu$F, $R = 100\,000\,\Omega$ and $V = 1000$ V, the voltage to which the capacitor has been charged when the charging current has decreased to 90 per cent of its initial value, and the time taken for the current to decrease to 90 per cent of its initial value. (App. El., L.U.)

45. Derive an expression for the stored electrostatic energy of a charged capacitor.

    A 10-μF capacitor in series with a 10-kΩ resistor is connected across a 500-V d.c. supply. The fully charged capacitor is disconnected from the supply and discharged by connecting a 1000-Ω resistor across its terminals. Calculate: (*a*) the initial value of the charging current; (*b*) the initial value of the discharge current; and (*c*) the amount of heat, in joules, dissipated in the 1000-Ω resistor. (App. El., L.U.)

46. A 20-μF capacitor is found to have an insulation resistance of 50 MΩ, measured between the terminals. If this capacitor is charged off a d.c. supply of 230 V, find the time required after disconnection from the supply for the p.d. across the capacitor to fall to 60 V. Prove any formula used. (App. El., L.U.)

47. A circuit consisting of a 6-μF capacitor, an electrostatic voltmeter and a resistor in parallel, is connected across a 140-V d.c. supply. It is then disconnected and the reading on the voltmeter falls to 70 V in 127 s. When the test is performed without the resistor, the time taken for the same fall in voltage is 183 s. Calculate the resistance of the resistor.

48. A constant direct voltage of V volts is applied across two plane parallel electrodes. Derive expressions for the electric field strength and flux density in the field between the electrodes and the charge on the electrodes. Hence, or otherwise, derive an expression for the capacitance of a parallel-plate capacitor.

    An electronic flash tube requires an energy input of 8.5 J which is obtained from a capacitor charged from a 2000-V d.c. source. The capacitor is to consist of two parallel plates, 11 cm in width, separated by a dielectric of thickness 0.1 mm and relative permittivity 5.5. Calculate the necessary length of each capacitor plate. (U.E.I., O.2)

49. An electrostatic device consists of two parallel conducting plates, each of area 1000 cm^2. When the plates are 10 mm apart in air, the attractive force between them is 0.1 N. Calculate the potential difference between the plates. Find also the energy stored in the system.

    If the device is used in a container filled with a gas of relative permittivity 4, what effect does this have on the force between the plates? (U.L.C.I., O.2)

50. The energy stored in a certain capacitor when connected across a 400-V d.c. supply is 0.3 J. Calculate: (*a*) the capacitance, and (*b*) the charge on the capacitor.

51. A variable capacitor having a capacitance of 800 pF is charged to a p.d. of 100 V. The plates of the capacitor are then separated until the capacitance is reduced to 200 pF. What is the change of p.d. across the capacitor? Also, what is the energy stored in the capacitor when its capacitance is: (*a*) 800 pF; (*b*) 200 pF? How has the increase of energy been supplied?

52. A 200-pF capacitor is charged to a p.d. of 50 V. The dielectric has a cross-sectional area of 300 cm^2 and a relative permittivity of 2.5. Calculate the energy density (in J/m^3) of the dielectric.

53. A parallel-plate capacitor, with the plates 20 mm apart, is immersed in oil having a relative permittivity of 3. The plates are charged to a p.d. of 25 kV. Calculate the force between the plates (in newtons per square metre of plate area) and the energy density (in J/m^3) within the dielectric.

54. A capacitor consists of two metal plates, each 600 mm × 500 mm, spaced 1.0 mm apart. The space between the metal plates is occupied by a dielectric having a relative permittivity of 6, and a p.d. of 3 kV is maintained between the plates. Calculate: (*a*) the capacitance in picofarads; (*b*) the electric field strength and the electric flux density in the dielectric; and (*c*) the force of attraction, in newtons, between the plates.

6 Alternating Voltage and Current

6.1 Generation of an alternating e.m.f.

Fig. 6.1 shows a loop AB carried by a spindle DD rotated at
a constant speed in an anticlockwise direction in a uniform
magnetic field due to poles NS. The ends of the loop are
brought out to two slip-rings C_1 and C_2, attached to but
insulated from DD. Bearing on these rings are carbon brushes
E_1 and E_2, which are connected to an external resistor R.
When the plane of the loop is horizontal, the two sides A and
B are moving parallel to the direction of the magnetic flux;
therefore, no flux is being cut and no e.m.f. is being generated
in the loop.

Fig. 6.1 Generation of an
alternating e.m.f.

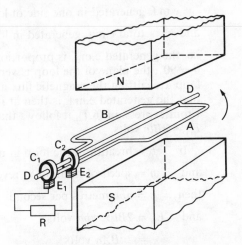

In fig. 6.2(a), the vertical dotted lines represent lines of
magnetic flux and loop AB is shown after it has rotated
through an angle θ from the horizontal position, namely the
position of zero e.m.f. Suppose the peripheral velocity of each
side of the loop to be v metres per second; then at the instant
shown in fig. 6.2, this peripheral velocity can be represented
by the length of a line AL drawn at right-angles to the plane
of the loop. We can resolve AL into two components AM
and AN, perpendicular and parallel respectively, to the
direction of the magnetic flux, as shown in fig. 6.2(b).

Since $\qquad \angle \text{MLA} = 90° - \angle \text{MAL} = \angle \text{MAO} = \theta,$

$$\therefore \qquad \text{AM} = \text{AL} \sin \theta = v \sin \theta.$$

Fig. 6.2 Instantaneous value
of generated e.m.f.

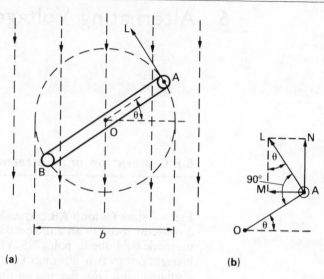

(a)

(b)

The e.m.f. generated in A is due entirely to the component
of the velocity perpendicular to the magnetic field. Hence, if B
is the flux density in teslas and if l is the length in metres of
each of the parallel sides A and B of the loop, it follows from
expression (2.3) that:

e.m.f. generated in one side of loop $= Blv \sin \theta$ volts

and   total e.m.f. generated in loop $= 2Blv \sin \theta$ volts   (6.1)

i.e. the generated e.m.f. is proportional to $\sin \theta$. When
$\theta = 90°$, the plane of the loop is vertical and both sides of the
loop are cutting the magnetic flux at the maximum rate, so
that the generated e.m.f. is then at its maximum value E_m.
From expression (6.1), it follows that when $\theta = 90°$,
$E_m = 2Blv$ volts.

If $b =$ breadth of the loop in metres

and $n =$ speed of rotation in revolutions per second,

then $v = \pi bn$ metres per second

and $E_m = 2Bl \times \pi bn$ volts

$$= 2\pi BAn \text{ volts}$$

where $A = lb =$ area of loop in square metres.

If the loop is replaced by a coil of N turns in series, each
turn having an area of A square metres,

maximum value of e.m.f. generated in coil

$$= E_m = 2\pi BAnN \text{ volts} \qquad (6.2)$$

and instantaneous value of e.m.f. generated in coil

$$= e^* = E_m \sin \theta = 2\pi BAnN \sin \theta \text{ volts} \qquad (6.3)$$

---

*Lower-case letters are used to represent instantaneous values
and upper-case letters represent definite values such as maximum,
average or r.m.s. values. In a.c. circuits, capital I and V without any
subscript represent r.m.s. values.

Fig. 6.3 Sine wave of e.m.f.

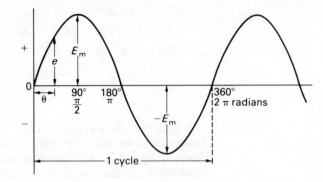

This e.m.f. can be represented by a sine wave as in fig. 6.3, where E_m represents the maximum value of the e.m.f. and e is the value after the loop has rotated through an angle θ from the position of zero e.m.f. When the loop has rotated through 180° or π radians, the e.m.f. is again zero. When θ is varying between 180° and 360° (π and 2π radians), side A of the loop is moving towards the right in fig. 6.1 and is therefore cutting the magnetic flux in the opposite direction to that during the first half-revolution. Hence, if we regard the e.m.f. as positive while θ is varying between 0 and 180°, it is negative while θ is varying between 180° and 360°; i.e. when θ varies between 180° and 270°, the value of the e.m.f. increases from zero to $-E_m$ and then decreases to zero as θ varies between 270° and 360°. Subsequent revolutions of the loop merely produce a repetition of the e.m.f. wave.

Each repetition of a variable quantity, recurring at equal intervals, is termed a *cycle*, and the duration of one cycle is termed its *period* (or *periodic time*). The number of such cycles that occur in one second is termed the *frequency* of that quantity. The unit of frequency is the *hertz* (Hz) in memory of Heinrich Rudolf Hertz, who, in 1888, was the first to demonstrate experimentally the existence and properties of electromagnetic radiation predicted by Maxwell in 1865.

Example 6.1
A coil of 100 turns is rotated at 1500 r/min in a magnetic field having a uniform density of 0.05 T, the axis of rotation being at right angles to the direction of the flux. The mean area per turn is 40 cm².
Calculate (a) the frequency, (b) the period, (c) the maximum value of the generated e.m.f. and (d) the value of the generated e.m.f. when the coil has rotated through 30° from the position of zero e.m.f.

(a) Since the e.m.f. generated in the coil undergoes one cycle of variation when the coil rotates through one revolution,

$$\therefore \qquad \text{frequency} = \text{no. of cycles per second}$$

$$= \text{no. of revolutions per second}$$

$$= 1500/60 = 25 \text{ Hz}.$$

(b) $\qquad$ Period $= \text{time of 1 cycle}$

$$= 1/25 = 0.04 \text{ s}.$$

(c) From expression (6.2),

$$E_{\mathrm{m}} = 2\pi \times 0.05 \times 0.004 \times 100 \times 1500/60 = 3.14 \text{ volts}.$$

(d) For $\theta = 30°$, $\sin 30° = 0.5$,

$$\therefore \qquad e = 3.14 \times 0.5 = 1.57 \text{ volts}.$$

In practice, the waveform of the e.m.f. generated in a conductor of an a.c. generator is usually similar to that shown in fig. 15.2. In an actual machine, however, the conductors are distributed in slots around the periphery of the machine and the effect is to make the waveform of the resultant e.m.f. nearly sinusoidal (section 22.4). In most a.c. calculations, the waveforms of both the voltage and the current are assumed to be sinusoidal.

6.2 Relationship between frequency, speed and number of pole pairs

The waveform of the e.m.f. generated in an a.c. generator undergoes one complete cycle of variation when the conductors move past a N and a S pole; and the shape of the wave over the negative half is exactly the same as that over the positive half. This symmetry of the positive and negative half-cycles does not necessarily hold for waveforms of voltage and current in circuits incorporating rectifiers or transistors.

If an a.c. generator has *p pairs* of poles and if its speed is *n* revolutions per second, then

$$\text{frequency} = f = \text{no. of cycles per second}$$

$$= \text{no. of cycles per revolution}$$

$$\times \text{ no. of revolutions per second}$$

$$= pn \text{ hertz}. \qquad (6.4)$$

Thus if a two-pole machine is to generate an e.m.f. having a frequency of 50 Hz, then from expression (6.4),

$$50 = 1 \times n$$

$$\therefore \qquad \text{speed} = 50 \text{ revolutions per second}$$

$$= 50 \times 60 = 3000 \text{ r/min}.$$

Since it is not possible to have fewer than two poles, the highest speed at which a 50-Hz a.c. generator can be operated is 3000 r/min.

6.3 Average and r.m.s. values of an alternating current

Let us first consider the general case of a current the waveform of which cannot be represented by a simple mathematical expression. For instance, the wave shown in fig. 6.4 is typical of the current taken by a transformer on no load. If n equidistant midordinates, i_1, i_2, etc., are taken over

Fig. 6.4 Average and r.m.s. values

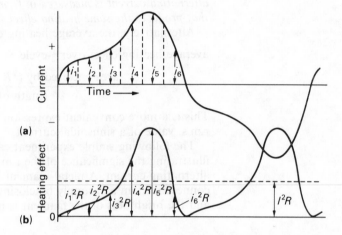

(a)

(b)

either the positive or the negative half-cycle, then:

$$\left.\begin{array}{l} \textit{average value of current} \\ \text{over half a cycle} \end{array}\right\} = I_{av} = \frac{i_1 + i_2 + \cdots + i_n}{n} \quad (6.5)$$

Or, alternatively,

$$\text{average value of current} = \frac{\text{area enclosed over half-cycle}}{\text{length of base over half-cycle}} \quad (6.6)$$

This method of expressing the average value is the more convenient when we come to deal with sinusoidal waves.

In a.c. work, however, the average value is of comparatively little importance. This is due to the fact that it is the power produced by the electric current that usually matters. Thus, if the current represented in fig. 6.4(a) is passed through a resistor having resistance R ohms, the heating effect of i_1 is i_1^2R, that of i_2 is i_2^2R, etc., as shown in fig. 6.4(b). The variation of the heating effect during the second half-cycle is exactly the same as that during the first half-cycle,

$$\therefore \quad \text{average heating effect} = \frac{i_1^2R + i_2^2R + \cdots + i_n^2R}{n}$$

Suppose I to be the value of *direct* current through the same resistance R to produce a heating effect equal to the average heating effect of the alternating current, then:

$$I^2R = \frac{i_1^2R + i_2^2R + \cdots + i_n^2R}{n}$$

$$\therefore \quad I = \sqrt{\left(\frac{i_1^2 + i_2^2 + \cdots + i_n^2}{n}\right)} \tag{6.7}$$

= square *root* of the *mean* of the *squares* of the curren

= root-mean-square (or r.m.s.) value of the current.

This quantity is also termed the *effective* value of the current. It will be seen that the r.m.s. or *effective value of an alternating current is measured in terms of the* direct *current that produces the same heating effect in the same resistance*.

Alternatively, the average heating effect can be expressed:

average heating effect over $\frac{1}{2}$-cycle

$$= \frac{\text{area enclosed by } i^2R \text{ curve over } \frac{1}{2}\text{-cycle}}{\text{length of base}} \tag{6.8}$$

This is a more convenient expression to use when deriving the r.m.s. value of a sinusoidal current.

The following simple experiment can be found useful in illustrating the significance of the r.m.s. value of an alternating current. A metal-filament lamp L (fig. 6.5) is connected to an a.c. supply by closing switch S on contact *a* and the brightness of the filament is noted. Switch S is then

Fig. 6.5 An experiment to demonstrate the r.m.s. value of an alternating current

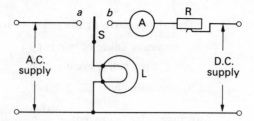

moved to *b* and the slider on resistor R is adjusted to give the same brightness.* The reading on a moving-coil ammeter A then gives the value of the direct current that produces the same heating effect as that produced by the alternating current. If the reading on ammeter A is, say, 0.3 ampere when equality of brightness has been attained, the r.m.s. value of the alternating current is 0.3 ampere.

The r.m.s. value is always greater than the average except for a rectangular wave, in which case the heating effect remains constant so that the average and the r.m.s. values are the same.

---

* For more precise adjustment, an illumination photometer (section 28.7) can be placed at a convenient distance from the lamp, and R adjusted to give the same reading when S is moved over from *a* to *b*.

$$\textit{Form factor} \text{ of a wave} = \frac{\text{r.m.s. value}}{\text{average value}} \qquad (6.9)$$

$$\left.\begin{array}{c}\textit{Peak} \text{ or } \textit{crest factor}\\ \text{of a wave}\end{array}\right\} = \frac{\text{peak or maximum value}}{\text{r.m.s. value}} \quad (6.10)$$

6.4 Average and r.m.s. values of sinusoidal currents and voltages

If I_m is the maximum value of a current which varies sinusoidally as shown in fig. 6.6(a), the instantaneous value i is represented by:

$$i = I_m \sin \theta$$

where θ = angle in radians from instant of zero current.

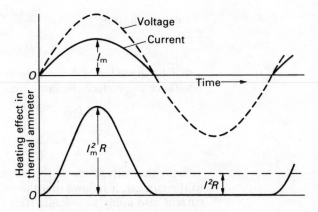

Fig. 6.6 Average and r.m.s. values of a sinusoidal current

For a very small interval $d\theta$ radians, the area of the shaded strip is $i \cdot d\theta$ ampere radians. The use of the unit 'ampere radian' avoids converting the scale on the horizontal axis from radians to seconds,

$\therefore$ total area enclosed by the current wave over $\frac{1}{2}$-cycle

$$= \int_0^\pi i \cdot d\theta = I_m \int_0^\pi \sin \theta \cdot d\theta = -I_m \left[\cos \theta \right]_0^\pi$$

$$= -I_m[-1-1]$$

$$= 2I_m \text{ ampere-radians.}$$

From expression (6.6),

$$\left.\begin{array}{c}\text{average value of current}\\ \text{over half a cycle}\end{array}\right\} = \frac{2I_m \text{ (ampere-radians)}}{\pi \text{ (radians)}}$$

i.e. $\qquad\qquad\qquad I_{av} = 0.637 I_m \text{ amperes} \qquad (6.11)$

If the current be passed through a resistor having

resistance R ohms, instantaneous heating effect $= i^2 R$ watts.

The variation of $i^2 R$ during a complete cycle is shown in fig. 6.6(b). During interval $d\theta$ radians, heat generated is $i^2 R \cdot d\theta$ watt radians and is represented by the area of the shaded strip. Hence:

$$\left.\begin{array}{r}\text{heat generated during}\\ \text{the first half-cycle}\end{array}\right\} = \text{area enclosed by the } i^2 R \text{ curve}$$

$$= \int_0^\pi i^2 R \cdot d\theta = I_m^2 R \int_0^\pi \sin^2 \theta \cdot d\theta$$

$$= \frac{I_m^2 R}{2} \int_0^\pi (1 - \cos 2\theta) \cdot d\theta$$

$$= \frac{I_m^2 R}{2} \left[\theta - \tfrac{1}{2}\sin 2\theta\right]_0^\pi$$

$$= \frac{\pi}{2} I_m^2 R \text{ watt radians}.$$

From expression (6.8),

$$\text{average heating effect*} = \frac{(\pi/2)I_m^2 R \text{ (watt radians)}}{\pi \text{ (radians)}}$$

$$= \tfrac{1}{2}I_m^2 R \text{ watts} \qquad (6.12)$$

If I be the value of direct current through the same resistance to produce the same heating effect,

$$I^2 R = \tfrac{1}{2}I_m^2 R$$

$\therefore$
$$I = \frac{I_m}{\sqrt{2}} = 0.707 I_m \qquad (6.13)$$

Since the voltage across the resistor is directly proportional to the current, it follows that the relationships derived for current also apply to voltage. Hence, in general:

average value of a sinusoidal current or voltage

$$= 0.637 \times \text{maximum value} \quad (6.14)$$

r.m.s. value of a sinusoidal current or voltage

$$= 0.707 \times \text{maximum value} \quad (6.15)$$

From expressions (6.9) and (6.10),

$$\text{form factor of a sine wave} = \frac{0.707 \times \text{maximum value}}{0.637 \times \text{maximum value}}$$

$$= 1.11 \qquad (6.16)$$

---

* This result is really obvious from the fact that $\sin^2 \theta = \tfrac{1}{2} - \tfrac{1}{2}\cos 2\theta$. In words, this means that the square of a sine wave may be regarded as being made up of two components: (a) a constant quantity equal to half the maximum value of the $\sin^2 \theta$ curve, and (b) a cosine curve having twice the frequency of the $\sin \theta$ curve. From fig. 6.6 it is seen that the curve of the heating effect undergoes two cycles of change during one cycle of current. The average value of component (b) over a complete cycle is zero; hence the average heating effect is $\tfrac{1}{2}I_m^2 R$.

and $\left.\begin{array}{r} \text{peak or crest factor} \\ \text{of a sine wave} \end{array}\right\} = \dfrac{\text{maximum value}}{0.707 \times \text{maximum value}}$

$$= 1.414 \qquad (6.17)$$

Example 6.2 *A moving-coil ammeter, a thermal* ammeter and a rectifier are connected in series with a resistor across a 110-V a.c. supply. The circuit has a resistance of 50 Ω to current in one direction and an infinite resistance to current in the reverse direction. Calculate:* (a) *the readings on the ammeters, and* (b) *the form and peak factors of the current wave. Assume the supply voltage to be sinusoidal.*

(a)   Maximum value of the voltage = $110/0.707 = 155.5$ V

∴   maximum value of the current = $155.5/50 = 3.11$ A.

During the positive half-cycle the current is proportional to the voltage and is therefore sinusoidal, as shown in fig. 6.7(a),

∴ $\left.\begin{array}{r} \text{average value of current over} \\ \text{the positive half-cycle} \end{array}\right\} = 0.637 \times 3.11$

$$= 1.98 \text{ A}.$$

During the negative half-cycle, the current is zero. Owing, however, to the inertia of the moving system, the moving-coil ammeter reads the average value of the current over the *whole* cycle,

∴   reading on M.C. ammeter $= \dfrac{1.98}{2} = 0.99$ A.

The variation of the heating effect in the thermal ammeter is shown in fig. 6.7(b), the maximum power being $I_m^2 R$, where R is the resistance of the instrument.

Fig. 6.7 Waveforms of voltage, current and power for example 6.2

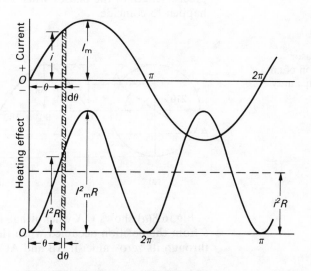

From expression (6.12) it is seen that the average heating effect over the positive half-cycle is $\frac{1}{2}I_m^2 R$; and since no heat is generated during the second half-cycle, it follows that the average heating effect over a complete cycle is $\frac{1}{4}I_m^2 R$.

---

* A thermal ammeter is an instrument the operation of which depends upon the heating effect of a current (see section 26.7).

If I be the direct current to produce the same heating effect,

$$I^2R = \tfrac{1}{4}I_m^2R$$

$$\therefore \qquad I = \tfrac{1}{2}I_m = 3.11/2 = 1.555\,\text{A},$$

i.e. reading on thermal ammeter $= 1.555\,\text{A}$.

A mistake that can very easily be made is to calculate the r.m.s. value of the current over the positive half-cycle as 0.707×3.11, namely 2.2 A, and then say that the reading on thermal ammeter is half this value, namely 1.1 A. The importance of working such a problem from first principles will therefore be evident.

(*b*) From (6.9), form factor $= 1.555/0.99 = 1.57$

and from (6.10), peak factor $= 3.11/1.555 = 2.0$.

6.5 Representation of an alternating quantity by a phasor*

Suppose OA in fig. 6.8(*a*) to represent to scale the maximum value of an alternating quantity, say, current; i.e. OA $= I_m$. Also, suppose OA to rotate counter-clockwise about O at a uniform angular velocity. This is purely a conventional direction which has been universally adopted. An arrowhead is drawn at the outer end of the phasor, partly to indicate which end is assumed to move and partly to indicate the precise length of the phasor when two or more phasors happen to coincide.

Fig. 6.8 Phasor representation of an alternating quantity

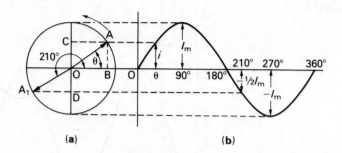

(a) (b)

Fig. 6.8(*a*) shows OA when it has rotated through an angle θ from the position occupied when the current was passing through its zero value. If AB and AC are drawn perpendicular

---

*It has been the practice to refer to a line, such as OA in fig. 6.8(*a*), as a *vector* when it represents the magnitude and phase of a sinusoidal alternating quantity. Strictly, however, a vector should only be used to represent the magnitude and direction of a quantity *in space*, e.g. the magnitude and direction of a force in mechanics. The term *phasor* has been adopted to replace the term *vector* for representing graphically the magnitude and phase of a sinusoidal alternating current or voltage.

to the horizontal and vertical axes respectively:

$$OC = AB = OA \sin \theta$$

$$= I_\mathrm{m} \sin \theta$$

$$= i, \text{ namely the value of the current at that instant.}$$

Hence the projection of OA on the vertical axis represents to scale the instantaneous value of the current. Thus when $\theta = 90°$, the projection is OA itself; when $\theta = 180°$, the projection is zero and corresponds to the current passing through zero from a positive to a negative value; when $\theta = 210°$, the phasor is in position OA_1, and the projection $= OD = \frac{1}{2}OA_1 = -\frac{1}{2}I_\mathrm{m}$; and when $\theta = 360°$, the projection is again zero and corresponds to the current passing through zero from a negative to a positive value. It follows that OA rotates through one revolution or 2π radians in one cycle of the current wave.

If f is the frequency in hertz, then OA rotates through f revolutions of $2\pi f$ radians in 1 second. Hence the angular velocity of OA is $2\pi f$ radians per second and is denoted by the symbol ω (omega):

i.e. $\qquad \omega = 2\pi f$ radians per second.

If the time taken by OA in fig. 6.8 to rotate through an angle θ radians be t seconds, then:

$$\theta = \text{angular velocity} \times \text{time}$$

$$= \omega t = 2\pi f t \text{ radians.}$$

We can therefore express the instantaneous value of the current thus:

$$i = I_\mathrm{m} \sin \theta = I_\mathrm{m} \sin \omega t = I_\mathrm{m} \sin 2\pi f t.$$

Let us next consider how two quantities such as voltage and current can be represented by a phasor diagram. Fig. 6.9(b) shows the voltage leading the current by an angle ϕ. In fig. 6.9(a), OA represents the maximum value of the current and OB that of the voltage. The angle between OA and OB must be the same angle ϕ as in fig. 6.9(b). Consequently when OA is along the horizontal axis, the current at that instant is zero and the value of the voltage is represented by the projection of OB on the vertical axis. These values correspond to instant O in fig. 6.9(b).

After the phasors have rotated through an angle θ, they occupy positions OA_1 and OB_1 respectively, with OB_1 still

Fig. 6.9 Phasor representation of quantities differing in phase

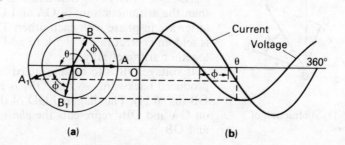

(a)   (b)

leading OA_1 by the same angle ϕ; and the instantaneous values of the current and voltage are again given by the projections of OA_1 and OB_1 on the vertical axis, as shown by the horizontal dotted lines.

If the instantaneous value of the current is represented by

$$i = I_m \sin \theta,$$

then the instantaneous value of the voltage is represented by

$$v = V_m \sin (\theta + \phi)$$

where $I_m = OA$ and $V_m = OB$ in fig. 6.9(a).

The current in fig. 6.9 is said to *lag* the voltage by an angle ϕ or the voltage is said to *lead* the current by an angle ϕ. The *phase difference* ϕ between the two phasors remains constant, irrespective of their position.

6.6 Addition and subtraction of sinusoidal alternating quantities

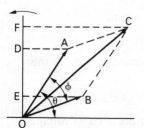

Fig. 6.10 Addition of phasors

Suppose OA and OB in fig. 6.10 to be phasors representing to scale the maximum values of, say, two alternating voltages having the same frequency but differing in phase by an angle ϕ. Complete the parallelogram OACB and draw the diagonal OC. Project OA, OB and OC on to the vertical axis. Then for the positions shown in fig. 6.10:

instantaneous value of OA = OD

instantaneous value of OB = OE

and   instantaneous value of OC = OF.

Since AC is parallel and equal to OB, DF = OE,

$\therefore$ $OF = OD + DF = OD + OE$

i.e.   the instantaneous value of OC $\Big\} = \Big\{$ sum of the instantaneous values of OA and OB.

Hence OC represents the maximum value of the resultant voltage to the scale that OA and OB represent the maximum values of the separate voltages. OC is therefore termed the *phasor sum* of OA and OB; and it is evident that OC is less than the arithmetic sum of OA and OB except when the latter are in phase with each other. This is the reason why it is seldom correct in a.c. work to add voltages or currents together arithmetically.

If voltage OB is to be subtracted from OA, then OB is produced backwards so that OB_1 is equal and opposite to OB (fig. 6.11). The diagonal OD of the parallelogram drawn on OA and OB_1 represents the *phasor difference* of OA and OB.

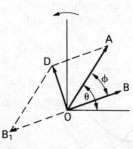

Fig. 6.11 Subtraction of phasors

Example 6.3   *The instantaneous values of two alternating voltages are represented respectively by $v_1 = 60 \sin \theta$ volts and $v_2 = 40 \sin (\theta - \pi/3)$ volts. Derive an expression for the instantaneous value of* (a) *the sum and* (b) *the difference of these voltages.*

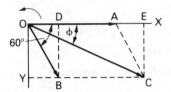

Fig. 6.12 Addition of phasors for example 6.3

(*a*) It is usual to draw the phasors in the position corresponding to $\theta = 0$,* i.e. OA in fig. 6.12 is drawn to scale along the X axis to represent 60 volts, and OB is drawn $\pi/3$ radians or 60° behind OA to represent 40 volts. The diagonal OC of the parallelogram drawn on OA and OB represents the phasor sum of OA and OB. By measurement, OC = 87 volts and angle ϕ between OC and the X axis is 23.5°, namely 0.41 radians; hence:

instantaneous sum of the two voltages = $87 \sin (\theta - 0.41)$ volts.

Alternatively, this expression can be found thus:

horizontal component of OA = 60 V

horizontal component of OB = OD = $40 \cos 60° = 20$ V.

∴ resultant horizontal component = OA + OD = 60 + 20

= 80 V = OE in fig. 6.12.

Vertical component of OA = 0

vertical component of OB = BD = $-40 \sin 60°$

= -34.64 V.

∴ resultant vertical component = -34.64 V = CE.

The minus sign merely indicates that the resultant vertical component is *below* the horizontal axis and that the resultant voltage must therefore lag relative to the reference phasor OA.

Hence maximum value of resultant voltage $\Big\}$ = OC = $\sqrt{\{(80)^2 + (-34.64)^2\}}$

= 87.2 V.

If ϕ is the phase difference between OC and OA,

$\tan \phi$ = EC/OE = $-34.64/80 = -0.433$

∴ $\phi = -23.4° = -0.41$ radian

and instantaneous value of resultant voltage $\Big\}$ = $87.2 \sin (\theta - 0.41)$ volts.

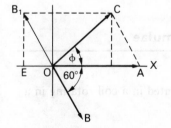

Fig. 6.13 Subtraction of phasors for example 6.3

(*b*) The construction for subtracting OB from OA is obvious from fig. 6.13. By measurement, OC = 53 volts and $\phi = 41° = 0.715$ radian.

∴ instantaneous difference of the two voltages $\Big\}$ = $53 \sin (\theta + 0.715)$ volts.

Alternatively, resultant horizontal component $\Big\}$ = OA − OE = 60 − 20

= 40 V = OD in fig. 6.13,

---

*The idea of a phasor rotating continuously serves to establish its physical significance, but its application in circuit analysis is simplified by *fixing* the phasor in position corresponding to $t = 0$, as in fig. 6.12, thereby eliminating the time function. Such a phasor represents the magnitude of the sinusoidal quantity and its phase relative to a reference quantity, e.g. in fig. 6.12, phasor OB lags the reference phasor OA by 60°.

and   resultant vertical component $= B_1 E = 34.64$ V

$= DC$ in fig. 6.13.

$\therefore$ $\left.\begin{array}{c}\text{maximum value of}\\\text{resultant voltage}\end{array}\right\} = OC = \sqrt{\{(40)^2 + (34.64)^2\}}$

$= 52.9$ V

and $\tan \phi = DC/OD = 34.64/40$

$= 0.866$

$\therefore$ $\phi = 40.9° = 0.714$ radians

and $\left.\begin{array}{c}\text{instantaneous value of}\\\text{resultant voltage}\end{array}\right\} = 52.9 \sin(\theta + 0.714)$ volts.

6.7 Phasor diagrams drawn with r.m.s. values instead of maximum values

It is important to note that when alternating voltages and currents are represented by phasors it is assumed that their waveforms are sinusoidal. It has already been shown that for sine waves the r.m.s. or effective value is 0.707 times the maximum value. Furthermore, ammeters and voltmeters are almost invariably calibrated to read the r.m.s. values. Consequently it is much more convenient to make the length of the phasors represent r.m.s. rather than maximum values. If the phasors of fig. 6.12, for instance, were drawn to represent to scale the r.m.s. instead of the maximum values of the voltages, the shape of the diagram would remain unaltered and the phase relationships between the various quantities would remain unaffected. Hence in all phasor diagrams from now onwards, the lengths of the phasors will, for convenience, represent the r.m.s. values.

Summary of important formulae

Instantaneous value of e.m.f. generated in a coil rotating in a uniform magnetic field

$$= e = E_m \sin \theta$$

$$= 2\pi BANn \sin \theta \text{ volts} \qquad (6.3)$$

$$f = np \qquad (6.4)$$

For n equidistant mid-ordinates over half a cycle,

$$\text{average value} = \frac{i_1 + i_2 + \cdots + i_n}{n} \qquad (6.5)$$

and

$$\text{r.m.s. or effective value} = \sqrt{\left(\frac{i_1^2 + i_2^2 + \cdots + i_n^2}{n}\right)} \qquad (6.7)$$

For sinusoidal waves,

$$\text{average value} = 0.637 \times \text{maximum value} \qquad (6.11)$$

and

$$\text{r.m.s. or effective value} = 0.707 \times \text{maximum value} \qquad (6.13)$$

$$\text{Form factor} = \frac{\text{r.m.s. value}}{\text{average value}} \qquad (6.9)$$

$$= 1.11 \text{ for a sine wave.}$$

$$\text{Peak or crest factor} = \frac{\text{peak or maximum value}}{\text{r.m.s. value}} \qquad (6.10)$$

$$= 1.414 \text{ for a sine wave.}$$

EXERCISES 6

Note. In questions reproduced from examination papers, the term 'vector' has been replaced by 'phasor' by kind permission of the respective institutions.

1. A coil is wound with 300 turns on a square former having sides 50 mm in length. Calculate the maximum value of the e.m.f. generated in the coil when it is rotated at 2000 r/min in a uniform magnetic field of density 0.8 T. What is the frequency of this e.m.f.?

2. Explain what is meant by the terms *wave form, frequency* and *average value*.
   A square coil of side 10 cm, having 100 turns, is rotated at 1200 r/min about an axis through the centre and parallel with two sides in a uniform magnetic field of density 0.4 T. Calculate: (*a*) the frequency; (*b*) the root-mean-square value of the induced e.m.f.; (*c*) the instantaneous value of the induced e.m.f. when the coil is at a position 40 degrees after passing its maximum induced voltage. (U.E.I., O.1)

3. A rectangular coil, measuring 30 cm by 20 cm and having 40 turns, is rotated about an axis coinciding with one of its longer sides at a speed of 1500 r/min in a uniform magnetic field of flux density 0.075 T. Find, from first principles, an expression for the instantaneous e.m.f. induced in the coil, if the flux is at right-angles to the axis of rotation.
   Evaluate this e.m.f. at an instant 0.002 s after the plane of the coil has been perpendicular to the field. (U.L.C.I., O.2)

4. Calculate the speed at which an eight-pole a.c. generator must be driven in order that it may generate an e.m.f. having a frequency of 60 Hz.

5. An a.c. generator driven at 375 r/min generates an e.m.f. having a frequency of 50 Hz. Calculate the number of poles.

6. If a four-pole a.c. generator is driven at a speed of 1800 r/min, what is the frequency of the generated e.m.f.?

7. Define the following terms: *period, phase difference.*
   An alternating voltage had the following instantaneous values, in volts, measured at equal intervals of time over half-

cycle:

$$0, \quad 30, \quad 40, \quad 45, \quad 55, \quad 80, \quad 90, \quad 56, \quad 0.$$

Draw the waveform to scale over half a cycle and determine the average and the root-mean-square values of the voltage.

(U.E.I., O.1)

8. The following ordinates were taken during a half-cycle of a symmetrical alternating-current wave, the current varying in a linear manner between successive points:

Phase angle, in degrees	0	15	30	45	60	75	90
Current, in amperes	0	3.6	8.4	14.0	19.4	22.5	25.0

Phase angle, in degrees	105	120	135	150	165	180
Current, in amperes	25.2	23.0	15.6	9.4	4.2	0

Determine: (*a*) the mean value; (*b*) the r.m.s. value; (*c*) the form factor. (N.C.T.E.C., O.1)

9. Explain the significance of the root-mean-square value of an alternating current or voltage waveform. Define the form factor of such a waveform.

Calculate from first principles the r.m.s. value and form factor of an alternating voltage having the following values over half a cycle, both half-cycles being symmetrical about the zero axis:

Time, in milliseconds	0	1	2	3	4
Voltage, in volts	0	100	100	100	0

These voltage values are joined by straight lines. (U.L.C.I., O.1)

10. A triangular voltage wave has the following values over one-half cycle, both half-cycles being symmetrical about the zero axis:

Time (ms)	0	10	20	30	40	50	60	70	80	90	100
Voltage (V)	0	2	4	6	8	10	8	6	4	2	0

Plot half-cycle of the waveform and hence determine: (*a*) the average value; (*b*) the r.m.s. value; (*c*) the form factor.

(U.L.C.I., O.1)

11. Describe, and explain the action of, an ammeter suitable for measuring the r.m.s. value of a current.

An alternating current has a periodic time $2T$. The current for a time one-third of T is 50 A; for a time one-sixth of T, it is 20 A; and zero for a time equal to one-half of T. Calculate the r.m.s. and average values of this current. (E.M.E.U., O.2)

12. A triangular voltage wave has a periodic time of $\frac{3}{100}$ s. For the first $\frac{2}{100}$ s of each cycle it increases uniformly at the rate of 1000 V/s, while for the last $\frac{1}{100}$ s it falls away uniformly to zero. Find, graphically or otherwise: (*a*) its average value; (*b*) its r.m.s. value; (*c*) its form factor. (E.M.E.U., O.2)

13. Define the root-mean-square value of an alternating current. Explain why this value is more generally employed in a.c. measurements than either the average or the peak value.

Under what circumstances would it be necessary to know (*a*) the average and (*b*) the peak value of an alternating current or voltage?

Calculate the ratio of the peak values of two alternating currents which have the same r.m.s. values, when the waveform of one is sinusoidal and that of the other triangular. What effect would lack of symmetry of the triangular wave about its peak value have upon this ratio? (App. El., L.U.)

14. A voltage, $100 \sin 314t$ volts, is maintained across a circuit consisting of a half-wave rectifier in series with a 50-Ω resistor. The resistance of the rectifier may be assumed to be negligible in the forward direction and infinity in the reverse direction. Calculate the average and the r.m.s. values of the current.

15. State what is meant by the root-mean-square value of an alternating current and explain why the r.m.s. value is usually more important than either the maximum or the mean value of the current.

    A moving-coil ammeter and a moving-iron ammeter are connected in series with a rectifier across a 110-V (r.m.s.) a.c. supply. The total resistance of the circuit in the conducting direction is 60 Ω and that in the reverse direction may be taken as infinity. Assuming the waveform of the supply voltage to be sinusoidal, calculate from first principles the reading on each ammeter. (App. El., L.U.)

16. If the waveform of a voltage has a form factor of 1.15 and a peak factor of 1.5, and if the peak value is 4.5 kV, calculate the average and the r.m.s. values of the voltage.

17. An alternating current was measured by a d.c. milliammeter in conjunction with a full-wave rectifier. The reading on the milliammeter was 7 mA. Assuming the waveform of the alternating current to be sinusoidal, calculate (a) the r.m.s. value and (b) the maximum value of the alternating current.

18. An alternating current, when passed through a resistor immersed in water for 5 minutes, just raised the temperature of the water to boiling point. When a direct current of 4 A was passed through the same resistor under identical conditions, it took 8 minutes to boil the water. Find the r.m.s. value of the alternating current. Neglect other factors than heat given to the water.

    If a rectifier type of ammeter connected in series with the resistor read 5.2 A when the alternating current was flowing, find the form factor of the alternating current.

19. Explain what is meant by the *r.m.s. value* of an alternating current.

    In a certain circuit supplied from 50-Hz mains, the potential difference has a maximum value of 500 V and the current has a maximum value of 10 A. At the instant $t = 0$, the instantaneous values of the p.d. and the current are 400 V and 4 A respectively, both increasing positively. Assuming sinusoidal variation, state trigonometrical expressions for the instantaneous values of the p.d. and the current at time t.

    Calculate the instantaneous values at the instant $t = 0.015$ s and find the angle of phase difference between the p.d. and the current.

    Sketch the phasor diagram. (S.A.N.C., O.1)

20. Explain with the aid of a sketch how the r.m.s. value of an alternating current is obtained.

    An alternating current i is represented by:

    $$i = 10 \sin 942t \text{ amperes.}$$

    Determine (a) the frequency, (b) the period, (c) the time taken from $t = 0$ for the current to reach a value of 6 A for a first and second time, (d) the energy dissipated when the current flows through a 20-Ω resistor for 30 minutes. (S.A.N.C., O.1)

21. (a) Explain the term *r.m.s. value* as applied to an alternating current.

    (b) An alternating current flowing through a circuit has a maximum value of 70 A, and lags the applied voltage by 60°.

The maximum value of the voltage is 100 V, and both current and voltage waveforms are sinusoidal. Plot the current and voltage waveforms in their correct relationship for the positive half of the voltage. What is the value of the current when the voltage is at a positive peak?   (W.J.E.C., O.1)

22. Two sinusoidal e.m.f.s of peak values 50 V and 20 V respectively but differing in phase by 30° are induced in the same circuit. Draw the phasor diagram and find the peak and r.m.s. values of the resultant e.m.f.

23. Two impedances are connected in parallel to the supply, the first takes a current of 40 A at a lagging phase angle of 30°, and the second a current of 30 A at a leading phase angle of 45°. Draw a phasor diagram to scale to represent the supply voltage and these currents. From this diagram, by construction, determine the total current taken from the supply and its phase angle.

24. Two circuits connected in parallel take alternating currents which can be expressed trigonometrically as:

$$i_1 = 13 \sin 314t \text{ amperes} \quad \text{and} \quad i_2 = 12 \sin (314t + \pi/4) \text{ amperes}.$$

Sketch the waveforms of these currents to illustrate maximum values and phase relationships.

By means of a phasor diagram drawn to scale, determine the resultant of these currents, and express it in trigonometric form. Give also the r.m.s. value and the frequency of the resultant current.   (E.M.E.U., O.1)

25. The voltage drops across two components, when connected in series across an a.c. supply, are: $v_1 = 180 \sin 314t$ volts and $v_2 = 120 \sin (314t + \pi/3)$ volts respectively. Determine with the aid of a phasor diagram: (a) the voltage of the supply in trigonometric form; (b) the r.m.s. voltage of the supply; (c) the frequency of the supply.   (E.M.E.U., O.1)

26. Three e.m.f.s, $e_A = 50 \sin \omega t$, $e_B = 80 \sin (\omega t - \pi/6)$ and $e_C = 60 \cos \omega t$ volts, are induced in three coils connected in series so as to give the phasor sum of the three e.m.f.s. Calculate the maximum value of the resultant e.m.f. and its phase relative to e.m.f. e_A. Check the results by means of a phasor diagram drawn to scale.

If the connections to coil B were reversed, what would be the maximum value of the resultant e.m.f. and its phase relative to e_A?   (App. El., L.U.)

27. Find graphically or otherwise the resultant of the following four voltages:

$$e_1 = 25 \sin \omega t, \qquad e_2 = 30 \sin (\omega t + \pi/6),$$
$$e_3 = 30 \cos \omega t, \qquad e_4 = 20 \sin (\omega t - \pi/4)$$

Express the answer in a similar form.   (U.L.C.I., O.2)

28. Four e.m.f.s, $e_1 = 100 \sin \omega t$, $e_2 = 80 \sin (\omega t - \pi/6)$, $e_3 = 120 \sin (\omega t + \pi/4)$ and $e_4 = 100 \sin (\omega t - 2\pi/3)$, are induced in four coils connected in series so that the sum of the four e.m.f.s is obtained. Find graphically or by calculation the resultant e.m.f. and its phase difference with (a) e_1 and (b) e_2.

If the connections to the coil in which the e.m.f. e_2 is induced are reversed, find the new resultant e.m.f.   (E.M.E.U., O.2)

29. The currents in three circuits connected in parallel to a voltage source are: (i) 4 A in phase with the applied voltage; (ii) 6 A lagging the applied voltage by 30°; (iii) 2 A leading the applied voltage by 45°. Represent these currents to scale on a phasor diagram, showing their correct relative phase displacement with each other.

Determine, graphically or otherwise, the total current taken from the source, and its phase angle with respect to the supply voltage.   (U.L.C.I., O.1)

7 Single-Phase Circuits

7.1 Alternating current in a circuit possessing resistance only

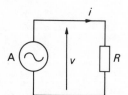

Fig. 7.1 Circuit with resistance only

Consider a circuit having a resistance R ohms connected across the terminals of an a.c. generator A, as in fig. 7.1, and suppose the alternating voltage to be represented by the sine wave of fig. 7.2. If the value of the voltage at any instant B is v volts, the value of the current* at that instant is given by:

$$i = v/R \text{ amperes.}$$

When the voltage is zero, the current is also zero; and since the current is proportional to the voltage, the waveform of the current is exactly the same as that of the voltage. Also the two quantities are *in phase* with each other; that is, they pass through their zero values at the same instant and attain their maximum values in a given direction at the same instant. Hence the current wave is as shown dotted in fig. 7.2.

If V_m and I_m be the maximum values of the voltage and current respectively, it follows that:

$$I_m = V_m/R \qquad (7.1)$$

But the r.m.s. value of a sine wave is 0.707 times the maximum value, so that:

$$\text{r.m.s. value of voltage} = V = 0.707 \, V_m$$

and $\qquad$ r.m.s. value of current $= I = 0.707 \, I_m$.

Substituting for I_m and V_m in (7.1) we have:

$$\frac{I}{0.707} = \frac{V}{0.707 \, R}$$

$\therefore \qquad\qquad\qquad I = V/R \qquad (7.2)$

Hence Ohm's Law can be applied without any modification to an a.c. circuit possessing resistance only.

If the instantaneous value of the applied voltage is

---

* An arrow or arrowhead, as in fig. 7.1, is used to indicate the direction in which the current flows when it is regarded positive. It is immaterial which direction is chosen as positive; but once it has been decided upon for a given circuit or network, the same direction must be adhered to for all the currents and voltages involved in that circuit or network.

Fig. 7.2 Voltage and current waveforms for a resistive circuit

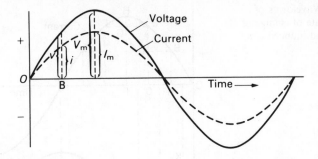

represented by:

$$v = V_m \sin \theta,$$

then instantaneous value of current in a resistive circuit $\Bigg\} = i = \dfrac{V_m}{R} \sin \theta$ (7.3)

Fig. 7.3 Phasor diagram for a resistive circuit

The phasors representing the voltage and current in a resistive circuit are shown in fig. 7.3. The two phasors are actually coincident, but are drawn slightly apart so that the identity of each may be clearly recognized. As mentioned on p. 173, it is usual to draw the phasors in the position corresponding to $\theta = 0$. Hence the phasors representing the voltage and current of expression (7.3) are drawn along the X axis.

7.2 Alternating current in a circuit possessing inductance only

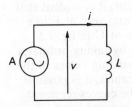

Fig. 7.4 Circuit with inductance only

Let us consider the effect of a sinusoidal current flowing through a coil having an inductance of L henrys and a negligible resistance, as in fig. 7.4. For instance, let us consider what is happening during the first quarter-cycle of fig. 7.5. This quarter-cycle has been divided into three equal intervals, OA, AC and CF seconds. During interval OA, the current increases from zero to AB; hence the average rate of change of current is AB/OA amperes per second, and is represented by ordinate JK drawn midway between O and A. From expression (4.2), the e.m.f., in volts, induced in a coil

$$= -L \times \text{rate of change of current in amperes per second};$$

consequently, the average value of the induced e.m.f. during interval OA is $-L \times$ AB/OA, namely $-L \times$ JK volts, and is represented by ordinate JQ in fig. 7.5.

Similarly, during interval AC, the current increases from AB to CE, so that the average rate of change of current is DE/AC amperes per second, which is represented by ordinate LM in fig. 7.5; and the corresponding induced e.m.f. is $-L \times$ LM volts and is represented by LR. During the third interval CF, the average rate of change of current is GH/CF, namely NP amperes per second; and the corresponding

Fig. 7.5 Waveforms of
current, rate of change of
current and induced e.m.f.

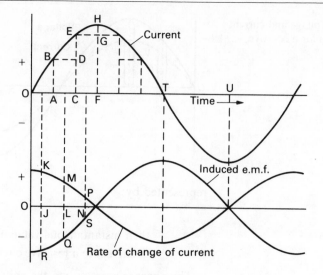

induced e.m.f. is $-L \times$ NP volts and is represented by NS.
At instant F, the current has ceased growing but has not yet
begun to decrease; consequently the rate of change of current
is then zero. The induced e.m.f. will therefore have decreased
from a maximum at O to zero at F. Curves can now be
drawn through the derived points, as shown in fig. 7.5.

During the second quarter-cycle, the current decreases, so
that the rate of change of current is negative and the induced
e.m.f. becomes positive, tending to prevent the current
decreasing. Since the sine wave of current is symmetrical
about ordinate FH, the curves representing the rate of change
of current and the e.m.f. induced in the coil will be symmetrical with those derived for the first quarter-cycle. Since
the rate of change of current at any instant is proportional to
the slope of the current wave at that instant, it is evident that
the value of the induced e.m.f. increases from zero at F to a
maximum at T and then decreases to zero at U in fig. 7.5.

By using shorter intervals, for example by taking ordinates
at intervals of 10° and noting the corresponding values of the
ordinates with the air of trigonometrical tables, it is possible
to derive fairly accurately the shapes of the curves representing the rate of change of current and the induced e.m.f.

From fig. 7.5, it will be seen that the induced e.m.f. attains
its maximum positive value a quarter of a cycle *after* the
current has done the same thing — in fact, it goes through all
its variations a quarter of a cycle after the current has gone
through similar variations. Hence the induced e.m.f. is said to
lag the current by a quarter of a cycle or the current is said
to *lead* the induced e.m.f. by a quarter of a cycle.

Since the resistance of the coil is assumed negligible, we
can regard the whole of the applied voltage as being absorbed
in neutralizing the induced e.m.f. Hence the curve of applied
voltage in fig. 7.6 can be drawn exactly equal and opposite to
that of the induced e.m.f.; and since the latter is sinusoidal,
the wave of applied voltage must also be a sine curve.

Fig. 7.6 Voltage and current waveforms for a purely inductive circuit

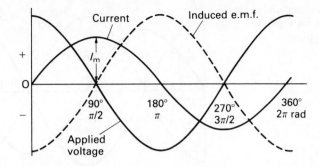

From fig. 7.6 it is seen that the applied voltage attains its maximum positive value a quarter of a cycle earlier than the current; in other words, the voltage applied to a purely inductive circuit leads the current by a quarter of a cycle or 90°, or the current lags the applied voltage by a quarter of a cycle or 90°.

The student might quite reasonably ask: If the applied voltage is neutralized by the induced e.m.f., how can there be any current? The answer is that if there were no current there would be no flux, and therefore no induced e.m.f. The current has to vary at such a rate that the e.m.f. induced by the corresponding variation of flux is equal and opposite to the applied voltage. Actually there is a slight difference between the applied voltage and the induced e.m.f., this difference being the voltage required to send the current through the low resistance of the coil.

7.3 Relationship between current and voltage in a purely inductive circuit

Suppose the instantaneous value of the current through a coil having inductance L henrys and negligible resistance to be represented by

$$i = I_m \sin \theta = I_m \sin 2\pi f t \qquad (7.4)$$

where t is the time, in seconds, after the current has passed through zero from negative to positive values, as shown in fig. 7.7.

Suppose the current to increase by di amperes in dt seconds, then

$$\left.\begin{array}{r}\text{instantaneous value}\\ \text{of induced e.m.f.}\end{array}\right\} = e = -L \cdot di/dt$$

$$= -LI_m \frac{d}{dt}(\sin 2\pi f t)$$

$$= -2\pi f L I_m \cos 2\pi f t$$

$$= 2\pi f L I_m \sin (2\pi f t - \pi/2) \qquad (7.5)$$

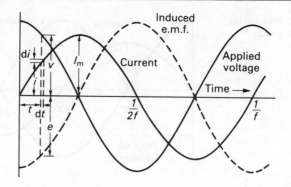

Fig. 7.7 Voltage and current waveforms for a purely inductive circuit

Since f represents the number of cycles per second, the duration of 1 cycle $= 1/f$ seconds. Consequently:

when $t = 0$, $\cos 2\pi ft = 1$

and $\qquad\qquad$ induced e.m.f. $= -2\pi fLI_m$.

When $t = 1/2f$, $\cos 2\pi ft = \cos \pi = -1$,

and $\qquad\qquad$ induced e.m.f. $= 2\pi fLI_m$.

Hence the induced e.m.f. is represented by the dotted curve in fig. 7.7, lagging the current by a quarter of a cycle.

Since the resistance of the circuit is assumed negligible, the whole of the applied voltage is absorbed in neutralizing the induced e.m.f.,

$$\left.\begin{array}{r}\text{instantaneous value of} \\ \text{applied voltage}\end{array}\right\} = v$$

$$= 2\pi fLI_m \cos 2\pi ft$$
$$= 2\pi fLI_m \sin(2\pi ft + \pi/2) \qquad (7.6)$$

Comparison of expressions (7.4) and (7.6) shows that the applied voltage leads the current by a quarter of a cycle. Also, from expression (7.6), it follows that the maximum value V_m of the applied voltage is $2\pi fLI_m$,

i.e. $\qquad V_m = 2\pi fLI_m$, so that $V_m/I_m = 2\pi fL$.

If I and V be the r.m.s. values, then:

$$\frac{V}{I} = \frac{0.707\,V_m}{0.707\,I_m} = 2\pi fL$$

$$= \textit{inductive reactance.}$$

The inductive reactance is expressed in ohms and is represented by the symbol X_L.

Hence $\qquad\qquad I = \dfrac{V}{2\pi fL} = \dfrac{V}{X_L} \qquad (7.7)$

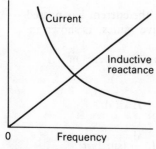

Fig. 7.8 Variation of reactance and current with frequency for a purely inductive circuit

The inductive reactance is proportional to the frequency and the current produced by a given voltage is inversely proportional to the frequency, as shown in fig. 7.8.

The phasor diagram for a purely inductive circuit is given

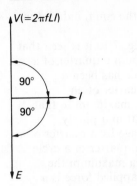

$V(=2\pi fLI)$

90°

90°

I

E

Fig. 7.9 Phasor diagram for a purely inductive circuit

in fig. 7.9, where E represents the r.m.s. value of the e.m.f. induced in the circuit; and V, equal and opposite to E, represents the r.m.s. value of the applied voltage.

7.4 Mechanical analogy of an inductive circuit

One of the most puzzling things to a student commencing the study of alternating currents is the behaviour of a current in an inductive circuit. For instance, why should the current in fig. 7.6 be at its maximum value when there is no applied voltage? Why should there be no current when the applied voltage is at its maximum? Why should it be possible to have a voltage applied in one direction and a current flowing in the reverse direction, as is the case during the second and fourth quarter-cycles in fig. 7.6?

It might therefore be found helpful to consider a simple mechanical analogy — the simpler the better. In mechanics, the *inertia* of a body opposes any change in the *speed* of that body. The effect of inertia is therefore analogous to that of *inductance* in opposing any change in the *current*.

Suppose we take a heavy metal cylinder C (fig. 7.10), such as a pulley or an armature, and roll it backwards and forwards on a horizontal surface between two extreme positions A and B. Let us consider the forces and the speed

Fig. 7.10 Mechanical analogy of a purely inductive circuit

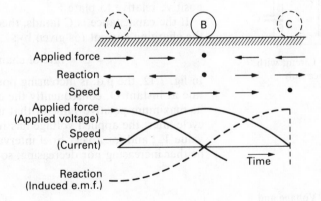

Applied force

Reaction

Speed

Applied force
(Applied voltage)

Speed
(Current)

Time

Reaction
(Induced e.m.f.)

while C is being rolled from A to B. At first the speed is zero, but the force applied to the body is at its maximum, causing C to accelerate towards the right. This applied force is reduced — as indicated by the length of the arrows in fig. 7.10 — until it is zero when C is midway between A and B. C ceases to accelerate and will therefore have attained its maximum speed from left to right.

Immediately after C has passed the mid-point, the direction of the applied force is reversed and increased until the body is brought to rest at B and then begins its return movement.

The reaction of C, on the other hand, is equal and opposite

to the applied force and corresponds to the e.m.f. induced in the inductive circuit.

From an inspection of the arrows in fig. 7.10 it is seen that the speed in a given direction is a maximum a quarter of a complete oscillation after the applied force has been a maximum in the same direction, but a quarter of an oscillation before the reaction reaches its maximum in that direction. This is analogous to the current in a purely inductive circuit lagging the applied voltage by a quarter of a cycle and leading the induced e.m.f. by a quarter of a cycle. Also it is evident that when the speed is a maximum the applied force is zero, and that when the applied force is a maximum the speed is zero; and that during the second half of the movement indicated in fig. 7.10, the direction of motion is opposite to that of the applied force. These relationships correspond exactly to those found for a purely inductive circuit.

7.5 Alternating current in a circuit possessing capacitance only

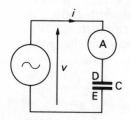

Fig. 7.11 Circuit with capacitance only

Fig. 7.11 shows a capacitor C connected in series with an ammeter A across the terminals of an a.c. source; and the alternating voltage applied to C is represented in fig. 7.12. Suppose this voltage to be positive when it makes plate D positive relative to plate E.

If the capacitance is C farads, then from expression (5.23), the charging current i is given by:

$$i = C \times \text{rate of change of p.d.}$$

In fig. 7.12, the p.d. is increasing positively at the maximum rate at instant O; consequently the charging current is also at its maximum positive value at that instant. A quarter of a cycle later, the applied voltage has reached its maximum value V_m; and for a very brief interval of time the p.d. is neither increasing nor decreasing, so that there is no current.

Fig. 7.12 Voltage and current waveforms for a purely capacitive circuit

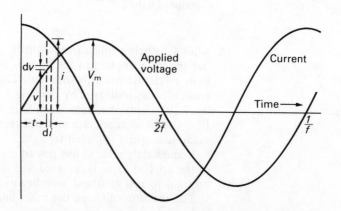

During the next quarter of a cycle, the applied voltage is decreasing. Consequently the capacitor discharges, the discharge current being in the negative direction.

When the voltage is passing through zero, the slope of the voltage curve is at its maximum, i.e. the p.d. is varying at the maximum rate; consequently the current is also a maximum at that instant.

7.6 Relationship between current and voltage in a purely capacitive circuit

Suppose the instantaneous value of the voltage* applied to a capacitor having capacitance C farads to be represented by:

$$v = V_m \sin \theta = V_m \sin 2\pi f t \qquad (7.8)$$

If the applied voltage increases by dv volts in dt seconds (fig. 7.12), then from (5.23),

$$\left.\begin{array}{l}\text{instantaneous value} \\ \text{of current}\end{array}\right\} = i = C \cdot dv/dt$$

$$= C \frac{d}{dt}(V_m \sin 2\pi f t)$$

$$= 2\pi f C V_m \cos 2\pi f t$$

$$= 2\pi f C V_m \sin(2\pi f t + \pi/2) \qquad (7.9)$$

Comparison of expressions (7.8) and (7.9) shows that the current leads the applied voltage by a quarter of a cycle, and the current and voltage can be represented by phasors as in fig. 7.13.

From expression (7.9) it follows that the maximum value I_m of the current is $2\pi f C V_m$,

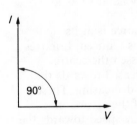

Fig. 7.13 Phasor diagram for a purely capacitive circuit

$$\therefore \qquad \frac{V_m}{I_m} = \frac{1}{2\pi f C}.$$

Hence, if I and V be the r.m.s. values,

$$\frac{V}{I} = \frac{1}{2\pi f C} = capacitive\ reactance \qquad (7.10)$$

The capacitive reactance is expressed in ohms and is represented by the symbol X_C.

Hence, $\qquad\qquad I = 2\pi f C V = V/X_C \qquad (7.11)$

The capacitive reactance is inversely proportional to the

---

* It will be noted that we start with the voltage wave in this case, whereas with inductance we started with the current wave. The reason for this is that in the case of inductance, we derive the induced e.m.f. by differentiating the current expression; whereas with capacitance, we derive the current by differentiating the voltage expression.

frequency, and the current produced by a given voltage is proportional to the frequency, as shown in fig. 7.14.

Example 7.1

A 30-μF capacitor is connected across a 400-V, 50-Hz supply. Calculate (a) the reactance of the capacitor and (b) the current.

(*a*) From expression (7.10):

$$\text{reactance} = \frac{1}{2 \times 3.14 \times 50 \times 30 \times 10^{-6}} = 106.2 \, \Omega.$$

(*b*) From expression (7.11):

$$\text{current} = 400/106.2 = 3.77 \, \text{A}.$$

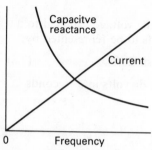

Fig. 7.14 Variation of reactance and current with frequency for a purely capacitive circuit

7.7 Analogies of a capacitance in an a.c. circuit

If the piston P in fig. 5.7. be moved backwards and forwards, the to-and-fro movement of the water causes the diaphragm to be distended in alternate directions. This hydraulic analogy, when applied to capacitance in an a.c. circuit, becomes rather complicated owing to the inertia of the water and of the piston; and as we do not want to take the effect of inertia into account at this stage, it is more convenient to consider a very light flexible strip L (fig. 7.15), such as a metre rule, having one end rigidly clamped. Let us apply an alternating force comparatively slowly by hand so as to oscillate L between positions A and B.

When L is in position A, the applied force is at its maximum towards the *left*. As the force is reduced, L moves towards the *right*. Immediately L has passed the centre position, the applied force has to be increased towards the right, while the speed in this direction is decreasing. These variations are indicated by the lengths of the arrows in fig. 7.15. From the latter it is seen that the speed towards the right is a maximum a quarter of a cycle before the applied force is a maximum in the same direction. The speed is therefore the analogue of the alternating current, and the applied force is that of the applied voltage. Hence capacitance in an electrical circuit is analogous to elasticity in mechanics, whereas inductance is analogous to inertia (section 7.4).

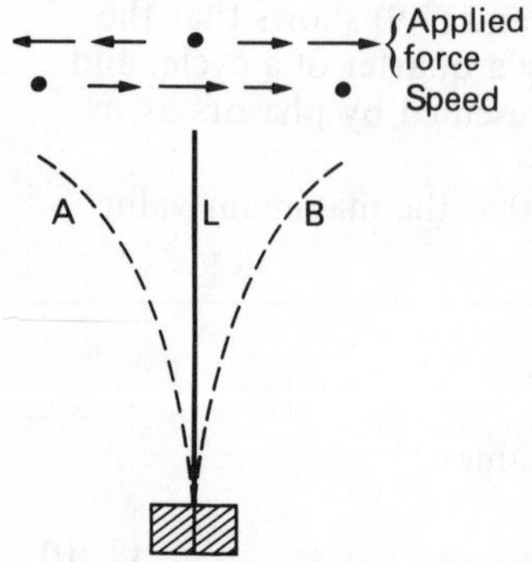

Fig. 7.15 Mechanical analogy of a capacitive circuit

7.8 Alternating current in a circuit possessing resistance, inductance and capacitance in series

We have already considered resistive, inductive and capacitive circuits separately. An actual circuit, however, may have

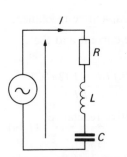

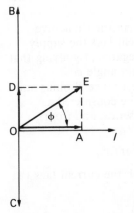

Fig. 7.16 Circuit with R, L and C in series

Fig. 7.17 Phasor diagram for fig. 7.16

resistance and inductance,* or resistance and capacitance, or resistance, inductance and capacitance in series. Hence, if we first consider the general case of R, L and C in series, we can adapt the results to the other two cases by merely omitting the capacitive or the inductive reactance from the expressions derived for the general case.

Fig. 7.16 shows a circuit having resistance R ohms, inductance L henrys and capacitance C farads in series, connected across an a.c. supply of V volts (r.m.s.) at a frequency of f hertz. Let I be the r.m.s. value of the current in amperes.

From section 7.1, the p.d. across R is RI volts in phase with the current and is represented by phasor OA in phase with OI in fig. 7.17.† From section 7.3, the p.d. across L is $2\pi fLI$, and is represented by phasor OB, leading the current by 90°; and from section 7.6, the p.d. across C is $I/(2\pi fC)$ and is represented by phasor OC lagging the current by 90°.

Since OB and OC are in direct opposition, their resultant is OD = OB − OC, OB being assumed greater than OC in fig. 7.17; and the supply voltage is the phasor sum of OA and OD, namely OE. From fig. 7.17,

$$OE^2 = OA^2 + OD^2 = OA^2 + (OB - OC)^2$$

$$\therefore \quad V^2 = (RI)^2 + \left(2\pi fLI - \frac{I}{2\pi fC} \right)^2$$

so that

$$I = \frac{V}{\sqrt{\left\{ R^2 + \left(2\pi fL - \frac{1}{2\pi fC} \right)^2 \right\}}} = \frac{V}{Z}. \qquad (7.12)$$

where Z = *impedance* of circuit in ohms

$$= \frac{V}{I} = \sqrt{\left\{ R^2 + \left(2\pi fL - \frac{1}{2\pi fC} \right)^2 \right\}} \qquad (7.13)$$

---

*A coil possesses resistance and inductance, and its resistance can be considered as part of the total resistance of the circuit. If the coil has a ferromagnetic core, the hysteresis and eddy-current losses in the core are equivalent to an increase in the effective resistance of the coil. Thus, if the core loss is 40 W when the current in the coil is 5 A, the resistor to give the same loss when connected in series with the circuit has a resistance of $40/5^2$, namely 1.6 Ω. Hence if the winding has a resistance of, say, 4 Ω, the effective resistance of the coil is (4 + 1.6), namely 5.6 Ω.

† When drawing a phasor diagram of this type, one should always start with the phasor representing the quantity that is common to the various components of the circuit. Thus with a series circuit it is the current that is the same in all the components, so that in fig. 7.17, OI is the first phasor to be drawn. For parallel circuits it is the voltage that is common to the various components, so that the voltage phasor is then the first to be drawn.

From this expression it is seen that:

$$\text{result reactance} = 2\pi f L - \frac{1}{2\pi f C}$$

$$= \text{inductive reactance} - \text{capacitive reactance}.$$

If ϕ = phase difference between the current and the supply voltage

$$\tan \phi = \frac{AE}{OA} = \frac{OD}{OA} = \frac{OB - OC}{OA} = \frac{2\pi f L I - I/(2\pi f C)}{RI}$$

$$= \frac{\text{inductive reactance} - \text{capacitive reactance}}{\text{resistance}} \quad (7.14)$$

$$\cos \phi = \frac{OA}{OE} = \frac{RI}{ZI} = \frac{\text{resistance}}{\text{impedance}} \quad (7.15)$$

$$\text{and} \quad \sin \phi = \frac{AE}{OE} = \frac{\text{resultant reactance}}{\text{impedance}} \quad (7.16)$$

If the inductive reactance is greater than the capacitive reactance, $\tan \phi$ is positive and the current lags the supply voltage by an angle ϕ; if less, $\tan \phi$ is negative, signifying that the current leads the supply voltage by an angle ϕ.

If a circuit consists of a coil having resistance R ohms and inductance L henrys, such a circuit can be considered as possessing resistance and inductance in series; and from (7.13),

$$\text{impedance} = \sqrt{\{R^2 + (2\pi f L)^2\}}$$

and from (7.14) the phase angle by which the current lags the supply voltage is given by:

$$\phi = \tan^{-1}(2\pi f L / R).$$

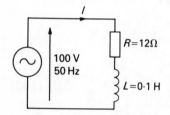

Fig. 7.18 Circuit diagram for example 7.2

Example 7.2

A coil having a resistance of 12 Ω and an inductance of 0.1 H is connected across a 100-V, 50-Hz supply. Calculate: (a) the reactance and the impedance of the coil; (b) the current; and (c) the phase difference between the current and the applied voltage.

When solving problems of this kind, students should first of all draw a circuit diagram (fig. 7.18) and insert all the known quantities. They should then proceed with the phasor diagram, fig. 7.19. It is not essential to draw the phasor diagram to exact scale, but it is helpful to draw it approximately correctly since it is then easy to make a rough check of the calculated values.

(a)   Reactance = $X_L = 2\pi f L$

$$= 2\pi \times 50 \times 0.1 = 31.4 \, \Omega.$$

Impedance = $Z = \sqrt{(R^2 + X_L^2)}$

$$= \sqrt{(12^2 + 31.4^2)} = 33.6 \, \Omega.$$

(b)   Current = $I = V/Z = 100/33.6 = 2.97 \, A$.

(c)   Tan $\phi = X/R = 31.4/12 = 2.617$

∴ $\phi = 69°$.

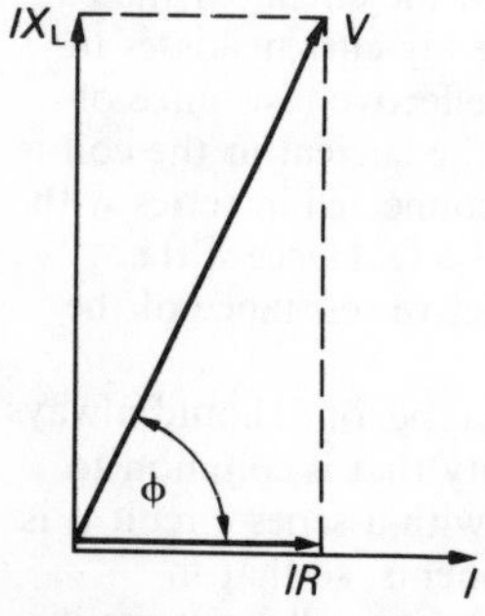

Fig. 7.19 Phasor diagram for example 7.2

Example 7.3

A metal-filament lamp, rated at 750 W, 100 V, is to be connected in series with a capacitor across a 230-V, 60-Hz supply. Calculate (a) the capacitance required and (b) the phase angle between the current and the supply voltage.

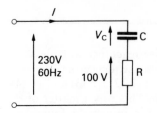

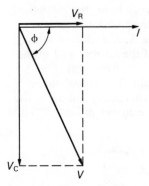

Fig. 7.20 Circuit diagram for example 7.3

Fig. 7.21 Phasor diagram for example 7.3

(*a*) The circuit is given in fig. 7.20, where R represents the lamp. In the phasor diagram of fig. 7.21, the voltage V_R across R is in phase with the current I, while the voltage V_C across C lags I by 90°. The resultant voltage V is the phasor sum of V_R and V_C, and from the diagram:

$$V^2 = V_R^2 + V_C^2$$

$$\therefore \quad (230)^2 = (100)^2 + V_C^2$$

$$\therefore \quad V_C = 207 \text{ V}.$$

$$\text{Rated current of lamp} = \frac{750 \text{ watts}}{100 \text{ volts}} = 7.5 \text{ A}.$$

From (7.11) $\quad 7.5 = 2 \times 3.14 \times 60 \times C \times 207$

$$\therefore \quad C = 96 \times 10^{-6} \text{ F} = 96 \, \mu\text{F}.$$

(*b*) If ϕ = phase angle between the current and the supply voltage,

$$\cos \phi = V_R/V \text{ (from fig. 7.21)}$$

$$= 100/230 = 0.435$$

$$\therefore \quad \phi = 64° \, 12'.$$

Example 7.4

A circuit having a resistance of 12 Ω, an inductance of 0.15 H and a capacitance of 100 μF in series, is connected across a 100-V, 50-Hz supply. Calculate: (a) the impedance; (b) the current; (c) the voltages across R, L and C; (d) the phase difference between the current and the supply voltage.

The circuit diagram is the same as that of fig. 7.16.
(*a*) From (7.13),

$$Z = \sqrt{\left\{(12)^2 + \left(2 \times 3.14 \times 50 \times 0.15 - \frac{10^6}{2 \times 3.14 \times 50 \times 100}\right)^2\right\}}$$

$$= \sqrt{\{144 + (47.1 - 31.85)^2\}} = 19.4 \, \Omega.$$

(*b*) Current $= V/Z = 100/19.4 = 5.15$ A.

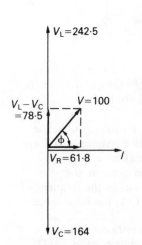

Fig. 7.22 Phasor diagram for example 7.4

(*c*) Voltage across $R = V_R = 12 \times 5.15 = 61.8$ V,

voltage across $L = V_L = 47.1 \times 5.15 = 242.5$ V

and voltage across $C = V_C = 31.85 \times 5.15 = 164.0$ V.

These voltages and current are represented by the respective phasors in fig. 7.22.

(*d*) Phase difference between current and supply voltage

$$= \phi = \cos^{-1}(V_R/V) = \cos^{-1} 61.8/100 = 51° \, 50'.$$

Or, alternatively, from (7.14),

$$\phi = \tan^{-1}(47.1 - 31.85)/12 = \tan^{-1} 1.271 = 51° \, 48'.$$

7.9 Circuit with R, L and C in series: effect of frequency variation

In fig. 7.17, we assumed the capacitive reactance to be less than the inductive reactance, and in consequence the expressions for capacitive reactance in (7.12), (7.13) and (7.14) have a negative sign in front of them. This was a matter of chance in the above discussion, but it is, in general, found convenient to regard inductive reactance as positive and capacitive reactance as negative.*

Fig. 7.23 shows the effect of frequency upon the inductive and capacitive reactances and upon the resultant reactance and the impedance of a circuit having R, L and C in series; and fig. 7.24 shows how the values of the current and of the voltages across R, L and C vary with the frequency, the applied voltage being assumed constant. The actual shapes and the relative magnitudes of these curves depend upon the values chosen for R, L and C; and the curves in figs. 7.23 and

Fig. 7.23 Variation of reactances and impedance with frequency

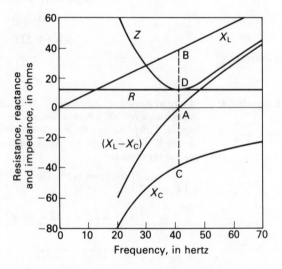

7.24 have been derived for the circuit of example 7.4, where $R = 12\,\Omega$, $L = 0.15\,\text{H}$, $C = 100\,\mu\text{F}$ and applied voltage $= 100\,\text{V}$.

It will be seen from fig. 7.23 that for frequency OA, the inductive reactance AB and the capacitive reactance AC are equal in magnitude so that their resultant is zero. Consequently the impedance is then the same as the resistance AD of the circuit. Furthermore, as the frequency is reduced below OA or increased above OA, the impedance

---

*These positive and negative signs are merely conventions. A 100-Ω capacitive reactance has exactly the same effect as 100-Ω inductive reactance as far as the *magnitude* of the current for a given voltage is concerned. It is only the *phase* of the current that is affected.

Fig. 7.24 Effect of frequency variation upon voltages across *R, L* and *C* in series

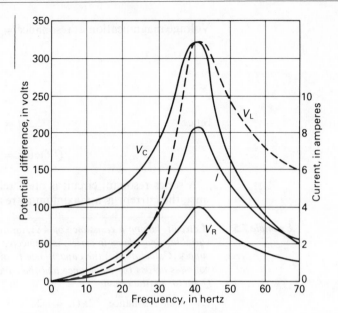

increases and therefore the current decreases. Also, it will be seen from fig. 7.24 that when the frequency is OA, the voltages across *L* and *C* are equal and each is much greater than the supply voltage, namely 100 V. Such a condition is referred to as *resonance*, an effect that is extremely important in communications, e.g. radio, partly because it provides a simple method of increasing the sensitivity of a receiver and partly because it gives selectivity, i.e. it enables a signal of given frequency to be considerably magnified so that it can be separated from signals of other frequencies.

7.10 Resonance in a circuit having *R, L* and *C* in series

From the preceding section, it is evident that, for a series circuit, resonance occurs when:

inductive reactance − capacitive reactance = 0

i.e. when

$$2\pi f L = \frac{1}{2\pi f C}$$

or

$$f = \frac{1}{2\pi\sqrt{(LC)}} \qquad (7.17)$$

At this frequency, $Z = R$ and $I = V/R$. If the resistance is small compared with the inductive and capacitive reactances, the p.d.s across the latter, namely $2\pi f L I$ and $I/2\pi f C$, are many times the supply voltage, as represented by the phasors in fig. 7.25.

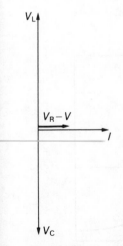

Fig. 7.25 Phasor diagram for series circuit at resonance

$$\text{Voltage magnification at resonance} = \frac{\text{voltage across } L \text{ (or } C)}{\text{supply voltage}}$$

$$= \frac{2\pi fLI}{RI} = \frac{2\pi fL}{R} \quad (7.18)$$

$$= Q \text{ factor of the circuit.}$$

Since $\qquad 2\pi f = 1/\sqrt{(LC)}$ at resonance,

$$\therefore \qquad Q \text{ factor} = \frac{2\pi fL}{R} = \frac{1}{R}\sqrt{\left(\frac{L}{C}\right)} \quad (7.19)$$

A series resonant circuit is often referred to as an *acceptor*, since the current is a maximum at resonance.

Example 7.5   *A circuit, having a resistance of 4 Ω, an inductance of 0.5 H and a variable capacitance in series, is connected across a 100-V, 50-Hz supply. Calculate:* (a) *the capacitance to give resonance;* (b) *the voltages across the inductance and the capacitance; and* (c) *the Q factor of the circuit.*

(*a*)   For resonance, $2\pi fL = 1/2\pi fC$

$$\therefore \qquad\qquad C = \frac{1}{(2 \times 3.14 \times 50)^2 \times 0.5}$$

$$= 20.3 \times 10^{-6} \text{ F} = 20.3 \,\mu\text{F}.$$

(*b*) At resonance, $I = V/R = 100/4 = 25 \text{ A}$.

$\therefore$ p.d. across inductance $= V_L = 2 \times 3.14 \times 50 \times 0.5 \times 25$

$$= 3925 \text{ V}$$

and   p.d. across capacitor $= V_C = 3925 \text{ V}$.

Or alternatively, from (7.11),

$$V_C = \frac{25 \times 10^6}{2 \times 3.14 \times 50 \times 20.3} = 3925 \text{ V}.$$

(*c*) From (7.18),

$$Q \text{ factor} = \frac{2 \times 3.14 \times 50 \times 0.5}{4} = 39.25.$$

Fig. 7.26 Variation of current with frequency for circuit of example 7.5

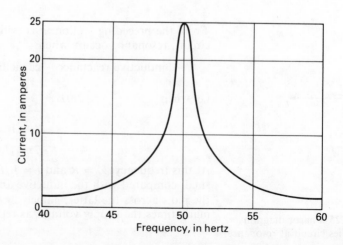

The voltages and current are represented by the respective phasors, but not to scale, in fig. 7.25, and fig. 7.26 shows how the current taken by this circuit varies with frequency, the applied voltage being assumed constant at 100 V.

In example 7.5 the voltages across the inductance and the capacitance at resonance are each nearly forty times the supply voltage. A fuller explanation, however, is necessary to understand the physical significance of such a result.

7.11 Natural frequency of oscillations in a circuit possessing inductance and capacitance

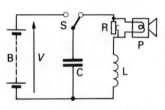

Fig. 7.27 A capacitor discharged through an inductor

Suppose C in fig. 7.27 to be a capacitor whose capacitance can be varied from, say, 1 to 20 μF, and suppose L to be an inductor whose inductance is variable between, say, 0.01 and 0.1 H. A loudspeaker P is connected across a resistor R having a low resistance. A two-way switch S enables C to be charged from a battery B and discharged through L, and R is adjusted to give a convenient volume of sound in P. It is found that each time C is discharged through L, a pizzicato note is emitted by P, similar to the sound produced by plucking the string of a violin or 'cello. Further, the pitch of the note can be varied by varying either C or L — the larger the capacitance and the inductance, the lower the pitch. But it is well known that a musical sound requires the vibration of some medium for its production and that the pitch is dependent upon the number of vibrations per second. Hence it follows that the discharge current through P is an alternating current of diminishing amplitude as shown by the full line in fig. 7.28, the p.d. across the capacitor being represented by the dotted line.

Fig. 7.28 Voltage and current waveforms for fig. 7.27

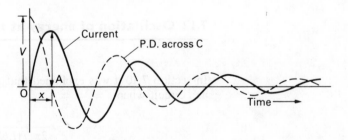

From expression (5.40), the energy stored in C at instant O is $\frac{1}{2}CV^2$ joules. At instant A there is no p.d. across and therefore no energy in C; but the current is I amperes, so that the energy stored in L is $\frac{1}{2}LI^2$ joules (section 4.9). If we neglect the energy wasted in the resistance of the circuit during interval OA, then:

$$\tfrac{1}{2}LI^2 = \tfrac{1}{2}CV^2$$

and
$$I = V\sqrt{(C/L)} \qquad (7.20)$$

If x seconds be the duration of the quarter-cycle OA, average e.m.f. induced in L during OA

$$= -L \times \text{average rate of change of current}$$

$$= -L \times I/x \text{ volts.}$$

If the small voltage drop due to the resistance of the circuit be neglected, the average p.d. applied to L is the same as the average p.d. across C. Hence — assuming a sine wave — we have

$$\frac{LI}{x} = \frac{2}{\pi} V \text{ (section 6.4).}$$

Substituting for I the value given in (7.20):

$$\frac{LV}{x} \times \sqrt{\frac{C}{L}} = \frac{2}{\pi} V,$$

$$\therefore \qquad x = \frac{\pi}{2} \sqrt{(LC)} \text{ seconds,}$$

so that the duration of one cycle $= 4x = 2\pi\sqrt{(LC)}$ seconds

and $\qquad$ frequency $= \dfrac{1}{2\pi\sqrt{(LC)}}$ $\qquad$ (7.21)

This quantity is termed the *natural frequency* of the circuit and represents the frequency with which energy is oscillating backwards and forwards between the capacitor and the inductor, the energy being at one moment stored as electrostatic energy in the capacitor, and a quarter of a cycle later as magnetic energy in the inductor. Owing to loss in the resistance of the circuit, the net amount of energy available to be passed backwards and forwards between L and C gradually decreases.

7.12 Oscillation of energy at resonance

In section 7.10 it was explained that resonance occurs in an a.c. circuit when

$$f_r = \frac{1}{2\pi\sqrt{(LC)}} \qquad (7.17)$$

From expressions (7.17) and (7.21) it follows that the condition for resonance is that the frequency of the applied alternating voltage is the same as the natural frequency of oscillation of the circuit. This condition enables a large amount of energy to be maintained in oscillation between L and C; thus, if the current and the p.d. across the capacitor in a resonant circuit be represented by the curves of fig. 7.29, the magnetic energy stored in L at instant A is $\frac{1}{2}LI_m^2$ joules, and the electrostatic energy in C at instant B is $\frac{1}{2}CV_{cm}^2$ joules,

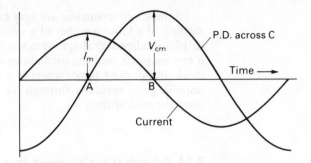

Fig. 7.29 Waveforms of current and p.d. for a capacitor

where V_{cm} represents the maximum value of the voltage across the capacitor.

Since $I_m = 2\pi f C V_{cm}$, and from (7.17), $L = \dfrac{1}{(2\pi f)^2 C}$,

$$\therefore \qquad \tfrac{1}{2}LI_m^2 = \tfrac{1}{2} \times \frac{1}{(2\pi f)^2 C} \times (2\pi f C V_{cm})^2$$

$$= \tfrac{1}{2}C V_{cm}^2,$$

i.e. the magnetic energy in L at instant A is equal to the electrostatic energy in C at instant B, and the power taken from the supply is simply that required to move this energy backwards and forwards between L and C through the resistance of the circuit.

7.13 Mechanical analogy of a resonant circuit

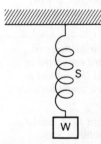

Fig. 7.30 Mechanical analogy of a resonant circuit

It was pointed out in section 7.4 that inertia in mechanics is analogous to inductance in the electric circuit, and in section 7.7 that elasticity is analogous to capacitance. A very very simple mechanical analogy of an electrical circuit possessing inductance, capacitance and a very small resistance can therefore be obtained by attaching a mass W (fig. 7.30) to the lower end of a helical spring S, the upper end of which is rigidly supported. If W is pulled down a short distance and then released, it will oscillate up and down with gradually decreasing amplitude. By varying the mass of W and the length of S it can be shown that the greater the mass and the more flexible the spring, the lower is the natural frequency of oscillation of the system.

If we set W into a slight oscillation and then give it a small downward tap each time it is moving downwards, the oscillations may be made to grow to a large amplitude. In other words, when the frequency of the applied force is the same as the natural frequency of oscillation, a small force can build up large oscillations, the work done by the applied force being that required to supply the losses involved in the transference of energy backwards and forwards between the kinetic and potential forms of energy.

Examples of resonance are very common; for instance, the rattling of a loose member of a vehicle at a particular speed or of a loudspeaker diaphragm when reproducing a sound of a certain pitch, and the oscillations of the pendulum of a clock and of the balance wheel of a watch due to the small impulse given regularly through the escapement mechanism from the mainspring.

7.14 Alternating current in a network possessing resistance, inductance and capacitance in parallel

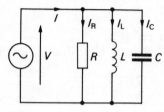

Fig. 7.31 *R, L* and *C* in parallel

Suppose three branches possessing resistance R ohms, inductance L henrys and capacitance C farads respectively to be connected in parallel, as in fig. 7.31, across a supply voltage V of frequency f hertz.

Current through $R = I_R = V/R$, in phase with V,

$$\text{current through } L = I_L = \frac{V}{2\pi fL}, \text{ lagging } V \text{ by } 90°$$

and   current through $C = I_C = 2\pi fCV$, leading V by $90°$.

These currents and the supply voltage are represented by the phasors in fig. 7.32, where it is assumed that I_L is greater than I_C.

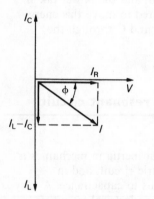

Fig. 7.32 Phasor diagram for fig. 7.31

Resultant of I_L and $I_C = I_L - I_C$,

and current from supply $= I = $ *phasor sum of* I_R and $(I_L - I_C)$

$$= \sqrt{\{I_R^2 + (I_L - I_C)^2\}}$$

If $\phi = $ phase difference between the supply voltage and the resultant current,

$$\tan \phi = (I_L - I_C)/I_R.$$

If $\tan \phi$ is positive, the resultant current lags the supply voltage by an angle ϕ; if negative, the resultant current leads.

Example 7.6   *Three branches possessing a resistance of 50 Ω, an inductance of 0.15 H, and a capacitance of 100 μF respectively, are connected in parallel across a 100-V, 50-Hz supply. Calculate: (a) the current in each branch; (b) the supply current; and (c) the phase angle between the supply current and the supply voltage.*

(a) The circuit diagram is given in fig. 7.33, where I_R, I_L and I_C represent the currents through the resistance, inductance and capacitance respectively.

$$I_R = 100/50 = 2 \text{ A},$$

$$I_L = \frac{100}{2 \times 3.14 \times 50 \times 0.15} = 2.125 \text{ A}$$

and $I_C = 2 \times 3.14 \times 50 \times 100 \times 10^{-6} \times 100 = 3.14 \text{ A}.$

In the case of parallel branches, the first phasor (fig. 7.34) to be

Fig. 7.33 Circuit diagram for example 7.6

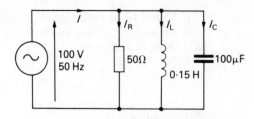

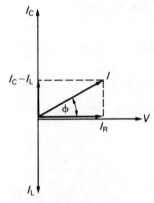

Fig. 7.34 Phasor diagram for example 7.6

drawn is that representing the quantity that is common to those circuits, namely the voltage. I_R is then drawn in phase with V, I_L lagging 90° and I_C leading 90°.

(b) The resultant of I_C and I_L

$$= I_C - I_L = 3.14 - 2.125$$
$$= 1.015 \, \text{A, leading by } 90°.$$

The current I taken from the supply is the resultant of I_R and $(I_C - I_L)$, and from fig. 7.34:

$$I^2 = I_R{}^2 + (I_C - I_L)^2 = 2^2 + (1.015)^2 = 5.03$$
$$\therefore \qquad I = 2.24 \, \text{A}.$$

(c) From fig. 7.34:

$$\cos \phi = \frac{I_R}{I} = \frac{2}{2.24} = 0.893$$

$$\phi = 26° \, 45'.$$

Since I_C is greater than I_L, the supply current leads the supply voltage by 26° 45'.

7.15 Resonance in parallel networks

If the values of L and C in fig. 7.31 were such that $I_L = I_C$, $(I_L - I_C)$ would be zero and the resultant current $I = I_R$; so that, theoretically, current could be flowing backwards and forwards round L and C without any current having to be supplied from the mains. In practice, the inductive circuit has some resistance, however small, so that the parallel inductive and capacitive branches are actually as represented in fig. 7.35.

Fig. 7.35 Coil and capacitor in parallel

Current through inductive branch $= I_1 = \dfrac{V}{\sqrt{\{R^2 + (2\pi fL)^2\}}}$

$\left. \begin{array}{l} \text{and phase angle} \\ \text{between } I_1 \text{ and } V \end{array} \right\} = \phi = \tan^{-1} \dfrac{2\pi fL}{R}$

Since R is assumed to be very small compared with $2\pi fL$, ϕ is nearly 90°.

Current taken by capacitor $= I_C = 2\pi fCV$, leading V by 90°.

If I_1 and I_C be such that the resultant current I is in phase with the supply voltage as shown in fig. 7.36, the network is said to be in resonance.

From fig. 7.36,

$$I_C = OA = I_1 \sin \phi \qquad (7.22)$$

But $\quad \sin \phi = \dfrac{\text{reactance of coil}}{\text{impedance of coil}} = \dfrac{2\pi fL}{\sqrt{\{R^2 + (2\pi fL)^2\}}}$

Substituting for I_C, I_1 and $\sin \phi$ in (7.22), we have:

$$2\pi fCV = \dfrac{2\pi fLV}{R^2 + (2\pi fL)^2}.$$

so that $\qquad f = \dfrac{1}{2\pi} \sqrt{\left(\dfrac{1}{LC} - \dfrac{R^2}{L^2} \right)}$

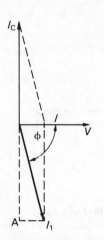

Fig. 7.36 Phasor diagram
for fig. 7.35

If R is very small compared with $2\pi fL$, as in communications circuits (see example 7.7),

$$C \simeq \dfrac{1}{(2\pi f)^2 L}$$

$\therefore \qquad\qquad f_r = \dfrac{1}{2\pi\sqrt{(LC)}} \qquad (7.23)$

which is the same as the resonance frequency of a series circuit.

From fig. 7.36, it is seen that when resonance occurs in a parallel circuit, the current circulating in L and C can be many times greater than the resultant current; in other words, by means of a parallel resonant network, the current taken from the supply can be greatly magnified, thus:

$$\dfrac{I_C}{I} = \dfrac{I_1 \sin \phi}{I} = \dfrac{\sin \phi}{\cos \phi} = \tan \phi = \dfrac{2\pi fL}{R}$$

$$= Q \text{ factor of circuit.}$$

It will be noted that the Q factor is a measure of voltage magnification in a series circuit and of current magnification in a parallel circuit.

The resultant current in a resonant parallel network is in phase with the supply voltage,

and $\quad \left. \begin{matrix} \text{impedance of} \\ \text{such a network} \end{matrix} \right\} = \dfrac{V}{I} = \dfrac{V}{OA \cot \phi}$

$$= \dfrac{V}{I_C} \tan \phi = \dfrac{1}{2\pi fC} \cdot \dfrac{2\pi fL}{R}$$

$$= L/CR \qquad (7.24)$$

This means that a resonant parallel network is equivalent to a non-reactive resistor of $L/(CR)$ ohms. This quantity is termed the *dynamic impedance* of the circuit; and it is obvious that the lower the resistance of the coil, the higher is the dynamic impedance of the parallel circuit. This type of circuit, when used in communications work, is referred to as a *rejector*,

Fig. 7.37 Resonance curve
for a rejector

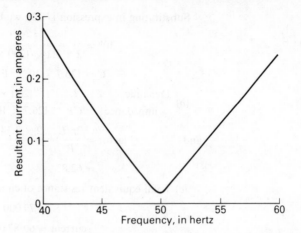

since its impedance is a maximum and the resultant current a
minimum at resonance.

Fig. 7.37 represents the variation of current with frequency
in a parallel resonant circuit consisting of an inductor of 4 Ω
resistance and 0.5 H inductance connected in parallel with a
20.3-μF capacitor across a constant voltage of 100 V, the
values being the same as those of the series circuit of
example 7.5. Since figs. 7.26 and 7.37 refer to circuits having
the same values of R, L and C, it is of interest to note that a
change of 10 per cent in the frequency from the resonance
value reduces the current to about an eighth in the series
circuit and increases the resultant current about eight times
when the circuits are in parallel.

Example 7.7   *A tuned circuit consisting of a coil having an inductance of* 200 μH
and a resistance of 20 Ω *in parallel with a variable capacitor is
connected in series with a resistor of* 8000 Ω *across a 60-V supply
having a frequency of* 1 *MHz. Calculate:* (a) *the value of C to give
resonance;* (b) *the dynamic impedance and the Q factor of the tuned
circuit; and* (c) *the current in each branch.*

(*a*) 1 MHz = 10^6 Hz,

∴   reactance of $L = 2 \times 3.14 \times 10^6 \times 200 \times 10^{-6} = 1256$ Ω,

so that the resistance of the coil is very small compared with its
reactance.

Fig. 7.38 Circuit diagram for
example 7.7

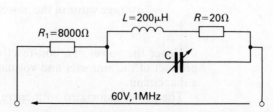

Substituting in expression (7.23), we have:

$$10^6 = \frac{1}{2 \times 3.14 \sqrt{(200 \times 10^{-6} \times C)}}$$

$$\therefore \qquad C = 126.7 \times 10^{-12}\,\text{F} = 126.7\,\text{pF}.$$

(b) $\left.\begin{array}{l}\text{Dynamic} \\ \text{impedance}\end{array}\right\} = \dfrac{L}{CR} = \dfrac{200 \times 10^{-6}}{126.7 \times 10^{-12} \times 20} = 79\,000\,\Omega$

and $\qquad Q \text{ factor} = \dfrac{2\pi f L}{R} = \dfrac{2 \times 3.14 \times 10^6 \times 200 \times 10^{-6}}{20}$

$$= 62.8.$$

(c) Total equivalent resistance of circuit

$$= 79\,000 + 8000 = 87\,000\,\Omega,$$

$$\therefore \qquad \text{current} = 60/87\,000\,\text{A} = 0.69\,\text{mA}.$$

P.D. across tuned circuit $= 0.69 \times 10^{-3} \times 79\,000 = 54.5\,\text{V}$

$\therefore$ current through inductive branch of tuned circuit

$$= \frac{54.5}{\sqrt{\{(20)^2 + (1256)^2\}}} = \frac{54.5}{1256}\,\text{A}$$

$$= 43.4\,\text{mA},$$

and current $\left.\begin{array}{l}\\ \text{through } C\end{array}\right\} = 2 \times 3.14 \times 10^6 \times 126.7 \times 10^{-12} \times 54.5\,\text{A}$

$$= 43.4\,\text{mA}.$$

Actually there is a very slight difference between the values of the currents in the two parallel branches, but it is so small as to be negligible. The results show that the current in each of the parallel circuits is about 62.8 times the resultant current taken from the supply.

7.16 Power in a non-reactive circuit having constant resistance

In section 6.3 it was explained that when an alternating current flows through a resistor of R ohms, the average heating effect over a complete cycle is I^2R watts, where I is the r.m.s. value of the current in amperes.

If V volts be the r.m.s. value of the applied voltage, then for a non-reactive circuit having constant resistance R ohms, $V = IR$:

$$\therefore \qquad \text{average value of the power} = I^2R = I \times IR$$

$$= IV \text{ watts}.$$

Hence the power in a non-reactive circuit is given by the product of the ammeter and voltmeter readings, exactly as in a d.c. circuit.

The power associated with energy transfer from the

electrical system to another system such as heat, light or mechanical drives is termed active power, thus the average power given by I^2R is the active power of the arrangement. There are other powers, e.g. when the energy does not transfer away from the electrical system but remains temporarily stored and returns almost immediately to the electrical system.

7.17 Power in a purely inductive circuit

Consider a coil wound with such thick wire that the resistance is negligible in comparison with the inductive reactance X_L ohms. If such a coil is connected across a supply voltage V, the current is given by $I = V/X_L$ amperes. Since the resistance is very small, the heating effect and therefore the active power are also very small, even though the voltage and the current be large. Such a curious conclusion — so different from anything we have experienced in d.c. circuits — requires fuller explanation if its significance is to be properly understood. Let us therefore consider fig. 7.39, which shows the applied voltage and the current for a purely inductive circuit, the current lagging the voltage by a quarter of a cycle.

Fig. 7.39 Power curve for a purely inductive circuit

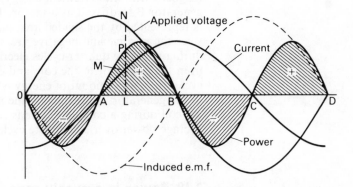

The power at any instant is given by the product of the voltage and the current at that instant; thus at instant L, the applied voltage is LN volts and the current is LM amperes, so that the power at that instant is LN × LM watts and is represented to scale by LP.

By repeating this calculation at various instants we can deduce the curve representing the variation of power over one cycle. It is seen that during interval OA the applied voltage is positive, but the current is negative, so that the power is negative; and that during interval AB, both the current and the voltage are positive, so that the power is positive.

The power curve is found to be symmetrical about the horizontal axis OD. Consequently the shaded areas marked

'−' are exactly equal to those marked '+', so that the mean value of the power over the complete cycle OD is zero.

It is necessary, however, to consider the significance of the positive and negative areas if we are to understand what is really taking place. So let us consider an a.c. generator P (fig. 7.40) connected to a coil Q whose resistance is negligible, and let us assume that the voltage and current are represented by the graphs in fig. 7.39. At instant A, there is no current and therefore no magnetic field through and around Q. During interval AB, the growth of the current is accompanied by a growth of flux as shown by the dotted lines in fig. 7.40. But the existence of a magnetic field involves some kind of a strain in the space occupied by the field and the storing up of energy in that field, as already dealt with in section 4.9. The current, and therefore the magnetic energy associated with it, reach their maximum values at instant B; and, since the loss in the coil is assumed negligible, it follows that at that instant the whole of the energy supplied to the coil during interval AB, and represented by the shaded area marked '+', is stored up in the magnetic field.

During the interval BC the current and its magnetic field are decreasing; and the e.m.f. induced by the collapse of the magnetic flux is in the same direction as the current. But any circuit in which the current and the induced or generated e.m.f. are in the same direction acts as a generator of electrical energy (see section 4.2). Consequently the coil is now acting as a generator transforming the energy of its magnetic field into electrical energy, the latter being sent to generator P to drive it as a motor. The energy thus returned is represented by the shaded area marked '−' in fig. 7.39; and since the positive and negative areas are equal, it follows that during alternate quarter-cycles electrical energy is being sent from the generator to the coil, and during the other quarter-cycles the same amount of energy is sent back from the coil to the generator. Consequently the net energy absorbed by the coil during a complete cycle is zero; in other words, the average power over a complete cycle is zero.

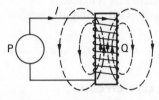

Fig. 7.40 Magnetic field of an inductive circuit

7.18 Power in a purely capacitive circuit

In this case, the current leads the applied voltage by a quarter of a cycle, as shown in fig. 7.41; and by multiplying the corresponding instantaneous values of the voltage and current, we can derive the curve representing the variation of power. During interval OA, the voltage and current are both positive so that the power is positive, i.e. power is being supplied from the generator to the capacitor; and the shaded area enclosed by the power curve during interval OA represents the value of the electrostatic energy stored in the capacitor at instant A.

During interval AB, the p.d. across the capacitor decreases from its maximum value to zero and the whole of the energy

Fig. 7.41 Power curve for a purely capacitive circuit

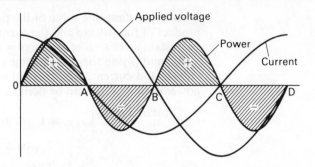

stored in the capacitor at instant A is returned to the generator; consequently the net energy absorbed during the half cycle OB is zero. Similarly, the energy absorbed by the capacitor during interval BC is returned to the generator during interval CD. Hence the average power over a complete cycle is zero.

7.19 Power in a circuit possessing resistance and reactance in series

Let us consider the general case of the current differing in phase from the applied voltage; thus in fig. 7.42(a), the current is shown lagging the voltage by an angle ϕ.

Let   instantaneous value of voltage $= v = V_m \sin \theta$

then   instantaneous value of current $= i = I_m \sin (\theta - \phi)$.

Fig. 7.42 Voltage, current and power curves

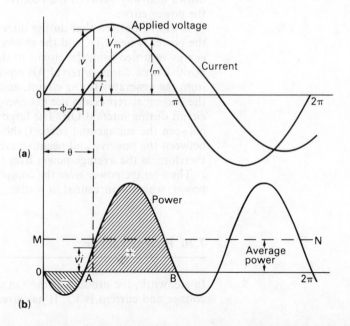

At any instant, the value of the power is given by the product of the voltage and the current at that instant, i.e. instantaneous value of power $= vi$ watts.

By multiplying the corresponding instantaneous values of voltage and current, the curve representing the variation of power in fig. 7.42(*b*) can be derived:

i.e. $\left.\begin{array}{c}\text{instantaneous}\\ \text{power}\end{array}\right\} = vi = V_m \sin \theta \cdot I_m \sin (\theta - \phi)$

$$= \tfrac{1}{2}V_m I_m \{\cos \phi - \cos (2\theta - \phi)\}$$

$$= \tfrac{1}{2}V_m I_m \cos \phi - \tfrac{1}{2}V_m I_m \cos (2\theta - \phi).$$

From this expression, it is seen that the instantaneous value of the power consists of two components:

(1) $\tfrac{1}{2}V_m I_m \cos \phi$, which contains no reference to θ and therefore remains constant in value.

(2) $\tfrac{1}{2}V_m I_m \cos (2\theta - \phi)$, the term 2θ indicating that it varies at twice the supply frequency; thus in fig. 7.42(*b*), it is seen that the power undergoes two cycles of variation for one cycle of the voltage wave. Furthermore, since the average value of a cosine curve over a *complete* cycle is zero, it follows that this component does not contribute anything towards the *average* value of the power taken from the generator.

Hence, $\left.\begin{array}{c}\text{average power}\\ \text{over one cycle}\end{array}\right\} = \tfrac{1}{2}V_m I_m \cos \phi$

$$= \frac{V_m}{\sqrt{2}} \cdot \frac{I_m}{\sqrt{2}} \cdot \cos \phi$$

$\therefore \qquad\qquad\qquad P = VI \cos \phi \qquad\qquad (7.25)$

where V and I are the r.m.s. values of the voltage and current respectively. In fig. 7.42(*b*), the average power is represented by the height above the horizontal axis of the dotted line MN drawn mid-way between the positive and negative peaks of the power curve.

It will be noticed that during interval OA in fig. 7.42(*b*), the power is negative, and the shaded negative area represents energy returned from the circuit to the generator. The shaded positive area during interval AB represents energy supplied from the generator to the circuit; and the difference between the two areas represents the net energy absorbed by the circuit during interval OB. The larger the phase difference between the voltage and current, the smaller is the difference between the positive and negative areas and the smaller, therefore, is the average power over the complete cycle.

The average power over the complete cycle is the active power, which is measured in watts.

7.20 Power factor

In a.c. work, the product of the r.m.s. values of the applied voltage and current is VI. It has already been shown that the active power $P = VI \cos \phi$ and the value of $\cos \phi$ has to lie between 0 and 1. It follows that the active power P can be either equal to or less than the product VI, which is termed the apparent power and is measured in voltamperes (VA).

The ratio of the active power P to the apparent power S is termed the power factor,

i.e.
$$\frac{\text{active power } P \text{ in watts}}{\text{apparent power } S \text{ in voltamperes}} = \text{power factor} \quad (7.26)$$

$$\therefore \qquad \cos \phi = \frac{P}{S} = \frac{P}{VI}$$

or active power P = apparent power S × power factor (7.27)

Comparison of expressions (7.25) and (7.26) shows that for *sinusoidal* voltage and current:

$$\text{power factor} = \cos \phi$$

From the general phasor diagram of fig. 7.17 for a *series* circuit, it follows that:

$$\cos \phi = \frac{IR}{V} = \frac{IR}{IZ} = \frac{\text{resistance}}{\text{impedance}} \quad (7.15)$$

It has become the practice to say that the power factor is *lagging* when the *current lags the supply voltage* and *leading* when the *current leads the supply voltage*. This means that the supply voltage is regarded as the reference quantity.

7.21 Active and reactive currents

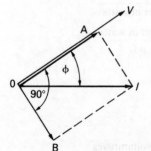

Fig. 7.43 Active and reactive components of current

If a current I lags the applied voltage V by an angle ϕ, as in fig. 7.43, it can be resolved into two components,* OA in phase with the voltage and OB lagging by 90°.

Since power $= IV \cos \phi = V \times \text{OI} \cos \phi = V \times \text{OA}$ watts, therefore OA is termed the *active* component of the current:

i.e. active component of current $= I \cos \phi$ (7.29)

Power due to component OB $= V \times \text{OB} \cos 90° = 0$, so that OB is termed the *reactive* component of the current:

i.e. reactive component of current $= I \sin \phi$ (7.30)

and reactive power Q in vars† $= VI \sin \phi$ (7.31)

Example 7.8 *A coil having a resistance of 6 Ω and an inductance of 0.03 H is connected across a 50-V, 60-Hz supply. Calculate: (a) the current; (b) the phase angle between the current and the applied voltage; (c) the*

---

* If the phasor diagram of fig. 7.43 refers to a circuit possessing resistance and inductance in series, OA and OB must *not* be labelled I_R and I_L respectively. This is an error frequently made by beginners.
† The term 'var' is short for voltampere reactive.

power factor; (d) *the apparent power;* (e) *the active power; and* (f) *the reactive power.*

(a) The phasor diagram for such a circuit is given in fig. 7.19.

$$\text{Reactance of circuit} = 2\pi f L = 2 \times 3.14 \times 60 \times 0.03$$
$$= 11.31\,\Omega.$$

From (7.13),　　impedance $= \sqrt{\{6^2 + (11.31)^2\}} = 12.8\,\Omega$

and　　　　　　current $= 50/12.8 = 3.91\,\text{A}.$

(b) From (7.14)　　$\tan\phi = \dfrac{X}{R} = \dfrac{11.31}{6} = 1.885.$

$\therefore$　　　　　　　　$\phi = 62°\,3'.$

(c) From (7.28), power factor $= \cos 62°\,3' = 0.469$

or, from (7.15),　power factor $= 6/12.8 = 0.469.$

(d) Apparent power $S = 50 \times 3.91 = 195.5\,\text{VA}.$

(e)　　　Active power $=$ apparent power $\times \cos\phi$
$$= 195.5 \times 0.469 = 91.7\,\text{W}.$$

Or, alternatively:

(f)　Active power $= I^2 R = (3.91)^2 \times 6 = 91.7\,\text{W}.$

Reactive power $=$ apparent power $\times \sin\phi$
$$= 195.5 \times \sin 62°\,3'$$
$$= 172.7\,\text{var}.$$

Example 7.9　*A single-phase motor operating off a 400-V, 50-Hz supply is developing 10 kW with an efficiency of 84 per cent and a power factor of 0.7 lagging. Calculate:* (a) *the input apparent power;* (b) *the active and reactive components of the current; and* (c) *the reactive power (in kilovars).*

(a)　　　　　Efficiency $= \dfrac{\text{output power in watts}}{\text{input power in watts}}$

$$= \dfrac{\text{output power in watts}}{IV \times \text{p.f.}}$$

$\therefore$　　　　$0.84 = \dfrac{10 \times 1000}{IV \times 0.7}$

so that　　　　　　$IV = 17\,000\,\text{VA}.$

$\therefore$ input　　　　　　$= 17\,\text{kVA}.$

(b)　Current taken by motor $= \dfrac{\text{input voltamperes}}{\text{voltage}}$

$$= 17\,000/400 = 42.5\,\text{A}.$$

$\therefore$　$\left.\begin{array}{l}\text{active component}\\ \text{of current}\end{array}\right\} = 42.5 \times 0.7 = 29.75\,\text{A}.$

Since　$\sin\phi = \sqrt{(1 - \cos^2\phi)} = \sqrt{\{1 - (0.7)^2\}} = 0.714,$

$\therefore$　$\left.\begin{array}{l}\text{reactive component}\\ \text{of current}\end{array}\right\} = 42.5 \times 0.714 = 30.35\,\text{A}.$

(c)　　　Reactive power $= 400 \times 30.35/1000$
$$= 12.14\,\text{kvar}.$$

Example 7.10

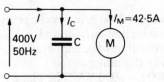

Fig. 7.44 Circuit diagram for example 7.10

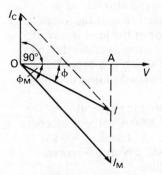

Fig. 7.45 Phasor diagram for fig. 7.44

Calculate the capacitance required in parallel with the motor of example 7.9 to raise the power factor to 0.9 lagging.

The circuit and phasor diagrams are given in figs. 7.44 and 7.45 respectively, M being the motor taking a current I_M of 42.5 A.

Current I_C taken by the capacitor must be such that when combined with I_M, the resultant current I lags the voltage by an angle ϕ, where $\cos\phi = 0.9$. From fig. 7.45,

$$\text{active component of } I_M = I_M \cos\phi_M = 42.5 \times 0.7$$
$$= 29.75 \text{ A}$$

and $\qquad$ active component of $I = I \cos\phi = I \times 0.9$.

These components are represented by OA in fig. 7.45,

$$\therefore \qquad I = 29.75/0.9 = 33.06 \text{ A}.$$

Reactive component of $I_M = I_M \sin\phi_M$
$$= 30.35 \text{ A (from example 7.9)}$$

and $\qquad$ reactive component of $I = I \sin\phi$
$$= 33.06\sqrt{\{1 - (0.9)^2\}}$$
$$= 33.06 \times 0.436 = 14.4 \text{ A}.$$

From fig. 7.45 it will be seen that:

$$I_C = \text{reactive component of } I_M - \text{reactive component of } I$$
$$= 30.35 - 14.4 = 15.95 \text{ A}.$$

But $\quad I_C = 2\pi f C V$

$$\therefore \quad 15.95 = 2 \times 3.14 \times 50 \times C \times 400$$

and $\quad C = 127 \times 10^{-6} \text{ F} = 127 \, \mu\text{F}.$

From example 7.10 it will be seen that the effect of connecting a 127-μF capacitor in parallel with the motor is to reduce the current taken from the supply from 42.5 to 33.06 A, without altering either the current or the power taken by the motor. This enables an economy to be effected in the size of the generating plant and in the cross-sectional area of conductor in the cable.

Example 7.11

An a.c. generator is supplying a load of 300 kW at a power factor of 0.6 lagging. If the power factor is raised to unity, how many more kilowatts can the generator supply for the same kVA loading?

Since the power in kW

$$= \text{number of kilovoltamperes} \times \text{power factor},$$

$\therefore$ number of kilovoltamperes

$$= 300/0.6 = 500 \text{ kVA}.$$

When the power factor is raised to unity:

number of kilowatts = number of kilovoltamperes = 500 kW.

Hence increased power supplied by generator

$$= 500 - 300 = 200 \text{ kW}.$$

7.22 The practical importance of power factor

If an a.c. generator is rated to give, say, 2000 A at a voltage of 400 V, it means that these are the highest current and voltage values the machine can give without the temperature exceeding a safe value. Consequently the rating of the generator is given as $400 \times 2000/1000 = 800$ kVA. The phase difference between the voltage and the current depends upon the nature of the load and not upon the generator. Thus if the power factor of the load is unity, the 800 kVA are also 800 kW; and the engine driving the generator has to be capable of developing this power together with the losses in the generator. But if the power factor of the load is, say, 0.5, the power is only 400 kW; so that the engine is developing only about one-half of the power of which it is capable, though the generator is supplying its rated output of 800 kVA.

Similarly, the conductors connecting the generator to the load have to be capable of carrying 2000 A without excessive temperature rise. Consequently they can transmit 800 kW if the power factor is unity, but only 400 kW at 0.5 power factor, for the same rise of temperature.

It is therefore evident that the higher the power factor of the load, the greater is the *active power* that can be generated by a given generator and transmitted by a given conductor.

The matter may be put another way by saying that, for a *given power*, the lower the power factor, the larger must be the size of the source to generate that power and the greater must be the cross-sectional area of the conductor to transmit it; in other words, the greater is the cost of generation and transmission of the electrical energy. This is the reason why supply authorities do all they can to improve the power factor of their loads, either by the installation of capacitors or special machines or by the use of tariffs which encourage consumers to do so.

7.23 Measurement of power in a single-phase circuit

Since the product of the voltage and current in an a.c. circuit must be multiplied by the power factor to give the active power in watts, the most convenient method of measuring the power is to use a watt-meter, the simplest form of which is the dynamometer type described in section 26.8.

7.24 Current locus for a circuit having constant inductance and variable resistance

Suppose a circuit having a constant inductive reactance of X ohms and a variable resistance to be supplied at a constant

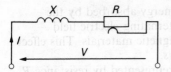

Fig. 7.46 Constant
inductance and variable
resistance in series

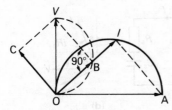

Fig. 7.47 Phasor diagram
for fig. 7.46

voltage V having a frequency f hertz, as in fig. 7.46.

When $R = 0$, current $= V/X$ and is represented by OA in fig. 7.47, lagging the supply voltage OV by 90°.

For the general case of R equal to some finite value,

$$I = \frac{V}{\sqrt{(R^2 + X^2)}}$$

and is represented by OI in fig. 7.47.

Draw VB perpendicular to OI and complete the rectangle OBVC; then, OB $= IR$, and BV $=$ OC $= IX$.

Also, since VB and OB are at right-angles to each other, the locus of B must be a semicircle on OV.

Join IA, as shown dotted in fig. 7.47.

For triangles IOA and VOB,

$$\angle \text{IOA} = 90° - \angle \text{VOB} = \angle \text{OVB}.$$

Also,

$$\frac{\text{OV}}{\text{BV}} = \frac{V}{XI} = \frac{V}{X} \cdot \frac{1}{I} = \frac{\text{OA}}{\text{OI}}.$$

Hence the two triangles are similar,

so that $\qquad \angle \text{OIA} = \angle \text{OBV} = 90°,$

and the locus of the extremity of the current phasor must therefore be a semicircle on OA. Also, it is evident from fig. 7.47 that the power component of the current is a maximum when the phase difference between the current and the supply voltage is 45°, i.e. when the value of the resistance is equal to that of the reactance.

7.25 Current locus for a circuit having constant resistance and variable inductance

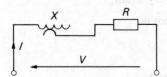

Fig. 7.48 Constant
resistance and variable
inductance in series

The circuit and phasor diagrams are given in figs. 7.48 and 7.49 respectively.

When $X = 0$, current $= V/R$ and is represented by OA in phase with OV.

If OI represents the current for the general case of X having some finite value, then OB and OC represent the corresponding voltages across R and X respectively. Since $\angle$ OBV is a right-angle, the locus of B is a semicircle on OV; and since OI is proportional to and in phase with OB, the locus of the extremity of the current phasor must be a semicircle on OA.

7.26 Imperfect capacitors and loss angle

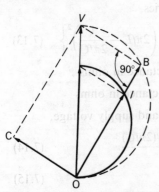

Fig. 7.49 Phasor diagram
for fig. 7.48

Losses in capacitors having a solid dielectric are due to:
(a) leakage current through the dielectric and along surface

paths between the terminals, (*b*) energy absorbed by the dielectric when subjected to an alternating electric field — analogous to hysteresis loss in magnetic materials. This effect is referred to as *dielectric hysteresis*.

An imperfect capacitor can be represented by resistance R_p in *parallel* with capacitor C_p as in fig. 7.50(*a*), or by resistance R_s in *series* with capacitor C_s, as in fig. 7.50(*c*). The values of

Fig. 7.50 Circuit and phasor diagrams for an imperfect capacitor

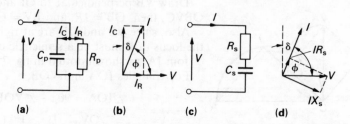

(a)   (b)   (c)   (d)

R_p and R_s must be such that the loss in each is equal to the total loss in the imperfect capacitor. The respective phasor diagrams are given in figs. 7.50(*b*) and (*d*). The angle by which the angle of lead of the current falls short of 90° is termed the *loss angle* and is represented by the Greek letter δ.

For the equivalent parallel circuit (fig. 7.50(*a*)),

$$\tan \delta = \frac{I_R}{I_c} = \frac{V}{R_p} \times \frac{1}{2\pi f C_p V} = \frac{1}{2\pi f C_p R_p}$$

and for the equivalent series circuit (fig. 7.50(*c*)),

$$\tan \delta = IR_s / IX_s = 2\pi f C_s R_s.$$

If δ is expressed in radians and is very small,

$$\tan \delta \simeq \delta \simeq \cos \phi.$$

Summary of important formulae

For a circuit with *R*, *L* and *C* in series,

$$\text{Impedance} = Z = \sqrt{\left\{ R^2 + \left(2\pi f L - \frac{1}{2\pi f C} \right)^2 \right\}} \quad (7.13)$$

where $2\pi f L$ = inductive reactance in ohms

and $1/2\pi f C$ = capacitive reactance in ohms

If ϕ = difference between current and supply voltage,

$$\tan \phi = \frac{2\pi f L - 1/(2\pi f C)}{R} \quad (7.14)$$

$$\cos \phi = R/Z \quad (7.15)$$

$$\sin \phi = \frac{2\pi f L - 1/(2\pi f C)}{Z} \quad (7.16)$$

For purely resistive circuit,

$$I = V/R \quad \text{and} \quad \phi = 0.$$

For purely inductive circuit,

$$I = V/2\pi fL$$

and $\qquad \phi = 90°$, current lagging.

For purely capacitive circuit,

$$I = 2\pi fCV$$

and $\qquad \phi = 90°$, current leading.

For resonance in a series circuit,

$$f = \frac{1}{2\pi\sqrt{(LC)}} \tag{7.17}$$

and $\qquad Q$ factor $= \dfrac{2\pi fL}{R} = \dfrac{1}{R}\sqrt{\left(\dfrac{L}{C}\right)}$ (7.19)

For resonance in a parallel circuit,

$$C = \frac{L}{R^2 + (2\pi fL)^2}$$

and $\qquad f = \dfrac{1}{2\pi\sqrt{(LC)}}$ when $R \ll 2\pi fL$ (7.23)

$$\left.\begin{array}{l}\text{Dynamic impedance of}\\ \text{resonant parallel circuit}\end{array}\right\} = \frac{L}{CR} \tag{7.24}$$

In an a.c. circuit,

$$\text{power factor} = \frac{\text{power in watts}}{\text{r.m.s. volts} \times \text{r.m.s. amperes}}$$

$$= \frac{\text{kilowatts}}{\text{kilovoltamperes}} \tag{7.26}$$

or $\qquad$ power in watts $= VI \times$ power factor (7.27)

For sinusoidal voltage and current having phase difference ϕ,

$$\text{power factor} = \cos\phi \tag{7.28}$$

and $\qquad$ power (in watts) $= VI\cos\phi$ (7.25)

Active or power component of current $= I\cos\phi$ (7.29)

Reactive or wattless component of current $= I\sin\phi$ (7.30)

Reactive power (in vars) $= VI\sin\phi$ (7.31)

EXERCISES 7

Note. In questions reproduced from examination papers, the term 'vector' has been replaced by 'phasor' by kind permission of the respective institutions.

1. A closed-circuit, 500-turn coil, of resistance $100\,\Omega$ and negligible inductance, is wound on a square frame of 40 cm side. The frame is pivoted at the mid-points of two opposite sides and is rotated at 250 r/min in a uniform magnetic field of 60 mT. The field direction is at right-angles to the axis of rotation.

   For the instant that the e.m.f. is maximum: (a) draw a diagram of the coil, and indicate the direction of rotation, and of the current flow, the magnetic flux, the e.m.f. and the force exerted by the magnetic field on the conductors; (b) calculate the e.m.f., the current and the torque, and hence verify that the mechanical power supplied balanced the electric power produced. (S.A.N.C., O.2)

2. An alternating p.d. of 100 V (r.m.s.), at 50 Hz, is maintained across a $20\text{-}\Omega$ non-reactive resistor. Plot to scale the waveforms of p.d. and current over one cycle. Deduce the curve of power and state its mean value.

3. An inductor having a reactance of $10\,\Omega$ and negligible resistance is connected to a 100-V (r.m.s.) supply. Draw to scale for one half-cycle, curves of voltage and current, and deduce and plot the power curve. What is the mean power over the half-cycle?

4. A coil having an inductance of 0.2 H and negligible resistance is connected across a 100-V a.c. supply. Calculate the current when the frequency is: (a) 30 Hz, and (b) 500 Hz.

5. A coil of inductance 0.1 H and negligible resistance is connected in series with a $25\text{-}\Omega$ resistor. The circuit is energized from a 250-V, 50-Hz source. Calculate: (a) the current in the circuit; (b) the p.d. across the coil; (c) the p.d. across the resistor; (d) the phase angle of the circuit. Draw to scale a phasor diagram representing the current and the component voltages. (U.L.C.I., O.1)

6. A coil connected to a 250-V, 50-Hz sinusoidal supply takes a current of 10 A at a phase angle of $30°$. Calculate the resistance and inductance of, and the power taken by the coil.

   Draw, for one half-cycle, curves of voltage and current and deduce and plot the power curve. Comment on the power curve. (S.A.N.C., O.1)

7. A $15\text{-}\Omega$ non-reactive resistor is connected in series with a coil of inductance 0.08 H and negligible resistance. The combined circuit is connected to a 240-V, 50-Hz supply. Calculate: (a) the reactance of the coil; (b) the impedance of the circuit; (c) the current in the circuit; (d) the power factor of the circuit; (e) the active power absorbed by the circuit. (N.C.T.E.C., O.1)

8. The potential difference measured across a coil is 20 V when a direct current of 2 A is passed through it. With an alternating current of 2 A at 40 Hz, the p.d. across the coil is 140 V. If the coil is connected to a 230-V, 50-Hz supply, calculate: (a) the current; (b) the active power; (c) the power factor. (S.A.N.C., O.1)

9. A non-inductive load takes a current of 15 A at 125 V. An inductor is then connected in series in order that the same current shall be supplied from 240-V, 50-Hz mains. Ignore the resistance of the inductor and calculate: (a) the inductance of the inductor; (b) the impedance of the circuit; (c) the phase difference between the current and the applied voltage. Assume the waveform to be sinusoidal. (U.L.C.I., O.1)

10. A series a.c. circuit, ABCD, consists of a resistor AB, an inductor BC, of resistance R and inductance L, and a resistor CD. When a current of 6.5 A flows through the circuit, the voltage drops across various points are: $V_{AB} = 65\,V$; $V_{BC} = 124\,V$; $V_{AC} = 149\,V$. The supply voltage is 220 V at 50 Hz.

Draw a phasor diagram to scale showing all the resistive and reactive volt-drops and, from the diagram, determine: (*a*) the volt-drop V_{BD} and the phase angle between it and the current; (*b*) the resistance and inductance of the inductor.

11. A coil of 0.5 H inductance and negligible resistance and a 200-Ω resistor are connected in series to a 50-Hz supply. Calculate the circuit impedance.

    An inductor in a radio set has to have a reactance of 11 kΩ at a frequency of 1.5 MHz. Calculate the inductance (in millihenrys). (N.C.T.E.C., O.2)

12. A coil takes a current of 10.0 A and dissipates 1410 W when connected to a 200-V, 50-Hz sinusoidal supply. When another coil is connected in parallel with it, the total current taken from the supply is 20.0 A at a power factor of 0.866.

    Determine the current and the overall power factor when the coils are connected in series across the same supply. (S.A.N.C., O.2)

13. When a steel-cored reactor and a non-reactive resistor are connected in series to a 150-V a.c. supply, a current of 3.75 A flows in the circuit. The potential differences across the reactor and across the resistor are then observed to be 120 V and 60 V respectively. If the d.c. resistance of the reactor is 4.5 Ω, determine the core loss in the reactor and calculate its equivalent series resistance. (N.C.T.E.C., O.2)

14. A single-phase network consists of three parallel branches, the currents in the respective branches being represented by:

$$i_1 = 20 \sin 314t \text{ amperes,}$$

$$i_2 = 30 \sin (314t - \pi/4) \text{ amperes,}$$

and $\qquad i_3 = 18 \sin (314t + \pi/2) \text{ amperes.}$

(*a*) Using a scale of 1 cm = 5 A, draw a phasor diagram and find the total maximum value of current taken from the supply and the overall phase angle.

(*b*) Express the total current in a form similar to that of the branch currents.

(*c*) If the supply voltage is represented by 200 sin 314t volts, find the impedance, resistance and reactance of the network. (W.J.E.C., O.1)

15. A non-inductive resistor is connected in series with a coil across a 230-V, 50-Hz supply. The current is 1.8 A and the potential differences across the resistor and the coil are 80 V and 170 V respectively. Calculate the inductance and the resistance of the coil, and the phase difference between the current and the supply voltage. Also draw the phasor diagram representing the current and the voltages. (App. El., L.U.)

16. An inductive circuit, in parallel with a non-inductive resistor of 20 Ω, is connected across a 50-Hz supply. The currents through the inductive circuit and the non-inductive resistor are 4.3 A and 2.7 A respectively, and the current taken from the supply is 5.8 A. Find (*a*) the active power absorbed by the inductive branch, (*b*) its inductance and (*c*) the power factor of the combined network. Sketch the phasor diagram. (App. El., L.U.)

17. A coil having a resistance of 15 Ω and an inductance of 0.2 H is connected in series with another coil having a resistance of 25 Ω and an inductance of 0.04 H to a 230-V, 50-Hz supply.

    Draw to scale the complete phasor diagram for the circuit and determine: (*a*) the voltage across each coil; (*b*) the active power dissipated in each coil; (*c*) the power factor of the circuit as a whole. (App. El., L.U.)

18. Two identical coils, each of 25-Ω resistance, are mounted coaxially a short distance apart. When one coil is supplied at 100 V, 50 Hz, the current taken is 2.1 A and the e.m.f. induced in the other coil on open-circuit is 54 V. Calculate the self inductance of each coil and the mutual inductance between them.

    What current will be taken if a p.d. of 100 V, 50 Hz, is supplied across the two coils in series? (App. El., L.U.)

19. Two similar coils have a coupling coefficient of 0.6. Each coil has a resistance of 8 Ω and a self inductance of 2 mH. Calculate the current and the power factor of the circuit when the coils are connected in series (a) cumulatively and (b) differentially, across a 10-V, 5-kHz supply.

20. A two-wire cable, 8 km long, has a capacitance of 0.3 μF/km. If the cable is connected to a 11-kV, 60-Hz supply, calculate the value of the charging current. The resistance and inductance of the conductors may be neglected.

21. Draw, to scale, phasors representing the following voltages, taking e_1 as the reference phasor:

$$e_1 = 80 \sin \omega t \text{ volts}; \quad e_2 = 60 \cos \omega t \text{ volts};$$

$$e_3 = 100 \sin (\omega t - \pi/3) \text{ volts}.$$

    By phasor addition, find the sum of these three voltages and express it in the form of $E_m \sin (\omega t \pm \phi)$.

    When this resultant voltage is applied to a circuit consisting of a 10-Ω resistor and a capacitor of 17.3 Ω reactance connected in series find an expression for the instantaneous value of the current flowing, expressed in the same form.

(N.C.T.E.C., O.2)

22. In order to use three 100-V, 60-W lamps on a 230-V, 50-Hz supply, they are connected in parallel and a capacitor is connected in series with the group. Find: (a) the capacitance required to give the correct voltage across the lamps; (b) the power factor of the network.

    If one of the lamps is removed, to what value will the voltage across the remaining two rise, assuming that their resistances remain unchanged? (U.L.C.I., O.2)

23. A 130-Ω resistor and a 30-μF capacitor are connected in parallel across a 200-V, 50-Hz supply. Calculate: (a) the current in each branch; (b) the resultant current; (c) the phase difference between the resultant current and the applied voltage; (d) the active power; and (e) the power factor. Sketch the phasor diagram.

24. A resistor and a capacitor are connected in series across a 150-V a.c. supply. When the frequency is 40 Hz the current is 5 A, and when the frequency is 50 Hz the current is 6 A. Find the resistance and capacitance of the resistor and capacitor respectively.

    If they are now connected in parallel across the 150-V supply, find the total current and its power factor when the frequency is 50 Hz. (U.L.C.I., O.2)

25. A series circuit consists of a non-inductive resistor of 10 Ω, an inductor having a reactance of 50 Ω and a capacitor having a reactance of 30 Ω. It is connected to a 230-V a.c. supply. Calculate: (a) the current; (b) the active power; (c) the power factor; (d) the voltage across each component. Draw to scale a phasor diagram showing the supply voltage and current and the voltage across each component. (N.C.T.E.C., O.2)

26. A coil having a resistance of 20 Ω and an inductance of 0.15 H is connected in series with a 100-μF capacitor across a 230-V,

50-Hz supply. Calculate: (*a*) the active and reactive components of the current; (*b*) the voltage across the coil; and (*c*) the power factor of the circuit. Sketch the phasor diagram.

27. A p.d. of 100 V at 50 Hz is maintained across a series circuit having the following characteristics: $R = 10\,\Omega$, $L = 100/\pi\,\text{mH}$, $C = 500/\pi\,\mu\text{F}$. Draw the phasor diagram and calculate: (*a*) the current; (*b*) the power factor of the circuit; (*c*) the active and reactive components of the current; (*d*) the reactive power (in vars).

28. A network consists of three branches in parallel. Branch A is a 10-Ω resistor, branch B is a coil of resistance $4\,\Omega$ and inductance 0.02 H, and branch C is an 8-Ω resistor in series with a 200-μF capacitor. The combination is connected to a 100-V, 50-Hz supply.

    Find the various branch currents and then, by resolving into in-phase and quadrature components, determine the total current taken from the supply and its power factor. A phasor diagram showing the relative positions of the various circuit quantities should accompany your solution. It need not be drawn to scale. (E.M.E.U., O.2)

29. A 50-Hz steel-cored reactor has an impedance of $10\,\Omega$ and a Q factor of 3 at 50 Hz. If the reactor is connected to a 25-V, 50-Hz supply, calculate the power loss.

    What value of capacitance must be connected in series with the reactor so that the current lags by $45°$ with respect to the supply voltage? (N.C.T.E.C., O.2)

30. A coil, having a resistance of $20\,\Omega$ and an inductance of 0.0382 H, is connected in parallel with a circuit consisting of a 150-μF capacitor in series with a 10-Ω resistor. The arrangement is connected to a 240-V, 50-Hz supply. Determine the current in each branch and, sketching a phasor diagram, the total supply current, power factor and active power.
    (E.M.E.U., O.2)

31. A 31.8-μF capacitor, a 127.5-mH inductor of resistance $30\,\Omega$ and a 100-Ω resistor are all connected in parallel to a 200-V, 50-Hz supply. Calculate the current in each branch. Draw a phasor diagram to scale to show these currents. Find the total current and its phase angle by drawing or otherwise.
    (N.C.T.E.C., O.2)

32. A 200-V, 50-Hz sinusoidal supply is connected to parallel network comprising three branches A, B and C, as follows:

    A — a coil of resistance $3\,\Omega$ and inductive reactance $4\,\Omega$,

    B — a series circuit of resistance $4\,\Omega$ and capacitive reactance $3\,\Omega$,

    C — a capacitor.

    Given that the power factor of the combined circuit is unity, find: (*a*) the capacitance of the capacitor in microfarads; (*b*) the current taken from the supply; (*c*) the active power absorbed.
    (S.A.N.C., O.2)

33. Two circuits, A and B, are connected in parallel to a 115-V, 50-Hz supply. The total current taken by the combination is 10 A at unity power factor. Circuit A consists of a 10-Ω resistor and a 200-μF capacitor connected in series; circuit B consists of a resistor and an inductive reactor in series.

    Determine the following data for circuit B: (*a*) the current; (*b*) the power factor; (*c*) the impedance; (*d*) the resistance; (*e*) the reactance. (N.C.T.E.C., O.2)

34. A parallel network consists of two branches A and B. Branch A has a resistance of $10\,\Omega$ and an inductance of 0.1 H in series. Branch B has a resistance of $20\,\Omega$ and a capacitance of $100\,\mu\text{F}$

in series. The network is connected to a single-phase supply of 250 V at 50 Hz.

Calculate the magnitude and phase angle of the current taken from the supply. Verify your answers by measurement from a phasor diagram drawn to scale. (U.L.C.I., O.2)

35. A series circuit comprises an inductor, of resistance 10 Ω and inductance 159 μH, and a variable capacitor connected to a 50-mV sinusoidal supply of frequency 1 MHz. What value of capacitance will result in resonant conditions and what will then be the current?

For what values of capacitance will the current at this frequency be reduced to 10 per cent of its value at resonance? (S.A.N.C., O.2)

36. A circuit consists of a 10-Ω resistor, a 30-mH inductor and a 1-μF capacitor, and is supplied from a 10-V variable-frequehcy source. Find the frequency for which the voltage developed across the capacitor is a maximum and calculate the magnitude of this voltage. (E.M.E.U., O.2)

37. Calculate the voltage magnification created in a resonant circuit connected to a 240-V, a.c. supply consisting of an inductor having inductance 0.1 H and resistance 2 Ω in series with a 100-μF capacitor. Explain the effects of increasing the above resistance value. (W.J.E.C., O.2)·

38. A series circuit consists of 0.5-μF capacitor, a coil of inductance 0.32 H and resistance 40 Ω and a 20-Ω non-inductive resistor. Calculate the value of the resonant frequency of the circuit.

When the circuit is connected to a 30-V a.c. supply at this resonant frequency, determine: (a) the p.d. across each of the three components; (b) the current flowing in the circuit; (c) the active power absorbed by the circuit. (N.C.T.E.C., O.2)

39. An e.m.f. whose instantaneous value at time t is given by 283 sin (314t + π/4) volts is applied to an inductive circuit and the current in the circuit is 5.66 sin(314t − π/6) amperes. Determine (a) the frequency of the e.m.f.; (b) the resistance and inductance of the circuit; (c) the active power absorbed.

If series capacitance is added so as to bring the circuit into resonance at this frequency and the above e.m.f. is applied to the resonant circuit, find the corresponding expression for the instantaneous value of the current. Sketch a phasor diagram for this conditions.

Explain why it is possible to have a much higher voltage across a capacitor than the supply voltage in a series circuit. (U.L.C.I., O.2)

40. A coil, of resistance R and inductance L, is connected in series with a capacitor C across a variable-frequency source. The voltage is maintained constant at 300 mV and the frequency is varied until a maximum current of 5 mA flows through the circuit at 6 kHz.

If, under these conditions, the Q factor of the circuit is 105, calculate: (a) the voltage across the capacitor; (b) the values of R, L and C. (U.E.I., O.2)

41. A constant voltage at a frequency of 1 MHz is maintained across a circuit consisting of an inductor in series with a variable capacitor. When the capacitor is set to 300 pF, the current has its maximum value. When the capacitance is reduced to 284 pF, the current is 0.707 of its maximum value. Find: (a) the inductance and the resistance of the inductor, and (b) the Q factor of the inductor at 1 MHz. Sketch the phasor diagram for each condition.

42. A coil of resistance 12 Ω and inductance 0.12 H is connected in

parallel with a 60-μF capacitor to a 100-V, variable-frequency supply. Calculate the frequency at which the circuit will behave as a non-reactive resistor, and also the value of the dynamic impedance. Draw for this condition the complete phasor diagram. (W.J.E.C., O.2)

43. Calculate, from first principles, the impedance at resonance of a circuit consisting of a coil of inductance 0.5 mH and effective resistance 20 Ω in parallel with a 0.0002-μF capacitor.
(N.C.T.E.C., O.2)

44. A coil has resistance of 400 Ω and inductance of 318 μH. Find the capacitance of a capacitor which, when connected in parallel with the coil, will produce resonance with a supply frequency of 1 MHz.
If a second capacitor of capacitance 23.5 pF is connected in parallel with the first capacitor, find the frequency at which resonance will occur. (S.A.N.C., O.2)

45. A single-phase motor takes 8.3 A at a power factor of 0.866 lagging when connected to a 230-V, 50-Hz supply. Two similar capacitors are connected in parallel with each other to form a capacitance bank. This capacitance bank is now connected in parallel with the motor to raise the power factor to unity. Determine the capacitance of each capacitor. (E.M.E.U., O.2)

46. (a) A single-phase load of 5 kW operates at a power factor of 0.6 lagging. It is proposed to improve this power factor to 0.95 lagging by connecting a capacitor across the load. Calculate the kVA rating of the capacitor.
(b) Give reasons why it is to a consumer's economic advantage to improve his power factor with respect to the supply, and explain the fact that the improvement is rarely made to unity in practice. (W.J.E.C., O.2)

47. A 25-kVA single-phase motor has a power factor of 0.8 lag. A 10-kVA capacitor is connected for power-factor correction. Calculate the input apparent power in kVA taken from the mains and its power factor when the motor is (a) on half load, (b) on full load. Sketch a phasor diagram for each case.
(N.C.T.E.C., O.2)

48. A single-phase motor takes 50 A at a power factor of 0.6 lagging from a 250-V, 50-Hz supply. What value of capacitance must a shunting capacitor have to raise the overall power factor to 0.9 lagging?
How does the installation of the capacitor affect the line and motor currents? (S.A.N.C., O.2)

49. A 240-V, single-phase supply feeds the following loads:
(a) incandescent lamps taking a current of 8 A at unity power factor; (b) fluorescent lamps taking a current of 5 A at 0.8 leading power factor; (c) a motor taking a current of 7 A at 0.75 lagging power factor. Sketch the phasor diagram and determine the total current, active power and reactive power taken from the supply and the overall power factor. (U.L.C.I., O.2)

50. The load taken from an a.c. supply consists of: (a) a heating load of 15 kW; (b) a motor load of 40 kVA at 0.6 power factor lagging; (c) a load of 20 kW at 0.8 power factor lagging.
Calculate the total load from the supply (in kW and kVA) and its power factor. What would be the kvar rating of a capacitor to bring the power factor to unity and how would the capacitor be connected? (U.E.I., O.2)

51. A coil of resistance 2 Ω and reactance 5 Ω is connected in series with a non-reactive resistor which is continuously variable between 0 and 8 Ω. Draw the current locus diagram when the circuit is connected to a 200-V supply, and hence find the values

of the maximum and minimum currents and the corresponding power factors and active powers. Find also the maximum active power.
(W.J.E.C., O.2)

52. A circuit having a constant resistance of 60 Ω and a variable inductance of 0–0.4 H is connected across a 100-V, 50-Hz supply. Derive from first principles the locus of the extremity of the current phasor. Find (*a*) the active power and (*b*) the inductance of the circuit when the power factor is 0.8.
(App. El., L.U.)

53. An a.c. circuit consists of a variable resistor in series with a coil for which $R = 20\,\Omega$ and $L = 0.1\,H$. Show that when this circuit is supplied at constant voltage and frequency, and the resistance is varied between zero and infinity, the locus of the current phasor is a circular arc. Calculate, when the supply voltage is 100 V and the frequency 50 Hz: (*a*) the radius (in amperes) of this arc; (*b*) the resistance of the variable resistor in order that the power taken from the mains may be a maximum.
(App. El., L.U.)

54. A variable capacitor and a resistor of 200 Ω are connected in series across a 200-V, 50-Hz supply. Draw a phasor locus diagram showing the variation of the current as the capacitance is changed from $10\,\mu F$ to $100\,\mu F$. Find: (*a*) the capacitance for a current of 0.8 A; (*b*) the current when the capacitance is $50\,\mu F$.

55. The equivalent series of circuit for a certain capacitor consists of a 2-Ω resistor in series with a 200-pF capacitor. Calculate the loss angle of the capacitor at a frequency of 5 MHz.
(U.L.C.I., O.2)

56. A certain capacitor has a loss angle of 0.03 radian and when it is connected across a 6.6-kV, 50-Hz supply, the power loss is 25 W. Calculate the component values of the equivalent parallel circuit.
(U.L.C.I., O.2)

8 Complex Notation

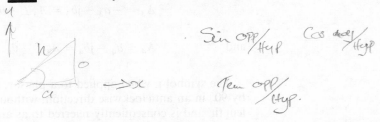

8.1 The j operator

In Chapter 7, problems on a.c. circuits were solved with the aid of phasor diagrams. So long as the circuits are fairly simple, this method is satisfactory; but with a more involved circuit, such as that of fig. 8.17, the calculation can be simplified by using complex algebra. This system enables equations representing alternating voltages and currents and their phase relationships to be expressed in simple algebraic form. It is based upon the idea that a phasor can be resolved into two components at right-angles to each other. For instance, in fig. 8.1(a), phasor OA can be resolved into components OB along the X axis and OC along the Y axis, where $OB = OA \cos \theta$ and $OC = OA \sin \theta$. It would obviously be incorrect to state that $OA = OB + OC$, since OA is actually $\sqrt{(OB^2 + OC^2)}$; but by introducing a symbol j to denote that OC is the component along the Y axis, we can represent the phasor thus:

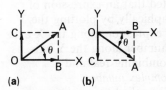

(a)   **(b)**

Fig. 8.1 Resolution of phasors

$$\mathbf{OA}^* = OB + jOC = OA(\cos \theta + j \sin \theta).$$

The phasor OA may alternatively be expressed thus:

$$\mathbf{OA} = OA \angle \theta.$$

If OA is occupying the position shown in fig. 8.1(b), the vertical component is negative, so that

$$\mathbf{OA} = OB - jOC = OA \angle -\theta.$$

Fig. 8.2 represents four phasors occupying different quadrants. These phasors can be represented thus:

$$\mathbf{A}_1 = a_1 + jb_1 = A_1 \angle \theta_1 \qquad (8.1)$$

where $A_1 = \sqrt{(a_1^2 + b_1^2)}$ and $\tan \theta_1 = b_1/a_1$.

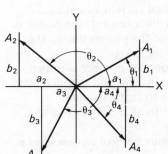

Fig. 8.2 Resolution of phasors

---

*Symbols representing phasors are printed in boldface, while those representing only the magnitudes are printed in italics, thus:

$\mathbf{V}, \mathbf{I}, \mathbf{Z}$ phasors or complex numbers,
V, I, Z magnitudes or moduli.

In manuscript, a phasor may be indicated by a horizontal bar above the symbol, and the magnitude or modulus by a vertical line on each side, thus: $|V|$.

Similarly
$$A_2 = -a_2 + jb_2 = A_2 \angle \theta_2$$
$$A_3 = -a_3 - jb_3 = A_3 \angle -\theta_3$$

and
$$A_4 = a_4 - jb_4 = A_4 \angle -\theta_4.$$

The symbol j, when applied to a phasor, alters its direction by 90° in an anticlockwise direction, without altering its length, and is consequently referred to as an *operator*. For example, if we start with a phasor A in phase with the X axis, as in fig. 8.3, then jA represents a phasor of the same length upwards along the Y axis. If we apply the operator j to jA, we turn the phasor anticlockwise through another 90°, thus giving jjA or j^2A in fig. 8.3. The symbol j^2 signifies that we have applied the operator j twice in succession, thereby rotating the phasor through 180°. This reversal of the phasor is equivalent to multiplying by -1; i.e. $j^2A = -A$, so that j^2 can be regarded as being numerically equal to -1 and $j = \sqrt{-1}$.

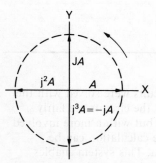

Fig. 8.3 Significance of the j operator

When mathematicians were first confronted by an expression such as $A = 3 + j4$, they thought of $j4$ as being an imaginary number rather than a real number; and it was Argand, in 1806, who first suggested that an expression of this form could be represented graphically by plotting the 3 units of *real* number in the above expression along the X axis and the 4 units of *imaginary* or j number along the Y axis, as in fig. 8.4. This type of number, combining real and imaginary numbers, is termed a *complex number*.

The term *imaginary*, though still applied to numbers containing j, such as $j4$ in the above expression, has long since lost its meaning of unreality, and the terms *real* and *imaginary* have become established as technical terms, like *positive* and *negative*. For instance, if a current represented by $I = 3 + j4$ amperes were passed through a resistor of, say, $10\,\Omega$, the power due to the 'imaginary' component of the current would be $4^2 \times 10 = 160\,W$, in exactly the same way as that due to the 'real' component would be $3^2 \times 10 = 90\,W$. From fig. 8.4 it is seen that the actual current would be 5 A. When this current flows through a $10\text{-}\Omega$ resistor, the power is $5^2 \times 10 = 250\,W$, namely the sum of the powers due to the real and imaginary components of the current.

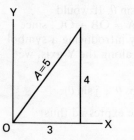

Fig. 8.4 Representation of $A = 3 + j4$

Since the real component of a complex number is drawn along the reference axis, namely the X axis, and the imaginary component is drawn at right-angles to that axis, these components are sometimes referred to as the *in-phase* and *quadrature* components respectively.

We can now summarize the various ways of representing a complex number algebraically:

$$A = a + jb \text{ (rectangular or Cartesian notation)},$$
$$= A(\cos \theta + j \sin \theta) \text{ (trigonometric notation)},$$
$$= A \angle \theta \text{ (polar notation)}.$$

8.2 Addition and subtraction of phasors

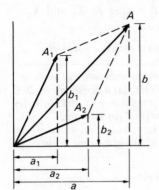

Fig. 8.5 Addition of phasors

Suppose $\mathbf{A}_1$ and $\mathbf{A}_2$ in fig. 8.5 to be two phasors to be added together. From this phasor diagram, it is evident that:

$$\mathbf{A}_1 = a_1 + jb_1 \quad \text{and} \quad \mathbf{A}_2 = a_2 + jb_2.$$

It was shown in section 8.6 that the resultant of $\mathbf{A}_1$ and $\mathbf{A}_2$ is given by $\mathbf{A}$, the diagonal of the parallelogram drawn on $\mathbf{A}_1$ and $\mathbf{A}_2$. If a and b be the real and imaginary components respectively of $\mathbf{A}$, then $\mathbf{A} = a + jb$.

But it is evident from fig. 8.5 that:

$$a = a_1 + a_2 \quad \text{and} \quad b = b_1 + b_2$$

$$\therefore \quad \mathbf{A} = a_1 + a_2 + j(b_1 + b_2)$$

$$= (a_1 + jb_1) + (a_2 + jb_2)$$

$$= \mathbf{A}_1 + \mathbf{A}_2.$$

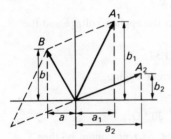

Fig. 8.6 Subtraction of phasors

Fig. 8.6 shows the construction for subtracting phasor $\mathbf{A}_2$ from phasor $\mathbf{A}_1$. If $\mathbf{B}$ is the phasor difference of these quantities and if a_1 is assumed less than a_2, then the real component of $\mathbf{B}$ is negative,

$$\therefore \quad \mathbf{B} = -a + jb = a_1 - a_2 + j(b_1 - b_2)$$

$$= (a_1 + jb_1) - (a_2 + jb_2) = \mathbf{A}_1 - \mathbf{A}_2.$$

8.3 Voltage, current and impedance

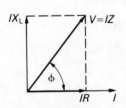

Fig. 8.7 Phasor diagram for R and L in series

Let us first consider a simple circuit possessing resistance R in *series* with an inductive reactance X_L. The phasor diagram is given in fig. 8.7, where the current phasor is taken as the reference quantity and is therefore drawn along the X axis,

i.e. $$\mathbf{I} = I + j0 = I \angle 0°.$$

From fig. 8.7 it is evident that:

$$\mathbf{V} = \mathbf{I}R + j\mathbf{I}X_L = \mathbf{I}(R + jX_L) = \mathbf{I}Z^* = \mathbf{I}Z \angle \phi$$

where $\mathbf{Z} = R + jX_L = Z \angle \phi$ and $\tan \phi = X_L/R$ (8.2)

Fig. 8.8 gives the phasor diagram for a circuit having a resistance R in series with a capacitive reactance X_C. From this diagram:

$$\mathbf{V} = \mathbf{I}R - j\mathbf{I}X_C = \mathbf{I}(R - jX_C) = \mathbf{I}Z = \mathbf{I}Z \angle -\phi$$

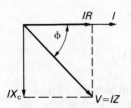

Fig. 8.8 Phasor diagram for R and C in series

---

* In the expression $\mathbf{V} = \mathbf{I}\mathbf{Z}$, $\mathbf{V}$ and $\mathbf{I}$ differ from $\mathbf{Z}$ in that they are associated with time-varying quantities, whereas $\mathbf{Z}$ is a complex number independent of time and is therefore not a phasor in the sense that $\mathbf{V}$ and $\mathbf{I}$ are phasors. It has, however, become the practice to refer to all complex numbers used in a.c. circuit calculations as phasors.

where

$$\mathbf{Z} = R - jX_C = Z \angle -\phi \quad \text{and} \quad \tan \phi = -X_C/R \quad (8.3)$$

Hence for the general case of a circuit having R, X_L and X_C in series,

$$\mathbf{Z} = R + j(X_L - X_C) \quad \text{and} \quad \tan \phi = (X_L - X_C)/R$$

Example 8.1   *Express in rectangular and polar notations, the impedance of each of the following circuits at a frequency of 50 Hz: (a) a resistance of 20 Ω in series with an inductance of 0.1 H; (b) a resistance of 50 Ω in series with a capacitance of 40 μF; (c) circuits (a) and (b) in series.*

If the terminal voltage is 230 V at 50 Hz, calculate the value of the current in each case and the phase of each current relative to the applied voltage.

(a) For 50 Hz, $\omega = 2\pi \times 50 = 314$ rad/s,

$$\therefore \qquad \mathbf{Z} = 20 + j314 \times 0.1 = 20 + j31.4 \, \Omega$$

Hence $\qquad Z = \sqrt{\{(20)^2 + (31.4)^2\}} = 37.2 \, \Omega$

and $\qquad I = 230/37.2 = 6.18$ A.

If ϕ be the phase difference between the applied voltage and the current,

$$\tan \phi = 31.4/20 = 1.57,$$

$$\therefore \qquad \phi = 57° \, 30', \text{ current lagging}.$$

The impedance can also be expressed:

$$\mathbf{Z} = 37.2 \angle 57° \, 30'^* \, \Omega.$$

If the applied voltage be taken as the reference quantity, then

$$\mathbf{V} = 230 \angle 0° \text{ volts},$$

$$\therefore \qquad \mathbf{I} = \frac{230 \angle 0°}{37.2 \angle 57° \, 30'} = 6.18 \angle -57° \, 30' \text{ A}.$$

(b) $\qquad \mathbf{Z} = 50 - j\dfrac{10^6}{314 \times 40} = 50 - j79.6 \, \Omega$

$$\therefore \qquad Z = \sqrt{\{(50)^2 + (79.6)^2\}} = 94 \, \Omega,$$

and $\qquad I = 230/94 = 2.447$ A.

$$\tan \phi = -79.6/50 = -1.592,$$

$$\therefore \qquad \phi = 57° \, 52', \text{ current leading}.$$

The impedance can also be expressed thus:

$$\mathbf{Z} = 94 \angle -57° \, 52' \, \Omega$$

$$\therefore \qquad \mathbf{I} = \frac{230 \angle 0°}{94 \angle -57° \, 52'} = 2.447 \angle 57° \, 52' \text{ A}.$$

(c) $\qquad \mathbf{Z} = 20 + j31.4 + 50 - j79.6$

$$= 70 - j48.2 \, \Omega$$

$$\therefore \qquad Z = \sqrt{\{(70)^2 + (48.2)^2\}} = 85 \, \Omega,$$

---

*This form is more convenient than that involving the j term when it is required to find the product or the quotient of two complex numbers; thus, $A \angle \alpha \times B \angle \beta = AB \angle (\alpha + \beta)$ and $A \angle \alpha / B \angle \beta = (A/B) \angle (\alpha - \beta)$.

and $I = 230/85 = 2.706$ A.

$$\tan \phi = -48.2/70 = -0.689,$$

$$\therefore \qquad \phi = 34° 34', \text{ current leading}.$$

The impedance can also be expressed:

$$\mathbf{Z} = 85 \angle -34° 34' \, \Omega$$

so that $\qquad \mathbf{I} = \dfrac{230 \angle 0}{85 \angle -34° 34'} = 2.706 \angle 34° 34' \text{ A}.$

Example 8.2 *Calculate the resistance and the inductance or capacitance in series for each of the following impedances:* (a) $10 + \text{j}15\,\Omega$; (b) $-\text{j}80\,\Omega$; (c) $50 \angle 30° \, \Omega$; *and* (d) $120 \angle -60° \, \Omega$. *Assume the frequency to be 50 Hz.*

(a) For $\mathbf{Z} = 10 + \text{j}15 \, \Omega$, resistance $= 10 \, \Omega$,

and $\qquad\qquad\qquad$ inductive reactance $= 15 \, \Omega$,

$\therefore \qquad\qquad\qquad\qquad$ inductance $= 15/314 = 0.0478$ H.

(b) For $\mathbf{Z} = -\text{j}80 \, \Omega$, resistance $= 0$,

and $\qquad\qquad\qquad$ capacitive reactance $= 80 \, \Omega$,

$\therefore \qquad\qquad\qquad\qquad$ capacitance $= \dfrac{1}{314 \times 80} \text{F} = 39.8 \, \mu\text{F}$.

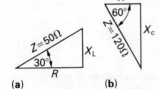

(a) $\qquad\qquad$ **(b)**

Fig. 8.9 Impedance triangles for example 8.2(c) and (d)

(c) Fig. 8.9(a) is an impedance triangle representing $50 \angle 30° \, \Omega$. From this diagram, it follows that the reactance is inductive and that $R = Z \cos \phi$ and $X_L = Z \sin \phi$,

$$\therefore \qquad \mathbf{Z} = R + \text{j}X_L = Z(\cos \phi + \text{j} \sin \phi)$$

$$= 50(\cos 30° + \text{j} \sin 30°) = 43.3 + \text{j}25 \, \Omega.$$

Hence, $\qquad\qquad$ resistance $= 43.3 \, \Omega$

and $\qquad\qquad$ inductive reactance $= 25 \, \Omega$,

so that $\qquad\qquad$ inductance $= 25/314 = 0.0796$ H.

(d) Fig. 8.9(b) is an impedance triangle representing $120 \angle -60° \, \Omega$. It will be seen that the reactance is capacitive, so that

$$\mathbf{Z} = R - \text{j}X_C = Z(\cos \phi - \text{j} \sin \phi)$$

$$= 120(\cos 60° - \text{j} \sin 60°) = 60 - \text{j}103.9 \, \Omega.$$

Hence, $\qquad\qquad$ resistance $= 60 \, \Omega$

and $\qquad$ capacitive reactance $= 103.9 \, \Omega$

$\therefore \qquad\qquad\qquad$ capacitance $= \dfrac{10^6}{314 \times 103.9} = 30.7 \, \mu\text{F}$.

8.4 Admittance, conductance and susceptance

When resistors having resistances R_1, R_2, etc., are in parallel, the equivalent resistance R is given by:

$$\frac{1}{R} = \frac{1}{R_1} + \frac{1}{R_2} + \cdots$$

In d.c. work the reciprocal of the resistance is known as *conductance* (section 1.8). It is represented by symbol G and the unit of conductance is the *siemen* (p. 9). Hence, if circuits having conductances G_1, G_2, etc., are in parallel, the total conductance G is given by:

$$G = G_1 + G_2 + \cdots$$

In a.c. work the conductance is the reciprocal of the resistance *only when the circuit possesses no reactance*. This matter is dealt with more fully in section 8.5.

If circuits having impedances Z_1, Z_2, etc., are connected in parallel across a supply voltage V, then:

$$I_1 = \frac{V}{Z_1}, \quad I_2 = \frac{V}{Z_2}, \quad \text{etc.}$$

If Z be the equivalent impedance of Z_1, Z_2, etc., in parallel and if I be the resultant current, then, using complex notation, we have:

$$\mathbf{I} = \mathbf{I}_1 + \mathbf{I}_2 + \cdots$$

$$\therefore \quad \frac{\mathbf{V}}{\mathbf{Z}} = \frac{\mathbf{V}}{\mathbf{Z}_1} + \frac{\mathbf{V}}{\mathbf{Z}_2} + \cdots$$

so that

$$\frac{1}{\mathbf{Z}} = \frac{1}{\mathbf{Z}_1} + \frac{1}{\mathbf{Z}_2} + \cdots \tag{8.4}$$

The reciprocal of impedance is termed *admittance* and is represented by the symbol Y, the unit being again the *siemens* (abbreviation, S). Hence, we may write expression (8.4) thus:

$$\mathbf{Y} = \mathbf{Y}_1 + \mathbf{Y}_2 + \cdots \tag{8.5}$$

It has already been shown that impedance can be resolved into a real component R and an imaginary component X, as in fig. 8.10(*a*). Similarly, an admittance may be resolved into a real component termed *conductance* and an imaginary component termed *susceptance*, represented by symbols G and B respectively as in fig. 8.10(*b*),

i.e. $\mathbf{Y} = G + jB$ and $\tan \phi = B/G$.

The significance of these terms will be more obvious when we consider their application to actual circuits.

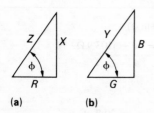

(a)   **(b)**

Fig. 8.10 Impedance and admittance triangles

8.5 Admittance of a circuit having resistance and inductive reactance in series

The phasor diagram for this circuit has already been given in fig. 8.7. From the latter it will be seen that the resultant voltage can be represented thus:

$$V = IR + jIX_L$$

$$\therefore \quad Z = \frac{V}{I} = R + jX_L$$

If Y be the admittance of the circuit, then:

$$Y = \frac{1}{Z} = \frac{1}{R + jX_L} = \frac{R - jX_L}{R^2 + X_L^2}^*$$

$$= \frac{R}{R^2 + X_L^2} - \frac{jX_L}{R^2 + X_L^2} = G - jB_L \qquad (8.6)$$

where $\quad G = \text{conductance} = \dfrac{R}{R^2 + X_L^2} = \dfrac{R}{Z^2} \qquad (8.7)$

and $\quad B_L = \text{inductive susceptance}$

$$= \frac{X_L}{R^2 + X_L^2} = \frac{X_L}{Z^2} \qquad (8.8)$$

From (8.7) it is evident that if the circuit has no reactance, i.e. if $X_L = 0$, then the conductance is $1/R$, namely the reciprocal of the resistance. Similarly, from (8.8) it follows that if the circuit has no resistance, i.e. if $R = 0$, the susceptance is $1/X_L$, namely the reciprocal of the reactance. In general, we define the *conductance* of a *series* circuit as the ratio of the resistance to the square of the impedance and the *susceptance* as the ratio of the reactance to the square of the impedance.

8.6 Admittance of a circuit having resistance and capacitive reactance in series

Fig. 8.8 gives the phasor diagram for this circuit. From this diagram it follows that:

$$V = IR - jIX_C$$

$$\therefore \quad Z = \frac{V}{I} = R - jX_C$$

and $\quad Y = \dfrac{1}{R - jX_C} = \dfrac{R + jX_C}{R^2 + X_C^2}$

$$= \frac{R}{R^2 + X_C^2} + \frac{jX_C}{R^2 + X_C^2} = G + jB_C \qquad (8.10)$$

---

*This method of transferring the j term from the denominator to the numerator is known as 'rationalizing'; thus,

$$\frac{1}{a + jb} = \frac{a - jb}{(a + jb)(a - jb)} = \frac{a - jb}{a^2 + b^2} \qquad (8.9)$$

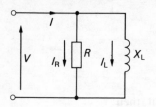

Fig. 8.11 *R and L in parallel*

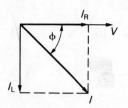

Fig. 8.12 *Phasor diagram for fig. 8.11*

where B_C = capacitive susceptance

$$= \frac{X_C}{R^2 + X_C^2} = \frac{X_C}{Z^2} \tag{8.11}$$

It will be seen that in the complex expression for an *inductive* circuit, the *impedance* has a *positive* sign in front of the imaginary component, whereas the imaginary component of the *admittance* is preceded by a negative sign. On the other hand, for a *capacitive* circuit, the imaginary component of the *impedance* has a *negative* sign and that of the *admittance* has a *positive* sign. Thus if the impedance of a circuit is represented by $(2 - j3)\,\Omega$, we know immediately that the circuit is capacitive; but if the admittance is $(2 - j3)\,S$ (siemens), the circuit must be inductive.

8.7 Admittance of a circuit having resistance and reactance in parallel

(a) Inductive reactance From the circuit and phasor diagrams of figs. 8.11 and 8.12 respectively, it follows that:

$$\mathbf{I} = I_R - jI_L = \frac{\mathbf{V}}{R} - \frac{j\mathbf{V}}{X_L}$$

$$\therefore \quad \mathbf{Y} = \frac{\mathbf{I}}{\mathbf{V}} = \frac{1}{R} - \frac{j}{X_L} = G - jB_L \tag{8.12}$$

(b) Capacitive reactance From figs. 8.13 and 8.14 it follows that:

$$\mathbf{I} = I_R + jI_C = \frac{\mathbf{V}}{R} + \frac{j\mathbf{V}}{X_C}$$

$$\therefore \quad \mathbf{Y} = \frac{\mathbf{I}}{\mathbf{V}} = \frac{1}{R} + \frac{j}{X_C} = G + jB_C \tag{8.13}$$

From expressions (8.12) and (8.13), it will be seen that if the admittance of a circuit is $(0.2 - j0.1)$ siemens, such a network can be represented as a resistance of $5\,\Omega$ in *parallel* with an inductive reactance of $10\,\Omega$; whereas if the impedance of a circuit is $(5 + j10)\,\Omega$, such a network can be represented as a resistance of $5\,\Omega$ in *series* with an inductive reactance of $10\,\Omega$.

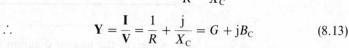

Fig. 8.13 *R and C in parallel*

Example 8.3   *Express in rectangular notation the admittance of circuits having the following impedances:* (a) $(4 + j6)\,\Omega$; (b) $20 \angle -30°\,\Omega$.

(a) $\mathbf{Z} = 4 + j6\,\Omega$,

$$\therefore \quad \mathbf{Y} = \frac{1}{4 + j6} = \frac{4 - j6}{16 + 36} = 0.0769 - j0.1154 \text{ siemens}.$$

(b) $\mathbf{Z} = 20 \angle -30° = 20(\cos 30° - j \sin 30°)$

$$= 20(0.866 - j0.5) = 17.32 - j10\,\Omega,$$

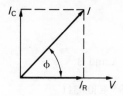

Fig. 8.14 Phasor diagram for fig. 8.13

$$\therefore \quad Y = \frac{1}{17.32 - j10} = \frac{17.32 + j10}{400}$$

$$= 0.0433 + j0.025 \, \text{S}.$$

Alternatively,

$$Y = \frac{1}{20 \angle -30} = 0.05 \angle 30$$

$$= 0.0433 + j0.025 \, \text{S}.$$

Example 8.4 *The admittance of a circuit is $(0.05 - j0.08)$ S. Find the values of the resistance and the inductive reactance of the circuit if they are (a) in parallel, (b) in series.*

(a) The conductance of the circuit is 0.05 S and its inductive susceptance is 0.08 S. From (8.12) it follows that if the circuit

Fig. 8.15 Circuit diagrams for example 8.4

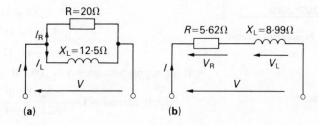

(a)   (b)

consists of a resistance in parallel with an inductive reactance, then:

$$\text{resistance} = \frac{1}{\text{conductance}} = \frac{1}{0.05} = 20 \, \Omega,$$

and inductive reactance $= \dfrac{1}{\text{inductive susceptance}} = \dfrac{1}{0.08} = 12.5 \, \Omega.$

(b) Since $Y = 0.05 - j0.08 \, \text{S}$,

$$\therefore \quad Z = \frac{1}{0.05 - j0.08} = \frac{0.05 + j0.08}{0.0089} = 5.62 + j8.99 \, \Omega.$$

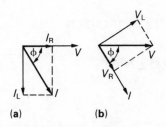

(a)   (b)

Fig. 8.16 Phasor diagrams for example 8.4

Hence if the circuit consists of a resistance in series with an inductance, the resistance is 5.62 Ω and the inductive reactance is 8.99 Ω. The two circuit diagrams are shown in figs. 8.15(a) and (b), and their phasor diagrams are given in figs. 8.16(a) and (b) respectively. The two circuits are equivalent in that they take the same current I for a given supply voltage V, and the phase difference ϕ between the supply voltage and current is the same in the two cases.

Example 8.5 *A network is arranged as indicated in fig. 8.17, the values being as shown. Calculate the value of the current in each branch and its phase relative to the supply voltage. Draw the complete phasor diagram.*

$$Z_A = 20 + j0 \, \Omega,$$

$$\therefore \quad Y_A = 0.05 + j0 \, \text{S}.$$

$$Z_B = 5 + j314 \times 0.1 = 5 + j31.4 \, \Omega,$$

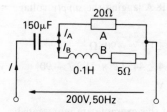

Fig. 8.17 Circuit diagram for example 8.5

$$\therefore \quad Y_B = \frac{1}{5 + j31.4} = \frac{5 - j31.4}{1010} = 0.004\,95 - j0.0311 \, \text{S}.$$

If Y_{AB} be the combined admittance of circuits A and B,

$$\mathbf{Y}_{AB} = 0.05 + 0.004\,95 - j0.0311$$
$$= 0.054\,95 - j0.0311\,\text{S}.$$

If Z_{AB} be the equivalent impedance of circuits A and B,

$$\mathbf{Z}_{AB} = \frac{1}{0.054\,95 - j0.0311} = \frac{0.054\,95 + j0.0311}{0.003\,987}$$
$$= 13.78 + j7.8\,\Omega.$$

Hence the circuit of fig. 8.17 can be replaced by that shown in fig. 8.18.

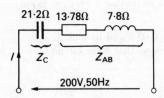

$$\mathbf{Z}_C = -j\frac{10^6}{314 \times 150} = -j21.2\,\Omega,$$

$\therefore$ total impedance $= \mathbf{Z}$

$$= 13.78 + j7.8 - j21.2 = 13.78 - j13.4$$
$$= \sqrt{\{(13.78)^2 + (13.4)^2\}} \angle \tan^{-1} - 13.4/13.78$$
$$= 19.22 \angle -44°\,12'\,\Omega.$$

Fig. 8.18 Equivalent circuit of fig. 8.17

If the supply voltage $= \mathbf{V} = 200 \angle 0°$ volts,

$\therefore$ supply current $= \mathbf{I} = \dfrac{200 \angle 0°}{19.22 \angle -44°\,12'} = 10.4 \angle 44°\,12'\,\text{A},$

i.e. the supply current is 10.4 A leading the *supply* voltage by 44° 12'. The p.d. across circuit AB $= \mathbf{V}_{AB} = \mathbf{I}\mathbf{Z}_{AB}.$

But $\mathbf{Z}_{AB} = 13.78 + j7.8$

$$= \sqrt{\{(13.78)^2 + (7.8)^2\}} \angle \tan^{-1} 7.8/13.78$$
$$= 15.85 \angle 29°\,30'\,\Omega,$$

$\therefore$ $\mathbf{V}_{AB} = 10.4 \angle 44°\,12' \times 15.85 \angle 29°\,30'$

$$= 164.8 \angle 73°\,42'\,\text{V}.$$

Since $\mathbf{Z}_A = 20 + j0 = 20 \angle 0°\,\Omega,$

$\therefore$ $\mathbf{I}_A = \dfrac{164.8 \angle 73°\,42'}{20 \angle 0°} = 8.24 \angle 73°\,42'\,\text{A},$

i.e. the current through branch A is 8.24 A leading the *supply* voltage by 73° 42'.

Similarly $\mathbf{Z}_B = 5 + j31.4 = 31.8 \angle 80°\,58'\,\Omega,$

$\therefore$ $\mathbf{I}_B = \dfrac{164.8 \angle 73°\,42'}{31.8 \angle 80°\,58'} = 5.18 \angle -7°\,16'\,\text{A},$

i.e. the current through branch B is 5.18 A lagging the *supply* voltage by 7° 16'.

Impedance of C $= \mathbf{Z}_C = -j21.2 = 21.2 \angle -90°\,\Omega,$

$\therefore$ p.d. across C $= \mathbf{V}_C = \mathbf{I}\mathbf{Z}_C = 10.4 \angle 44°\,12' \times 21.2 \angle -90°$

$$= 220 \angle -45°\,48'\,\text{V}.$$

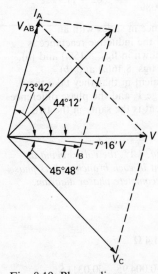

Fig. 8.19 Phasor diagram for example 8.5

The various voltages and currents of this example are represented by the respective phasors in fig. 8.19.

8.8 Calculation of power, using complex notation

Suppose the alternating voltage across and the current in a circuit to be represented respectively by:

$$\mathbf{V} = V \angle \alpha = V(\cos \alpha + j \sin \alpha) = a + jb \qquad (8.14)$$

and $\qquad \mathbf{I} = I \angle \beta = I(\cos \beta + j \sin \beta) = c + jd \qquad (8.15)$

Since the phase difference between the voltage and current is $(\alpha - \beta)$,

$$\text{active power} = VI \cos(\alpha - \beta)$$
$$= VI(\cos \alpha \cdot \cos \beta + \sin \alpha \cdot \sin \beta)$$
$$= ac + bd \qquad (8.16)$$

i.e. the power is given by the sum of the products of the real components and of the imaginary components.

$$\text{Reactive power} = VI \sin(\alpha - \beta)$$
$$= VI(\sin \alpha \cdot \cos \beta - \cos \alpha \cdot \sin \beta)$$
$$= bc - ad \qquad (8.17)$$

If we had proceeded by multiplying (8.14) by (8.15), the result would have been:

$$(ac - bd) + j(bc + ad).$$

The terms within the brackets represent neither the active power nor the reactive power. The correct expressions for these quantities are derived by multiplying the voltage by the *conjugate* of the current, the conjugate of a complex number being a quantity that differs only in the sign of the imaginary component; thus the conjugate of $c + jd$ is $c - jd$. Hence:

$$(a + jb)(c - jd) = (ac + bd) + j(bc - ad)$$
$$= (\text{active power}) + j(\text{reactive power})$$
$$= P + jQ.$$

Example 8.6   *The p.d. across and the current in a circuit are represented by* $(100 + j200)$ *V and* $(10 + j5)$ *A respectively. Calculate the active power and the reactive power.*

From (8.16),   active power $= (100 \times 10) + (200 \times 5)$
$$= 2000 \text{ W}.$$

From (8.17),   reactive power $= (200 \times 10) - (100 \times 5)$
$$= 1500 \text{ var}.$$

Alternatively, $100 + j200 = 223.6 \angle 63\;26' \text{ V}$

and $\qquad\qquad\qquad 10 + j5 = 11.18 \angle 26\;34' \text{ A},$

$\therefore\quad$ phase difference between $\Big\}$ $= 63\;26' - 26\;34' = 36\;52'$.
voltage and current

Hence,   active power $= 223.6 \times 11.18 \cos 36\;52'$
$$= 2000 \text{ W}$$

and   reactive power $= 223.6 \times 11.18 \sin 36\;52'$
$$= 1500 \text{ var}.$$

Summary of important formulae

$$\mathbf{A} = a + jb = A(\cos\theta + j\sin\theta) = A \angle \theta$$

where
$$A = \sqrt{(a^2 + b^2)} \quad \text{and} \quad \theta = \tan^{-1} b/a$$

$$A \angle \alpha \times B \angle \beta = AB \angle (\alpha + \beta)$$

$$\frac{A \angle \alpha}{B \angle \beta} = \frac{A}{B} \angle (\alpha - \beta)$$

$$\frac{1}{a + jb} = \frac{a - jb}{a^2 + b^2} \tag{8.9}$$

For a circuit having R and L in *series*,

$$\mathbf{Z} = R + jX_L = Z \angle \phi \tag{8.2}$$

$$\text{Admittance} = \mathbf{Y} = \frac{1}{\mathbf{Z}} = \frac{R}{Z^2} - \frac{jX_L}{Z^2} = G - jB_L$$

$$= Y \angle -\phi \tag{8.6}$$

For a circuit having R and C in *series*,

$$\mathbf{Z} = R - jX_C = Z \angle -\phi \tag{8.3}$$

and
$$\mathbf{Y} = \frac{R}{Z^2} + \frac{jX_C}{Z^2} = G + jB_C = Y \angle \phi \tag{8.10}$$

$$\text{Conductance} = G = R/Z^2 \text{ and is } 1/R \text{ only when } X = 0.$$

$$\text{Susceptance} = B = X/Z^2 \text{ and is } 1/X \text{ only when } R = 0.$$

For impedances $\mathbf{Z}_1 = R_1 + jX_1$ and $\mathbf{Z}_2 = R_2 + jX_2$ in *series*,

$$\text{total impedance} = \mathbf{Z} = \mathbf{Z}_1 + \mathbf{Z}_2 = (R_1 + R_2) + j(X_1 + X_2)$$

$$= Z \angle \phi$$

where
$$Z = \sqrt{\{(R_1 + R_2)^2 + (X_1 + X_2)^2\}}$$

and
$$\phi = \tan^{-1}(X_1 + X_2)/(R_1 + R_2).$$

For a circuit having R and L in *parallel*,

$$\mathbf{Y} = \frac{1}{R} - \frac{j}{X_L} = G - jB_L = Y \angle -\phi \tag{8.12}$$

For a circuit having R and C in *parallel*,

$$\mathbf{Y} = \frac{1}{R} + \frac{j}{X_C} = G + jB_C = Y \angle \phi \tag{8.13}$$

For admittances $\mathbf{Y}_1 = G_1 + jB_1$ and $\mathbf{Y}_2 = G_2 + jB_2$ in *parallel*,

$$\text{total admittance} = \mathbf{Y} = \mathbf{Y}_1 + \mathbf{Y}_2 = (G_1 + G_2) + j(B_1 + B_2)$$

$$= Y \angle \phi$$

where
$$Y = \sqrt{\{(G_1 + G_2)^2 + (B_1 + B_2)^2\}}$$

and
$$\phi = \tan^{-1}(B_1 + B_2)/(G_1 + G_2).$$

If $\qquad\qquad \mathbf{V} = a + jb$

and $\qquad\qquad \mathbf{I} = c + jd,$

$$\text{active power} = ac + bd \qquad\qquad (8.16)$$

and reactive power $= bc - ad \qquad\qquad (8.17)$

EXERCISES 8

1. Express in rectangular and polar notations the phasors for the following quantities: (*a*) $i = 10 \sin \omega t$; (*b*) $i = 5 \sin (\omega t - \pi/3)$; (*c*) $v = 40 \sin (\omega t + \pi/6)$.

   Draw a phasor diagram representing the above voltage and currents.

2. With the aid of a simple diagram, explain the j-notation method of phasor quantities.

   Four single-phase generators whose e.m.f.s can be represented by: $e_1 = 20 \sin \omega t$; $e_2 = 40 \sin (\omega t + \pi/2)$; $e_3 = 30 \sin (\omega t - \pi/6)$; $e_4 = 10 \sin (\omega t - \pi/3)$; are connected in series so that their resultant e.m.f. is given by $e = e_1 + e_2 + e_3 + e_4$. Express each e.m.f. and the resultant in the form $a \pm jb$. Hence find the maximum value of e and its phase angle relative to e_1.

   (U.L.C.I., O.2)

3. Express each of the following phasors in polar notation and draw the phasor diagram: (*a*) $10 + j5$; (*b*) $3 - j8$.

4. Express each of the following phasors in rectangular notation and draw the phasor diagram: (*a*) $20 \angle 60°$; (*b*) $40 \angle -45°$.

5. Add the two phasors of Q. 3 and express the result in: (*a*) rectangular notation; (*b*) polar notation. Check the values by drawing a phasor diagram to scale.

6. Subtract the second phasor of Q. 3 from the first phasor and express the result in: (*a*) rectangular notation; (*b*) polar notation. Check the values by means of a phasor diagram drawn to scale.

7. Add the two phasors of Q. 4 and express the result in: (*a*) rectangular notation; (*b*) polar notation. Check the values by means of a phasor diagram drawn to scale.

8. Subtract the second phasor of Q. 4 from the first phasor and express the result in: (*a*) rectangular notation; (*b*) polar notation. Check the values by a phasor diagram drawn to scale.

9. Calculate the resistance and inductance or capacitance in *series* for each of the following impedances, assuming the frequency to be 50 Hz: (*a*) $50 + j30 \,\Omega$; (*b*) $30 - j50 \,\Omega$; (*c*) $100 \angle 40° \,\Omega$; (*d*) $40 \angle -60° \,\Omega$.

10. Derive expressions, in rectangular and polar notations, for the admittances of the following impedances: (*a*) $10 + j15 \,\Omega$; (*b*) $20 - j10 \,\Omega$; (*c*) $50 \angle 20° \,\Omega$; (*d*) $10 \angle -70° \,\Omega$.

11. Derive expressions, in rectangular and polar notations, for the impedances of the following admittances: (*a*) $0.2 + j0.5$ siemens; (*b*) $0.08 \angle -30°$ siemens.

12. Calculate the resistance and inductance or capacitance in *parallel* for each of the following admittances, assuming the frequency to be 50 Hz: (*a*) $0.25 + j0.6 \,\text{S}$; (*b*) $0.05 - j0.1 \,\text{S}$; (*c*) $0.8 \angle 30° \,\text{S}$; (*d*) $0.5 \angle -50° \,\text{S}$.

13. A voltage, $v = 150 \sin (314t + 30°)$ volts, is maintained across a coil having a resistance of $20 \,\Omega$ and an inductance of $0.1 \,\text{H}$. Derive expressions for the r.m.s. values of the voltage and

current phasors in: (*a*) rectangular notation; (*b*) polar notation. Draw the phasor diagram.

14. A voltage, $v = 150 \sin(314t + 30°)$ volts, is maintained across a circuit consisting of a 20-Ω non-reactive resistor in series with a loss-free 100-μF capacitor. Derive an expression for the r.m.s. value of the current phasor in: (*a*) rectangular notation; (*b*) polar notation. Draw the phasor diagram.

15. Calculate the values of resistance and reactance which, when in parallel, are equivalent to a coil having a resistance of 20 Ω and a reactance of 10 Ω.

16. The impedances of two parallel branches can be represented by $(24 + j18)$ Ω and $(12 - j22)$ Ω respectively. If the supply frequency is 50 Hz, find the resistance and inductance or capacitance of each circuit. Also, derive a symbolic expression in polar form for the admittance of the combined circuits, and thence find the phase angle between the applied voltage and the resultant current. (W.J.E.C., O.2)

17. A coil of resistance 25 Ω and inductance 0.044 H is connected in parallel with a branch made up of a 50-μF capacitor in series with a 40-Ω resistor, and the whole is connected to a 200-V, 50-Hz supply. Calculate, using symbolic notation, the total current taken from the supply and its phase angle, and draw the complete phasor diagram. (W.J.E.C., O.2)

18. The current in a circuit is given by $4.5 + j12$ A when the applied voltage is $100 + j150$ V. Determine: (*a*) the complex expression for the impedance, stating whether it is inductive or capacitive; (*b*) the active power; (*c*) the phase angle between voltage and current.

19. Explain how alternating quantities can be represented by complex numbers.

    If the potential difference across a circuit is represented by $40 + j25$ V, and the circuit consists of a coil having a resistance of 20 Ω and an inductance of 0.06 H and the frequency is 79.5 Hz, find the complex number representing the current in amperes. (App. El., L.U.)

20. The impedances of two parallel branches can be represented by $(20 + j15)$ Ω and $(10 - j60)$ Ω respectively. If the supply frequency is 50 Hz, find the resistance and the inductance or capacitance of each branch. Also, derive a complex expression for the admittance of the combined network, and thence find the phase angle between the applied voltage and the resultant current. State whether this current is leading or lagging relatively to the voltage. (App. El., L.U.)

21. An alternating e.m.f. of 100 V is induced in a coil of impedance $10 + j25$ Ω. To the terminals of this coil there is joined a circuit consisting of two parallel impedances, one of $30 - j20$ Ω and the other of $50 + j0$ Ω. Calculate the current in the coil in magnitude and phase with respect to the induced voltage. (U.L.C.I., Adv. El. Tech.)

22. A circuit consists of a 30-Ω non-reactive resistor in series with a coil having an inductance of 0.1 H and a resistance of 10 Ω. A 60-μF loss-free capacitor is connected in parallel with the *coil*. The network is connected across a 200-V, 50-Hz supply. Calculate the value of the current in each branch and its phase relative to the supply voltage.

23. An impedance of $2 + j6$ Ω is connected in series with two impedances of $10 + j4$ Ω and $12 - j8$ Ω, which are in parallel. Calculate the magnitude and power factor of the main current when the combined circuit is supplied at 200 V. (U.L.C.I., Adv. El. Tech.)

24. The arms of an a.c. bridge (Maxwell) are arranged thus:

    AB: a non-reactive resistor of 300 Ω.
    BC: a variable resistance R in series with a variable inductance L.
    CD: a coil, the resistance and reactance of which are required.
    DA: a non-reactive resistor of 100 Ω.

    An alternating voltage is applied across AC. Deduce the conditions for zero p.d. across BD, and calculate the resistance and the inductance of the coil if balance is obtained with $R = 64\,Ω$ and $L = 0.28\,\text{H}$.
      Note. The solution makes use of the fact that if $a + jb = c + jd$, then $a = c$ and $b = d$.

25. The arms of an a.c. bridge (de Sauty) are arranged thus:

    AB: a non-reactive resistor of 1000 Ω.
    BC: a variable no-loss capacitor having capacitance C.
    CD: a capacitor X, the capacitance of which is required.
    DA: a non-reactive resistor of 100 Ω.

    If an alternating voltage is maintained across AC, deduce the condition for zero p.d. across BD. If the value of capacitance C be 0.068 μF when the bridge is balanced, calculate the capacitance of X.

26. The arms of an a.c. bridge (Owen) have the following impedances:

    AB: a coil in series with a variable resistor P.
    BC: a no-loss 0.5-μF capacitor in series with a variable resistor Q.
    CD: a no-loss 0.3-μF capacitor.
    DA: a 500-Ω non-reactive resistor.

    If an alternating voltage be maintained across AC, deduce the conditions for zero p.d. across BD. If the values of P and Q to give a balance be 126 Ω and 534 Ω respectively, calculate the resistance and the inductance of the coil.

27. The p.d. across and the current in a given circuit are represented by $(200 + j30)$ V and $(5 - j2)$ A respectively. Calculate the active power and the reactive power. State whether the reactive power is leading or lagging.

28. Had the current in Q. 27 been represented by $(5 + j2)$ A, what would have been the active power and the reactive power? Again state whether the reactive power is leading or lagging.

29. A p.d. of 200 ∠ 30° V is applied to two branches connected in parallel. The currents in the respective branches are 20 ∠ 60° A and 40 ∠ 30° A. Find the apparent power (in kVA) and the active power (in kW) in each branch and in the main network. Express the current in the main network in the form $A + jB$.
                                                                        (W.J.E.C., O.2)

9 Electronic Systems

9.1 Analogue and digital systems

The subject of electronics is one that is difficult to define since it refers to an extremely wide range of electrical technology having only the unifying aspect that electronics deals principally with the communication of information and/or data handling. Electronic systems fall into two broad categories: (*a*) analogue, and (*b*) digital.

In analogue systems, the information or data is given as an electrical signal that varies in direct proportion to the information or data. It follows that the variation must be continuous and, between the limits of operation of the system, the variation can have one of an infinite number of values. Such variation is associated with, for example, the production of sound, in radio receivers, and vision, in television sets.

An analogue system in its most basic form has an input electrical signal which is either a voltage or a current varying directly in proportion to the input information. The input information is converted into the electrical signal by a transducer. Typical transducers include a microphone (converting sound to electrical e.m.f.), pressure transducers (converting pressure, say air or water pressure, to electrical signals), tachometers (converting speed to an e.m.f. using the principle $e = Blv$) and light detectors (converting intensity of light to e.m.f.).

The analogue system generally takes the input signal and enlarges it. In practice, this is achieved by a variety of means but all have the common feature that the input signal is used to control the flow of energy from a more powerful source. Such a process is called amplification. It can be likened to pressing the accelerator of a car in order to make it go faster. The driver is controlling the release of energy, which comes from the engine and not from the driver.

Once the signal has been increased sufficiently, it can then be converted to another useful form, e.g. a radio converts its electrical signal into sound to which we can listen. This is achieved by the loudspeaker transducing the electrical signal into sound.

Unlike analogue systems, digital systems can have signals that have one of a limited number of discrete values. The most common digital systems are binary systems in which the signals can have only one of two values. These values are referred to as 0 and 1, being the absence and presence of the supply current or voltage.

Digital systems are most commonly associated with data handling devices such as calculators, computers, watches and microprocessors. However, digital systems are also increasingly used in conjunction with analogue systems. This development arises from the difficulty of producing analogue amplifiers which amplify equally over the full range of operation. This problem can be avoided by digital systems, hence it is often advantageous to introduce analogue-to-digital converters and, following the digital system, digital-to-analogue converters to reproduce the analogue signal.

At this introductory stage, it is preferable to consider analogue systems separately from digital systems due to the difference in the fundamental modes of operation.

9.2 Basic amplifier principles

The purpose of an amplifier is to produce gain. That is to say, a small input signal power controls a larger output signal power. Certain devices not normally referred to as amplifiers do nevertheless come into this category by definition. In a relay, for example, the power required by the coil to close the contacts can be considerably less than that involved in the circuit switched by the contacts. Another example is the separately excited d.c. generator being driven at constant speed. Here the power being fed to a load connected across the output terminals can be controlled by a relatively small power fed to its field winding.

In the basic amplifier, there is a further characteristic that it must exhibit. The waveform of the input signal voltage or current must be maintained to a fairly high degree of accuracy in the output signal. This can be illustrated by comparing two signals representing the same note played by two different musical instruments. They are both of the same fundamental frequency representing the pitch but the waveforms differ, representing different tones. This is illustrated in fig. 9.1.

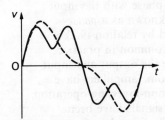

Fig. 9.1 Signal waveforms

Fig. 9.2 shows an amplifier with a resistive load R_L connected across the output terminals. The basic parameters of the amplifier are:

$$\text{voltage gain } (G_v) = \frac{\text{output signal voltage}}{\text{input signal voltage}}$$

$$G_v = \frac{V_2}{V_1} \qquad (9.1)$$

$$\text{current gain } (G_i) = \frac{\text{output signal current}}{\text{input signal current}}$$

$$G_i = \frac{I_2}{I_1} \qquad (9.2)$$

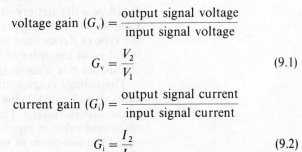

Fig. 9.2 Amplifier block diagram

$$\text{power gain } (G_{\mathrm{p}}) = \frac{\text{output signal power}}{\text{input signal power}}$$

$$G_{\mathrm{p}} = \frac{V_2 \times I_2}{V_1 \times I_1}$$

$$G_{\mathrm{p}} = G_{\mathrm{v}}G_{\mathrm{i}} \qquad (9.3)$$

Fig. 9.3 shows typical waveforms where the signals are assumed to vary sinusoidally with time. The waveforms met

Fig. 9.3 Amplifier signal waveforms

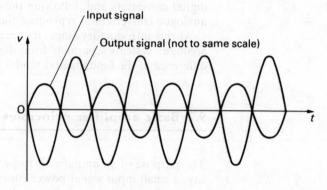

in practice tend to be more complex, as illustrated in fig. 9.1, but it can be shown that such waves are formed from a series of pure sine waves the frequencies of which are exact multiples of the basic or fundamental frequency. These sine waves are known as harmonics. The use of sine waves in the analysis and testing of amplifiers is therefore justified.

Examination of the waveforms in fig. 9.3 shows that the output signal voltage is 180° out of phase with the input signal voltage. Such an amplifier is known as a *phase-inverting* one and the gain as defined by relation (9.1) is negative. These amplifiers are very common in practice although non-inverting types, where the output and input signals are in phase, are also met. Some amplifiers have alternative inputs for inverting and non-inverting operation.

So far only the input and output signals have been considered in the operation of the amplifier. It is necessary, however, to provide a source of power from which is obtained the output signal power fed to the load; the magnitude of this signal power being controlled by the magnitude of the input signal. The source itself has to be a direct current one and could be a dry battery or a rectified supply as described in Chapter 11. The magnitude of this supply voltage depends on the type of device used in the amplifier. For thermionic valves it will be of the order of hundreds of volts while for transistors it is often in the range 3 to 30 volts.

The voltage range within which the output signal voltage can vary is limited, being very much dependent on the value of the supply voltage. This means therefore that there is a maximum value of input signal that will produce an output signal the waveform of which remains an acceptable replica of the input signal waveform. Increasing the input signal beyond

Fig. 9.4 Clipping of output
signal waveform

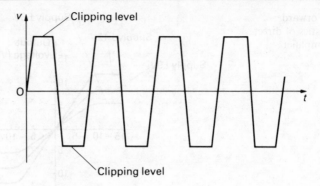

this level will produce 'clipping' of the input signal waveform
as illustrated in fig. 9.4, although the clipping levels need not
be symmetrical about the zero level.

Examination of the gain of an amplifier will show that it
does not remain constant with the frequency of the input
signal. Some amplifiers exhibit a reduction in gain at both
high and low frequencies while others have a reduction at
high frequencies only. In both cases there is a considerable
frequency range over which the gain remains essentially

Fig. 9.5 Gain/frequency
characteristics
(*a*) Capacitance-coupled
amplifier (*b*) Direct-coupled
amplifier

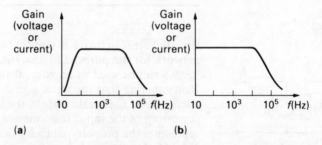

constant. It is within this frequency range that the amplifier is
designed to operate. The effect of the unequal gain is to
produce another form of waveform distortion since the
harmonics present in a complex input signal waveform may
not be amplified by the same amount. Fig. 9.5 shows typical
gain/frequency characteristics where it should be noted that
the frequency scales are logarithmic. The advantage of an
amplifier with a characteristic as illustrated in fig. 9.5(*b*) is
that it is capable of amplifying signals at the very low
frequencies met with in many industrial applications. It is
known as a direct-coupled (d.-c.) amplifier and integrated-
circuit amplifiers are of this type. Fig. 9.6 shows the responses
obtained from both types of amplifier for an input signal
which changes instantaneously from one level to another.
Lack of low frequency gain in the first type prevents faithful
reproduction of the steady parts of the input signal, the
output being merely a 'blip' as the input signal changes from
one level to the other. The characteristics showing output
voltage against input voltage for several d.c. supply voltages
as given in fig. 9.7 are representative of integrated-circuit d.-c.
amplifiers. Output voltage variation on either side of zero

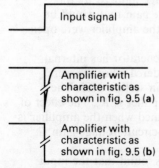

Fig. 9.6 Response to step
change

Fig. 9.7 Forward
characteristics of direct-
coupled amplifier

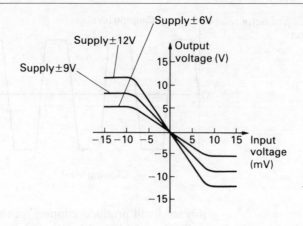

volts is obtained by the use of power supply voltages which
are spaced on both sides of zero, e.g. $+12$ V and -12 V as
opposed to $+24$ V and 0 V.

9.3 Amplifier constant-voltage equivalent networks

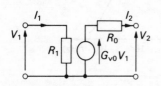

Fig. 9.8 Amplifier constant-
voltage equivalent network

An equivalent network is one that can replace the actual
network for the purpose of analysis. The network shown in
fig. 9.8 can be used as an equivalent network for an amplifier.
Remembering that the input signal only controls the flow of
energy from a separate source, the network has a circuit
consisting of the input resistance of the amplifier R_i and this
represents the property whereby the input circuit of the
amplifier loads the source of the input signal. It follows that

$$R_i = \frac{V_1}{I_1} \qquad (9.4)$$

The ability of the amplifier to produce gain requires a
source of controlled energy and this is represented by the
voltage generator. The generator amplifies the input voltage
V_1 by a gain of G_{vo} so that the voltage induced by the
generator is $G_{vo}V_1$. G_{vo} is the voltage gain that would be
obtained if the output terminals of the amplifier were open-
circuited.

However, any practical voltage generator has internal
resistance, a principle already considered when we were
introduced to Thévenin's Theorem in section 1.15. The
internal output resistance of the amplifier is R_o. The effect of
R_o causes the voltage gain G_v, obtained when the amplifier is
loaded, which is less than the open-circuit voltage gain G_{vo}
due to the voltage drop across R_o.

In most equivalent networks, it is found that the input
resistance R_i has a relatively high value while the output
resistance R_o has a relatively low value. Ideally, R_i would be

infinitely large thus reducing the input current and hence the input power to zero. In practice, amplifier designs generally try to ensure that the input power is as small as possible, usually because the input signal has little power available to it. The output resistance R_o gives rise to a loss of the power produced by the equivalent generator and hence it is desirable that the loss is reduced as far as possible, which effectively requires that R_o be as small as possible.

The use of such equivalent networks is restricted to the signal quantities only. They should not be used in calculations concerned with the direct quantities associated with the amplifier. Moreover, their use assumes an exact linear relationship between input and output signals, i.e. the amplifier produces no waveform distortion.

Fig. 9.9 shows an amplifier, represented by its equivalent circuit, being supplied by a source of signal E_s volts and source resistance R_s, and feeding a load resistance R_L.

Fig. 9.9 Amplifier with signal source and load

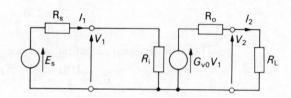

Expressions for the important parameters are derived.

Input signal voltage to amplifier $= V_1 = \dfrac{R_i E_s}{R_s + R_i}$ (9.5)

Output signal voltage $= V_2 = \dfrac{R_L}{R_o + R_L} G_{vo} V_1$

$\therefore$ Voltage gain $= G_v = \dfrac{V_2}{V_1} = \dfrac{G_{vo} R_L}{R_o + R_L}$ (9.6)

Output signal current $= I_2 = \dfrac{G_{vo} V_1}{R_o + R_L} = \dfrac{G_{vo} I_1 R_i}{R_o + R_L}$

$\therefore$ Current gain $= G_i = \dfrac{I_2}{I_1} = \dfrac{G_{vo} R_i}{R_o + R_L}$ (9.7)

Power gain $= G_p = \dfrac{\text{signal power into load}}{\text{signal power into amplifie}}$

$\therefore$ $G_p = \dfrac{V_2^2/R_L}{V_1^2/R_i} = \dfrac{I_2^2 R_L}{I_1^2 R_i} = \dfrac{VI}{V_1^2 I_2^2} = G_v G_i$ (9.8)

Example 9.1   *An integrated-circuit amplifier has an open-circuit voltage gain of 5000, an input resistance of 15 kΩ and an output resistance of 25 Ω. It is supplied from a signal source of internal resistance 5.0 kΩ and it feeds a resistive load of 175 Ω. Determine the magnitude of the signal source voltage to produce an output signal voltage of 1.0 V. What value of load resistance would halve the signal voltage output for the same input?*

Fig. 9.10 Part of
example 9.1

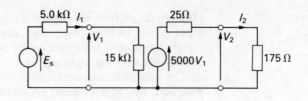

The network is shown in fig. 9.10

$$V_2 = \frac{175}{25 + 175} \times 5000 V_1$$

$$\text{Voltage gain} = G_v = \frac{V_2}{V_1} = \frac{175 \times 5000}{200} = 4350$$

$$\therefore \quad V_1 = \frac{1.0}{4350} = 2.30 \times 10^{-4}\,\text{V} = 230\,\mu\text{V}$$

$$V_1 = \frac{15}{5 + 15} E_s$$

$$\therefore \quad E_s = \frac{20 \times 230}{15} = 307\,\mu\text{V}$$

For half the signal output the voltage gain must be halved.

$$\therefore \quad G_v = \frac{4350}{2} = 2175 = \frac{R_L \times 5000}{25 + R_L}$$

$$\therefore \quad 2175 \times 25 + 2175 R_L = 5000 R_L$$

$$\therefore \quad R_L = \frac{2175 \times 25}{2825} = 19.2\,\Omega.$$

9.4 Amplifier constant-current equivalent networks

The constant-voltage equivalent network lends itself to the analysis of amplifiers in which the signals are primarily considered in terms of voltage. However, some amplifiers, especially those using epitaxial or junction transistors, are better considered in terms of current. In order to do this, it is first of all necessary to introduce the constant-current generator. It was shown in section 1.15 that a source of electrical energy could be represented by a source of e.m.f. in series with a resistance. This is not, however, the only form of representation. Consider such a source feeding a load resistor R_L as shown in fig. 9.11.

From this circuit:

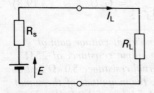

Fig. 9.11 Energy source
feeding load

$$I_L = \frac{E}{R_s + R_L} = \frac{\dfrac{E}{R_s}}{\dfrac{R_s + R_L}{R_s}}$$

$$I_{L} = \frac{R_{s}}{R_{s} - R_{L}} \times I_{s} \qquad (9.9)$$

where $I_s = E/R_s =$ the current which would flow in a short circuit across the output terminals of the source.

Fig. 9.12 Equivalence of constant-voltage generator and constant-current generator forms of representation

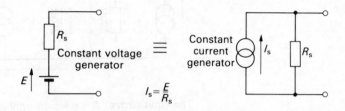

$$I_s = \frac{E}{R_s}$$

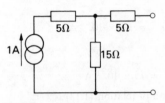

Fig. 9.13 Network for example 9.2

Comparing relation (9.9) with relation (1.13) it can be seen that, when viewed from the load, the source appears as a source of current (I_s) which is dividing between the internal resistance (R_s) and the load resistor (R_L) connected in parallel. For the solution of problems either form of representation can be used. In many practical cases an easier solution is obtained using the current form. Fig. 9.12 illustrates the equivalence of the two forms.

The internal resistance of the constant-current generator must be taken as infinite, since the resistance of the complete source must be R_s as is obtained with the constant-voltage form.

The ideal constant-voltage generator would be one with zero internal resistance so that it would supply the same voltage to all loads. Conversely, the ideal constant-current generator would be one with infinite internal resistance so that it supplied the same current to all loads. These ideal conditions can be approached quite closely in practice.

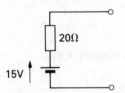

Fig. 9.14 Part of example 9.2

Example 9.2

Represent the network shown in fig. 9.13 by one source of e.m.f. in series with a resistance.

Potential difference across output terminals $= V_0 = 1 \times 15 = 15 \text{ V}$.

Resistance looking into output terminals $= 5 + 15 = 20 \, \Omega$

therefore the circuit can be represented as shown in fig. 9.14.

By using the constant-current generator principle, we can create the alternative form of amplifier equivalent network which is shown in fig. 9.15. Here the input circuit remains as before but the output network consists of a current generator producing a current $G_{is}I_1$ where G_{is} is the current gain that would be obtained with the output terminals short-circuited.

Since the voltage and current equivalent network are also to be equivalent to each other, it follows that

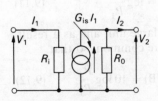

Fig. 9.15 Amplifier constant-current equivalent network

$$G_{vo} V_1 = G_{is} I_1 R_0 = G_{is} \frac{V_1}{R_i} R_0$$

$$\therefore \quad G_{vo} = \frac{R_0}{R_i} G_{is} \qquad (9.10)$$

Example 9.3   *For the amplifier shown in fig. 9.16, determine the current gain and hence the voltage gain.*

Fig. 9.16 Part of example 9.3

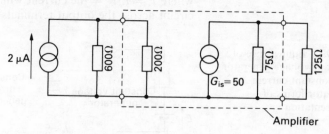

The input current I_i is given by application of relation (1.13), hence

$$I_i = \frac{600}{600 + 200} \times 2 \times 10^{-6} = 1.5 \times 10^{-6} \, \text{A}.$$

Hence the generator current is

$$G_{is}I_i = 50 \times 1.5 \times 10^{-6} = 75 \times 10^{-6} \, \text{A}$$

and the output current I_o by application of relation (1.13) is

$$I_o = \frac{75}{75 + 125} \times 75 \times 10^{-6} = 28.1 \times 10^{-6} \, \text{A}$$

$$\therefore \quad G_I = \frac{I_o}{I_i} = \frac{28.1 \times 10^{-6}}{1.5 \times 10^{-6}} = 18.8$$

Note that the amplifier gain is the ratio of the output terminal current to the input terminal current.

$$G_v = \frac{V_o}{V_i} = \frac{28.1 \times 10^{-6} \times 125}{1.5 \times 10^{-6} \times 200} = 11.7$$

9.5 Logarithmic units

It is sometimes found convenient to express the ratio of two powers P_1 and P_2 in logarithmic units known as *bels* as follows:

$$\text{Power ratio in bels} = \log \frac{P_2}{P_1} \qquad (9.11)$$

It is found that the *bel* is rather a large unit and as a result the *decibel* (one-tenth of a bel) is more common, so that

$$\text{Power ratio in decibels (dB)} = 10 \log \frac{P_2}{P_1} \qquad (9.12)$$

If the two powers are developed in the same resistance or equal resistances then

$$P_1 = \frac{V_1^2}{R} = I_1^2 R \quad \text{and} \quad P_2 = \frac{V_2^2}{R} = I_2^2 R$$

where V_1, I_1, V_2 and I_2 are the voltages across and the currents in the resistance. Therefore

$$10 \log \frac{P_2}{P_1} = 10 \log \frac{V_2^2/R}{V_1^2/R} = 10 \log \frac{V_2^2}{V_1^2}$$

$$\text{Power ratio in dB} = 20 \log \frac{V_2}{V_1} \qquad (9.13)$$

Similarly, $\qquad$ Power ratio in dB $= 20 \log \dfrac{I_2}{I_1}$ $\qquad (9.14)$

The relationships defined by relations (9.13) and (9.14) although expressed by ratios of voltages and currents respectively still represent power ratios. They are, however, used extensively, although by fundamental definition erroneously, to express voltage and current ratios where common resistance values are not involved. For example, the voltage gain of an amplifier is expressed often in decibels. Care should be taken in this use of decibels since the expression of power gain in decibels would be in complete agreement with the basic definition.

Fig. 9.17 shows two amplifiers connected in cascade, in which case the input of the second amplifier is the load on the first one. The overall voltage gain is given by

$$G_v = \frac{V_3}{V_2} \times \frac{V_2}{V_1}$$

Expressing this in decibels

$$\text{Voltage gain in dB} = 20 \log \frac{V_3}{V_2} \times \frac{V_2}{V_1}$$

$$= 20 \log \frac{V_3}{V_2} + 20 \log \frac{V_2}{V_1}$$

Thus the overall voltage gain in decibels is equal to the sum of the voltage gains in decibels of the individual amplifiers. This is a most useful result and can be extended to any number of amplifiers in cascade. It is obviously applicable to current and power gains.

The use of decibels gives a representation of one power (or voltage) with reference to another. If P_2 is greater than P_1 then P_2 is said to be $10 \log P_2/P_1$ dB *up* on P_1. For P_2 less than P_1, $\log P_2/P_1$ is negative. Since $\log P_1/P_2 = -\log P_2/P_1$, it is usual to determine the ratio greater than unity and P_2 is said to be $10 \log P_1/P_2$ *down* on P_1.

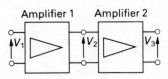

Amplifier 1 $\qquad$ Amplifier 2

Fig. 9.17 Amplifiers in cascade

Example 9.4 $\qquad$ *The voltage gain of an amplifier when it feeds a resistive load of 1.0 kΩ is 40 dB. Determine the magnitude of the output signal voltage and the signal power in the load when the input signal is 10 mV.*

$$20 \log \frac{V_2}{V_1} = 40$$

∴ $\qquad\qquad \log \dfrac{V_2}{V_1} = 2.0$

∴ $\qquad\qquad \dfrac{V_2}{V_1} = 100$

$$\therefore \quad V_2 = 100 \times 10 = 1000 \, \text{mV} = 1.0 \, \text{V}$$

$$P_2 = \frac{V_2^2}{R_L} = \frac{1.0^2}{1000} = \frac{1}{1000} \, \text{W} = 1 \, \text{mW}$$

Example 9.5 *Express the power dissipated in a 15-Ω resistor in decibels relative to 1.0 mW when the voltage across the resistor is 1.5 V r.m.s.*

$$P_2 = \frac{1.5^2}{15} \, \text{W} = 150 \, \text{mW}$$

Power level in dB relative to 1 mW

$$= 10 \log \frac{150}{1} = 10 \times 2.176 = 21.76 \, \text{dB}.$$

Example 9.6 *An amplifier has an open-circuit voltage gain of 70 dB and an output resistance of 1.5 kΩ. Determine the minimum value of load resistance so that the voltage gain is not more than 3.0 dB down on the open-circuit value. With this value of load resistance determine the magnitude of the output signal voltage when the input signal is 1.0 mV.*

$$20 \log G_{vo} = 70$$

$$\therefore \quad \log G_{vo} = 3.50$$

$$\therefore \quad G_{vo} = 3160$$

$$20 \log G_v = 70 - 3 = 67$$

$$\therefore \quad \log G_v = 3.35$$

$$\therefore \quad G_v = 2240$$

$$\therefore \quad \frac{R_L}{R_o + R_L} 3160 = 2240$$

$$\therefore \quad 3160 R_L = 2240 \times 1.5 + 2240 R_L$$

$$\therefore \quad R_L = \frac{2240 \times 1.5}{920} = 3.65 \, \text{k}\Omega.$$

Alternatively

since $20 \log G_{vo} - 20 \log G_v = 3.0$

$$\therefore \quad 20 \log \frac{G_{vo}}{G_v} = 3.0$$

$$\therefore \quad \frac{G_{vo}}{G_v} = 1.41$$

$$\therefore \quad \frac{R_L}{R_o + R_L} = \frac{1}{1.41}$$

$$\therefore \quad R_L = \frac{1.5}{0.41} = 3.65 \, \text{k}\Omega.$$

$$V_2 = 2240 \times 1.0 = 2240 \, \text{mV} = 2.24 \, \text{V}.$$

9.6 Frequency response

It has already been stated in section 9.2 that amplifier gain decreases at high frequencies and in some cases at low frequencies also. The equivalent circuits, as have been used so far, gives no indication of this since they contain pure resistance only. They have therefore to be modified for the frequency ranges in which the gain decreases.

One cause of loss of gain at high frequencies is the presence of shunt capacitance across the load. This can be due to stray capacitance in the external circuit and, what is usually more important, capacitance within the amplifier itself. The effective load on the amplifier tends to zero as the frequency tends to infinity. Thus the voltage gain decreases because of the decrease in load impedance, and the current gain decreases because the shunt capacitance path drains current away from the load. Another cause of loss of gain at high frequencies is the inherent decrease of available gain in the amplifier, i.e. G_{vo} decreases with frequency. The manner in which the gain decreases due to this effect is similar to that produced by shunt capacitance and both effffects can be represented on the equivalent circuit by the connection of capacitance across the output terminals assuming that, at the frequency being considered, only one of the effects is appreciable. In practice this is often a reasonable assumption over a considerable frequency range. Fig. 9.18 shows the output section of the equivalent circuit modified for high-frequency operation. Associated with the loss of gain will be a shift of phase between input and output signals from the nominal value. A more complex circuit would be required if both the effects considered above had to be taken into account simultaneously.

Loss of gain at low frequencies is due to the use of certain capacitors in the circuit. Their values are chosen such that the reactances are very small at the frequencies being used, so that little of the signal voltage is developed across them. At low frequencies, however, appreciable signal is developed across them, resulting in a loss of signal at the load. As with the high frequency response there is a corresponding shift of phase between input and output signals.

The *bandwidth* of an amplifier is defined as the difference in frequency between the lower and upper frequencies, f_1 and f_2 respectively, at which the gain is 3.0 dB down on its maximum value.

The frequency response that results in an amplifier with capacitors in its circuitry often takes the form shown in fig. 9.19. The loss of output signal and hence the loss of gain are clearly shown both at low frequencies and at high frequencies. Note the use of a logarithmic scale for the base — this is necessary to expand the characteristic at low frequencies as otherwise it would be crushed and the form of the characteristic would be lost.

The gain may be shown either in numerical form or in logarithmic form. In the latter case the bandwidth is determined by the 3-dB points which are the half-power points. If P_m is the power associated with the maximum gain, usually termed the mid-band gain, and the corresponding output voltage is V_o, then for the half-power condition let the

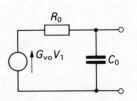

Fig. 9.18 Equivalent network for high-frequency operation

Fig. 9.19 Frequency
responses of a typical *R–C*
coupled amplifier and a
typical tuned amplifier

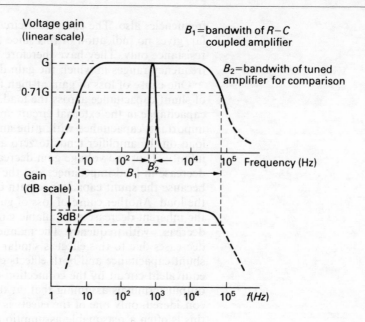

voltage be V_1.

$$\therefore \quad \frac{V_1^2}{R} = \tfrac{1}{2}P_m = \frac{V_o^2}{2R}$$

$$\therefore \quad V_1 = \frac{1}{\sqrt{2}} V_o = 0.71 V_o \qquad (9.15)$$

It follows that since the input voltage is assumed constant
then the voltage gain must also have fallen to 0.71 of the mid-
band gain. The frequency response bandwidth can therefore
also be determined by the points at which the voltage gain
has fallen to 0.71G. Although this has been argued for voltage
gain, it would equally have held for current gain.

For a direct-coupled amplifier the bandwidth will be simply
f_2 since the gain extends down to zero frequency. The
passband or working frequency range is that bounded by f_1
and f_2. If this lies within the range of frequencies normally
audible to the ear as sound waves, e.g. 30 Hz to 15 kHz, the
amplifier is referred to as an *audio amplifier*. Such amplifiers
are used in sound reproduction systems. The signals applied
to the cathode-ray tube in television receivers require greater
passbands extending from 0 to several megahertz and are
known as *video amplifiers*. In both audio and video amplifiers
approximately constant gain over a fairly wide range of
frequencies is required and they are collectively known as
broad-band amplifiers. Other types known as *narrow-band
amplifiers* are used where the bandwidth is considerably less
than the centre frequency. This type of amplifier provides
selectivity between signals of different frequency. Fig. 9.20
shows a typical frequency response curve for a broad-band
amplifier while the curve for a narrow-band or tuned
amplifier is incorporated into fig. 9.19.

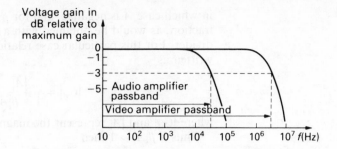

Fig. 9.20 Frequency response characteristic of a broad-band amplifier

9.7 Feedback

Feedback is the process whereby a signal derived in the output section of the amplifier is fed back into the input section. In this way the amplifier can be used to provide characteristics which differ from those of the basic amplifier. The signal fed back can be either a voltage or a current, being applied in series or shunt respectively with the input signal. Moreover, the feedback signal, whether voltage or current, can be directly proportional to the output signal voltage or current. This gives rise to four basic types of feedback, i.e. series-voltage, series-current, shunt-voltage and shunt-current. The characteristics produced by these four types of feedback are similar in some respects and differ in others. Series-voltage feedback will be considered here as a representative type and fig. 9.21 shows such a feedback amplifier.

The block marked β is that part of the network which provides the feedback voltage $V_f = \beta V_2$. In one of its simplest forms it could consist of two resistors connected across the output to form a voltage divider, the feedback voltage being the signal developed across one of the resistors. The voltage gain $G_v = A$ will be dependent on the load which the β network presents to the amplifier, although in many practical applications the values of the components used in the network are such as to present negligible loading.

From fig. 9.21,

input signal to basic amplifier $= V_a = V_1 + V_f = V_1 + \beta V_2$

$$V_2 = A V_a = A(V_1 + \beta V_2)$$

$$V_2(1 - \beta G_v) = A V_1$$

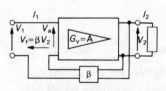

Fig. 9.21 Series-voltage feedback amplifier

Voltage gain with feedback $= G_{vf} = \dfrac{V_2}{V_1}$

$$G_{vf} = \frac{A}{1 - \beta A} \tag{9.16}$$

If the magnitude of $1 - \beta A$ is greater than unity then the magnitude of G_{vf} is less than that of A and the feedback is said to be *negative* or degenerative. The simplest means of accomplishing this is to provide a phase-inverting amplifier, in which case A is negative, and for β to be a positive fraction, as would be obtained with a simple resistive voltage divider. For this particular case relation (9.16) can be written as

$$|G_{vf}| = \frac{|A|}{1 + \beta|A|}$$

where $|G_{vf}|$ and $|A|$ represent the magnitudes of the quantities. Thus if $\beta|A| \gg 1$ then

$$G_{vf} \doteq \frac{|A|}{\beta|A|}$$

i.e.

$$|G_{vf}| \doteq \frac{1}{\beta} \qquad (9.17)$$

Relation (9.17) illustrates that the voltage gain with negative feedback is relatively independent of the voltage gain of the basic amplifier, provided that the product $\beta|A|$ remains large compared to unity. This is one of the most important characteristics of negative series-voltage feedback amplifiers. Thus the desired voltage gain can be obtained with a high degree of stability by selection of the component values in the feedback network.

To explain further the significance of such stability, we need to be aware that A can vary, either due to ageing of the components or due to the supply voltage not remaining constant. With the feedback gain being largely dependent on β, such changes in A become insignificant. For instance, consider an amplifier which has a gain of 2000 before the introduction of feedback and let there be negative series feedback with a feedback ratio of 0.25.

$$G_{vf} = \frac{A}{1 - \beta A} \quad \text{where } A = -2000$$

$$= \frac{-2000}{1 + (0.25 \times 2000)} = 3.99$$

Now let A fall to 1000, either due to ageing or change of supply voltage. Now

$$G_{vf} = \frac{-1000}{1 + (0.25 \times 1000)} = 3.98$$

Such a change in gain is almost negligible, yet the intrinsic amplifier feedback, i.e. without feedback, has fallen by 50 per cent. Thus although the overall gain is much reduced, it is now stabilized and will be able to give considerably more consistent performance.

Example 9.7   *An amplifier with an open-circuit voltage gain of −1000 and an output resistance of 100 Ω feeds a resistive load of 900 Ω. Negative feedback is provided by connection of a resistive voltage divider across the output and one-fiftieth of the output voltage fed back in series with the input signal. Determine the voltage gain with feedback. What percentage change in the voltage gain with feedback would be*

Fig. 9.22 Part of
example 9.7

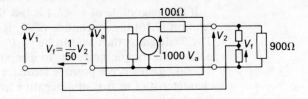

produced by a 50 per cent change in the voltage gain of the basic
amplifier due to a change in the load?

The loading effect of the feedback network can be neglected.

The network is shown in fig. 9.22.

$$A_1 = G_v = \frac{900}{100 + 900}(-1000) = -900$$

Voltage gain with feedback $= G_{vf} = \dfrac{A_1}{1 - \beta A_1}$

$$= \frac{-900}{1 - (\frac{1}{50})(-900)} = \frac{-900}{1 + 18} = -47.4$$

For $A_2 = -450$

$$\therefore \quad G_{vf} = \frac{-450}{1 - (\frac{1}{50})(-450)} = \frac{-450}{1 + 9} = -45.0$$

$$\Delta G_{vf} = 47.4 - 45.0 = 2.3.$$

Percentage change in $G_{vf} = \dfrac{2.4}{47.4} \times 100 = 5.1$ per cent.

Example 9.8   *An amplifier is required with an overall voltage gain of* 100 *and which
does not vary by more than* 1.0 *per cent. If it is to use negative
feedback with a basic amplifier, the voltage gain of which can vary by*
20 *per cent, determine the minimum voltage gain required and the
feedback factor.*

$$100 = \frac{A}{1 + \beta A} \qquad\qquad\qquad [1]$$

$$99 = \frac{0.8A}{1 + \beta\, 0.8A} \qquad\qquad\qquad [2]$$

$$\therefore \quad 100 + 100\beta A = A \qquad\qquad\qquad [3]$$

$$99 + 79.2\beta A = 0.8A \qquad\qquad\qquad [4]$$

$[3] \times 0.792$

$$\therefore \quad 79.2 + 79.2\beta A = 0.792A \qquad\qquad\qquad [5]$$

$[4] - [5]$

$$19.8 = 0.008A$$

$$\therefore \quad A = \frac{19.8}{0.008} = 2475$$

Substitute in [3]

$$100 + 100\beta \times 2475 = 2475$$

$$\therefore \quad \beta = \frac{2375}{100 \times 2475} = 0.009\,60$$

If the magnitude of $1 - \beta A$ is less than unity, then from relation (9.16) the magnitude of G_{vf} is greater than that of A. The feedback is then said to be *positive* or regenerative. It is not very common, however, to use positive feedback to increase gain since a positive feedback amplifier has opposite characteristics to that of a negative feedback amplifier. Thus the stability of the voltage gain with feedback will be worse than that of the basic amplifier.

There is one case of the use of positive feedback that is of considerable practical importance. That is the case where $\beta A = 1$, which gives, from relation (9.17),

$$G_{vf} = \frac{A}{0} = \infty$$

An amplifier with infinite gain is one that can produce an output signal with no externally applied input signal. It provides its own input signal via the feedback network. Such a circuit is known as an *oscillator*, and to produce an output signal at a predetermined frequency the circuitry is arranged so that $\beta A = 1$ at that frequency only. It must be stressed that the condition $\beta A = 1$ must be satisfied in magnitude and phase.

It is necessary to consider the condition $\beta A > 1$. This can only be a transient condition in a feedback amplifier since it means that the output signal amplitude would be increasing with time. This build-up of amplitude will eventually entail the amplifier operating in non-linear parts of its characteristics, perhaps even into the cut-off and/or saturation regions. This results in an effective decrease in A and the system settles down when $\beta A = 1$. Thus the substitution of values of βA greater than $+1$ in relation (9.16) has little practical significance.

In section 9.6 it was stated that loss of gain at high and low frequencies was accompanied by a change of phase shift between input and output signals. It is possible therefore for feedback, which is designed to be negative in the passband, to become positive at high or low frequencies and introduce the possibility of the condition $\beta A = 1$ existing. The amplifier would then oscillate at the appropriate frequency and it is said to be unstable. Much of the design work associated with nominally negative feedback amplifiers is concerned with maintaining stability against oscillation.

Example 9.9   *An amplifier has a voltage gain of -1000 within the passband. At a specific frequency f_x outwith the passband the voltage gain is $15\,dB$ down on the passband value and there is zero phase between input and output signal voltages. Determine the maximum amount of negative feedback that can be used so that the feedback amplifier will be stable.*

Let   voltage gain at $f_x = G_{vx}$

$\therefore$ $20 \log \dfrac{G_v}{G_{vx}} = 15$

$\therefore$ $\dfrac{G_v}{G_{vx}} = 5.62$

$$\therefore \qquad G_{vx} = \frac{1000}{5.62} = 178$$

Oscillation will occur if $\qquad \beta G_v = 1$

$$\therefore \qquad \text{Maximum value of } \beta = \frac{1}{178} = 0.0058$$

9.8 Effect of feedback on input and output resistances

So far, the effect of series-voltage feedback on voltage gain has been considered. It is necessary to consider its effect on the other characteristics of the amplifier. While the introduction of series feedback will affect the magnitude of the input current I_1 for a given value of input voltage V_1, the current gain will be unaffected, i.e. a given value of I_1 will still produce the same value of I_2 as specified by relation (9.7). It follows therefore that the power gain which is the product of the voltage and current gains will change by the same factor as the voltage gain.

The input resistance with feedback can be determined with reference to fig. 9.21 as follows:

$$\left. \begin{array}{c} \text{Input resistance} \\ \text{with feedback} \end{array} \right\} = R_{if} = \frac{V_1}{I_1}$$

$$= \frac{V_a - V_f}{I_1} = \frac{V_a - \beta V_2}{I_1} = \frac{V_a - \beta A V_a}{I_1}$$

$$= \frac{V_a(1 - \beta A)}{I_1}$$

$$R_{if} = R_i(1 - \beta A) \tag{9.18}$$

Thus the input resistance is increased by negative series-voltage feedback and hence decreased by positive series-voltage feedback.

The output resistance with feedback can be determined from the ratio of the open-circuit output voltage to the short-circuit output current.

$$V_{2oc} = A V_a = G_{vo}(V_1 + V_f) = A(V_1 + \beta V_{2oc})$$

$$V_{2oc} = \frac{A V_1}{1 - \beta A}$$

$$I_{2sc} = \frac{A V_a}{R_o} = \frac{A V_1}{R_o}$$

(there is no signal fed back in this case since there is no

output voltage)

$$\text{Output resistance with feedback} = R_{of} = \frac{V_{2o/c}}{I_{2s/c}}$$

$$R_{of} = \frac{R_o}{1 - \beta A} \qquad (9.19)$$

Thus the output resistance is decreased by negative series-voltage feedback and hence increased by positive series-voltage feedback.

Example 9.10   *An amplifier has an open-circuit voltage gain of* 1000, *an input resistance of* 2000 Ω *and an output resistance of* 1.0 Ω. *Determine the input signal voltage required to produce an output signal current of* 0.5 A *in a* 4.0-Ω *resistor connected across the output terminals. If the amplifier is then used with negative series-voltage feedback so that one-tenth of the output signal is fed back to the input, determine the input signal voltage to supply the same output signal current.*

From relation (9.7)

$$\frac{I_2}{I_1} = \frac{AR_i}{R_o + R_L} = \frac{1000 \times 2000}{1.0 + 4.0} = 4.0 \times 10^5$$

$$I_1 = \frac{0.5}{4.0 \times 10^5} = 1.25 \times 10^{-6}\,\text{A}$$

$$V_1 = I_1 R_i = 1.25 \times 10^{-6} \times 2 \times 10^3 = 2.5\,\text{mV}$$

With feedback

$$\frac{I_2}{I_1} = 4.0 \times 10^5 \text{ (as before)}$$

$$\therefore \qquad I_1 = 1.25 \times 10^{-6}\,\text{A}.$$

$$R_{if} = R_i(1 + \beta A) = 2000\left(1 + \frac{1}{10} \times \frac{4}{1+4} \times 1000\right)$$

$$= 2000(1 + 80) = 162\,000\,\Omega$$

$$\therefore \qquad V_1 = 1.25 \times 10^{-6} \times 1.62 \times 10^5$$

$$= 0.202\,\text{V} = 202\,\text{mV}.$$

9.9 Effect of feedback on bandwidth

Consider an amplifier which a frequency response characteristic of the form shown in fig. 9.23 and operating without feedback. The bandwidth limiting frequencies are f_1 and f_2. Now apply negative feedback so that the midband gain falls to G_{vf}. Due to feedback, G_{vf} is much smaller than A and the characteristic becomes flatter. It follows that the new cut-off frequencies $f_{1'}$ and $f_{2'}$ are further apart and hence the new bandwidth is increased. We may therefore conclude that the bandwidth of an amplifier is increased following the application of negative feedback.

Fig. 9.23 Effect of negative
feedback on bandwidth

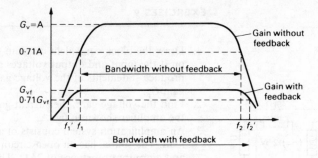

9.10 Distortion

Amplifiers produce three types of distortion:

(a) Phase distortion

In this distortion, consider a signal made from a number of alternating voltages of differing frequencies, During the amplification, the phase shifts experienced by the different frequencies can also be different, thus the total waveform being amplified becomes distorted. This is not too important in sound amplifiers, i.e. audio-frequency amplifiers, because our ears are not too sensitive, yet it would be most important to video-frequency amplifiers since our eyes are very much more sensitive to changes.

(b) Amplitude distortion

This is also known as frequency distortion and arises when signals of differing frequencies are not all amplified to the same extent. The cause of this can be observed by reference to the frequency response characteristic shown in fig. 9.23 in which we can observe that the gain is less at low and high frequencies.

(c) Harmonic distortion

This arises in amplifiers which are non-linear in their response. It follows that the input signal could be a pure sinusoidally varying voltage yet the output voltage would include multiple-frequency components, e.g. a 1-kHz input signal might be accompanied by other output signals at frequencies of 2 kHz, 3 kHz and so on. These additional frequencies have been generated within the amplifier and therefore give rise to distortion. The multiple-frequency signals are termed harmonics, hence the name for the distortion.

Each form of distortion is present in most amplifiers to a greater or lesser extent. However, the distortions are very much dependent on the basic gain of the amplifier, hence the introduction of feedback can and does reduce distortion to a considerable extent.

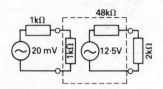

Fig. A

EXERCISES 9

1. Draw the block equivalent diagram of an amplifier and indicate on it the input and output voltages and currents. Hence produce statements of the voltage gain G_v and the power gain G_p.

2. Find the voltage gain, the current gain and the power gain of the amplifier shown in fig. A.

3. An amplification system consists of a course, an amplifier and a load. The source has an open-circuit output voltage of 25 mV and an output resistance of 2 kΩ. The amplifier has an open-circuit voltage gain of 975, an input resistance of 8 kΩ and an output resistance of 3 Ω. The load resistance is 3 Ω.

   Draw the equivalent circuit of this system and hence determine the voltage, current and power gains of the amplifier.

   If the load resistance is reduced to 2 Ω, determine the current gain of the amplifier.

4. (a) Explain the term *bandwidth* and describe the relationship between gain and bandwidth.

   (b) Describe the effect on the input and output resistance of series-voltage negative feedback.

   (c) An amplifier has the following parameters:

Input resistance (kΩ)	Output resistance (kΩ)	Short-circuit current gain
500	4.7	5000

   Draw the equivalent circuit for the amplifier and determine the open-circuit voltage gain.

5. (a) An amplifier has the following parameters:

Input resistance (kΩ)	Output resistance (Ω)	Open-circuit voltage gain
2	250	1000

   It is used with a load resistance of 750 Ω. Draw the Thévenin (constant voltage generator) equivalent circuit of the amplifier and its load. For this amplifier calculate: (i) the voltage gain; (ii) the current gain.

   (b) Draw the Norton (constant current generator) equivalent circuit of the amplifier and its load.

   (c) A 10 mV ideal voltage generator is applied to the amplifier input. Determine the current of the source generator for the amplifier equivalent circuit.

6. For the amplifier shown in fig. B, determine the current gain and the voltage gain.

Fig. B

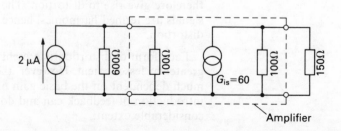

Amplifier

7. An amplifier has a voltage gain of 50 dB. Determine the output voltage when the input voltage is 2.0 mV.

8. Express in decibels the gain of an amplifier which gives an output of 10 W from an input of 0.1 W.

9. The output of a signal generator is calibrated in decibels for a resistive load of 600 Ω connected across its output terminals. Determine the terminal voltage to give (a) 0 dB corresponding to 1 mW dissipation in the load; (b) + 10 dB; (c) − 10 dB.

10. An amplifier system consists of a single source, an amplifier and a loudspeaker acting as a load.

   The source has an open-circuit voltage output of 20 mV and an output resistance of 1 kΩ. The amplifier has an input resistance of 9 kΩ, an open-circuit voltage gain of 60 dB and an output resistance of 8 Ω. The load has a resistance of 8 Ω.

   Draw the equivalent circuit of this system. Calculate: (a) the current, voltage and power gains; express these gains in dB; (b) the output power if an identical loudspeaker is added in parallel with the original loudspeaker.

11. Draw and explain the gain/frequency characteristic of a resistance-capacitance coupled amplifier. Reference should be made to mid-band frequencies and to cut-off points.

   An amplifier has an open-circuit voltage gain of 800, an output resistance of 20 Ω and an input resistance of 5 kΩ. It is supplied from a signal source of e.m.f. 10 mV and internal resistance 5 kΩ. If the amplifier supplies a load of 30 Ω, determine the magnitude of the output signal voltage and the power gain (expressed in decibels) of the amplifier.

12. The voltage gain of an amplifier is 62 dB when the amplifier is loaded by a 5-kΩ resistor. When the amplifier is loaded by a 10-kΩ resistor, the voltage gain is 63 dB. Determine the open-circuit voltage gain and the output resistance of the amplifier.

13. An amplifier has an input resistance of 20 kΩ and an output resistance of 15 Ω. The open circuit gain of the amplifier is 25 dB and it has a resistive load of 135 Ω.

   The amplifier is supplied from an a.c. signal source of internal resistance 5 kΩ and r.m.s. amplitude 15 dB above a reference level of 1 μV.

   Draw the equivalent circuit of this arrangement of source, amplifier and load.

   Calculate: (a) the voltage at the input terminals to the amplifier; (b) the voltage across the load; (c) the power gain in dB.

14. (a) An amplifier has series-voltage feedback. With the aid of a block diagram for such an amplifier, derive the general feedback equation:

$$G = \frac{A}{1 - \beta A}.$$

   (b) (i) In an amplifier with a constant input a.c. signal of 0.5 V, the output falls from 25 V to 15 V when feedback is applied. Determine the fraction of the output voltage which is fed back.

   (ii) If due to ageing the amplifier gain without feedback falls to 40, determine the new gain of the stage, assuming the same value of feedback.

15. The gain of an amplifier with feedback is 110. Given that the feedback fraction is + 1.5 per cent, determine the normal gain of the amplifier.

16. When a feedback fraction of 1/60 is introduced to an amplifier,

its gain changes by a factor of 2.5. Find the normal gain of the amplifier.

17. An amplifier has a gain of 120. What change in gain will occur if a -3.5 per cent feedback is introduced? If the overall gain had to be reduced to 10, what feedback would be required?

18. The voltage amplification of an amplifier is 65. If feedback fractions of 0.62 per cent, $-1/50$, $-1/80$ were introduced, express the new amplifier gain in dB.

19. The gain of an amplifier with feedback is 53, and without feedback the gain is 85. Express the feedback as a percentage and the change in gain in dB.

20. An amplifier with -4.5 per cent feedback has a gain of 12.04 dB. Determine the amplifier gain in dB without feedback.

21. An amplifier is required with a voltage gain of 100. It is to be constructed from a basic amplifier unit of voltage gain 500. Determine the necessary fraction of the output voltage that must be used as negative series voltage feedback. Hence determine the percentage change in the voltage gain of the feedback amplifier if the voltage gain of the basic amplifier (*a*) decreases by 10 per cent (*b*) increases by 10 per cent.

22. A series-voltage feedback amplifier has a feedback factor $\beta = 9.5 \times 10^{-4}$. If its voltage gain without feedback is 1000, calculate the voltage gain when the feedback is (*a*) negative and (*b*) positive. What percentage increase in voltage gain without feedback would produce oscillation in the positive feedback case?

23. An amplifier has the following characteristics:

$$\text{Open-circuit voltage gain} = 75 \,\text{dB},$$

$$\text{Input resistance} = 40 \,\text{k}\Omega,$$

$$\text{Output resistance} = 1.5 \,\text{k}\Omega.$$

It has to be used with negative series-voltage feedback to produce an output resistance of 5.0 Ω. Determine the necessary feedback factor. With this feedback factor and a resistive load of 10 Ω determine the output signal voltage when the feedback amplifier is supplied from a signal source of 10 mV and series resistance 10 kΩ.

10 Semiconductor Devices

10.1 Introduction

Having considered the general principles of amplifiers, we next have to consider some of the ways in which amplifier networks operate. Over the history of electronics, many devices have been used but, in the latter part of this century, the most common forms of device are based on semiconductor materials. This chapter is devoted to introducing such materials and explaining some simple applications, while the transistor, which is fundamental to most amplifiers, is described in the next chapter.

10.2 Atomic structure

It has already been stated in section 5.2 that an atom of a material consists of a nucleus carrying a positive charge surrounded by one or more electrons revolving around the nucleus. Electrons which are moving in orbits close to the nucleus are subject to relatively strong forces of attraction towards the protons of the nucleus, whereas those in the outer orbits are acted upon by progressively smaller forces, and the electrons in the outermost orbit can be easily detached from their atoms to become carriers of negative charges.

In semiconductor work, the materials with which we are principally concerned are germanium and silicon. These materials possess a crystalline structure, i.e. the atoms are arranged in an orderly manner. In both germanium and silicon, each atom has four electrons orbiting in the outermost shell and is therefore said to have a valency of four; or, alternatively, the atoms are said to be *tetravalent*. In the case of the silicon atom, the nucleus consists of 14 protons and 14 neutrons; and when the atom is neutral, the nucleus is surrounded by 14 electrons, 4 of which are *valence electrons*, one or more of which may be detached from the atom. If the four valence electrons were detached, the atom would be left with 14 units of positive charge on the protons and 10 units of negative charge on the 10 remaining electrons,

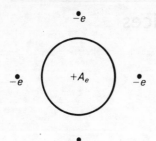

Fig. 10.1 An isolated tetravalent atom

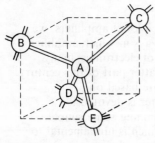

Fig. 10.2 Atomic structure of a lattice crystal

thus giving an *ion* (i.e. an atom possessing a net positive or negative charge) carrying a net positive charge of 4e, where **e** represents the magnitude of the charge on an electron, namely 1.6×10^{-19} C. The neutrons possess no resultant electric charge. A tetravalent atom, isolated from other atoms, can therefore be represented as in fig. 10.1, where the circle represents the ion **carrying** the net positive charge of 4e and the four dots represent the four valence electrons.

The cubic diamond lattice arrangement of the atoms in a perfect crystal of germanium or silicon is represented by the circles in fig. 10.2, where atoms B, C, D and E are located at diagonally opposite corners of the six surfaces of an imaginary cube, shown dotted, and atom A is located at the centre of the cube. The length of each side of the dotted cube is about 2.8×10^{-10} m for germanium and about 2.7×10^{-10} m for silicon.

10.3 Covalent bonds

When atoms are as tightly packed as they are in a germanium or a silicon crystal, the simple arrangement of the valence electrons shown in fig. 10.1 is no longer applicable. The four valence electrons of each atom are now shared with the adjacent four atoms: thus in fig. 10.2, atom A shares its four valence electrons with atoms B, C, D and E. In other words, one of A's valence electrons is linked with A and B, another with A and C, etc. Similarly, one valence electron from each of atoms B, C, D and E is linked with atom A. One can imagine the arrangement to be somewhat as depicted in fig. 10.3, where the four dots, marked $-\mathbf{e}_A$, represent the four

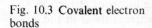

Fig. 10.3 **Covalent** electron bonds

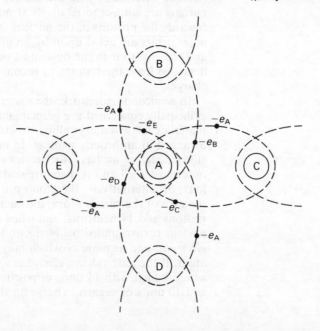

valence electrons of atom A, and dots $-e_B$, $-e_C$, $-e_D$ and $-e_E$ represent the valence electrons of atoms B, C, D and E respectively that are linked with atom A. The dotted lines are not intended to indicate the actual paths or the relative positions of these valence electrons but merely that the electrons on the various dotted lines move around the two atoms enclosed by a given dotted line.

It follows that each positive ion of germanium or silicon, carrying a net charge of 4e, has 8 electrons, i.e. 4 *electron-pairs*, surrounding it. Each electron-pair is referred to as a *covalent bond*; and in fig. 10.2, the covalent bonds are represented by the pairs of parallel lines between the respective atoms. An alternative two-dimensional method of representing the positive ions and the valence electrons forming the covalent bonds is shown in fig. 10.4, where the large circles represent the ions, each with a net positive charge of 4e, and the bracketed dots represent the valence electrons.

Fig. 10.4 Tetravalent atoms with covalent bonds

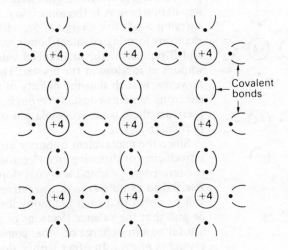

Covalent bonds

These covalent bonds serve to keep the atoms together in crystal formation and are so strong that at absolute zero temperature, i.e. $-273°C$, there are no free electrons. Consequently, at that temperature, pure* germanium and silicon behave as perfect insulators. At normal atmospheric temperature, some of the covalent bonds are broken, i.e. some of the valence electrons break away from their atoms. This effect is discussed in section 10.6, but as a first approximation, we can assume that pure germanium and silicon are perfect insulators and that the properties utilized in semiconductor rectifiers are produced by controlled amounts of impurities introduced into pure germanium and silicon crystals.

---

* A crystal can be regarded as 'pure' when impurities are less than 1 part in 10^{10}. Such a crystal is referred to as an *intrinsic* semiconductor.

10.4 An n-type semiconductor

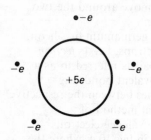

Fig. 10.5 An isolated pentavalent atom

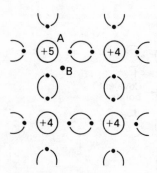

Fig. 10.6 An n-type semiconductor

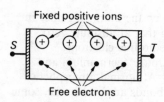

Fig. 10.7 An n-type semiconductor

Certain elements such as phosphorus, arsenic and antimony are pentavalent, i.e. each atom has 5 valence electrons, and an isolated pentavalent atom can be represented by an ion having a net positive charge of 5e and five valence electrons as in fig. 10.5. When a minute* trace of the order of 1 part in 10^8 of such an element is added to pure germanium or silicon, the conductivity is considerably increased. We shall now consider the reason for this effect.

When an atom of a pentavalent element such as antimony is introduced into a crystal of pure germanium, it enters into the lattice structure by replacing one of the tetravalent germanium atoms, but *only four of the five valence electrons of the antimony atom can join as covalent bonds*. Consequently the substitution of a pentavalent atom for a germanium atom provides a free electron. This state of affairs is represented in fig. 10.6, where A is the ion of, say, an antimony atom, carrying a positive charge of 5e, with four of its valence electrons forming covalent bonds with four adjacent atoms, and B represents the unattached valence electron free to wander at random in the crystal. This random movement, however, is such that the density of these free or mobile electrons remains constant throughout the crystal and therefore there is no accumulation of free electrons in any particular region.

Since the pentavalent impurity atoms are responsible for introducing or donating free electrons into the crystal, they are termed *donors*; and a crystal doped with such impurity is referred to as an *n-type* (i.e. negative-type) semiconductor. It will be noted that each antimony ion has a *positive* charge of 5e and that the valence electrons of each antimony atom have a total *negative* charge of $-5e$; consequently the doped crystal is *neutral*. In other words, donors provide fixed positively-charged ions and an equal number of electrons free to move about in the crystal, as represented by the circles and dots respectively in fig. 10.7.

The greater the amount of impurity in a semiconductor, the greater is the number of free electrons per unit volume and therefore the greater is the conductivity of the semiconductor.

When there is no p.d. across the metal electrodes, S and T, attached to opposite ends of the semiconductor, the paths of the random movement of one of the free electrons may be represented by the full lines AB, BC, CD, etc., in fig. 10.8, i.e. the electron is accelerated in direction AB until it collides with an atom with the result that it may rebound in direction

---

*It may assist us in realizing how minute this impurity is if we were to imagine a portion of a crystal magnified to such an extent that there was one atom for every cubic centimetre of a room, $8\,m \times 4\,m \times 3\,m$, then an impurity of 1 in 10^8 would correspond to $1\,cm^3$ being occupied by an atom of the impurity.

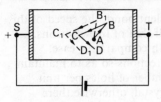

Fig. 10.8 Movement of a free electron

BC, etc. Different free electrons move in different directions so as to maintain the density constant; in other words, there is no resultant drift of electrons towards either S or T.

Let us next consider the effect of connecting a cell across S and T, the polarity being such that S is positive relative to T. The effect of the electric field (or potential gradient) in the semiconductor is to modify the random movement of the electron as shown by the dotted lines AB_1, B_1C_1 and C_1D_1 in fig. 10.8, i.e. there is superimposed on the random movement a drift of the electron towards the positive electrode S, and the number of electrons entering electrode S from the semiconductor is the same as that entering the semiconductor from electrode T.

10.5 A p-type semiconductor

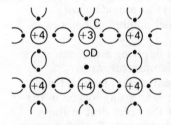

Fig. 10.9 A p-type semiconductor

Materials such as indium, gallium, boron and aluminium, are trivalent, i.e. each atom has only three valence electrons and may therefore be represented as an ion having a positive charge of $3e$ surrounded by 3 valence electrons. When a trace of, say, indium is added to pure germanium, the indium atoms replace the corresponding number of germanium atoms in the crystal structure, but each indium atom can provide only three valence electrons to join with the four valence electrons of adjacent germanium atoms, as shown in fig. 10.9, where C represents the indium atom. Consequently there is an incomplete valence bond, i.e. there is a vacancy represented by the small circle D in fig. 10.9. This vacancy is referred to as a *hole* — a term that is peculiar to semiconductors. This incomplete valent bond has the ability to attract a covalent electron from a near-by germanium atom, thereby filling the vacancy at D but creating another hole at, say, E as in fig. 10.10(a). Similarly the incomplete valent bond due to hole E attracts a covalent electron from another germanium atom, thus creating a new hole at, say, F as in fig. 10.10(b).

If the semiconductor is not being subjected to an external electric field, the position of the hole moves at random from

Fig. 10.10 Movement of a hole in a p-type semiconductor

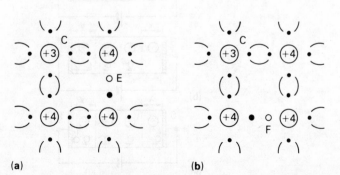

(a) (b)

one covalent bond to another covalent bond, the speed of this random movement being about half that of free electrons, since the latter can move about with comparative ease. Different holes move in different directions so as to maintain the density of the holes (i.e. the number of holes per unit volume) uniform throughout the crystal, otherwise there would be an accumulation of positive charge in one region with a corresponding negative charge in another region.

It will be seen that each of the *germanium* atoms associated with holes E and F in fig. 10.10(*a*) and (*b*) respectively has a nucleus with a resultant positive charge 4**e** and three valence electrons having a total negative charge equal to −3**e**. Hence each atom associated with a hole is an ion possessing a net positive charge **e**, and the movement of a hole from one atom to another can be regarded as the movement of a positive charge **e** within the structure of the p-type semiconductor.

Germanium and silicon, doped with an impurity responsible for the formation of holes, are referred to as *p-type* (positive-type) semiconductors; and since the trivalent impurity atoms can accept electrons from adjacent germanium or silicon atoms, they are termed *acceptors*.

It will be seen from fig. 10.10 that the trivalent impurity atom, with the *four covalent bonds complete*, has a nucleus carrying a positive charge 3**e** and four electrons having a total negative charge −4**e**. Consequently such a trivalent atom is an ion carrying a net *negative* charge −**e**. For each such ion, however, there is a *hole* somewhere in the crystal; in other words, the function of an acceptor is to provide *fixed* negatively-charged ions and an equal number of holes as in fig. 10.11.

Let us next consider the effect of applying a p.d. across the two opposite faces of a p-type semiconductor, as in fig. 10.12(*a*), where S and T represent metal plates, S being positive relative to T. The negative ions, being locked in the crystal structure, cannot move, but the holes drift in the direction of the electric field, namely towards T. Consequently region *x* of the semiconductor acquires a net negative charge and region *y* acquires an equal net positive charge, as in

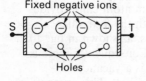

Fixed negative ions

Holes

Fig. 10.11 A p-type semiconductor

Fig. 10.12 Drift of holes in a p-type semiconductor

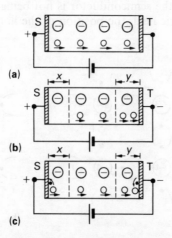

fig. 10.12(*b*). These charges attract electrons from T into region *y* and repel electrons from region *x* into S, as indicated in fig. 10.12(*c*). The electrons attracted from T combine with holes in region *y* and electrons from covalent bonds enter S, thus creating in region *x* new holes which move from that region towards electrode T. The rate at which holes are being neutralized near electrode T is the same as that at which they are being created near electrode S. Hence, in a p-type semiconductor, we can regard the current as being due to the drift of holes in the conventional direction, namely from the positive electrode S to the negative electrode T.

10.6 Junction diode

Let us now consider a crystal, one-half of which is doped with p-type impurity and the other half with n-type impurity. Initially, the p-type semiconductor has mobile holes and the same number of fixed negative ions carrying exactly the same total charge as the total positive charge represented by the holes. Similarly the n-type semiconductor has mobile electrons and the same number of fixed positive ions carrying the same total charge as the total negative charge on the mobile electrons. Hence each region is initially neutral.

Owing to their random movements, some of the holes will diffuse* across the boundary into the n-type semiconductor and some of the free electrons will similarly diffuse into the p-type semiconductor, as in fig. 10.13(*a*). Consequently region A acquires an excess negative charge which repels any more electrons trying to migrate from the n-type into the p-type semiconductor. Similarly, region B acquires a surplus of positive charge which prevents any further migration of holes across the boundary. These positive and negative charges are concentrated near the junction, somewhat as indicated in fig. 10.13(*b*), and thus form a potential barrier between the two regions.

Forward bias. Let us next consider the effect of applying a p.d. across metal electrodes S and T, S being positive relative to T, as in fig. 10.13(*c*). The direction of the electric field in the semiconductor is such as to produce a drift of holes towards the right in the p-type semiconductor and of free electrons towards the left in the n-type semiconductor. In the region of the junction, free electrons and holes combine, i.e. free electrons fill the vacancies represented by the holes. For each combination, an electron is liberated from a covalent bond in the region near positive plate S and enters that plate, thereby creating a new hole which moves through

---

* It will be noted that *diffusion* takes place when there is a difference in the concentration of carriers in adjacent regions of a crystal; but *drift* of carriers takes place only when there is a difference of potential between two regions.

Fig. 10.13 Junction diode

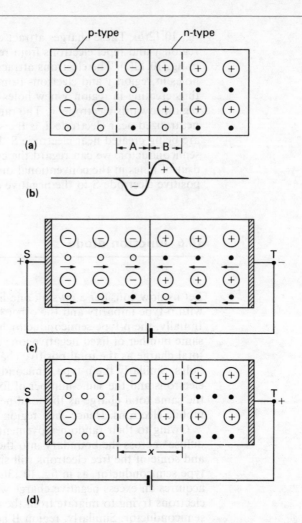

p-type   n-type

(a)

← A →|← B →

(b)

+

−

+S   T−

(c)

−S   T+

← x →

(d)

the p-type material towards the junction, as described in section 10.5. Simultaneously, an electron enters the n-region from the negative plate T and moves through the n-type semiconductor towards the junction, as described in section 10.4. The current in the diode is therefore due to hole-flow in the p-region, electron-flow in the n-region and a combination of the two in the vicinity of the junction.

Reverse bias. When the polarity of the applied voltage is reversed, as shown in fig. 10.13(*d*), the holes are attracted towards the negative electrode S and the free electrons towards the positive electrode T. This leaves a region *x*, known as a *depletion layer*, in which there are no holes or free electrons, i.e. there are no charge carriers in this region apart from the relatively few that are produced spontaneously by thermal agitation, as mentioned below. Consequently the junction behaves as an insulator.

In practice, there is a small current due to the fact that at room temperature, thermal agitation or vibration of atoms takes place in the crystal and some of the valence electrons acquire sufficient velocity to break away from their atoms, thereby producing *electron-hole pairs*. An electron-hole pair has a life of about 100 microseconds in germanium and about 50 microseconds in silicon. The generation and recombination of electron-hole pairs is a continuous process and is a function of the temperature. The higher the temperature, the greater is the rate at which generation and recombination of electron-hole pairs take place and therefore the lower the *intrinsic resistance* of a crystal of pure germanium or silicon.

These thermally liberated holes and free electrons are referred to as *minority carriers* because, at normal temperature, their number is very small compared with the number of *majority carriers* due to the doping of the semiconductor with donor and acceptor impurities. Hence, in a p-type semiconductor, holes form the majority carriers and electrons the minority carriers, whereas in an n-type crystal, the majority carriers are electrons and holes are the minority carriers.

When a germanium junction diode is biased in the reverse direction, the current remains nearly constant for a bias varying between about 0.1 volt and the breakdown voltage. This constant value is referred to as the *saturation current* and is represented by I_s in fig. 10.14. In practice, the reverse

Fig. 10.14 Static characteristic for a germanium junction diode having negligible surface leakage

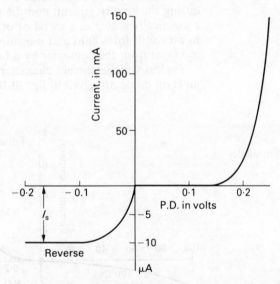

current increases with increase of bias, this increase being due mainly to surface leakage. In the case of a germanium junction diode in which the surface leakage is negligible, the current is given by the expression:

$$i = I_s(e^{ev/kT} - 1) \tag{10.1}$$

where I_s = saturation current with negative bias,

       e = charge on electron = 1.6×10^{-19} C,

       v = p.d., in volts, across junction,

      k = Boltzmann's constant = 1.38×10^{-23} J/K

and T = thermodynamic temperature = $(273.15 + \theta)$°C

Let us assume a saturation current of, say, $10 \mu A$; then for a temperature of 300 K ($= 27°C$), we have from expression (10.1),

$$i = 10(e^{38.6v} - 1) \text{ microamperes} \qquad (10.2)$$

Values of current i, calculated from expression (10.2) for various values of v, are plotted in fig. 10.14.

10.7 Construction and static characteristics of a junction diode

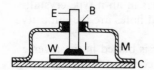

Fig. 10.15 A germanium junction diode

Fig. 10.15 shows one arrangement of a germanium junction diode. A thin wafer or sheet W is cut from an n-type germanium crystal, the area of the wafer being proportional to the current rating of the diode. The lower surface of the wafer is soldered to a copper plate C and a bead of indium I is placed centrally on the upper surface. The unit is then heat-treated so that the indium forms a p-type alloy with the germanium. A copper electrode E is soldered to the bead during the heat treatment, and the whole element is hermetically sealed in a metal or other opaque container M to protect it from light and moisture. The electrode E is insulated from the container by a bush B.

Typical voltage/current characteristics of a germanium junction diode are given in fig. 10.16, the full lines being for a

Fig. 10.16 Static characteristics of a germanium junction diode

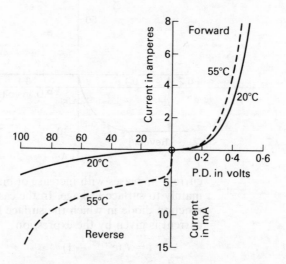

temperature of the surrounding air (i.e. ambient temperature) of 20°C and the dotted lines for 55°C. For a given reverse bias, the reverse current roughly doubles for every 10°C rise of temperature. This rectifier can withstand a peak inverse voltage of about 100 V at an ambient temperature of 20°C.

The silicon junction diode is similar in appearance to the

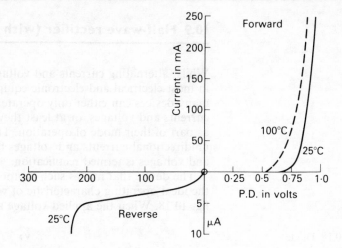

Fig. 10.17 Static
characteristics of a silicon
junction diode

germanium diode and typical voltage/current characteristics
are given in fig. 10.17. The properties of silicon junction
diodes differ from those of germanium junction diodes in the
following respects:

(*a*) the forward voltage drop is roughly double that of the
corresponding germanium diode;

(*b*) the reverse current at a given temperature and voltage
is approximately a hundredth of that of the corresponding
germanium diode, but there is little sign of current saturation
as is the case with germanium — in fact, the reverse current of
a silicon diode is roughly proportional to the square root of
the voltage until breakdown is approached;

(*c*) it can withstand a much higher reverse voltage and can
operate at temperatures up to about 150–200°C, compared
with about 75–90°C for germanium;

(*d*) the reverse current of a silicon diode, for a given
voltage, practically doubles for every 8°C rise of temperature,
compared with 10°C for germanium.

10.8 Rectifier circuits

Since a diode has the characteristic of having a much greater
conductivity in one direction than in the other, it will produce
a direct component of current when connected in series with
an alternating voltage and a load. This process is known as
rectification and is the main use to which diodes are put.
There are numerous applications for rectification, e.g. driving
a d.c. motor from a.c. mains and the production of direct-
voltage supplies for electronic amplifiers.

10.9 Half-wave rectifier (with resistive load)

Whilst alternating currents and voltages play the leading roles in most electrical and electronic equipment, nevertheless many devices can either only operate on unidirectional currents and voltages, or at least they require such a supply as part of their mode of operation. The process of obtaining unidirectional currents and voltages from alternating currents and voltages is termed rectification.

The device that makes such a process possible is a diode, the ideal operating characteristic of which is given in fig. 10.18. When the applied voltage acts in the forward

Fig. 10.18 Diode characteristics

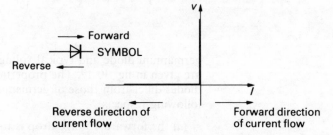

direction, there is no voltage drop across the diode and a current flows unimpeded. However, when the applied voltage acts in the reverse direction, a voltage drop appears across the diode and no current flows.

It is possible to obtain rectification by means of a single diode as indicated in fig. 10.19. The current can only flow

Fig. 10.19 Half-wave rectification

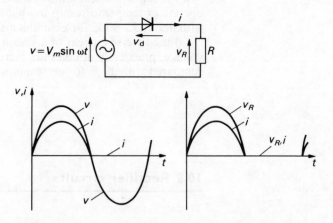

through the diode in one direction and thus the load current can only flow during alternate half-cycles. For this reason, the system is known as half-wave rectification. The load current, and hence the voltage drop across the load, is unidirectional and could be described as direct, although this term is more usually reserved for steady unidirectional quantities.

With reference to the circuit shown in fig. 10.19.

v = instantaneous supply voltage,

v_d = instantaneous voltage across the diode,

v_R = instantaneous voltage across the load resistance,

i = instantaneous diode current.

Using Kirchhoff's Second Law in the closed loop:

$$v = v_d + v_R = v_d + iR$$

$$\therefore \quad i = -\frac{1}{R}v_d + \frac{v}{R} \tag{10.3}$$

Examination of the characteristics shown in figs. 10.16 and 10.17 indicates that v_d has a small value during the positive half-cycle when the diode conducts but is equal to v during the negative half-cycle, there being negligible current and v_R consequently also being negligible. However, if we simplify the voltage/current characteristic to the idealized one shown in fig. 10.18 then the value of v_d is zero and the effective resistance (termed the forward resistance) of the diode is zero. If the supply voltage is

$$v = V_m \sin \omega t$$

then
$$i = \frac{v}{R} = \frac{V_m \sin \omega t}{R} = I_m \sin \omega t.$$

Thus if the mean value of the current (neglecting the reverse current) = I_{dc} then:

$$I_{dc} = \frac{1}{2\pi} \int_0^\pi I_m \sin \omega t \, d(\omega t) = \frac{I_m}{2\pi}\left[-\cos \omega t \right]_0^\pi$$

$$= \frac{I_m}{2\pi}\left[-\cos \pi + \cos 0 \right] = \frac{I_m}{2\pi}[1 + 1]$$

$$\therefore \quad I_{dc} = \frac{I_m}{\pi} = 0.318 I_m \tag{10.4}$$

Similarly the r.m.s. value of the current = I_{rms} then:

$$I_{rms} = \sqrt{\frac{1}{2\pi} \int_0^\pi I_m^2 \sin^2 \omega t \, d(\omega t)}$$

$$= \sqrt{\frac{I_m^2}{2\pi} \int_0^\pi \tfrac{1}{2}(1 - \cos 2\omega t) \, d(\omega t)}$$

$$= \sqrt{\frac{I_m^2}{2\pi} \times \frac{1}{2}\left[\omega t + \tfrac{1}{2}\sin 2\omega t \right]_0^\pi} = \sqrt{\frac{I_m^2}{4\pi}[\pi]}$$

$$\therefore \quad I_{rms} = \frac{I_m}{2} = 0.5 I_m \tag{10.5}$$

The voltage across the load is given by:

$$v_R = Ri = RI_m \sin \omega t = V_m \sin \omega t.$$

Therefore, in the same way as derived for the current:

$$\text{Mean value of the load voltage} = V_{dc} = \frac{V_m}{\pi} \qquad (10.6)$$

$$\text{R.M.S. value of the load voltage} = V_{rms} = \frac{V_m}{2} \qquad (10.7)$$

The maximum voltage, which occurs across the diode in the reverse direction, is known as the peak inverse voltage (P.I.V.). This must be less than the breakdown voltage of the diode if it is not to conduct appreciably in the reverse direction. The peak inverse voltage for the diode in this circuit occurs when the potential of B is positive with respect to A by its maximum amount. The reverse resistance of the diode will, in the great majority of practical cases, be very much greater than the load resistance and most of the applied voltage will appear across the diode. Thus the peak inverse voltage equals approximately the peak value of the supply voltage.

Since the production of direct current from an a.c. supply is the object of the circuit, the useful power output is that produced in the load by the d.c. component of the load current. The efficiency of a rectifier circuit is defined as:

$$\text{Efficiency } \eta = \frac{\text{power in the load due to d.c. component of current}}{\text{total power dissipated in the circuit}}$$

$$= \frac{I_{dc}^2 R}{I_{rms}^2 R} = \left[\frac{I_m}{\pi}\right]^2 \times \left[\frac{2}{I_m}\right]^2 = 0.405$$

This efficiency is based on $r_d = 0$. If it had a greater value, the efficiency would be reduced since the total power would be $I_{rms}^2(r_d + R)$,

hence $$\eta_m = 0.405 \qquad (10.8)$$

Waveforms for the half-wave rectifier circuit are shown in fig. 10.20.

Fig. 10.20 Waveforms for half-wave rectifier circuit

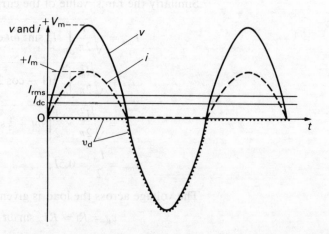

10.10 Full-wave rectifier network (with resistive load)

The half-wave rectifier gave rise to an output which had a unidirectional current but the resulting current does not compare favourably with the direct current that one would expect from, say, a battery. One reason is that the half-wave rectifier is not making use of the other half of the supply waveform. It would be technically and economically advantageous if both halves were rectified and this may be achieved by a full-wave rectifier network.

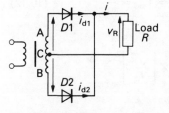

Fig. 10.21 Full-wave rectifier with resistive load

The basic full-wave rectifier network is shown in fig. 10.21. C is a centre tap on the secondary of the transformer, thus the e.m.f.s induced in each section of the secondary are equal, and when the potential of A is positive with respect to C, so is that of C positive with respect to B. With these polarities, diode D_1 will conduct while diode D_2 is non-conducting. When these polarities reverse, diode D_2 will conduct and D_1 will be non-conducting. In this way, each diode conducts on alternate half-cycles, passing current through the load in the same direction.

For
$$v_{AC} = v = V_m \sin \omega t$$

then
$$v_{BC} = -v = -V_m \sin \omega t.$$

For identical diodes with a forward resistance $r_d = 0$ and an infinite reverse resistance, then during the period that v is positive:

Diode 1 current $= i_{d1} = \dfrac{V_m \sin \omega t}{R} = I_m \sin \omega t.$

Diode 2 current $= i_{d2} = 0.$

During the period that v is negative:

Diode 1 current $= i_{d1} = 0.$

Diode 2 current $= i_{d2} = \dfrac{-V_m \sin \omega t}{R} = -I_m \sin \omega t.$

At any instant the load current is given by:

$$i = i_{d1} + i_{d2}.$$

Thus in this circuit the current will repeat itself twice every cycle of the supply voltage, therefore:

Mean value of the load current $= I_{dc}$

$$= \frac{1}{\pi} \int_0^\pi I_m \sin \omega t \, d(\omega t)$$

$$\therefore \quad I_{dc} = \frac{2I_m}{\pi} = 0.637 I_m \qquad (10.9)$$

R.M.S. value of the load current $= I_{rms}$

$$= \sqrt{\frac{1}{\pi} \int_0^\pi I_m^2 \sin^2 \omega t \, d(\omega t)}$$

$$\therefore \quad I_{rms} = \frac{I_m}{\sqrt{2}} = 0.707 I_m \qquad (10.10)$$

Similarly, for the load voltage:

Mean value of the load voltage $= V_{dc} = \dfrac{2V_m}{\pi} \qquad (10.11)$

R.M.S. value of the load voltage $= V_{rms} = \dfrac{V_m}{\sqrt{2}} \qquad (10.12)$

For this circuit:

$$\eta = \frac{I_{dc}^2 R}{I_{rms}^2 R} = \left[\frac{2I_m}{\pi}\right]^2 \times \left[\frac{\sqrt{2}}{I_m}\right]^2 = \frac{8}{\pi^2}$$

Again since $r_d = 0$

$$\eta = 0.81 \qquad (10.13)$$

Waveforms for the full-wave rectifier network are shown in fig. 10.22.

Fig. 10.22 Waveforms for full-wave rectifier circuit

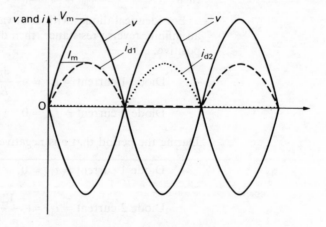

10.11 Bridge rectifier network (with resistive load)

The full-wave rectifier only makes use of each half of the transformer winding for half of the time. The winding can be used all of the time, or the transformer can be omitted in certain cases, by the application of the bridge rectifier network. Such a network is shown in fig. 10.23(a).

When the potential of A is positive with respect to B, diodes D_1 and D_3 conduct and a current flows in the load. When the potential of B is positive with respect to A diodes

Fig. 10.23 Full-wave bridge rectification

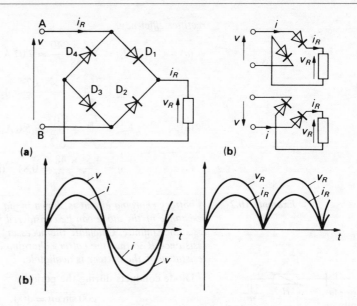

(a)

(b)

(b)

D_2 and D_4 conduct and the current in the load is in the same direction as before. Thus a full-wave type of output is obtained. The expressions derived for current and load voltage for the full-wave circuit will be applicable here also. Note should be taken, however, that, at any instant, two diodes are conducting: fig. 10.23(*b*) shows the active circuits during each half-cycle.

The waveform diagrams shown in fig. 10.23(*c*) indicate that the current *i* taken from the supply is purely alternating yet the load current i_R is unidirectional and therefore basically a direct current.

The efficiency of the network is given by:

$$\eta = \frac{I_{dc}^2 R}{I_{rms}^2 R} = \left[\frac{2}{\pi} I_m\right]^2 \times \left[\frac{\sqrt{2}}{I_m}\right]^2 = \frac{8}{\pi^2}$$

$$\therefore \qquad \eta = 0.81 \qquad\qquad (10.14)$$

Neglecting the forward resistances of the diodes, the voltage which appears across the non-conducting diodes is equal to the supply voltage. Thus the peak inverse voltage is equal to the peak value of the supply voltage.

There are certain advantages to be obtained by using the bridge circuit as opposed to the centre-tapped secondary full-wave circuit. For a given mean load voltage, the bridge circuit uses only half the number of secondary turns, and for a diode, of given maximum peak-inverse-voltage rating, twice the mean output voltage can be obtained from it since the peak inverse voltage encountered is V_m compared to $2V_m$ for the full-wave circuit.

Example 10.1   *The four diodes used in a bridge rectifier circuit have forward resistances which may be considered negligible, and infinite reverse resistances. The alternating supply voltage is 240 V r.m.s. and the resistive load is 48.0 Ω. Calculate:* (a) *the mean load current;* (b) *the rectifier efficiency.*

(a) $$I_m = \frac{\sqrt{2} \times 240}{48} = 7.07 \text{ A}.$$

$$I_{dc} = \frac{2I_m}{\pi} = \frac{2 \times 7.07}{\pi} = 4.5 \text{ A}.$$

(b) $$I_{rms} = \frac{I_m}{\sqrt{2}} = \frac{7.07}{\sqrt{2}} = 5.0 \text{ A}$$

$$\therefore \qquad \eta = \frac{4.5^2 \times 48}{5.0^2 \times 48} = 0.81 \quad \text{(i.e. 81 per cent).}$$

Example 10.2   *A battery charging circuit is shown in fig. 10.24. The forward resistance of the diode can be considered negligible and the reverse resistance infinite. Calculate the necessary value of the variable resistance R so that the battery charging current is 1.0 A. The internal resistance of the battery is negligible.*

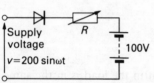

Supply voltage
R
100V
$v = 200 \sin\omega t$

Fig. 10.24 Circuit for example 10.2

Diode conducts during the period $v > 100$ V

$$200 \sin \omega t = 100,$$

i.e. $$\sin \omega t = 0.5$$

i.e. when $$\omega t = \frac{\pi}{6} \text{ and } \frac{5\pi}{6}$$

$\therefore$ diode conducts when $\frac{\pi}{6} < \omega t < \frac{5\pi}{6}$.

During conduction $i = \dfrac{v - 100}{R} = \dfrac{200 \sin \omega t - 100}{R}$

Mean value of current

$$= 1.0 = \frac{1}{2\pi} \int_{\pi/6}^{5\pi/6} \frac{200 \sin \omega t - 100}{R} \, d(\omega t)$$

$$= \frac{1}{2\pi R} \Big[-200 \cos \omega t - 100\omega t \Big]_{\pi/6}^{5\pi/6}$$

$$= \frac{1}{2\pi R} \left[-200 \cos \frac{5\pi}{6} - 100 \times \frac{5\pi}{6} + 200 \cos \frac{\pi}{6} + 100 \times \frac{\pi}{6} \right]$$

$$= \frac{1}{2\pi R} \left[-200 \left(-\frac{\sqrt{3}}{2} \right) - \frac{500\pi}{6} + 200 \left(\frac{\sqrt{3}}{2} \right) + \frac{100\pi}{6} \right]$$

$$= \frac{1}{2\pi R} \left[200\sqrt{3} - \frac{400\pi}{6} \right]$$

$$R = \frac{200\sqrt{3}}{2\pi} - \frac{400\pi}{2\pi \times 6}$$

$$\therefore \quad R = 55.1 - 33.3 = 21.8 \ \Omega.$$

10.12 Smoothing

The rectifier circuits so far described have produced, as required, a direct component of current in the load. There remains, however, a large alternating component. In a large number of applications it is desirable to keep this latter component small. This can be accomplished by the use of smoothing circuits, the simplest of which consists of a capacitor in parallel with the load. Fig. 10.25 shows such an arrangement.

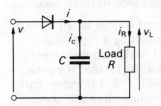

Fig. 10.25 Half-wave rectifier with capacitor input filter

The diode conducts when the supply voltage v is more positive than the load voltage v_L. During this conduction period, if the diode forward resistance is neglected, then the load voltage is equal to the supply voltage. Therefore if:

$$v = V_m \sin \omega t$$

$$i_R = \frac{v}{R} = \frac{V_m \sin \omega t}{R}$$

$$i_C = C \frac{dv_s}{dt} = C\omega V_m \cos \omega t.$$

$$\text{Diode current} = i = i_R + i_C = \frac{V_m}{R} \sin \omega t + \omega C V_m \cos \omega t$$

$$= \sqrt{\left(\frac{V_m}{R}\right)^2 + (\omega C V_m)^2} \sin (\omega t + \phi)$$

$$\therefore \quad i = \frac{V_m}{R} \sqrt{1 + \omega^2 C^2 R^2} \sin (\omega t + \phi) \qquad (10.15)$$

where

$$\tan \phi = \frac{\omega C V_m}{\dfrac{V_m}{R}} = \omega CR.$$

If $i = 0$, i.e. diode cuts off, when $t = t_2$ then

$$\sin (\omega t_2 + \phi) = 0$$

$$\therefore \quad \omega t_2 + \phi = \pi$$

$$\therefore \quad \omega t_2 = \pi - \phi \qquad (10.16)$$

If $\omega CR \gg 1$ then

$$\phi \doteqdot \frac{\pi}{2}$$

and

$$\omega t_2 \doteqdot \pi - \frac{\pi}{2} = \frac{\pi}{2}$$

i.e. the diode ceases to conduct near the instant at which v has its positive maximum value.

While the diode is non-conducting, C will discharge through R and the load voltage will be given by:

$$v_L = V e^{-(t - t_2)/CR}$$

where V = the capacitor voltage at the instant the diode cuts off. This will equal approximately the peak value of the supply voltage if $\omega CR \gg 1$. The diode will start to conduct again during the period that the supply voltage is positive and increasing. This instant can be determined by equating $V e^{-(t - t_2)/CR}$ and $V_m \sin \omega t$; one solution of which will give

$T + t_1$ where:

T = period of the supply voltage,

and t_1 = instant of time at which the diode starts to conduct.

In order to keep the variation in load voltage down during the period when the diode is non-conducting a long time constant CR_L compared to the period of the supply voltage is required. Note, however, from relation (10.15) that the peak value of the diode current increases with C and therefore care must be taken to ensure that the maximum allowable diode peak current is not exceeded. Waveforms for the circuit are shown in fig. 10.26.

Fig. 10.26 Waveforms for half-wave rectifier with capacitor input filter

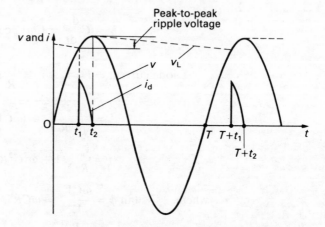

A similar analysis could be carried out for the full-wave circuit. The operation is identical to that of the half-wave circuit during the charging period, but the capacitor discharges into the load resistance for a shorter period, giving less amplitude of ripple for a given time constant during the non-conducting period. Fig. 10.27 shows waveforms for a full-wave circuit with a capacitor input filter.

Fig. 10.27 Waveforms for full-wave rectifier with capacitor input filter

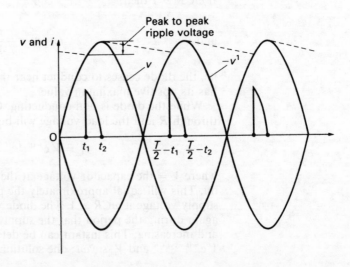

Fig. 10.28 Smoothing circuits (*a*) Series inductor filter (*b*) *L–C* filter (*c*) π-Filter (*d*) Resistive filter

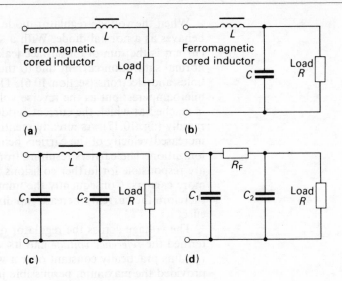

(a)

(b)

(c)

(d)

Fig. 10.28 shows some other types of filter circuits used in practice for smoothing purposes. The ideal arrangement is to connect a low valued impedance at the frequency being used and which presents a high resistance to direct current, e.g. a capacitor, in parallel with the load, and/or a high valued impedance at the frequency being used and which presents a low resistance to direct current, e.g. an inductor, in series with the load. This has the effect of minimizing the ripple across the load while having little effect on the direct voltage developed at the input to the filter circuit. The circuit of fig. 10.28(*d*) does not meet these requirements fully in so much as the resistor *R* will reduce the direct voltage as well as the ripple. The circuit is useful, however, in low current applications where the direct voltage drop across *R* can be kept small; a resistor has an economic advantage over a ferromagnetic-cored inductor.

10.13 Zener diode

If the reverse voltage across a p–n junction is gradually increased, a point is reached where the energy of the current carriers is sufficient to dislodge additional carriers. These carriers, in turn, dislodge more carriers and the junction goes into a form of avalanche breakdown characterized by a rapid increase in current as shown in fig. 10.17. The power due to a relatively large reverse current, if maintained for an appreciable time, can easily ruin the device.

Special junction diodes, often known as *zener diodes*, but more appropriately termed *voltage regulator diodes*, are available in which the reverse breakdown voltage is in the range of about 4 V to about 75 V, the actual voltage depending upon the type of diode.

When the voltage regulator diode is forward-biased, it behaves as a normal diode. With a small reverse voltage, the current is the sum of the surface leakage current and the normal saturation current due to the thermally generated holes and electrons (section 10.6). This current is only a few microamperes; but as the reverse voltage is increased, a value is reached at which the current suddenly increases very rapidly (fig. 10.17). As already mentioned, this is due to the increased velocity of the carriers being sufficient to cause ionization. The carriers resulting from ionization by collision are responsible for further collisions and thus produce still more carriers. Consequently the numbers of carriers, and therefore the current, increase rapidly due to this avalanche effect.

The voltage across the regulator diode after breakdown is termed the *reference voltage* and its value for a given diode remains practically constant over a wide range of current, provided the maximum permissible junction temperature is not exceeded.

Fig. 10.29 shows how a voltage regulator diode (or zener diode) can be used as a voltage stabilizer to provide a

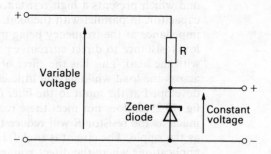

Fig. 10.29 A voltage stabilizer

constant voltage from a source whose voltage may vary appreciably. A resistor R is necessary to limit the reverse current through the diode to a safe value.

10.14 Three-phase rectifier networks

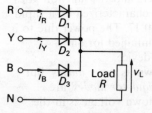

Fig. 10.30 Simple 3-phase rectifier network

Diodes are used with polyphase supplies to give a rectified output which has basically less ripple than rectifiers operating from single-phase supplies. Fig. 10.30 shows a simple 3-phase rectifier network. Each diode will conduct for one-third of a cycle, the conduction path changing instantaneously from one diode to another as one phase voltage becomes more positive than another. For example when v_{RN} becomes less positive than v_{YN} diode D_1 ceases to conduct while D_2 starts to conduct.

The mean load current can be determined as follows:

$$I_{dc} = \frac{1}{2\pi/3} \int_{\pi/6}^{5\pi/6} I_m \sin \omega t \, d(\omega t)$$

$$= \frac{3I_m}{2\pi} \left[-\cos \omega t \right]_{\pi/6}^{5\pi/6} = \frac{3I_m}{2\pi} \left[-\cos \frac{5\pi}{6} + \cos \frac{\pi}{6} \right]$$

$$= \frac{3I_m}{2\pi} \left[\frac{\sqrt{3}}{2} + \frac{\sqrt{3}}{2} \right]$$

$$I_{dc} = \frac{3\sqrt{3}I_m}{2\pi} \tag{10.17}$$

The mean load voltage will be given by:

$$V_{dc} = I_{dc} R$$

$$\therefore \qquad V_{dc} = \frac{3\sqrt{3}I_m R}{2\pi} \tag{10.18}$$

where $I_m R = V_m$, the peak value of the phase voltage if the diode forward resistance is neglected. Fig. 10.31 shows waveforms for the network.

Fig. 10.31 Waveforms for network shown in fig. 10.30

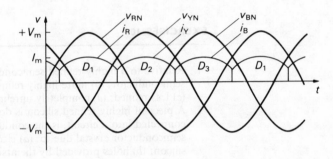

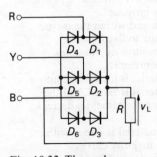

Fig. 10.32 Three-phase bridge rectifier network

The rectifier circuit shown in fig. 10.32 has the advantage that there is no necessity to have a neutral point available. Diodes D_1, D_2, D_4 and D_5 form a bridge rectifier network, similar to that described in section 10.11, between the R and Y lines, similarly D_2, D_3, D_5 and D_6 between the B and R lines and D_3, D_1, D_6 and D_4 between the B and R lines. When v_{RY} is at its maximum positive value diodes D_1 and D_5 will be conducting. These two diodes will continue conducting until v_{RB} is more positive than v_{RY} and diodes D_1 and D_6 will then take over the conduction.

In this way a pair of diodes will conduct for one-sixth of a cycle at a time, each diode conducting for one-third of a cycle. The waveforms for the network are shown in fig. 10.33 where it can be seen that the ripple frequency is twice that obtained in the previous circuit. The amplitude of ripple is decreased also since the conduction periods about the peaks of the supply is decreased.

Fig. 10.33 Waveforms for network shown in fig. 10.32

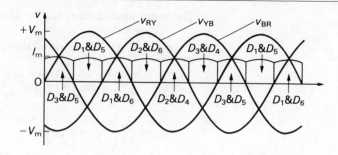

The mean load current can be determined as follows:

$$I_{dc} = \frac{1}{\pi/3} \int_{\pi/3}^{2\pi/3} I_m \sin \omega t \, d(\omega t)$$

$$= \frac{3I_m}{\pi} \left[-\cos \omega t \right]_{\pi/3}^{2\pi/3}$$

$$= \frac{3I_m}{\pi} \left[-\cos \frac{2\pi}{3} + \cos \frac{\pi}{3} \right] = \frac{3I_m}{\pi} \left[\tfrac{1}{2} + \tfrac{1}{2} \right]$$

$$\therefore \qquad I_{dc} = \frac{3I_m}{\pi} \tag{10.19}$$

EXERCISES 10

1. Relative to a highly refined semiconductor, is a doped semiconductor: (*a*) more highly refined; (*b*) similarly refined; (*c*) less refined; (*d*) completely unrefined?

2. A piece of highly refined silicon is doped with arsenic. When connected into a circuit, is the conduction process in the semiconductor crystal due to: (*a*) electrons provided by the silicon; (*b*) holes provided by the arsenic; (*c*) holes provided by the silicon; (*d*) electrons provided by the silicon and the arsenic?

3. With reference to the semiconductor material specified in Q. 2, is the conduction process: (*a*) intrinsic; (*b*) extrinsic; (*c*) both but mainly extrinsic; (*d*) both but mainly intrinsic?

4. In the barrier layer of a p–n junction and within the p-type material, is the charge: (*a*) positive due to fixed atoms; (*b*) negative due to fixed atoms; (*c*) positive due to holes present; (*d*) negative due to electrons present?

5. A piece of doped semiconductor material is introduced into a circuit. If the temperature of the material is raised, will the circuit current: (*a*) increase; (*b*) remain the same; (*c*) decrease; (*d*) cease to flow?

6. Some highly refined semiconductor material is doped with indium. For such a material, are the majority carriers: (*a*) electrons from the semiconductor; (*b*) electrons from the indium; (*c*) holes from the semiconductor; (*d*) holes from the indium?

7. With the aid of a suitable sketch, describe the essential features of a semiconductor diode and also sketch a graph showing the forward and reverse characteristics of a typical silicon diode.

8. Explain, with reference to a semiconductor material, what is meant by: (*a*) intrinsic conductivity; (*b*) extrinsic conductivity.

9. Discuss the phenomenon of current flow in (i) intrinsic, (ii) p-type, and (iii) n-type semiconductors, and hence explain the rectifying action of a p–n junction.

10. A silicon diode has a reverse-bias resistance which can be considered as infinitely large. In the forward-bias direction, no current flows until the voltage across the device exceeds 600 mV. The junction then behaves as a constant resistance of 20 Ω. From this information draw to scale the characteristic of this device.

11. If sine waves of peak values (*a*) 0.1 V, (*b*) 1 V, (*c*) 5 V are connected in turn in series with the device described in Q. 10, what is the peak current in each case, assuming that the diode is still operating within its rated limits?

12. What are the peak currents in Q. 11 if in addition a 50-Ω resistor is included in series with the sinewave source in each case. (You might need some assistance with this one.)

13. Sketch one form of full-wave rectifier circuit together with smoothing components. If the supply frequency is 400 Hz, what is the ripple frequency?

14. A silicon diode has forward characteristics given in the following table:

V (V)	0	0.50	0.75	1.00	1.25	1.50	1.75	2.00
I (A)	0	0.1	0.3	1.5	3.3	5.0	6.7	8.5

The diode is connected in series with a 1.0 Ω resistor across an alternating voltage supply of peak value 10.0 V. Determine the peak value of the diode forward current and the value of the series resistor which would limit the peak current to 5.0 A.

15. A semiconductor diode, the forward and reverse characteristics of which can be considered ideal, is used in a half-wave rectifier circuit supplying a resistive load of 1000 Ω. If the r.m.s. value of the sinusoidal supply voltage is 250 V determine: (*a*) the peak diode current; (*b*) the mean diode current; (*c*) the r.m.s. diode current; and (*d*) the power dissipated in the load.

16. Two semiconductor diodes used in a full-wave rectifier circuit have forward resistances which will be considered constant at 1.0 Ω and infinite reverse resistances. The circuit is supplied from a 300–0–300 V r.m.s. secondary winding of a transformer and the mean current in the resistive load is 10 A. Determine the resistance of the load, the maximum value of the voltage which appears across the diodes in reverse and the efficiency of the circuit.

17. (*a*) Sketch network diagrams showing how diodes may be connected to the secondary winding of a suitable transformer in order to obtain unsmoothed: (i) half-wave rectification, (ii) full-wave rectification with two diodes, (iii) full-wave rectification with four diodes. For each network, show the waveform of the output voltage.

(*b*) Explain the necessity of smoothing the output voltage before applying it to a transistor amplifier.

(*c*) Sketch a typical network for smoothing such a supply to a transistor amplifier.

18. A full-wave rectifier circuit supplies a 2000-Ω resistive load. The characteristics of the diodes used can be considered ideal and each half of the secondary winding of the transformer develops an output voltage of 250 V. If the mean current in the load is to be limited to 100 mA by the connection of equal resistors in

series with the diodes, determine the value of these resistors and the power dissipated in them.

19. Draw the circuit diagram of a full-wave bridge rectifier network supplying a resistive load.

   By reference to the circuit diagram, explain the operation of the network, including in your answer diagrams illustrating the input and output current waveforms.

   What would be the effect on the operation of the other three diodes if one of the component diodes were to become short-circuited?

20. Describe, with the aid of suitable diagrams, the rectifier action of a semiconductor diode. (Reference should be made to the principle of conduction in the semiconductor materials and the potential difference at the barrier layer.)

   The rectifier diodes shown in fig. A are assumed to be ideal. Calculate the peak current in each of the resistors, given that the applied voltage is sinusoidal.

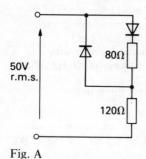

50V r.m.s.

80Ω

120Ω

Fig. A

21. The four semiconductors diodes used in a bridge rectifier circuit have forward resistances which can be considered constant at 0.1 Ω and infinite reverse resistances. They supply a mean current of 10 A to a resistive load from a sinusoidally varying alternating supply of 20 V r.m.s. Determine the resistance of the load and the efficiency of the circuit.

22. A half-wave rectifier circuit is used to charge a battery of e.m.f. 12 V and negligible internal resistance. The sinusoidally varying alternating supply voltage is 24 V peak. Determine the value of resistance to be connected in series with the battery to limit the charging current to 1.0 A. What peak current would flow in the diode if the battery were reversed?

23. A stabilized power supply is tested and the following results are obtained:

   Unstabilized input voltage to the zener circuit constant at 25 V.
   Output voltage from the stabilizer on no-load 16 V.
   Output voltage from the stabilizer on full load (50 mA) 14 V.
   The output stabilized voltage on full load falls to 13.5 V when the unstabilized voltage is reduced to 18 V.

   Find (a) the output resistance; and (b) the stabilization factor.

24. A zener diode has a characteristic which may be considered as two straight lines, one joining the points

$$I = 0, V = 0 \quad \text{and} \quad I = 0, V = -9\,\text{V};$$

and the other joining the points

$$I = 0, V = -9\,\text{V} \quad \text{and} \quad I = -45\,\text{mA}, V = -10\,\text{V}.$$

Find: (a) its resistance after breakdown; (b) the value of the stabilization series resistor if the unstabilized voltage is 15 V and the load voltage is 9.5 V, when the load current is 30 mA; (c) the range of load current this circuit can deal with if the maximum dissipation in the zener is 0.5 W; (d) the range of unstabilized voltage this circuit can deal with on a constant load current of 20 mA; (e) the stabilization factor.

11 Simple Transistor Amplifiers

11.1 Introduction

Having become familiar with the semiconductor diode in Chapter 10, it is now possible to progress to an understanding of the transistor, which is fundamental to most amplifier arrangements as well as many switching arrangements associated with digital systems.

There are two basic types of transistor: (a) the bipolar junction transistor, (b) the field effect transistor (FET).

Early transistor circuits depended entirely on the junction transistor and for that reason it will be considered first.

11.2 Bipolar junction transistor

A bipolar junction transistor is a combination of two junction diodes and consists of either a thin layer of p-type semiconductor sandwiched between two n-type semiconductors, as in fig. 11.1(a), and referred to as an n–p–n transistor, or a thin layer of an n-type semiconductor sandwiched between two p-type semiconductors, as in fig. 11.1(b), and referred to as a p–n–p transistor. The thickness of the central layer, known as the *base*, is of the order of 25 μm (or 25×10^{-6} m).

Fig. 11.1 Arrangement of a bipolar junction transistor

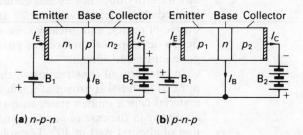

(a) n-p-n   (b) p-n-p

The junction diode formed by n_1–p in fig. 11.1(a) is biased in the forward direction by a battery B_1 so that free electrons are urged from n_1 towards p. Hence n_1 is termed an *emitter*. On the other hand, the junction diode formed by n_2–p in fig. 11.1(a) is biased in the reverse direction by battery B_2 so that if battery B_1 were disconnected, i.e. with zero emitter current, no current would flow between n_2 and p apart from that due to the thermally generated minority carriers referred to in section 10.6. However, with B_1 connected as in fig. 11.1(a), the electrons from emitter n_1 enter p and diffuse through the base until they come within the influence of n_2, which is connected to the positive terminal of battery B_2. Consequently the electrons which reach n_2 are collected by the metal electrode attached to n_2; hence n_2 is termed a *collector*.

Some of the electrons, in passing through the base, combine with holes; others reach the base terminal. The electrons which do not reach collector n_2 are responsible for the current at the base terminal, and the distribution of electron flow in an n–p–n transistor can be represented diagrammatically as in fig. 11.2. The *conventional* directions of the currents are represented by the arrows marked I_E, I_B and I_C.*

By making the thickness of the base very small and the impurity concentration in the base much less than in the emitter and collector, the free electrons emerging from the emitter have little opportunity of combining with holes in the base, with the result that about 98 per cent of these electrons reach the collector.

By Kirchhoff's First Law, $I_E = I_B + I_C$, so that if $I_C = 0.98 I_E$, then $I_B = 0.02 I_E$; thus, when $I_E = 1$ mA, $I_C = 0.98$ mA and $I_B = 0.02$ mA, as in fig. 11.3.

In the above explanation, we have dealt with the n–p–n transistor, but exactly the same explanation applies to the p–n–p transistor of fig. 11.1(b) except that the movement of electrons is replaced by the movement of holes.

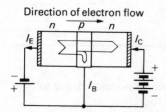

Fig. 11.2 Electron flow in an n–p–n transistor

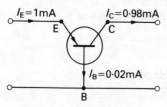

Fig. 11.3 Currents in emitter, collector and base circuits of a p–n–p transistor

11.3 Construction of a bipolar transistor

The first step is to purify the germanium or silicon so that any impurity does not exceed about 1 part in 10^{10}. Various methods have been developed for attaining this exceptional degree of purity, and intensive research is still being carried out to develop new methods of purifying and of doping germanium and silicon.

In one form of construction of the p–n–p transistor, the purified material is grown as a single crystal, and while the material is in a molten state, an n-type impurity (e.g. antimony in the case of germanium) is added in the proportion of about 1 part in 10^8. The solidified crystal is then sawn into slices about 0.1 mm thick. Each slice is used to form the base region of a transistor; thus in the case of n-type germanium, a pellet of p-type impurity such as indium is

---

* See note on subscripts, p. xvii.

placed on each side of the slide, the one which is to form the collector being about three times the size of that forming the emitter. One reason for the larger size of the collector bead is that the current carriers from the emitter spread outwards as they pass through the base, and the larger area of the collector enables the latter to collect these carriers more effectively. Another reason is that the larger area assists in dissipating the greater power loss at the collector-base junction. This greater loss is due to the p.d. between collector and base being greater than that between emitter and base.

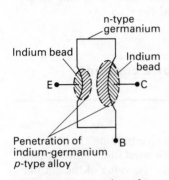

Fig. 11.4 Construction of a p–n–p germanium transistor

The assembly is heated in a hydrogen atmosphere until the pellets melt and dissolve some of the germanium from the slice, as shown in fig. 11.4. Leads for the emitter and collector are soldered to the surplus material in the pellets to make non-rectifying contacts, and a nickel tab is soldered to make connection to the base. The assembly is then hermetically sealed in a metal or glass container, the glass being coated with opaque paint. The transistor is thus protected from moisture and light.

11.4 Common-base and common-emitter circuits

In fig. 11.1 the transistor is shown with the base connected directly to both the emitter and collector circuits; hence the arrangement is referred to as a *common-base* circuit. The conventional way of representing this arrangement is shown in fig. 11.5(a) and (b) for n–p–n and p–n–p transistors respectively, where E, B and C represent the emitter, base and collector terminals. The arrowhead on the line joining the emitter terminal to the base indicates the conventional direction of the current in that part of the circuit.

Fig. 11.6 shows the *common-emitter* method of connecting a transistor. In these diagrams it will be seen that the emitter is

Fig. 11.5 Common-base circuits

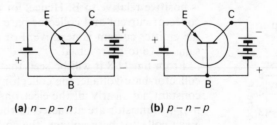

(a) *n – p – n* **(b)** *p – n – p*

Fig. 11.6 Common-emitter circuits

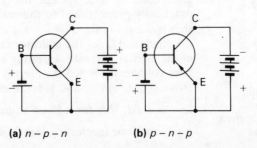

(a) *n – p – n* **(b)** *p – n – p*

connected directly to the base and collector circuits. This method is more commonly used than the common-base circuit owing to its higher input resistance and the higher current and power gains.

The common-collector circuit is used only in special cases, and will therefore not be considered here.

11.5 Static characteristics for a common-base circuit

Fig. 11.7 shows an arrangement for determining the static characteristics of an n–p–n transistor used in a common-base circuit. The procedure is to maintain the value of the emitter

Fig. 11.7 Determination of static characteristics for a common-base n–p–n transistor circuit

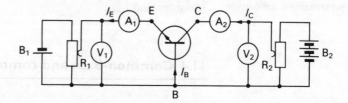

current, indicated by A_1, at a constant value, say 1 mA, by means of the slider on R_1, and note the readings on A_2 for various values of the collector-base voltage given by voltmeter V_2. The test is repeated for various values of the emitter current and the results are plotted as in fig. 11.8.

In accordance with B.S. 3363, the current is assumed to be positive when its direction is from the external circuit towards the transistor terminal, and the voltage V_{CB} is positive when C is positive relative to B. Hence, for an n–p–n transistor, the collector current and collector-base voltage are positive but the emitter current is negative. For a p–n–p transistor, all the signs have to be reversed.

From fig. 11.8 it will be seen that for positive values of the collector-base voltage, the collector current remains almost constant, i.e. nearly all the electrons entering the base of an n–p–n transistor are attracted to the collector. Also, for a given collector-base voltage, the collector current is practically proportional to the emitter current. This relationship is shown in fig. 11.9 for V_{CB} equal to 4 V. The ratio of the change, ΔI_C, of the collector current to the change, ΔI_E, of the emitter current (neglecting signs), for a given collector-base voltage, is termed the *current amplification factor for a common-base circuit* and is represented by the symbol α,

i.e. $\quad \alpha = \Delta I_C / \Delta I_E$ for a given value of V_{CB} (11.1)

$\qquad$ = slope (neglecting signs) of I_C / I_E graph in fig. 11.9.

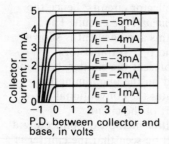

Fig. 11.8 Static characteristics for a common-base n–p–n transistor circuit

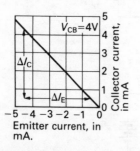

Fig. 11.9 Relationship between collector and emitter currents for a given collector-base voltage

11.6 Static characteristics for a common-emitter circuit

Fig. 11.10 shows an arrangement for determining the static characteristics of an n–p–n transistor used in a common-emitter circuit. Again, the procedure is to maintain the *base*

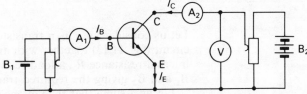

Fig. 11.10 Determination of static characteristics for a common-emitter n–p–n transistor circuit

current, I_B, through a microammeter A_1 constant at, say, $25\,\mu A$, and to note the collector current, I_C, for various values of the collector–emitter voltage V_{CE}, the test being repeated for several values of the base current, and the results are plotted as shown in fig. 11.11. For a given voltage between collector and emitter, e.g. for $V_{CE} = 4$ volts, the relationship between the collector and base currents is practically linear as shown in fig. 11.12.

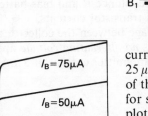

Fig. 11.11 Static characteristics for a common-emitter n–p–n transistor circuit

The ratio of the change, ΔI_C, of the collector current to the change, ΔI_B, of the base current, for a given collector–emitter voltage, is termed the *current amplification factor for a common-emitter circuit* and is represented by the symbol β,

i.e. $\beta = \Delta I_C/\Delta I_B$ for a given value of V_{CE} (11.2)

= slope of graph in fig. 11.12.

11.7 Relationship between α and β

From figs. 11.7 and 11.10 it is seen that:

$$I_E = I_C + I_B$$
$$\therefore \quad \Delta I_E = \Delta I_C + \Delta I_B.$$

From expression (11.1),

$$\alpha = \Delta I_C/\Delta I_E$$
$$= \Delta I_C/(\Delta I_C + \Delta I_B)$$
$$\therefore \quad 1/\alpha = 1 + \Delta I_B/\Delta I_C$$
$$= 1 + 1/\beta = (1 + \beta)/\beta$$

Hence $\alpha = \beta/(1 + \beta)$ (11.3)

and $\beta = \alpha/(1 - \alpha)$ (11.4)

Thus,   if $\alpha = 0.98$, $\beta = 0.98/0.02 = 49$,

and   if $\alpha = 0.99$, $\beta = 0.99/0.01 = 99$,

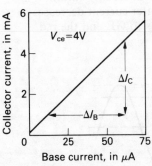

Fig. 11.12 Relationship between collector and base currents for a given collector-emitter voltage

i.e. a small variation in α corresponds to a large variation in β. It is therefore better to determine β experimentally and calculate therefrom the corresponding value of α by means of expression (11.3).

11.8 Load line for a transistor

Let us consider an n–p–n transistor used in a common-base circuit (fig. 11.13) together with an a.c. source S having an internal resistance R_s, a load resistance R and bias batteries B_1 and B_2 giving the required transistor currents.

Let us assume that the voltage between the collector and base is to be 3 V when there is *no alternating voltage* applied to the emitter, and that with an emitter current of 3 mA, the

Fig. 11.13 An n–p–n transistor with load resistance R

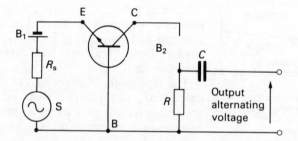

corresponding collector current is 2.9 mA. This corresponds to point D in fig. 11.14. Also suppose the resistance R of the load to be 1000 Ω. Consequently the corresponding p.d. across R is 0.0029 × 1000, namely 2.9 V, and the total bias supplied by battery B_2 must be 3 + 2.9, namely 5.9 V. This is represented by point P in fig. 11.14. The straight line PDQ drawn through P and D is the *load line*, the inverse of the slope of which is equal to the resistance of the load, e.g. HP/DH = 2.9/0.0029 = 1000 Ω. The load line is the locus of the variation of the collector current for any variation in the emitter current. Thus, if the emitter current decreases to 1 mA, the collector current decreases to GE and the p.d.

Fig. 11.14 Graphical determination of the output current of a transistor in a common-base circuit

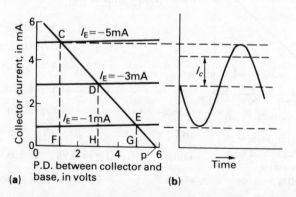

across *R* decreases to GP. On the other hand, if the emitter current increases to 5 mA, the collector current increases to FC and the p.d. across *R* increases to FP. Hence, if the emitter current varies sinusoidally between 1 and 5 mA, the collector current varies as shown at (*b*) in fig. 11.14. This curve will be sinusoidal if the graphs are linear and are equally spaced over the working range.

The function of C is to eliminate the d.c. component of the voltage across *R* from the output voltage.

If I_c* be the r.m.s. value of the *alternating* component of the collector current, then $I_c R$ gives the r.m.s. value of the alternating component of the output voltage, and $I_c^2 R$ gives the output power due to the alternating e.m.f. generated in S. Hence the larger the value of *R*, the greater the voltage and power gains, the maximum value of *R* for a given collector supply voltage being limited by the maximum permissible distortion of the output voltage.

Similar procedure can be used to determine the output voltage and power for a transistor used in a common-emitter circuit.

11.9 Transistor as an amplifier

Consider again the basic action of an amplifier. The input signal shown in fig. 11.15 controls the amount of power that the amplifier takes from the power source and converts into power in the load. We have also seen that in the transistor the collector current is controlled by the emitter or base currents. By connecting a load effectively between the collector and the common terminal, the transistor can produce gain.

Again the input signal is generally of an alternating quantity. However, the transistor requires to operate in a unidirectional mode otherwise the negative parts of the alternating quantity would cause, say, the emitter-base junction to be reverse biased and this would prevent normal transistor action occurring. As a result, it is necessary to introduce a bias.

Fig. 11.15 Mode of operation of an amplifier

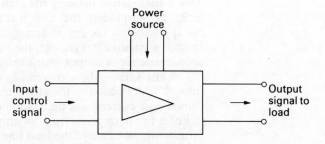

* See note on subscripts, p. xvii.

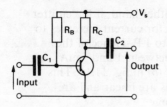

Fig. 11.16 Simple common-emitter transistor amplifier

Fig. 11.16 shows a practical transistor amplifier circuit utilizing an n–p–n transistor in the common-emitter mode. The resistor R_C is connected between the collector and the positive supply from a battery or other d.c. source. The resistor R_B is included to provide the bias current to the base of the transistor. In order to separate the direct current of the transistor arrangement from the alternating signal entering and leaving the amplifier, capacitors are included (it will be recalled that capacitors appear to pass an alternating current but do not pass a direct current). The input a.c. signal is fed into the transistor via C_1 which prevents the signal source having any effect on the steady component of the base current. Similarly, the coupling capacitor C_2 prevents the load connected across the output terminals from affecting the steady conditions of the collector. The capacitances of C_1 and C_2 are selected to ensure that their reactances are negligible at the operating frequencies and consequently they have negligible effect on the amplifier operation so far as the a.c. signal is concerned.

Let I_B be the steady base current, usually termed the quiescent current, and let V_{BE} be the quiescent base-emitter voltage; then

$$V_S = I_B R_B + V_{BE}$$

$$\therefore \qquad R_B = \frac{V_S - V_{BE}}{I_B} \qquad (11.5)$$

Often V_{BE} is significantly smaller than V_S and, approximately,

$$R_B = \frac{V_S}{I_B} \qquad (11.6)$$

Let the applied signal current to the base be given by

$$i_b = I_{b_m} \sin \omega t$$

whereby the total base current is

$$I_B + I_{b_m} \sin \omega t.$$

At any instant

$$v_{ce} = V_S - i_c R_C$$

and

$$i_c = -\frac{1}{R_C} \cdot v_{ce} + \frac{V_S}{R_C} \qquad (11.7)$$

This is the relation defining the load line of the form shown in fig. 11.14(a) except that now it is expressed in terms that are applicable to the circuit arrangement shown in fig. 11.16. In such a circuit arrangement, the load line applied to an appropriate set of output characteristics is shown in fig. 11.17.

For any value of base current, i_C and v_{CE} can be taken from the intersection of the characteristic appropriate to the chosen base current and the load line.

For a linear response from an amplifier, it is necessary to operate on that part of the load line which meets the family of characteristics at regular intervals. It can be clearly seen that a value greater than A would incur a much smaller

Fig. 11.17 Load line for a
bipolar transistor amplifier

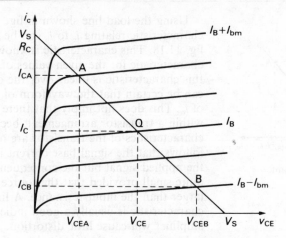

interval than the others of lesser value, all of which are
reasonably equal. B is the lowest acceptable point of
operation on the load line and Q is more or less half-way
between A and B. The nearer it is to being exactly half-way
then the better will be the linear amplification.

A, Q and B correspond to the maximum, mean and
minimum values of the base current, which are $I_B + I_{b_m}$, I_B
and $I_B - I_{b_m}$ respectively. The corresponding values of the
collector current are I_{CA}, I_C and I_{CB} respectively, and the
collector-emitter voltages are V_{CEA}, V_{CE} and V_{CEB} respectively.

The current gain, $$G_i = \frac{\Delta I_o}{\Delta I_i} = \frac{I_{CA} - I_{CB}}{2 I_{b_m}} \qquad (11.8)$$

The voltage gain, $$G_v = \frac{\Delta V_o}{\Delta V_i} \qquad (11.9)$$

and the power gain, $$G_p = \frac{P_o}{P_i} \qquad (11.10)$$

where P_o is the signal power in the load and P_i is the signal
power into the transistor.

In practice, the voltage gain can only be determined
accurately if we know the change in input voltage that has
brought about the change in base current. However, little
error is introduced if the collector-emitter voltage is assumed
constant when calculating the base-emitter voltage producing
the base current. This simplification is often taken a stage
further by assuming a linear input characteristic, which leads
to an equivalent circuit similar to that considered in fig. 9.15.

Care should be taken in interpreting relation (11.10). The
powers P_i and P_o are those related to the signal frequency
and they exclude the powers associated with the direct
conditions, i.e. the quiescent conditions. For instance, the
signal power in the load is the product of the r.m.s. signal
load voltage and the r.m.s. signal load current.

It follows that

$$G_p = G_v G_i \qquad (11.11)$$

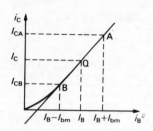

Fig. 11.18 Dynamic characteristic

Using the load line shown in fig. 11.17, then the characteristic relating i_c to i_b can be derived, as shown in fig. 11.18. This characteristic is known as the dynamic characteristic for the given values of R_C and V_S. Provided that this characteristic is linear over the operating range then we can be certain that the waveform of i_c will be identical to that of i_b. This does not imply that there will be no distortion within a transistor arrangement, because the input characteristics of the transistor are not themselves linear. It follows that the signal base current is not an exact replica of the applied signal but the consequent distortion is generally quite small provided that the source resistance is considerably larger than the input resistance. A linear dynamic characteristic is usually a good indicator that the transistor amplifier will cause little distortion.

In considering the circuit shown in fig. 11.16, we took R_C to be effectively the load, there being no other load connected across the output terminals of the amplifier. If a resistor R_L was connected across the output terminals, then the total load presented to the transistor amplifier would effectively be R_C and R_L in parallel (assuming the reactance of C_2 to be negligible). The effective load is therefore

$$R_P = \frac{R_C \cdot R_L}{R_C + R_L}$$

and the amplifier performance can be determined by drawing a load line for R_P and passing through the quiescent point on the characteristics. Note that the quiescent point Q is the same for both the load line associated with R_C and the load line associated with R_P because in both cases it refers to the zero-signal condition. The load line associated with R_C is termed the d.c. load line, while that associated with R_P is the a.c. load line.

The use of the a.c. load line is better illustrated by means of the following example.

Example 11.1

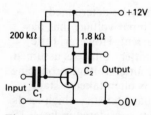

Fig. 11.19 Bipolar transistor amplifier stage for example 11.1

A bipolar transistor amplifier stage is shown in fig. 11.19 and the transistor has characteristics, shown in fig. 11.20, which may be considered linear over the working range. A 2.2-kΩ resistive load is connected across the output terminals and a signal source of sinusoidal e.m.f. 0.6 V peak and internal resistance 10 kΩ is connected to the input terminals. The input resistance of the transistor is effectively constant at 2.7 kΩ. Determine the current, voltage and power gains of the stage. The reactances of the coupling capacitors C_1 and C_2 may be considered negligible.

First determine the extremities of the d.c. load line.

If $\qquad\qquad i_c = 0 \quad$ then $\quad V_s = 12 \text{ V} = v_{ce}$

If $\qquad v_{ce} = 0 \quad$ then $\quad i_c = \dfrac{V_s}{R_c} = \dfrac{12}{1.8 \times 10^3} \equiv 6.7 \text{ mA}$

For the a.c. load line,

$$R_p = \frac{1.8 \times 2.2}{1.8 + 2.2} = 1.0 \text{ k}\Omega$$

$$= \frac{\Delta v}{\Delta I}$$

Fig. 11.20 Bipolar transistor characteristics for example 11.1

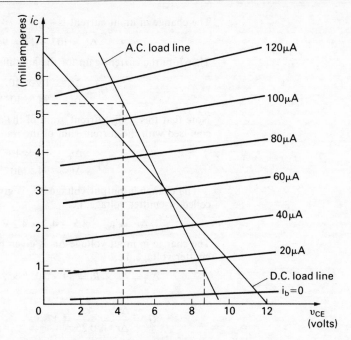

Hence the slope of the a.c. load line

$$= -\frac{1000}{1} \equiv 1.0 \text{ mA/V}$$

and the quiescent base current

$$I_{B_o} = \frac{12}{200 \times 10^3} \equiv 60 \,\mu\text{A}.$$

The a.c. load line is therefore drawn with a slope of -1.0 mA/V through the quiescent point Q, which is given by the intersection of the 60 μA characteristic and the d.c. load line.

The input circuit consists of the base-emitter junction in parallel with the bias resistor, i.e. the signal passes through both in parallel. However, the transistor input resistance is 2.7 kΩ hence the shunting effect of 200 kΩ is negligible, and effectively the entire input circuit can be represented by fig. 11.21.

The peak signal base current

$$= \frac{0.6}{(10 + 2.7) \times 10^3} \equiv 47 \,\mu\text{A}$$

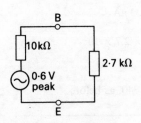

Fig. 11.21 Equivalent input circuit for example 11.1

and hence the maximum signal base current

$$= 60 + 47 = 107 \,\mu\text{A}$$

and the minimum signal base current

$$= 60 - 47 = 13 \,\mu\text{A}$$

From the a.c. load line in fig. 11.20,

$$\Delta i_c = 5.1 - 0.9 = 4.2 \text{ mA}.$$

This change in collector current is shared between the 1.8-kΩ collector resistor and the load resistor of 2.2 kΩ. Hence the change in output current is

$$\Delta i_o = 4.2 \times \frac{1.8 \times 10^3}{(1.8 + 2.2) \times 10^3} = 1.9 \text{ mA}.$$

The change in input current is

$$\Delta i_i = 107 - 13 = 94 \,\mu A.$$

Therefore the current gain for the amplifier stage is

$$G_i = \frac{\Delta i_o}{\Delta i_i} = \frac{1.9 \times 10^{-3}}{94 \times 10^{-6}} = 20.$$

Note that this is the current grain of the amplifier and should not be confused with the current gain of the transistor, which is given by

$$G_i = \frac{\Delta i_c}{\Delta i_b} = \frac{4.2 \times 10^{-3}}{94 \times 10^{-6}} = 45.$$

The change in output voltage Δv_o is given by the change in collector-emitter voltage, i.e.

$$\Delta v_o = v_{ce} = 8.5 - 4.3 = 4.2 \text{ V (peak to peak)}$$

The change in input voltage Δv_i is given by the change in base-emitter voltage, i.e.

$$\Delta v_i = \Delta i_i \times R_i = 39 \times 10^{-6} \times 2.7 \times 10^3 = 0.25 \text{ V}$$

(peak to peak)

$$\therefore \qquad G_v = \frac{\Delta v_o}{\Delta v_i} = \frac{4.2}{0.25} = 17$$

$$G_p = G_v G_i = 17 \times 20 = 340.$$

Alternatively,

$$\text{R.M.S. output voltage} = \frac{4.2}{2\sqrt{2}} = 1.5 \text{ V}$$

$$\therefore \quad \text{R.M.S. output current} = \frac{1.5}{2.2 \times 10^3} = 0.68 \text{ mA}$$

and

$$P_o = (0.68 \times 10^{-3})^2 \times 2.2 \times 10^3$$

$$\equiv 1.0 \text{ mW}$$

$$\text{R.M.S. input voltage} = \frac{0.25}{2\sqrt{2}} \equiv 88 \text{ mV}$$

$$\therefore \quad \text{R.M.S. input current} = \frac{88 \times 10^{-3}}{2.7 \times 10^3} \equiv 33 \,\mu A$$

and

$$P_i = (33 \times 10^{-6})^2 \times 2.7 \times 10^3$$

$$\equiv 2.9 \,\mu W$$

$$\therefore \qquad G_p = \frac{1.0 \times 10^{-3}}{2.9 \times 10^{-6}} = 340, \text{ as before.}$$

11.10 Circuit component selection

The simple transistor circuit shown in fig. 11.16 was translated into an equivalent circuit for the purpose of example 11.1. However, what would we have done if simply asked to make an amplifier from a given transistor? For instance, given the transistor, the characteristics of which were shown in fig. 11.20, how were the resistances of the other circuit components selected?

Normally, given the transistor characteristics, it is the practice first to select a quiescent point. This requires selecting a curve somewhere in the middle of the family of characteristics, which in fig. 11.20 immediately suggests that we should take the curve for $i_b = 60 \,\mu\text{A}$.

In order to obtain such a value, we require to forward-bias the base-emitter junction, i.e. being an n–p–n transistor, the base should be positive with respect to the emitter. This is achieved by R_B connected between the positive supply V_S and the base. If we neglect the base-emitter voltage drop, then

$$R_B = V_S/I_b = 12/60 \times 10^{-6} = 200 \,\text{k}\Omega.$$

To take the base-emitter voltage as negligible might be rather optimistic since generally it is in the order of 1 V. Even so, the voltage across the bias resistor would be 11 V in which case R_B would require a value of about 183 kΩ.

Having determined R_B, it is also possible to determine an approximate value for R_C. The quiescent point normally takes a value of about half the supply voltage V_S, which in this case suggests a quiescent collector-emitter voltage of about 6 V. From the characteristics, the corresponding collector current is 2.9 mA. If v_{CE} is 6 V then the voltage across R_C is 6 V and the current in R_C is 2.9 mA, hence $R_C = 6/2.9 \times 10^3 = 2 \,\text{k}\Omega$.

In example 11.1, it happens that the resistance of R_C was 1.8 kΩ, but a value of 2 kΩ, which is reasonably similar in magnitude, would be as acceptable.

This description of the determination of R_B and R_C seems rather off-hand in the manner with which the values have been accepted. Having sought high correlation of input and output waveforms and a linear relationship, it subsequently seems strange that 1.8 kΩ can be taken as much the same as 2 kΩ. However, the circuit components of many electronic circuits are manufactured to a tolerance of ± 10 per cent; thus the approximations made above would lie within such limits. The variations in practice result in considerable ranges in performance between supposedly identical amplifiers, although most tend to have the same performance.

When the source of the input alternating signal is connected to the amplifier, consideration has to be given to the problem that the source normally has a low internal resistance. If it were not for the coupling capacitor C_1, some of the current through the bias resistor would be diverted away from the base-emitter junction and thus the bias conditions would be affected. However, the coupling capacitor does not permit any of the direct bias current to pass through the source. Provided the capacitance of C_1 is so chosen that the reactance $1/2\pi f C_1$ is small compared with the input resistance of the transistor then its effect on the alternating signal is negligible. In this case, we would be wanting the reactance to be about 10 per cent of the input resistance in magnitude.

Finally, it should not be thought that the simple bias

arrangement considered is the only one used. Among the many other arrangements found in practice, the bias can be effected by a potential divider across the supply or a resistor connected in parallel between the emitter and the 0-V line.

11.11 Equivalent circuits of a transistor

So long as a transistor is operated over the linear portions of its characteristics, the actual values of the bias voltages are of no consequence, hence these voltages can be omitted from an equivalent circuit used to calculate the current, voltage and power gains. Instead, the values of the voltages and currents shown on an equivalent circuit diagram refer only to the r.m.s. values of the alternating components of these quantities.

Since the transistor itself is an amplifier, it can be represented by an equivalent circuit of the form shown in fig. 9.15. In order to indicate that the circuit components and gain ratios apply specifically to the transistor, the values are specified as *h*-parameters as shown in fig. 11.22.

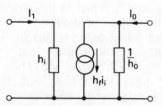

h_i is equivalent to the general input resistance R_i, h_o is equivalent to the general output resistance R_o except that h_o is expressed as a conductance in siemens rather than as a resistance in ohms, and h_f is the current gain of the transistor. Because there is a variety of connections for the transistor, we add a letter to the parameter subscript to indicate the mode of operation, thus h_{ie} is the input resistance of a transistor connected in the common-emitter mode and h_{ib} is the input resistance in the common-base mode.

Fig. 11.22 *h*-parameters for a transistor equivalent circuit

Consider now the transistor in the amplifier arrangement shown in fig. 11.23(*a*). The equivalent circuit for the amplifier is shown in fig. 11.23(*b*) and here again the bias resistor is seen to shunt the input to the transistor while the output consists of the load resistor shunted by the collector resistor.

In converting the actual circuit arrangement to the equivalent circuit, it will also be noted that the capacitors have been omitted, it being assumed that their reactances are negligible. In both diagrams, the source has been taken as a voltage source of E_S in series with its source resistance R_S. This contrasts with the equivalent circuit of the transistor which has a current generator, but it has to be remembered that the action of the transistor depends on the input current and therefore it is essentially a current-operated device.

From fig. 11.23(*b*), for the transistor

$$G_i = \frac{I_c}{I_b}$$

and

$$I_c = \frac{1/h_{oe}}{1/h_{oe} + R_p} \times h_{fe} I_b$$

$$\therefore \quad G_i = \frac{h_{fe}}{1 + h_{oe} R_p} \tag{11.12}$$

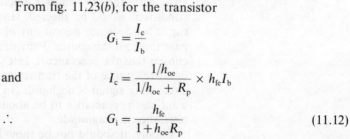

Fig. 11.23 Common-emitter
amplifier stage and its
equivalent circuit

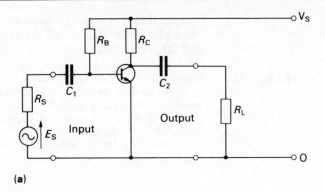

(a)

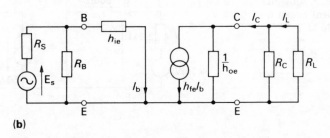

(b)

This is the current gain of the transistor but not of the
amplifier stage. The stage gain is given by

$$G_i = \frac{I_L}{I_b}$$

and
$$I_L = \frac{R_c}{R_c + R_L} I_c$$

$\therefore$
$$G_i = \frac{h_{fe}}{1 + h_{oe}P_p} \cdot \frac{R_c}{R_c + R_L} \tag{11.13}$$

The voltage gain for the transistor is given by

$$G_v = \frac{v_{ce}}{v_{be}} = \frac{-I_c R_p}{I_b h_{ie}}$$

The minus sign is a consequence of the direction of the
collector current.

From relation (11.12)

$$G_v = \frac{-h_{fe} R_p}{h_{ie}(1 + h_{oe} R_p)} \tag{11.14}$$

Since the voltage that appears across the transistor's collector
and emitter terminals is the same as that which appears
across the load, the voltage gain expression applies both to
the transistor and to the amplifier stage. Again the minus sign
results from the current directions and also this implies that
there is a 180° phase shift between input and output, as
shown in fig. 11.24.

Having examined the approach to analysing the
performance of the transistor and also of the amplifier stage,
we should now consider a numerical example in which we

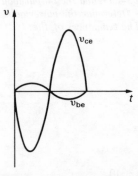

Fig. 11.24 Common-emitter
transistor input and output
voltage waveforms

will see that it is easier to calculate the performance from first principles rather than recall the analyses which resulted in the last three relations.

Example 11.2

A transistor amplifier stage comprises a transistor, of parameters $h_{ie} = 800\,\Omega$, $h_{fe} = 50$ and $h_{oe} = 20\,\mu S$, and bias components and coupling capacitors, of negligible effect. The input signal consists of an e.m.f. of 60 mV from a source of internal resistance 2.2 kΩ, and the total load on the stage output is 4 kΩ.

Determine the current, voltage and power gains of the amplifier stage.

Fig. 11.25 Amplifier equivalent circuit for example 11.2

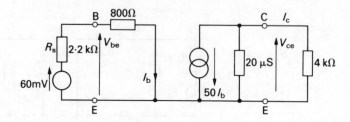

The equivalent circuit is shown in fig. 11.25.

$$I_b = \frac{60 \times 10^{-3}}{2200 + 800} = 20 \times 10^{-6}\,\text{A}$$

$$\frac{1}{h_{oe}} = \frac{1}{20 \times 10^{-6}} \equiv 50\,\text{k}\Omega$$

$$I_c = \frac{50 \times 10^3}{(50 + 4) \times 10^3} \times 50 \times 20 \times 10^{-6} = 926 \times 10^{-6}\,\text{A}$$

$$G_i = \frac{926 \times 10^{-6}}{20 \times 10^{-6}} = 46.3$$

$$G_v = \frac{-926 \times 10^{-6} \times 4 \times 10^3}{20 \times 10^{-6} \times 800} = 231.5.$$

$$G_p = 46.3 \times 231.5 = 10\,720.$$

Example 11.3

A transistor amplifier is shown in fig. 11.26. The parameters of the transistor are $h_{ie} = 2\,\text{k}\Omega$, $h_{oe} = 25\,\mu S$ and $h_{fe} = 55$, and the output load resistor dissipates a signal power of 10 mW. Determine the power gain of the stage and the input signal e.m.f. E. The reactances of the capacitors may be neglected.

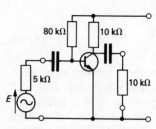

Fig. 11.26 Amplifier for example 11.3

The equivalent circuit is shown in fig. 11.27.

$$\frac{1}{h_{oe}} = \frac{1}{25 \times 10^{-6}} \equiv 40\,\text{k}\Omega$$

$$P = 10 \times 10^{-3}\,\text{W} = \frac{v_{ce}^2}{10 \times 10^3}$$

$$\therefore \qquad v_{ce} = 10\,\text{V}$$

$$\therefore \qquad 55\,I_b = \frac{10}{40 \times 10^3} + \frac{10}{10 \times 10^3} + \frac{10}{10 \times 10^3}$$

$$\therefore \qquad I_b = 40.9 \times 10^{-6}\,\text{A}$$

Fig. 11.27 Equivalent circuit for example 11.3

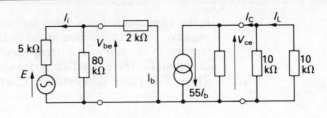

$$\therefore \quad I_i = 40.9 \times 10^{-6} + 40.9 \times 10^{-6} \cdot \frac{2 \times 10^3}{80 \times 10^3}$$

$$= 41.9 \times 10^{-6}\,\text{A}$$

$$E = (41.9 \times 10^{-6} \times 5 \times 10^3)$$
$$+ (40.9 \times 10^{-6} \times 2 \times 10^3)$$
$$= 0.29\,\text{V}.$$

$$P_i = V_{be}I_i = (40.9 \times 10^{-6} \times 2 \times 10^3)$$
$$\times (41.9 \times 10^{-6})$$
$$= 3.43 \times 10^{-6}\,\text{W}$$

$$P_o = 10 \times 10^{-3}\,\text{W}$$

$$G_p = \frac{10 \times 10^{-3}}{3.43 \times 10^{-6}} = 2920.$$

Before turning our attention to the common-base transistor amplifier, it is worth noting that

$$h_{fe} = \beta \tag{11.15}$$

where β is the current amplification factor for a common-emitter circuit as defined in relation (11.2).

The circuit of a common-base transistor amplifier and the corresponding equivalent circuit diagram are shown in fig. 11.28. The expressions for the gains are identical in form to those derived for the common-emitter arrangement except that the appropriate parameters have to be applied, thus

$$G_i = \frac{I_c}{I_e} = \frac{h_{fb}}{1 + h_{ob}R_p} \tag{11.16}$$

$$G_v = \frac{V_{cb}}{V_{eb}} = \frac{-h_{fb}R_p}{h_{ib}(1 + h_{ob}R_p)^2} \tag{11.17}$$

Fig. 11.28 Common-base amplifier stage and its equivalent circuit

$$G_p = G_v G_i.$$

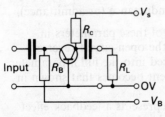

(a)

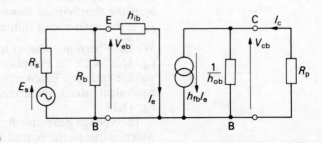

(b)

$h_{fb} (= \alpha)$ normally has a value just under unity and is negative, since an increase in the emitter current produces an increase in the collector current which is in the opposite direction, i.e. there is a $180°$ phase shift between I_e and I_c. There is no phase shift between V_{eb} and V_{cb}.

11.12 Hybrid parameters

For any bipolar junction transistor, there are six possible variables, these being i_b, i_c, i_e, v_{be}, v_{ce} and v_{cb}. However,

$$i_b + i_c = i_e$$

and

$$v_{cb} + v_{be} = v_{ce}.$$

From these relations, the operation of a transistor can be predicted provided we have the characteristics specifying the relationships between two of the voltages and two of the currents.

It is most convenient to consider input and output quantities. At this point, let us restrict our consideration to a common-emitter transistor in which case the input and output quantities are v_{be}, i_b, v_{ce} and i_c. In general mathematical terms, these relations can be expressed in the forms

$$v_{be} = f_1(i_b, v_{ce}) \tag{11.18}$$

$$i_c = f_2(i_b, v_{ce}) \tag{11.19}$$

Provided that all the variable quantities are sinusoidal in nature, then these expressions can be expanded as

$$V_{be} = h_{ie}I_b + h_{re}V_{ce} \tag{11.20}$$

$$i_c = h_{fe}I_b + h_{oe}V_{ce} \tag{11.21}$$

These parameters are the small-signal hybrid parameters which are generally termed the h-parameters. Note that the signals are assumed small in order to ensure that the operation remains within the linear sections of the transistor characteristics.

For the common-emitter transistor,

h_{ie} is the short-circuit input resistance (or impedance)
h_{re} is the open-circuit reverse voltage ratio
h_{fe} is the short-circuit forward current ratio
h_{oe} is the open-circuit output conductance (or admittance).

We have been introduced to all of these parameters in fig. 11.22 with the exception of the open-circuit reverse voltage ratio h_{re}. When introduced into the full h-parameter equivalent circuit, the arrangement becomes that shown in fig. 11.29.

The voltage generator $h_{re}V_{ce}$ represents a feedback effect which is due to the output signal voltage controlling, to a

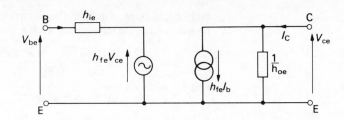

Fig. 11.29 *h*-parameter equivalent circuit for a common-emitter transistor

very limited extent, the input signal current. Due to the normal voltage gain of a transistor being quite large, it might be expected that as a consequence the feedback effect could be appreciable, but h_{re} is always very small (less than 1×10^{-3}) with the result that the feedback voltage generator can be omitted from the equivalent circuit without introducing any important error. Again we may recall that transistor circuit components are liable to considerable variation in their actual values compared with their anticipated values, and, within the expected range of tolerances, the omission of the feedback generator will have a negligible effect on any circuit analysis. It follows that the analyses undertaken in section 11.11 correspond satisfactorily to the full *h*-parameter equivalent circuit.

Similar equivalent circuits can be used for the common-base and common-collector configurations for a junction transistor.

11.13 Limitations to the bipolar junction transistor

For most bipolar transistors, the input resistance, as given by h_{ie}, tends to be about $1\,k\Omega$, which for electronic circuits tends to be rather a small value. It follows that there is always a significant input current which has two effects: (*a*) the amplifier gain is somewhat limited; (*b*) the power levels increase the size and rating of circuit components more than might be desirable.

In order to understand the limitation of gain, consider an amplifier with a gain of 40 and an input resistance of $1\,k\Omega$. A source, of e.m.f. $100\,mV$ and internal resistance $50\,k\Omega$, is connected to the input. The circuit is shown in fig. 11.30. The input voltage is

$$V_i = \frac{1 \times 10^3 \times 100 \times 10^{-3}}{(50 + 1) \times 10^3} \equiv 2\,mV$$

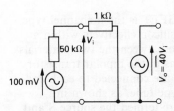

Fig. 11.30 Effect of input resistance on grain

and the output voltage

$$V_o = 40 \times 2 = 80\,mV.$$

The result is that, in spite of the amplification, the output voltage is less than the input e.m.f.— this is due to the low input resistance when compared with the source internal

resistance. Now consider the situation had the input
resistance been greater, say $1 \, \text{M}\Omega$, in which case,

$$V_i = \frac{1 \times 10^6 \times 100 \times 10^{-3}}{(50 + 1000) \times 10^3} = 95.2 \, \text{mV}$$

and

$$V_o = 40 \times 95.2 \times 10^{-3} = 3.8 \, \text{V}$$

It can therefore be seen that the increase of input resistance
has increased the gain by more than 47 times. Also, because
the input resistance is much higher, the input current is
substantially reduced and therefore the size and rating of the
input circuit components can be reduced.

In order to obtain the increase in input resistance, we have
to look for another type of transistor — the field effect
transistor, generally known as the FET.

11.14 Field effect transistor (FET)

Unlike the bipolar junction transistors, which are all basically
similar in spite of a variety of constructional forms, the FET
is more a collective term for a family of transistor devices, of
which there are two principal groups: (*a*) the junction-type
FET (JUGFET); and (*b*) the insulated-type FET (IGFET).

The JUGFET can be in two forms, p-channel and
n-channel, depending on the type of semiconductor forming
the basis of the transistor. The IGFET has two distinct
modes of operation, one known as the depletion mode and
the other as the enhancement mode, and again each mode
subdivides into p-channel and n-channel.

Like the bipolar junction transistor, all FETs are three-
electrode devices; the electrodes are the source, the gate and
the drain, which can be taken as corresponding to the
emitter, base and collector respectively.

11.15 JUGFET

The more common variety of JUGFET is the n-channel type.
This consists of a piece of n-type silicon effectively within a
tube of p-type silicon, the interface between the two materials
being the same intimate junction as in the bipolar transistor
junctions. Non-rectifying contacts are connected to each end
of the n-type piece and to the p-type tube, as shown in
fig. 11.31. The two end contacts are termed the source S and
the drain D while that to the tube or shroud is the gate G.

The principle of the action of a JUGFET can be explained
by first considering the operation of the source and drain
while the gate is left disconnected. Since we are considering

Fig. 11.31 The basic JUGFET

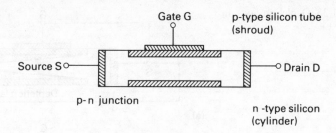

an n-channel JUGFET, consider a d.c. supply connected with the drain positive with respect to the source, as shown in fig. 11.32. The application of the drain-source voltage causes a conventional current to flow from the drain to the source: the current consists of electrons, which are the majority carriers in the n-channel, moving from the source to the drain. Being semiconductor material, the voltage/current relationship is almost linear and the n-channel more or less behaves as a resistor.

Fig. 11.32 Application of a drain-source voltage to a JUGFET

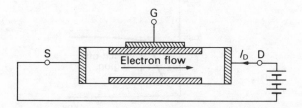

To bring the gate into action, first connect it to the source as shown in fig. 11.32(*a*). The drain-source current is still flowing and therefore the n-channel voltage becomes greater as electrons flow from the source towards the drain. It follows that electrons nearer the drain also experience a higher voltage with respect to the gate, hence the voltage across the p–n junction is greater at the righthand end than at the left. Also the p–n junction is reverse biased and this bias is greater nearer the drain than nearer the source. This causes a depletion layer which becomes greater as the bias increases, i.e. the depletion layer is greater nearer to the drain.

The depletion layer, which acts as an insulator, is shown in fig. 11.33(*a*). It reduces the effective cross-section of the n-channel and therefore restricts the flow of electrons. For this reason, the FET is said to be operating in the depletion mode.

If the drain-source voltage is increased, as shown in fig. 11.33(*b*), the current might increase but the increase in voltage also increases the depletion layer which restricts the increase. Eventually a point is reached at which the depletion layer completely absorbs the n-channel, and the drain-source current I_D reaches a limiting value called the saturation current I_{DSS}. The current/voltage characteristic is shown in fig. 11.33(*c*). The saturation point is termed the pinch-off point.   If the drain-source voltage is disconnected and a bias voltage applied to the gate, as shown in fig. 11.34, the

Fig. 11.33 Depletion layer in
a JUGFET

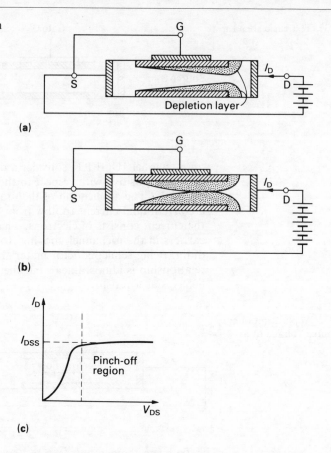

(a)

(b)

(c)

Fig. 11.34 Bias voltage
applied to gate of a JUGFET

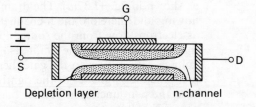

depletion layer is evenly set up in the n-channel. This
effectively reduces the cross-section and means that when a
drain-source voltage is reapplied, the pinch-off point will be
experienced at a lower voltage, and the currents, including the
saturation current, will all be less.

The combined effect of applying both drain-source voltage
and bias voltage is shown in fig. 11.35. Here the drain current
depends on both voltages, and the more negative the bias

Fig. 11.35 Combined
depletion effects in a JUGFET

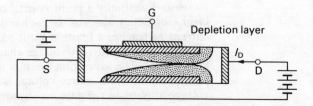

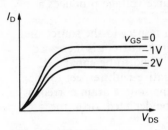

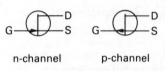

Fig. 11.36 Output characteristics of an n-channel JUGFET

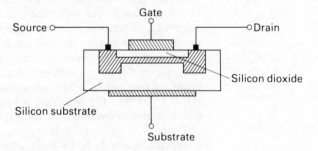

Fig. 11.37 JUGFET symbols

gate voltage the smaller is the saturation current. The gate voltage could stop the flow of drain current if increased sufficiently, but more importantly, the gate voltage has a more significant effect on the drain current than the drain voltage.

The resulting output characteristics of an n-channel JUGFET are shown in fig. 11.36. Operation occurs beyond pinch-off so that the control is by the gate voltage. The characteristics compare in form with those shown in fig. 11.12 for the junction transistor and the amplifier action can be obtained in the manner illustrated in fig. 11.17.

The symbol used in circuit diagrams for an n-channel JUGFET is shown along with that of the p-channel variety in fig. 11.37. The operation of the p-channel JUGFET is the converse of the n-channel action, with all polarities and current flow directions reversed.

The gate-source junctions present very high values of resistance (megohms) since they consist of reverse-biased junctions. It follows that the gate currents are very small but, due to the increase in minority charge carriers with temperature, these currents do vary with temperature.

11.16 IGFET

As its name indicates, the IGFET has its gate insulated from the channel. A simplified form of construction is shown in fig. 11.38, in which the main bulk of the material is low-conductivity silicon; this is termed the substrate. For an n-channel IGFET, the substrate is p-type silicon into which is

Fig. 11.38 Simple IGFET

introduced a thin n-channel terminated in the drain and source electrodes. The gate is separated from the channel by a layer of silicon dioxide which is an insulator ensuring that the gate is isolated from both source and drain. The silicon dioxide can be thought of as the dielectric in a capacitor consisting of two plates, one being the gate and the other the channel.

In the IGFET, the flow of electrons from drain to source is again the same as in the JUGFET and the pinch-off effect can also be achieved by connecting the gate to the source.

However, variation of the gate-source voltage produces a distinctly different effect.

If the gate is made positive with respect to the source and hence to the channel, the source-drain current is increased. Because of this increase, the FET is said to be enhanced. Conversely, if the gate is made negative with respect to the source and hence to the channel, the source-drain current is decreased and the FET is said to be depleted, or in the depletion mode.

The output characteristics of an n-channel IGFET are shown in fig. 11.39 and the symbols for IGFETs are shown in

Fig. 11.39 Output characteristics of an n-channel IGFET

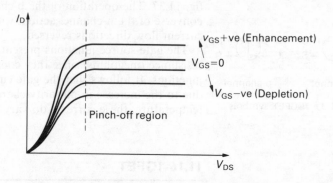

fig. 11.40. Again the p-channel IGFET is the converse of the n-channel. Note that a connection to the substrate is available and may be used as the other terminal for the bias arrangement.

The cause of the variation in drain current when bias voltages are applied to the gate lies in the capacitive effect between gate and channel. If the gate is positively charged then the channel is negatively charged and the channel therefore experiences an increase in charge carriers, i.e. it is enhanced. When the bias is reversed, there is a depletion in the number of charge carriers available.

The cause of the pinch-off effect is less obvious but stems from the p–n junction between channel and substrate, there being no p–n junction between channel and gate as in the JUGFET. Often the substrate is connected to the source, which is also convenient for the gate-source bias.

There is an important derivative of the IGFET which is the enhancement-mode IGFET for which the output characteristics are shown in fig. 11.41. In it, the channel is omitted and conduction can only take place when the gate is positive with respect to the substrate. This induces a conduction channel in the substrate and it follows that operation can only take place in the enhancement mode.

Although there are differences between the IGFET symbols, these are not applied in practice because this could be confusing. Finally, the IGFET is sometimes also termed a MOSFET due to the metal, oxide and silicon form of construction, and even MOSFET might be further contracted as MOST. IGFETs can easily be destroyed by the build-up of

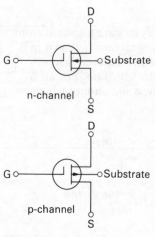

Fig. 11.40 IGFET symbols

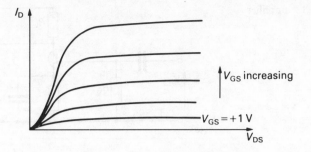

Fig. 11.41 Enhancement-mode n-channel IGFET

charge across the dielectric separating the gate and substrate. To avoid this, often these are linked at manufacture and the links have to be removed when the IGFETs are installed.

11.17 Static characteristics of a FET

There are three principal characteristics: the drain resistance, the mutual conductance and the amplification factor.

The ratio of the change, ΔV_{DS}, of the drain-source voltage to the change, ΔI_D, of the drain current, for a given gate-source voltage, is termed the drain resistance r_d,

i.e. $\qquad r_d = \Delta V_{DS}/\Delta I_D$ for a given value V_{GS} $\qquad$ (11.22)

The value of r_d is large in most FETs, say in the order of $10\,\text{k}\Omega$ to $500\,\text{k}\Omega$.

The ratio of the change, ΔI_D, of the drain current to the change, ΔV_{GS}, of the gate-source voltage, for a given drain-source voltage, is termed the mutual conductance g_m,

i.e. $\qquad g_m = \Delta I_D/\Delta V_{GS}$ for a given value V_{DS} $\qquad$ (11.23)

g_m is usually measured in milliamperes per volt.

The product of drain resistance and mutual conductance gives the amplification factor μ,

i.e. $\qquad \mu = r_d g_m$ $\qquad$ (11.24)

11.18 Equivalent circuit of a FET

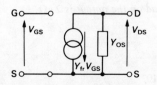

Fig. 11.42 *Y*-parameter equivalent circuit for a FET

Like the junction transistor, the FET can be represented by a circuit which is equivalent to its response to small signals causing it to operate over the linear part of its characteristics. The FET equivalent circuit, shown in fig. 11.42, is similar to that for the junction transistor except that the input resistance is so high that it can be considered infinite, i.e. equivalent to an open circuit.

The parameters are termed Y parameters. Y_{fs} is the forward transconductance (the s indicates that it is for the common-

Fig. 11.43 FET amplifier

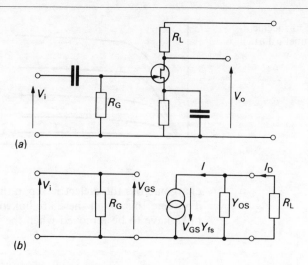

(a)

(b)

source mode) and is the ratio of I_D to V_{GS}. It follows that it is expressed in amperes per volt, or sometimes in siemens. Y_{os} is the output conductance (in the common-source mode) and is also measured in amperes per volt.

Consider the transistor in the amplifier arrangement illustrated in fig. 11.43(a). The equivalent circuit is shown in fig. 11.43(b) and again the capacitors are taken to have negligible effect.

From fig. 11.43(b),

$$I = I_D + V_o Y_{os}$$
$$= I_D + I_D R_L Y_{os}$$
$$\therefore \quad I_D = \frac{I}{1 + R_L Y_{os}}$$
$$= \frac{V_{GS} Y_{fs}}{1 + R_L Y_{os}}$$
$$V_o = I_D R_L$$
$$= \frac{V_{GS} Y_{fs} R_L}{1 + R_L Y_{os}}$$
$$\therefore \quad A_v = \frac{Y_{fs} R_L}{1 + R_L Y_{os}} \tag{11.25}$$

Example 11.4 *A transistor amplifier stage comprises a FET, of parameters $Y_{fs} = 2.2\,mA/V$ and $Y_{os} = 20\,\mu S$, and bias components and coupling capacitors of negligible effect. The total load on the output is $2\,k\Omega$. Determine the voltage gain.*

$$A_v = \frac{Y_{fs} R_L}{1 + Y_{os} R_L}$$
$$= \frac{2.2 \times 10^{-3} \times 2 \times 10^3}{1 + (2 \times 10^3 \times 20 \times 10^{-6})} = 4.23.$$

EXERCISES II

1. Explain the function of an equivalent circuit.
2. Explain the relationship between h_{ie} and a suitable device characteristic.
3. Define the term h_{ie}.
4. Define the term h_{fe}.
5. Explain the relationship between h_{fe} and a suitable device characteristic.
6. Define the term h_{oe}.
7. Explain the relationship between h_{oe} and a suitable device characteristic.
8. Draw the equivalent circuit for a common-emitter bipolar transistor amplifier and derive suitable formulae for the amplifier: current gain, voltage gain and power gain. Neglect bias, decoupling and coupling components.
9. Indicate suitable values of the components employed in the circuit of the previous question if:

$$V_{cc} = 15\,\text{V}, \qquad V_{be} = 0.8\,\text{V}, \qquad I_b = 200\,\mu\text{A},$$
$$I_c = 5\,\text{mA}, \qquad V_{ce} = 5\,\text{V}.$$

10. In the simplified amplifier network shown in fig. A the hybrid parameters of the transistor are $h_{ie} = 650\,\Omega$, $h_{fe} = 56$ and $h_{oe} = 100\,\mu\text{S}$. The input sinusoidal signal is $1.0\,\text{mV}$ r.m.s.

Fig. A

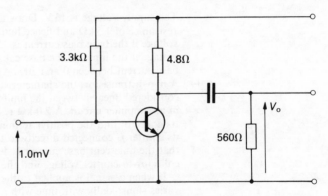

(*a*) Draw the small-signal equivalent network; (*b*) hence calculate:

    (i) the current in the 560-Ω output resistor,
    (ii) the voltage gain (in dB),
    (iii) the current gain (in dB).

11. An amplifier has the following parameters:

Input resistance (kΩ)	Output resistance (kΩ)	Open-circuit voltage gain
470	8.2	50

A signal source having an e.m.f. of $75\,\mu\text{V}$ and negligible internal resistance is connected to the input of the amplifier. The output is connected to a 10-kΩ resistive load. Draw the equivalent circuit and hence determine: (i) the output voltage developed across the load; (ii) the current gain; (iii) the voltage gain.

12. The equivalent circuit of an amplifier is shown in fig. B.
    When the load resistance is $R_1 = 175\,\Omega$, the voltage gain of
    the amplifier $A_v = 4375$. Calculate the amplifier output
    resistance R_o.

Fig. B

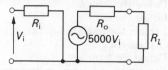

When the value of R_1 is changed to $275\,\Omega$, the current gain
of the amplifier is $A_i = 16.7 \times 10^4$. Calculate the value of the
amplifier input resistance R_i.

13. The collector characteristics of a p–n–p transistor can be
    considered as three straight lines through the following points:

   (a) $I_b = 300\,\mu A$ ($I_c = 8\,\text{mA}$, $V_{ce} = 2\,\text{V}$ and $I_c = 9.5\,\text{mA}$,
       $V_{ce} = 14\,\text{V}$)
   (b) $I_b = 150\,\mu A$ ($I_c = 4.5\,\text{mA}$, $V_{ce} = 2\,\text{V}$ and $I_c = 5.3\,\text{mA}$,
       $V_{ce} = 14\,\text{V}$)
   (c) $I_b = 0\,\mu A$ ($I_c = 1.5\,\text{mA}$, $V_{ce} = 2\,\text{V}$ and $I_c = 2\,\text{mA}$,
       $V_{ce} = 14\,\text{V}$)

   The supply voltage is 16 V. Draw a load line for a load
   resistance of $1.5\,\text{k}\Omega$ and hence find: (i) the collector current and
   voltage if the base bias current is $150\,\mu A$; (ii) the r.m.s. output
   voltage if the input to the base causes a sinusoidal variation of
   base current between 0 and $300\,\mu A$; (iii) the current gain.

14. A n–p–n transistor, the characteristics of which can be
    considered linear between the limits shown in the table, is used
    in an amplifier circuit. A 2.0-$k\Omega$ resistor is connected between
    its collector and the positive terminal of the 9-V d.c. supply and
    its emitter is connected directly to the negative terminal. Given
    that the quiescent base current is $40\,\mu A$ determine the quiescent
    collector-to-emitter voltage and the quiescent collector current.
    If when a signal is applied to the circuit the base current
    varies sinusoidally with time with a peak alternating component
    of $20\,\mu A$, determine the alternating component of the collector
    current and hence the current gain of the stage. The load across
    the output terminals of the circuit can be considered very high
    in comparison with $2.0\,\text{k}\Omega$.

	$I_b = 60\,\mu A$		$I_b = 40\,\mu A$		$I_b = 20\,\mu A$	
V_{ce}	1 V	10 V	1 V	10 V	1 V	10 V
I_c mA	2.7	3.2	1.8	2.1	0.9	1.1

15. The transistor used in the circuit shown in fig. C has
    characteristics which can be considered linear between the limits
    shown in the table. Determine the value of R_B to give a
    quiescent base current of $80\,\mu A$.
    If the signal base current varies sinusoidally with time and
    has a peak value of $40\,\mu A$ determine the r.m.s. value of the
    signal voltage across the load and hence the signal power in the

Fig. C

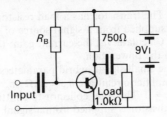

load. The reactances of the coupling capacitors can be considered zero.

I_b (μA)	120		100		80		60		40	
V_{ce} (V)	1	12	1	12	1	12	1	12	1	12
I_c (mA)	10.8	13.4	9.0	11.2	7.2	9.0	5.3	6.7	3.7	4.5

16. The output characteristics, which can be considered linear, for the n–p–n silicon transistor used in the amplifier circuit shown in fig. D are specified in the table below. The source of signal

Fig. D

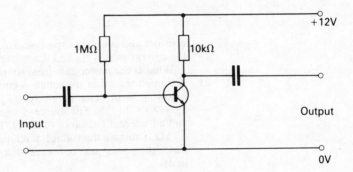

can be represented by a constant current generator of 24 μA peak and internal resistance 3.0 kΩ. The stage feeds an identical one. Determine the current, voltage and power gains of the stage. The input resistances of the transistors can be considered constant at 6.0 kΩ and the reactances of the coupling capacitors are negligible.

I_b (μA)	0		4		8		12		16		20	
V_{ce} (V)	1	10	1	10	1	10	1	10	1	10	1	10
I_c (mA)	0.0	0.0	0.14	0.18	0.33	0.39	0.49	0.61	0.68	0.83	0.86	1.04

17. A common-emitter, bipolar transistor has the following h-parameters: $h_{ie} = 1.5$ kΩ, $h_{fe} = -60$ and $h_{oe} = 12.5$ μS. If the load resistance can vary between 5 kΩ and 10 kΩ, calculate the maximum and minimum values of the amplifier's (a) current gain; (b) voltage gain; and (c) power gain.

18. The hybrid parameters for a transistor used in the common-emitter configuration are $h_{ie} = 1.5$ kΩ, $h_{fe} = 70$ and $h_{oe} = 100$ μS.

The transistor has a load resistor of 1 kΩ in the collector and is supplied from a signal source of resistance 800 Ω.

Calculate: (*a*) the current gain; (*b*) the voltage gain; and (*c*) the power gain.

19. The small-signal hybrid parameters for a transistor used in an amplifier circuit are $h_{ie} = 2\,k\Omega$, $h_{fe} = 60$, $h_{oe} = 20\,\mu S$ and $h_{fe} = 0$. The total collector to emitter load is 10 kΩ and the transistor is supplied from a signal source of e.m.f. 100 mV r.m.s. and internal resistance 3 kΩ. Determine the current, voltage and power gains for the stage and the signal power developed in the load.

20. The transistor in the given circuit (fig. E) has the following hybrid parameters: $h_{ie} = 1\,k\Omega$, $h_{fe} = 50$, $h_{oe} = 100\,\mu S$ and $h_{re} = 0$. Draw the equivalent circuit, neglecting the effect of bias

Fig. E

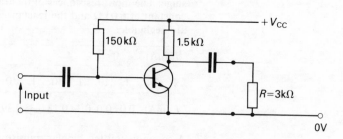

resistors and coupling capacitors, and obtain the magnitude of the current gain I_L/I_B and the voltage gain V_o/E_s.

What is the power gain from source to load?

21. A transistor used in the common-emitter configuration has the following small signal parameters: $h_{ie} = 1.0\,k\Omega$, $h_{fe} = 49$ and $h_{oe} = 80 \times 10^{-6}\,S$. The source of signal has an e.m.f. of 10 mV and an internal resistance of 600 Ω. The load resistance is 4.7 kΩ. Estimate the voltage developed across the load and hence the power gain of the transistor, also the power gain in dB.

22. The n–p–n transistor used in the circuit shown in fig. F has the following small signal parameters: $h_{ie} = 1.4\,k\Omega$, $h_{fe} = 50$ and $h_{oe} = 25\,\mu S$. The input signal is supplied from a signal source of 30 mV and internal resistance 3 kΩ.

Determine the output voltage at mid-band frequencies.

Fig. F

150 kΩ   1.5 kΩ   +V_CC

Input

R = 3 kΩ

0V

23. The common-emitter stage shown in fig. G feeds an identical stage, and is fed from a source of constant e.m.f. of 100 mV. $h_{ie} = 2.0\,k\Omega$, $h_{fe} = 45$, $h_{oe} = 30\,\mu S$.

Calculate:

(*a*) the input resistance of the stage;

Fig. G

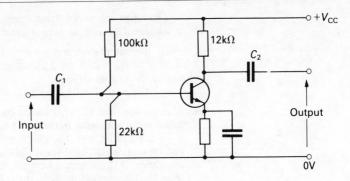

(b) the output resistance of the stage;
(c) the input current to the next stage;
(d) the input voltage to the next stage;
(e) the power gain;
(f) the power gain in dB.

24. The transistor of the amplifier shown in fig. H has an input resistance of 750 Ω, an output resistance of 100 Ω and a short-circuit current gain of 12 000. Draw the mid-band equivalent circuit of the amplifier.
   Determine the current, voltage and power gains of the amplifier.

Fig. H

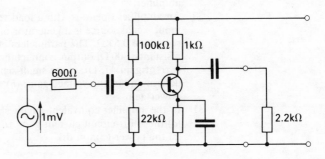

25. In the simplified transistor amplifier network shown in fig. I, the hybrid parameters are $h_{ie} = 700$ Ω, $h_{fe} = 48$ and $h_{oe} = 80$ μS.

Fig. I

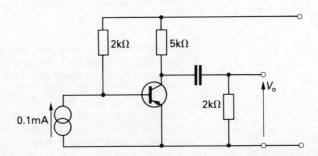

The input sinusoidal signal current is 0.1 mA r.m.s. Draw the small-signal equivalent network and hence calculate:

(i) the alternating component of the collector current,
(ii) the current in the 2-kΩ output resistor,
(iii) the output voltage,
(iv) the overall voltage amplification.

26. A transistor used in the common-base configuration has the following small signal parameters: $h_{ib} = 75\,\Omega$, $h_{fb} = -0.95$, $h_{ob} = 0\,S$ and $h_{rb} = 0$. The effect of the applied sinusoidal signal can be represented by a source of e.m.f. of 30 mV and internal resistance 75 Ω connected between the emitter and base, and the total load connected between the collector and base is 5.0 kΩ resistance. Determine the collector-to-base signal voltage in magnitude and phase relative to the source of e.m.f., and the power gain of the transistor.

27. An amplifier has a voltage gain of 15 at 50 Hz. Its response at 1.0 kHz gives a voltage gain of 40. What is the relative response at 50 Hz in dB to its response at 1.0 kHz?

28. An amplifier has an open-circuit voltage gain of 600 and an output resistance of 15 kΩ and input resistance of 5.0 kΩ. It is supplied from a signal source of e.m.f. 10 mV and internal resistance 2.5 kΩ and it feeds a load of 7.5 kΩ. Determine the magnitude of the output signal voltage and the power gain in dB of the amplifier.

29. An amplifier has a short-circuit current gain of 100, an input resistance of 2.5 kΩ and an output resistance of 40 kΩ. It is supplied from a current generator signal source of 12 μA in parallel with a 50-kΩ resistor. The amplifier load is 10 kΩ. Determine the current voltage and power gains in dB of the amplifier.

30. An amplifier operates with a load resistance of 2.0 kΩ. The input signal source is a generator of e.m.f. 50 mV and internal resistance 0.5 kΩ. The parameters of the amplifier are input resistance 800 Ω, output conductance 80 μS, and short-circuit current gain 47. Draw the small-signal equivalent circuit and determine: (a) the voltage gain; (b) the current gain; (c) the power gain.

31. In the amplifier equivalent circuit shown in fig. J determine: (a) the short-circuit current gain; (b) the current gain I_L/I_I; (c) the power gain in dB.

Fig. J

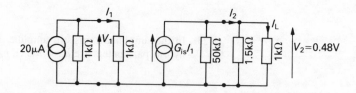

32. Without referring to the text, give the symbols for: (a) a p-channel junction FET; (b) an n-channel enhancement and depletion mode insulated gate FET; (c) a p-channel enhancement insulated gate FET; (d) a p–n–p bipolar transistor.

33. In each of the cases in Q. 32, give the polarities of the voltages normally applied to each electrode and state whether the flow of conventional current through each device is primarily by electrons or holes.

34. In a FET the following measurements are noted:
    With V_{GS} constant at -4 V the drain current is 5 mA when V_{DS} is $+4$ V, and 5.05 mA when V_{DS} is $+15$ V.
    Find r_d.
    The current is restored to 5 mA if the gate voltage is now changed to -4.02 V.
    Find g_m.
35. Using the characteristics for an n-channel FET shown in fig. K find at $V_{DS} = 12$ V and $V_{GS} = -1.5$ V

Fig. K

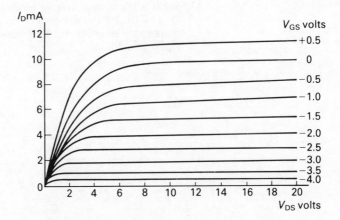

 (i) r_d,
 (ii) g_m,
 (iii) μ.

 For a value of $V_{DS} = 14$ V derive the transfer characteristic (I_D/V_{GS}).
 Using the characteristics shown in fig. K, derive the transfer characteristic for $V_{DS} = 10$ V. Compare the result with that of the previous question. What conclusions do you reach?
 Using the same characteristics find I_D, V_{DS} and V_{GS} at the points marked A, B, E, F and G.
36. A bipolar transistor used as a common-emitter amplifier has an input resistance of 2 kΩ and a current gain under operating conditions of 38. If the output current is supplied to a 5-kΩ load resistor, find the voltage gain. If the bipolar transistor is replaced by an FET having g_m 2.6 mA/V and the load resistor is changed to a value of 22 kΩ, find the voltage gain.
37. Find the output voltages for the two devices employed in Q. 36 if the input is derived from a low frequency generator of e.m.f. 50 mV and output resistance of 6 kΩ.
38. Find the approximate input impedance of an FET at (*a*) 10 Hz; (*b*) 1 MHz; if the input resistance is 5 MΩ shunted by a capacitor of 200 pF.
39. An insulated gate FET operating in the enhancement mode gives a saturation drain current of 4 mA when the gate source voltage is -2 V and the drain source voltage is $+25$ V. Is this a p- or n-channel device? What is the value of g_m?
40. The FET in Q. 39 uses as a load a coil of inductance 110 μH and resistance 15 Ω tuned to a frequency of 250 kHz by a capacitor connected in parallel with the coil. What is the value of capacitance needed and what is the stage gain at the resonant frequency if $g_m = 5.5$ mA/V?

41. The tuned load in the previous question is replaced by a transformer having 1000 primary turns and 2000 secondary turns and the secondary is connected to a resistor of 5 kΩ resistance. The input to the gate is a sine wave of peak to peak value 80 mV. Find the r.m.s. output voltage across the 5-kΩ resistor.

42. A FET is used as a switch. In the ON state the current flowing is 20 mA with a V_{DS} at 4 V. What is the effective ON resistance? If the FET is used as a series switch with a direct voltage of 40 V and a load resistor of 2.0 kΩ find the voltage across the load in the ON position.

43. What is the average voltage produced across the load in Q. 42 if the FET is switched on by a square wave of mark/space ratio 2:1 (the FET passes no current in the OFF (space) condition)?

12 Further Semiconductor Amplifiers

12.1 Cascaded amplifiers

Often a single-transistor amplifier stage is unable to provide the gain necessary for the function to be performed by the system in which the amplifier operates. Alternatively, the stage might be able to provide the necessary gain but only with the introduction of distortion. In order to overcome the difficulties associated with either of these problems, amplifier stages can be connected in cascade as shown in fig. 9.17. Cascaded amplifiers are connected in such a manner that the output of one amplifier stage is the input to a second stage, thus providing further gain within the system.

The advantage of cascading is the increase of gain, but this is offset by the increased distortion due to the non-linearity of each amplifier stage.

A simple two-stage n–p–n transistor amplifier could take the form shown in fig. 12.1.

Fig. 12.1 Two-stage n–p–n transistor amplifier

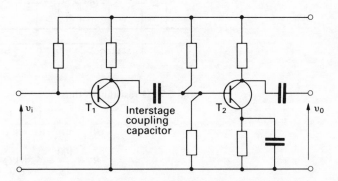

The input signal v_i is amplified in the usual manner by the transistor T_1 and the amplified signal is transferred to the second stage by the interstage coupling capacitor. It should be recalled that the coupling capacitor 'conducts' the a.c. signal but 'blocks' the d.c. component of the supply. The a.c. amplified signal is then further amplified by transistor T_2 and the enhanced output signal emerges as v_o by means of a further coupling capacitor.

The analysis of a two-stage amplifier is similar to that already described in section 11.11 and is best illustrated by means of an example.

Example 12.1   *A two-stage amplifier is shown in fig. 12.2. Draw the mid-band equivalent circuit and hence determine the overall current and power gains of the amplifier. Both transistors have input resistances of 1000 Ω, short-circuit current gains of 60 and output conductances of 20 μS.*

Fig. 12.2

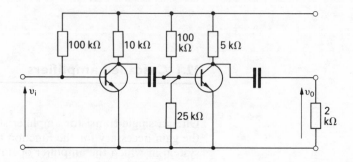

The mid-band equivalent circuit is shown in fig. 12.3.

$$\frac{1}{h_{oe}} = \frac{1}{20 \times 10^{-6}} \equiv 50 \text{ k}\Omega.$$

For the output of the first stage, combine the 50-kΩ, 10-kΩ, 100-kΩ

Fig. 12.3

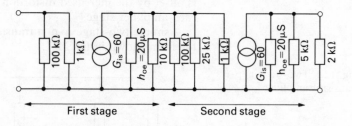

First stage   Second stage

and 25-kΩ resistances thus

$$\frac{1}{R'} = \frac{1}{50} + \frac{1}{10} + \frac{1}{100} + \frac{1}{25} = \frac{17}{100}$$

$$R' = \frac{100}{17} = 5.88 \text{ k}\Omega.$$

Also, for the output of the second stage, combine the 50-kΩ and the 5-kΩ resistances thus

$$\frac{1}{R''} = \frac{1}{50} + \frac{1}{5} = \frac{11}{50}$$

$$\therefore \qquad R'' = \frac{50}{11} = 4.55 \text{ k}\Omega.$$

The network shown in fig. 12.3 can now be reduced to that shown in fig. 12.4.

For the input current i_i to the amplifier, the current i_i to the first-stage transistor is obtained by applying the current-sharing rule

$$i_1 = \frac{100}{100 + 1} i_i = \frac{100}{101} i_i.$$

Fig. 12.4

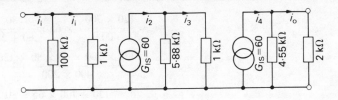

The output of the current generator associated with the first-stage transistor is

$$i_2 = 60 \times i_1 = 60 \times \frac{100}{101} i_{\mathrm{i}}.$$

The current i_3 to the second-stage transistor is given by

$$i_3 = \frac{5.88}{5.88 + 1} i_2 = \frac{5.88}{6.88} \times 60 \times \frac{100}{101} i_{\mathrm{i}}.$$

The output current of the current generator associated with the second-stage transistor is

$$i_4 = 60 \times i_3 = 60 \times \frac{5.88}{6.88} \times 60 \times \frac{100}{101} i_{\mathrm{i}}.$$

The amplifier output current is given by

$$i_{\mathrm{o}} = \frac{4.55}{4.55 + 2} i_4 = \frac{4.55}{6.55} \times 60 \times \frac{5.88}{6.88} \times 60 \times \frac{100}{101} i_{\mathrm{i}}$$

$$\therefore \quad A_{\mathrm{i}} = \frac{i_{\mathrm{o}}}{i_{\mathrm{i}}} = \frac{4.55}{6.55} \times 60 \times \frac{5.88}{6.88} \times 60 \times \frac{100}{101} = 2120$$

$$A_{\mathrm{p}} = A_{\mathrm{i}}^2 \times \frac{R_{\mathrm{o}}}{R_{\mathrm{i}}} \quad \text{where } R_{\mathrm{i}} = \frac{100 \times 1}{100 + 1} = 990$$

$$= 2120^2 \times \frac{2000}{990} = 9\,080\,000.$$

Example 12.2   *Determine the overall voltage gain, current gain and power gain of the two-stage amplifier shown in fig. 12.5.*

Fig. 12.5

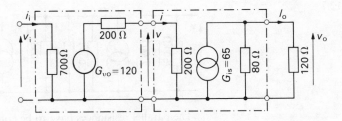

$$v_{\mathrm{i}} = i_{\mathrm{i}}\, 700$$

$$\therefore \quad v = 120 \times i_{\mathrm{i}}\, 700 \times \frac{200}{200 + 200}$$

$$\therefore \quad i = 120 \times i_{\mathrm{i}}\, 700 \times \frac{200}{200 + 200} \times \frac{1}{200} = \frac{120 \times 700}{400} i_{\mathrm{i}}$$

$$\therefore \quad i_o = \frac{120 \times 700}{400} i_i \times 65 \times \frac{80}{80 + 120}$$

$$\therefore \quad v_o = \frac{120 \times 700 \times 65 \times 80 \times 120}{400 \times 200}$$

$$\therefore \quad A_v = \frac{v_o}{v_i} = \frac{120 \times 700 \times 65 \times 80 \times 120 \times i_i}{400 \times 700 \times 200 \times i_i}$$

$$= 936.$$

$$A_i = \frac{i_o}{i_i} = \frac{120 \times 700 \times 65 \times 80 \times i_i}{400 \times 200 \times i_i} = 5460$$

$$\therefore \quad A_p = A_v A_i = 936 \times 5460$$

$$= 5\,110\,000.$$

FET amplifiers can also be connected in cascade. The
manner of coupling can either be achieved using the normal
resistance–capacitance coupling method or by means of a
transformer. The former method is illustrated in
example 12.3, while a typical transformer coupling
arrangement is shown in fig. 12.6.

Fig. 12.6 Transformer
coupling of two FET
amplifiers

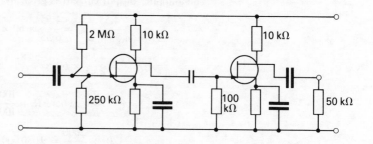

Example 12.3

*A two-stage amplifier is shown in fig. 12.7. Draw the mid-band
equivalent circuit and hence determine the overall voltage, current and
power gains of the amplifier. Both transistors have a mutual
conductance $g_m = 3\,mA/V$ and a drain resistance $r_d = 200\,k\Omega$.*

Fig. 12.7

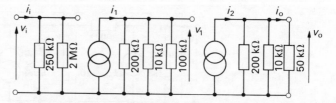

The parallel resistances have to be combined in order to simplify
the circuit analysis; thus

$$R' = \frac{250 \times 2000}{250 + 2000} = 222\,k\Omega$$

$$\frac{1}{R''} = \frac{1}{200} + \frac{1}{10} + \frac{1}{100} = \frac{23}{100}$$

$$\therefore \quad R'' = \frac{200}{23} = 8.7\,k\Omega$$

$$\therefore \qquad \frac{1}{R'''} = \frac{1}{200} + \frac{1}{10} + \frac{1}{50} = \frac{25}{200}$$

$$\therefore \qquad R''' = \frac{200}{25} = 8.0 \,\text{k}\Omega.$$

Thus the mid-band equivalent circuit simplifies to that shown in fig. 12.8.

Fig. 12.8

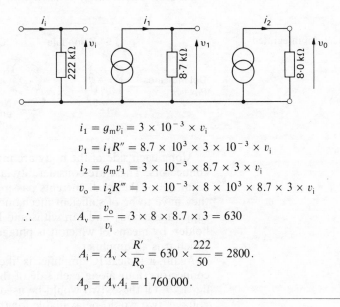

$$i_1 = g_m v_i = 3 \times 10^{-3} \times v_i$$

$$v_1 = i_1 R'' = 8.7 \times 10^3 \times 3 \times 10^{-3} \times v_i$$

$$i_2 = g_m v_1 = 3 \times 10^{-3} \times 8.7 \times 3 \times v_i$$

$$v_o = i_2 R''' = 3 \times 10^{-3} \times 8 \times 10^3 \times 8.7 \times 3 \times v_i$$

$$\therefore \qquad A_v = \frac{v_o}{v_i} = 3 \times 8 \times 8.7 \times 3 = 630$$

$$A_i = A_v \times \frac{R'}{R_o} = 630 \times \frac{222}{50} = 2800.$$

$$A_p = A_v A_i = 1\,760\,000.$$

12.2 Integrated circuits

Thus far, the treatment of semiconductor devices has led to the construction of networks made from a considerable number of components. The consideration of the cascaded amplifier stages indicates that such numbers can rise to a hundred or more, even for quite simple arrangements. Apart from the cost of connecting all these pieces together, the space taken up by them is also an important factor when designing electronic equipment.

However, the manufacture of transistors is achieved by a process that makes many transistors on a single slice, which is then cut up to form the individual transistors. Although not simple to effect, it was a logical development to keep the transistors together on the slice, thereby forming the complete amplifier arrangement in one step.

From this development was evolved the integrated circuit, often referred to as an IC. Integrated circuits can be produced to form single amplifier stages or multistage amplifiers, both with the advantage of automatic compensation for temperature drift. The outcome is that one component can replace up to a hundred or more with a dramatic reduction in the volume taken up, i.e. the outcome is the miniaturized microcircuit which is very small, cheap and reliable. It cannot be overemphasized that the cost of an IC is significantly less than the circuit components which it can replace.

Typically, an IC might comprise 20–30 transistors with possibly 50–100 other passive components such as resistors, all in the volume formerly taken up by a single bipolar transistor. The form of construction is shown in fig. 12.9.

Fig. 12.9 An integrated circuit

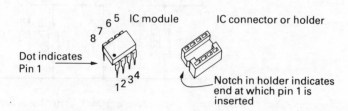

Along each side of the body are mounted a number of connectors. These are considerably larger than would be required for the small currents passing through them, but they have to be of sufficient mechanical strength that they can support the body and can withstand being pushed into the holder, by means of which it is plugged into the network for which it is an amplifier.

Again, a typical IC amplifier is likely to have eight connectors, four along each side of the body as shown in fig. 12.9. Of these, two would be used to provide the supply voltage, two would be available for the input signals, one would be available for the output signal while a sixth would be used to earth the IC. The remaining connectors can be used for frequency compensation but this use is optional.

The fact that there are two input connectors or terminals and there are also two connectors which need not be used highlights a significant difference between IC arrangements and the individual component networks we have previously considered. Because the complexity of the IC can be varied to a considerable extent without affecting the cost, many ICs are manufactured with optional facilities; thus a number of functions can be performed by the one IC.

It might appear strange that there are two input terminals but the versatility is such that the choice of terminal permits the output signal either to be inverted or non-inverted, i.e. out of phase or in phase with the input signal. Further, many ICs can accept two signals at these terminals and operate on the difference between them. This will be discussed later in this chapter.

The component amplifiers which make up a typical integrated-circuit amplifier are directly coupled rather than having a system within the semiconductor slice of introducing active capacitor components to be used for coupling. It follows that ICs can be used for amplifying direct signals as well as alternating signals.

Although integrated circuits have been discussed in the context of amplifiers, it should be noted that an integrated circuit is a form of construction rather than a device performing a specific function or range of functions. In Chapter 13, we shall look at logic systems, which also are generally manufactured in the form of integrated circuits. However, this chapter is limited to amplifier operation for which the basic integrated circuit usually takes the form of an operational amplifier, often abbreviated to op-amp.

12.3 Operational amplifiers

The common IC operational amplifier is one which has a very high gain and finds widespread use in many areas of electronics. Its applications are not limited to linear amplification systems but include digital logic systems as well.

There are certain properties common to all operational amplifiers as follows:

  (i) an inverting input;
  (ii) a non-inverting input;
  (iii) a high input impedance (usually assumed infinite) at both inputs;
  (iv) a low output impedance;
  (v) a large voltage gain when operating without feedback (typically 10^5);
  (vi) the voltage gain remains constant over a wide frequency range;
  (vii) relatively free of drift due to ambient temperature change, hence the direct voltage output is zero when there is no input signal;
  (viii) good stability, being free of parasitic oscillation.

The basic form of operational amplifier is shown in fig. 12.10.

Fig. 12.10 The basic operational amplifier

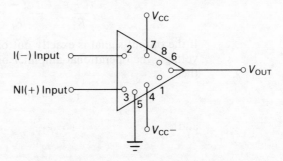

It will be noted that the input terminals are marked + and −. These are not polarity signs; rather the − indicates the inverting input terminal while the + indicates the non-inverting input terminal.

This basic unit performs in a variety of ways according to the manner of the surrounding circuitry. A number of general applications are now considered.

12.4 The inverting operational amplifier

The circuit of an inverting operational amplifier is shown in fig. 12.11.

Fig. 12.11 Inverting operational amplifier

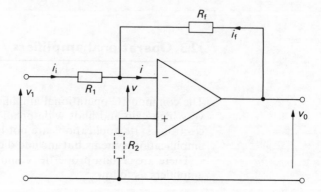

The open-loop gain of the op-amp is A, thus the output voltage $v_o = Av$. R_2 may be assumed negligible, hence

$$v_i - v = i_i R_1.$$

If the input impedance to the amplifier is very high then $i \simeq 0$, hence

$$i_i = -i_f$$

but

$$i_f = \frac{v_o - v}{R_f}$$

and

$$i_i = \frac{v_i - v}{R_1}$$

$\therefore$

$$\frac{v_i - v}{R_1} = -\frac{v_o - v}{R_f} = \frac{v - v_o}{R_f}.$$

If the output signal is exactly out of phase with the input voltage, the operational amplifier being in its inverting mode, then

$$v_o = -Av$$

and

$$v = -\frac{v_o}{A}$$

hence

$$\frac{v_i + \dfrac{v_o}{A}}{R_1} = \frac{-\dfrac{v_o}{A} - v_o}{R_f}$$

and

$$v_i + \frac{v_o}{A} = -\frac{v_o}{A} \cdot \frac{R_1}{R_f} - v_o \cdot \frac{R_1}{R_f}.$$

Generally, R_1 and R_f are of approximately the same range of resistance, e.g. $R_1 = 100\,\text{k}\Omega$ and $R_f = 1\,\text{M}\Omega$, and A is very large, e.g. $A = 10^5$, hence

$$\frac{v_o}{A} \quad \text{and} \quad \frac{v_o}{A} \cdot \frac{R_1}{R_f}$$

can be neglected,

$$\therefore \qquad v_i \simeq -v_o \frac{R_1}{R_f} \tag{12.1}$$

It follows that the overall gain is given approximately by

$$A_\text{v} = -\frac{R_f}{R_1} \tag{12.2}$$

From this relationship it is seen that the gain of the amplifier depends on the resistances of R_1 and R_f, and that the inherent gain of the op-amp, provided it is large, does not affect the overall gain.

Usually, in practice, the non-inverting input is earthed through R_2 thus minimizing the worst effects of the offset voltage and thermal drift. The offset voltage is the voltage difference between the op-amp input terminals required to bring the output voltage to zero. Finally, the output often includes a resistance of about 50 to 200 Ω in order to give protection in the event of the load being short-circuited.

12.5 The summing amplifier

This is a development of the inverting operational amplifier. Consider the arrangement shown in fig. 12.12. For the three

Fig. 12.12 Summing amplifier

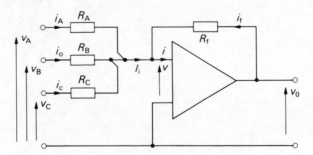

input signals v_A, v_B and v_C, the currents in the resistors R_A, R_B and R_C are:

$$i_A = \frac{v_A - v}{R_A}, \quad i_B = \frac{v_B - v}{R_B} \quad \text{and} \quad i_C = \frac{v_C - v}{R_C}$$

hence $\qquad i_i = \frac{v_A - v}{R_A} + \frac{v_B - v}{R_B} + \frac{v_C - v}{R_C}$

again $\qquad i_f = \frac{v_o - v}{R_f}$

and since $i \simeq 0$

$$\frac{v - v_o}{R_f} = \frac{v_A - v}{R_A} + \frac{v_B - v}{R_B} + \frac{v_C - v}{R_C}.$$

In a summing amplifier, usually v is very small compared with the other voltages, hence

$$-\frac{v_o}{R_f} = \frac{v_A}{R_A} + \frac{v_B}{R_B} + \frac{v_C}{R_C}.$$

If

$$R_f = R_A = R_B = R_C$$

then

$$-v_o = v_A + v_B + v_C$$

or

$$v_o = -(v_A + v_B + v_C).$$

It can now be seen that, apart from the phase reversal, the output voltage is the sum of the input voltages. From this comes the title — summing amplifier or summer.

It is this form of operation which leads to the general term — operational amplifier. The operation referred to is a mathematical operation and the basic op-amp can be made not only to add but to subtract, integrate, etc.

The summation can be illustrated by the following simple instance. If $v_A = 2\,V$, $v_B = -4\,V$ and $v_C = 6\,V$ then $v_o = -(2 - 4 + 6) = -4\,V$. Since instantaneous values have been chosen, it may be inferred that the operation works for alternating voltages as well as for steady voltages.

Example 12.4   *Two voltages, $+0.6\,V$ and $-1.4\,V$, are applied to the two input resistors of a summation amplifier. The respective input resistors are $400\,k\Omega$ and $100\,k\Omega$, and the feedback resistor is $200\,k\Omega$. Determine the output voltage.*

$$\frac{v_o}{R_f} = -\left[\frac{v_A}{R_A} + \frac{v_B}{R_B}\right] \qquad \text{from (12.4)}$$

$$\therefore \qquad v_o = -200\left[\frac{0.6}{400} + \frac{-1.4}{100}\right]$$

$$= 2.5\,V$$

12.6 The non-inverting amplifier

The circuit of a non-inverting amplifier is shown in fig. 12.13. It is shown in two common forms which are identical electrically, but the conversion from one diagram layout to the other can give many readers difficulty.

Due to the very high input resistance, the input current is negligible hence the voltage drop across R_2 is negligible and

$$v_i = v.$$

Fig. 12.13 Non-inverting amplifier

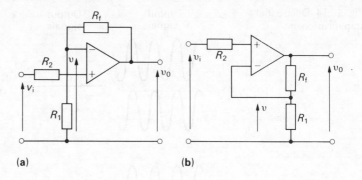

(a)   **(b)**

Especially using form (*b*) of fig. 12.13, it can be seen that

$$v = \frac{R_1}{R_1 + R_f} v_o$$

$$\therefore \qquad v_i = \frac{R_1}{R_1 + R_f} v_o$$

$$\therefore \qquad A_v = \frac{v_o}{v_i} = 1 + \frac{R_f}{R_1} \qquad (12.5)$$

Again we see that the gain of the amplifier is independent of the gain of the op-amp.

12.7 Differential amplifiers

The differential amplifier is the general case of the op-amp, which has been taken only in specific situations so far. The function of the differential amplifier is to amplify the difference between two signals. Being a linear amplifier, the output is proportional to the difference in signal between the two input terminals.

If we apply the same sine wave signal to both inputs, there will be no difference and hence no output signal, as shown in fig. 12.14. If one of the signals were inverted, the difference between the signals would be twice one of the signals — and hence there would be a considerable output signal.

This at first sight appears to be a complicated method of achieving amplification. Differential amplification has two advantages:

(*a*) The use of balanced input signals reduces the effect of interference as illustrated in fig. 12.15. Here a balanced signal is transmitted by two signals which are identical other than being out of phase. Interference at a lower frequency has distorted each signal but since the distortion effects are in phase, the amplifier output of the interference signals is zero. It follows that the amplified signal is devoid of interference.

(*b*) The differential amplifier can be used with positive or negative feedback. In an idealized differential amplifier, as

Fig. 12.14 Differential
amplification

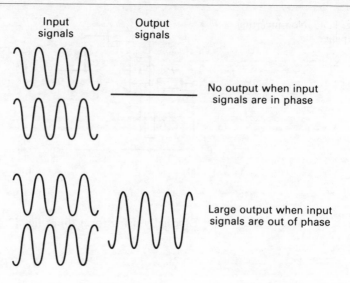

Input
signals

Output
signals

No output when input
signals are in phase

Large output when input
signals are out of phase

Fig. 12.15 Effect of
interference

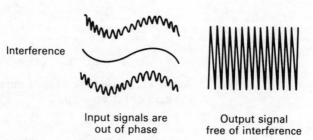

Interference

Input signals are
out of phase

Output signal
free of interference

shown in fig. 12.16,

$$v_o = A_v(v_1 - v_2) \tag{12.6}$$

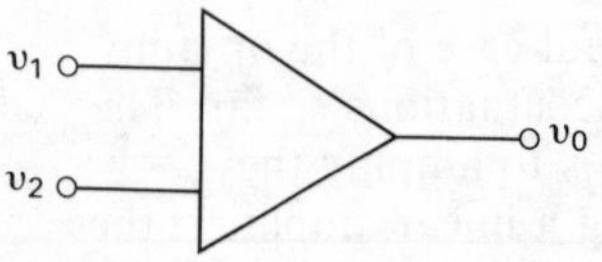

$v_1 \circ\!\!-\!\!$

$v_2 \circ\!\!-\!\!$ $\circ\, v_0$

Fig. 12.16 Simple differential
amplifier

In a practical differential amplifier, the relationship is more complicated because the output depends not only on the difference $v_d = v_1 - v_2$ but also upon the average signal — the common-mode signal v_{com} — where

$$v_{com} = \frac{v_1 + v_2}{2} \tag{12.7}$$

Nevertheless, the gain is still given by

$$A_v = \frac{v_o}{v_1 - v_2} \tag{12.8}$$

Example 12.5   *A differential amplifier has an open-circuit voltage gain of* 100. *The input signals are* 3.25 V *and* 3.15 V. *Determine the output voltage.*

$$v_o = A_v(v_1 - v_2) = 100(3.25 - 3.15) = 10\text{ V}$$

The 10-V output is therefore the amplified difference between the input signals. The mean input signal is

$$v_{com} = \frac{v_i + v_2}{2} = \frac{3.25 + 3.15}{2} = 3.20\text{ V}$$

It follows that the net input signal $v_1 = 3.25 - 3.20 = 0.05$ V and $v_2 = 3.15 - 3.20 = -0.05$ V.

Using the net signals

$$v_o = A_v(v_1 - v_2) = 100[0.05 - (-0.05)] = 10\,\text{V}$$

thus showing that the mean signal is not amplified. The mean signal is termed the common-mode signal.

12.8 Common-mode rejection ratio

This is a figure of merit for a differential amplifier, the name usually being abbreviated as C.M.R.R. It is defined as

$$\text{C.M.R.R.} = \frac{\text{Differential gain}}{\text{Common-mode gain}} = \frac{A_v}{A_{\text{com}}}$$

The C.M.R.R. should be large so that output errors are minimized. For instance, using the figures in example 12.5, consider the output if the two input signals had both been 3.20 V. In this case

$$v_o = A_v(v_1 - v_2) = 100(3.20 - 3.20) = 0.$$

Such a figure is idealistic because, in practice, the circuit components have manufacturing tolerances and it would almost certainly result in there being a very small output. As a result we define the common-mode gain as

$$A_{\text{com}} = \frac{v_o}{v_{\text{com}}} \tag{12.6}$$

Example 12.6　　*The differential amplifier used in example 12.5 has a common input signal of 3.20 V to both terminals. This results in an output signal of 26 mV. Determine the common-mode gain and the C.M.R.R.*

$$A_{\text{com}} = \frac{26 \times 10^{-3}}{3.20} = 0.0081$$

$$\text{C.M.R.R.} = \frac{A_v}{A_{\text{com}}} = \frac{100}{0.0081} = 12\,300$$

$$\equiv 20 \log 12\,300 = 81.8\,\text{dB}.$$

EXERCISES 12

1. Determine the overall voltage and current gains of the two-stage amplifier shown in fig. A and hence derive the corresponding

Fig. A

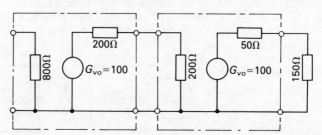

diagram of an equivalent single-stage amplifier.
2. Sketch and explain the gain/frequency characteristic of a CR-coupled amplifier.

   Determine the overall current and power gains of the two-stage amplifier shown in fig. B.

Fig. B

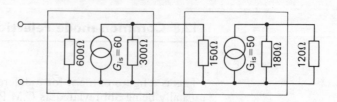

3. Fig. C shows the equivalent circuit of two identical amplifiers connected in cascade. The open-circuit voltage gain of each amplifier is 100.

Fig. C

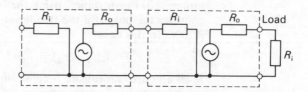

Prove that if the output load equals R_i, then the overall power gain of the cascaded system is given by

$$10^8 \times \left[\frac{R_i}{(R_o + R_i)} \right]^4.$$

Given $R_i = 1000\ \Omega$ and $R_o = 2000\ \Omega$, calculate: (a) the voltage, current and power gains of each stage; and (b) the overall power gain.
4. Determine the overall voltage and current gains of the two-stage amplifier shown in fig. D and hence determine the

Fig. D

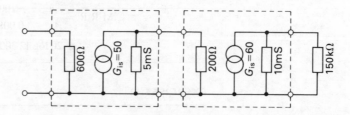

corresponding diagram of an equivalent single-stage amplifier, stating all essential values.
5. The input resistance of each of the identical amplifiers shown in fig. E is 600 Ω. Calculate V_1, V_2 and the input power.

Fig. E

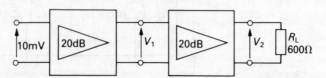

6. The input resistance of each of the amplifiers shown in fig. F is 600 Ω.
   Calculate V_i, V_o and the input power P_i.

Fig. F

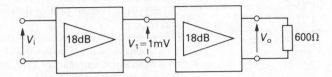

7. Two identical voltage amplifiers, whose input and output impedances are matched, are connected in cascade. If the input signal source is 20 mV and the output voltage of the arrangement is 20 V, determine: (a) the initial gain in dB; (b) the gain if the final stage impedance is reduced by 25 per cent but the power remains constant.

8. (a) Draw the circuit diagram of a non-inverting amplifier which uses an operational amplifier.
   (b) Derive the voltage gain. State any assumptions.
   (c) Comment on the input resistance.

9. (a) State the function of the circuit shown in fig. G.
   (b) Determine the output voltage for an input of 6 V r.m.s.

Fig. G

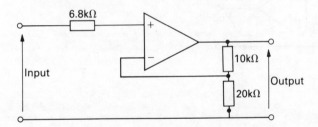

10. (a) Draw the circuit diagram of an op-amp connected as a voltage follower.
    (b) Derive the voltage gain. State any assumptions.
    (c) State the characteristics of such an amplifier.

11. (a) Draw the functional diagram of an operational amplifier connected in the inverting mode.
    (b) Derive the expression for voltage gain in terms of the circuit resistors.
    (c) Calculate the value of the feedback resistor given that the voltage gain is -1000 and the input resistor is 10 kΩ.
    (d) State FOUR necessary properties of an operational amplifier.
    (e) Draw a functional diagram of a voltage follower using an operational amplifier.

12. Two voltages, $+0.5$ V and -1.5 V, are applied to the two input resistors of an op-amp connected as a summer. The input resistors are 500 kΩ and 100 kΩ respectively. A feedback resistor of 200 kΩ is employed. What is the output voltage?

13. In Q. 12, what resistance of input resistor is needed in series with the $+0.5$-V supply to reduce the output to zero?

14. A differential amplifier employs resistors in the input and feedback paths, all of the same value. If two sine waves

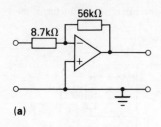

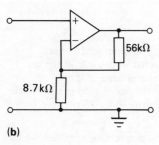

(a)

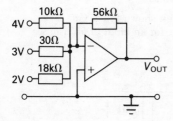

(b)

Fig. H

1.5 sin ωt and 2 sin ωt are applied simultaneously to the two inputs, what is the equation of the output voltage?

If the two waves have equations 1.5 sin ωt and 1.5 cos ωt what is the new output?

15. Calculate the voltage gain of the circuits shown in figs. H(a) and (b).
16. Determine the output voltage of the circuit shown in fig. I.
17. Two voltages, +0.1 V and +0.08 V, are applied simultaneously to a differential amplifier of gain 10 and C.M.R.R. 60 dB. What is the output voltage?

Fig. I

13 Digital Systems

13.1 Introduction to logic

It has been noted in Chapter 9 that an analogue amplifier has the inherent problem that the output is not a uniform amplification of the input, due to the non-linear nature of the circuit components and due to the variation of response with frequency. Such variation can be accepted in, say, sound amplification but, when it comes to dealing with signals carrying information, any variations would change the information. However, if the information can be reduced to a simple form of being, say, a circuit switched on and off, and, moreover, if the on condition could be represented by 10 V and the off condition by 0 V, it would not matter if after transmission the on condition were reduced to 9 V, as this could be applied to a new 10-V source thereby ensuring the output was exactly the same as the input; i.e. we can ensure that the signal is returned to its original condition. It is by such means that we can see television transmitted from the other side of the world with no apparent loss of quality.

The condition therefore for digital operation is that all information must be capable of being reduced to one or other of two states. In logic systems, these states are ON and OFF, which correspond in human terms to the responses yes and no.

13.2 Basic logic statements or functions

A logic circuit is one that behaves like a switch, i.e. a two-position device with ON and OFF states. This is termed a binary device, in which the ON state is represented by 1 and the OFF state by 0.

We require to devise a logic statement or communication which can be expressed in only one of two forms. For instance, consider the options on boarding a bus with the possibility of it taking you to two towns, A and B.

13.3 The OR function

'The bus will take me to either A or B.' The success of the bus taking you to one or other can be represented by F; thus F occurs when the bus goes to A or B or both. It might travel through A to get to B or vice-versa. Thus it can be stated as

$$F = A \text{ OR } B \qquad (13.1)$$

which in logic symbols is expressed as

$$F = A + B \qquad (13.2)$$

The positive sign is not the additive function but means OR in logic. In an electrical system, this statement is equivalent to two switches in parallel, as shown in fig. 13.1. The lamp F lights when either switch or both switches are closed.

Fig. 13.1 Switch arrangement equivalent to OR logic

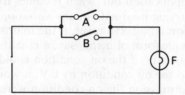

This operation can be better understood by considering the four combinations of switching options as given in part (a) of the table:

(a) *Switching circuit*

A	B	Lamp
open	open	off
open	closed	on
closed	open	on
closed	closed	on

(b) *Truth table*

A	B	F
0	0	0
0	1	1
1	0	1
1	1	1

The switching circuit table (a) can readily be reduced to the truth table in (b), using a 1 for the closed or ON condition and 0 for the open or OFF condition. A truth table permits a simple summary of the options for any logic function.

13.4 The AND function

'The bus will take me to A and B.' This success F occurs only when the bus goes to both A and B. This can be stated as

$$F = A \text{ AND } B \qquad (13.3)$$

which in logic symbols is expressed as

$$F = A \cdot B \tag{13.4}$$

The period sign is not the multiplicative function but means AND in logic. In an electric circuit, this statement is equivalent to two switches in series, as shown in fig. 13.2. The lamp F only lights when both are closed.

Fig. 13.2 Switch arrangement equivalent to AND logic

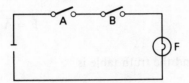

The truth table for the AND function is

A	B	F
0	0	0
0	1	0
1	0	0
1	1	1

13.5 The EXCLUSIVE-OR function

'The bus will take me either to A or to B but not to both.' In this case the success F can only be achieved either by A or by B but not by both, and the logic statement is

$$F = A \oplus B \tag{13.5}$$

This situation is typically represented in an electric circuit by the two-way switching associated with a stair light, as shown in fig. 13.3.

Fig. 13.3 Switch arrangement equivalent to the EXCLUSIVE-OR logic

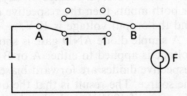

The truth table for the EXCLUSIVE-OR function is

A	B	F
0	0	0
0	1	1
1	0	1
1	1	0

13.6 The NOT function

This is a simple function in which the input is inverted; thus if the input represents the ON (1) condition then the output F is the OFF (0) condition, and vice-versa. It is stated in logic symbols as

$$F = \bar{A} \qquad (13.6)$$

and the truth table is

A	F
0	1
1	0

13.7 Logic gates

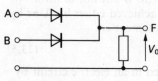

Fig. 13.4 Simple OR gate circuit

Circuits which perform logic functions are called gates. Generally, these are produced as circuit modules in the form associated with integrated circuits, or chips. In order to gain some understanding of the circuitry, consider the following simple OR and AND gates.

A simple diode OR gate is shown in fig. 13.4. If no voltage is applied to either input then the output voltage V_0 is also zero. If, however, a voltage of, say, 10 V is applied to either or both inputs then the respective diodes are forward biased and the output voltage is 10 V.

A simple diode AND gate is shown in fig. 13.5. If zero voltage is applied to either A or B or to both then the respective diodes are forward biased and current flows from the source. The result is that the output voltage V_0 is 0 V. However, if 10 V is applied to both A and B then the p.d.s across both diodes are zero and the output voltage rises to 10 V from the source. In practice most gates also incorporate switching transistors but these are practical refinements of the principles illustrated by the diode gates.

It is unusual to show the circuitry of a gate; rather a symbol is given to represent the entire gate. The symbols are summarized in fig. 13.6. Note that there are a number of systems of logic symbols in general use and only the two

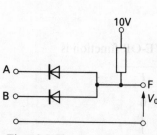

Fig. 13.5 Simple AND gate circuit

most common are shown. Gates can have more than two inputs; thus the truth tables in fig. 13.6 have been extended for three inputs.

13.8 The NOR function

This function is the combination of an OR function and a NOT function. It is realized by connecting a NOT gate to the output of an OR gate. This inverts the output, providing the truth table shown in fig. 13.6.

Fig. 13.6 Logic gate symbols and their respective truth tables

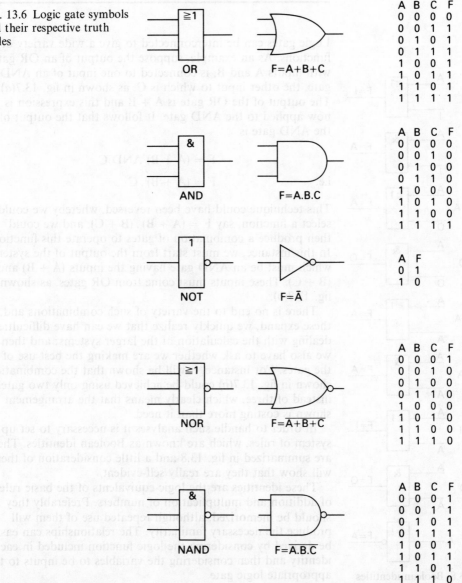

A	B	C	F
0	0	0	0
0	0	1	1
0	1	0	1
0	1	1	1
1	0	0	1
1	0	1	1
1	1	0	1
1	1	1	1

OR $F = A + B + C$

A	B	C	F
0	0	0	0
0	0	1	0
0	1	0	0
0	1	1	0
1	0	0	0
1	0	1	0
1	1	0	0
1	1	1	1

AND $F = A.B.C$

A	F
0	1
1	0

NOT $F = \bar{A}$

A	B	C	F
0	0	0	1
0	0	1	0
0	1	0	0
0	1	1	0
1	0	0	0
1	0	1	0
1	1	0	0
1	1	1	0

NOR $F = \overline{A + B + C}$

A	B	C	F
0	0	0	1
0	0	1	1
0	1	0	1
0	1	1	1
1	0	0	1
1	0	1	1
1	1	0	1
1	1	1	0

NAND $F = \overline{A.B.C}$

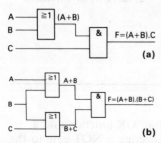

Fig. 13.7 Simple logic
networks

13.9 The NAND function

This function is realized by connecting a NOT gate to the
output of an AND gate, again inverting the output as
indicated by the truth tables shown in fig. 13.6. Both NOR
and NAND gates come as single gates, the combinations of
functions having been reduced in each case to a single circuit.

13.10 Logic networks

Logic gates can be interconnected to give a wide variety of
functions. As an example, suppose the output of an OR gate,
with inputs A and B, is connected to one input of an AND
gate, the other input to which is C, as shown in fig. 13.7(a).
The output of the OR gate is A + B and this expression is
now applied to the AND gate. It follows that the output of
the AND gate is

$$F = (A + B) \text{ AND } C$$

i.e. $$F = (A + B) . C$$

This technique could have been reversed, whereby we could
select a function, say F = (A + B) . (B + C), and we could
then produce a combination of gates to operate this function.
In this instance, we must start from the output of the system,
which must be an AND gate having the inputs (A + B) and
(B + C). These inputs must come from OR gates, as shown in
fig. 13.7(b).

There is no end to the variety of such combinations and, as
these expand, we quickly realize that we can have difficulties
dealing with the calculation of the larger systems; and then
we also have to ask whether we are making the best use of
the gates. For instance, it will be shown that the combination
shown in fig. 13.7(b) could be achieved using only two gates
instead of three, which clearly means that the arrangement
shown is costing more than it need.

In order to handle such analyses it is necessary to set up a
system of rules, which are known as Boolean identities. These
are summarized in fig. 13.8 and a little consideration of them
will show that they are really self-evident.

These identities are the logic equivalents of the basic rules
of addition and multiplication of numbers. Preferably they
should be memorized, although repeated use of them will
produce the necessary familiarity. The relationships can easily
be derived by considering the logic function included in each
identity and then considering the variables to be inputs to the
appropriate logic gate.

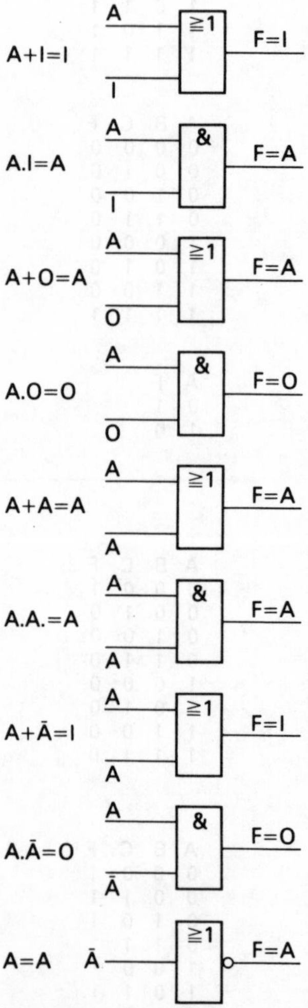

Fig. 13.8 Boolean identities

As an instance, consider the first identity which involves an OR function. The inputs are A and 1, and we know that the output of an OR gate is 1 provided that one of the inputs is 1. In this case, since one of the inputs is 1, it follows that the output must always be 1.

A similar consideration of the other eight identities will show them to be proved.

13.11 Combinational logic

In order to deal with more complex logic systems, we need to become familiar with a number of Boolean theorems.

(a) The Commutative Rules

$$A + B = B + A \tag{13.7}$$

$$A . B = B . A \tag{13.8}$$

i.e. the order of presentation of the terms is of no consequence.

(b) The Associative Rules

$$A + (B + C) = (A + B) + C \tag{13.9}$$

$$A . (B . C) = (A . B) . C \tag{13.10}$$

i.e. the order of association of the terms is of no consequence.

(c) The Distributive Rules

$$A + B . C = (A + B) . (A + C) \tag{13.11}$$

$$A . (B + C) = A . B + A . C \tag{13.12}$$

These rules provide the logic equivalents of factorization and expansion in algebra, although we do not use the terms factorization and expansion in Boolean logic. We can satisfy ourselves that relation (13.11) is correct by means of the following truth table.

A	B	C	$B . C$	$A + B . C$	$A + B$	$A + C$	$(A + B) . (A + C)$
0	0	0	0	0	0	0	0
0	0	1	1	0	0	0	0
0	1	0	1	0	0	0	0
0	1	1	1	0	0	0	0
1	0	0	0	0	0	0	0
1	0	1	1	1	0	1	1
1	1	0	1	1	1	0	1
1	1	1	1	1	1	1	1

Similarly, we can satisfy ourselves that (13.12) is also correct. There are two interesting cases of the Distributive Rules in which the inputs are limited to A and B; thus

$$A + A.B = A \qquad (13.13)$$

$$A.(A + B) = A \qquad (13.14)$$

These are particularly useful in reducing the complexity of a combinational logic system.

(d) de Morgan's Laws

$$\overline{A.B.C} = \bar{A} + \bar{B} + \bar{C} \qquad (13.15)$$

$$\overline{\bar{A}.\bar{B}.\bar{C}} = \overline{A + B + C} \qquad (13.16)$$

If we consider the truth table for a NAND gate, we can observe the proof of (13.15). A NAND gate with inputs A, B and C can be expressed as

$$F = \text{NOT (A AND B AND C)}$$
$$= \overline{A.B.C}$$

The truth table for the NAND gate is

A	B	C	A	B	C	F
0	0	0	1	1	1	1
0	0	1	1	1	0	1
0	1	0	1	0	1	1
0	1	1	1	0	0	1
1	0	0	0	1	1	1
1	0	1	0	1	0	1
1	1	0	0	0	1	1
1	1	1	0	0	0	0

From the table, it can be observed that

$$F = \bar{A} + \bar{B} + \bar{C}$$

hence

$$\overline{A.B.C} = \bar{A} + \bar{B} + \bar{C}$$

A similar proof of (13.16) can be observed from the NOR gate.

With the use of the various rules listed above, it is possible to reduce the numbers of gates, as indicated in fig. 13.7; there the three gates of fig. 13.7(b) were shown to be replaceable by the two gates in fig. 13.7(a). Consider a practical instance, as illustrated by the following example.

Example 13.1

An electrical control system uses three positional sensing devices, each of which produce 1 output when the position is confirmed. These devices are to be used in conjunction with a logic network of AND and OR gates and the output of the network is to be 1 when two or more of the sensing devices are producing signals of 1. Draw a network diagram of a suitable gate arrangement.

If we consider the possible combinations which satisfy the necessary conditions, it will be observed that there are four, i.e. any two devices or all three devices providing the appropriate signals; thus

$$F = A.B.\bar{C} + A.\bar{B}.C + \bar{A}.B.C + A.B.C.$$

Using the fifth identity illustrated in fig. 13.8, the term $A . B . C$ can be repeated as often as desired, hence

$$F = A . B . \bar{C} + A . B . C + A . \bar{B} . C + A . B . C + \bar{A} . B . C + A . B . C$$

Using the Second Distributive Rule

$$F = A . B . (\bar{C} + C) + A . C . (\bar{B} + B) + B . C . (\bar{A} + A)$$

but $\qquad A + \bar{A} = 1, \quad B + \bar{B} = 1 \quad \text{and} \quad C + \bar{C} = 1$

hence $\qquad\qquad F = A . B + B . C + C . A$

The network which would effect this function is shown in fig. 13.9.

Fig. 13.9

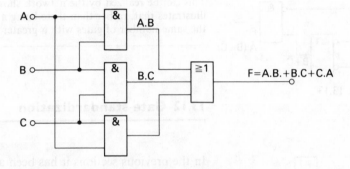

Example 13.2   *Draw the circuit of gates that could effect the function*

$$F = \overline{\overline{A . B} + \overline{A . C}}$$

Simplify this function and hence redraw the circuit that could effect it.

The gate circuit based on the original function is shown in fig. 13.10.

Fig. 13.10

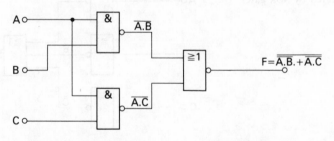

Using de Morgan's Theorems, $\quad F = \overline{\overline{A . B} + \overline{A . C}}$

$$= \overline{\overline{A . B}} . \overline{\overline{A . C}}$$

(Associative Rule) $\qquad = A . B . A . C$

$$= A . B . C.$$

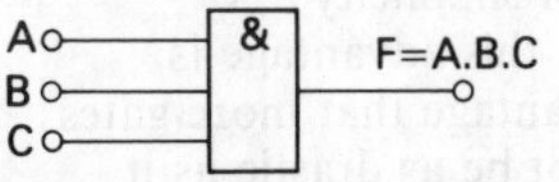

The simple circuit is shown in fig. 13.11.

Ao———
Bo——— [&] F=A.B.C
Co———

Fig. 13.11

In example 13.2 it is seen that the original function could be effected by a simple AND gate, which would result in a substantial saving of gate components.

Not all applications of combinational logic rules result in a saving, as shown by example 13.3.

Example 13.3 *Draw the circuit of gates that would effect the function*

$$F = \overline{A + B \cdot C}$$

Simplify this function and hence redraw the circuit that could effect it.

The gate circuit based on the original function is shown in fig. 13.12.

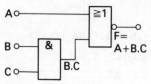

Fig. 13.12

Using de Morgan's Theorem, $F = \overline{A + B \cdot C}$

$$= \bar{A} \cdot \overline{(B \cdot C)}$$

$$= \bar{A} \cdot (\bar{B} + \bar{C})$$

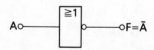

This can be realized by the network shown in fig. 13.13, which illustrates that, rather than there being a saving, we have involved the same number of gates with a greater number of inverters.

Fig. 13.13

13.12 Gate standardization

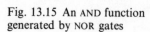

Fig. 13.14 A NOT function generated by a NOR gate

In the previous sections it has been assumed that we are free to use any combination of logic gates. Sometimes it is easier to standardize onto using a single form of gate. For example, any function can be effected using only NOR gates, and figs. 13.14, 13.15 and 13.16 illustrate how NOT, AND and OR gates can be realized using only NOR gates.

Fig. 13.15 An AND function generated by NOR gates

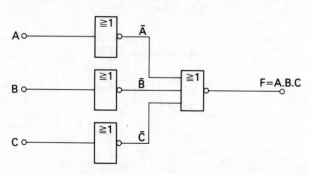

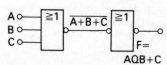

Fig. 13.16 An OR function generated by NOR gates

At first sight it would appear that restricting our choice of gates to one type has the advantage of simplicity — of avoiding using a range of gates — but this advantage is apparently offset by the major disadvantage that more gates are being used. However, this need not be as drastic as it might appear. Consider the instance of the NOT-EQUIVALENT gate illustrated by example 13.4.

Example 13.4 *Draw a logic circuit, incorporating any gates of your choice, which will produce an output 1 when its two inputs are different. Also draw a logic circuit, incorporating only NOR gates, which will perform the same function.*

For such a requirement, the function takes the form

$$F = \bar{A} \cdot B + A \cdot \bar{B}$$

This is the NOT-EQUIVALENT function and the logic circuit is shown in fig. 13.17.

This can be converted directly into NOR logic gate circuitry, as shown in fig. 13.18. Examination of the circuitry shows that two pairs of NOR gates are redundant since the output of each pair is the same as its input. These gates have been crossed out and the simplified circuit is shown in fig. 13.19.

Fig. 13.17

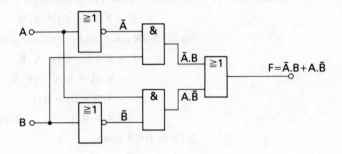

Fig. 13.18

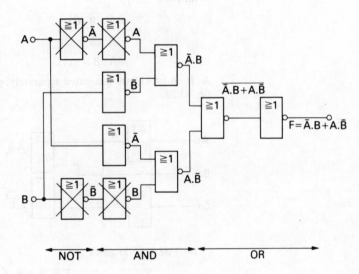

Fig. 13.19

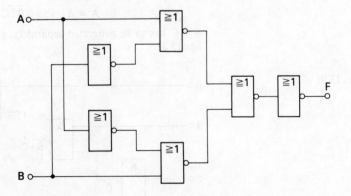

Similar systems to the one developed in example 13.4 can be created using only NAND gates. Methods specified for NOR or NAND gates will not guarantee the simplest form of circuit, but the methods used in the reduction of circuits to their absolute minimal forms are beyond the scope of an introductory text.

Example 13.5

Draw circuits which will generate the function

$$F = B.(\bar{A} + \bar{C}) + \bar{A}.\bar{B}$$

using (a) *NOR gates; and* (b) *NAND gates.*

$$
\begin{aligned}
F &= B.(\bar{A} + \bar{C}) + \bar{A}.\bar{B} \\
&= B.\bar{A} + B.\bar{C} + \bar{A}.\bar{B} && \text{(Second Distributive Rule)} \\
&= \bar{A}(B + \bar{B}) + B.\bar{C} && \text{(Second Distributive Rule)} \\
&= \bar{A} + B.\bar{C} && \text{(First Rule of Complementation)}
\end{aligned}
$$

(a) For NOR gates

$$\left.\begin{aligned}\text{Complement of}\\ \text{function}\end{aligned}\right\} = \bar{F} = \overline{\bar{A} + B.\bar{C}}$$

$$
\begin{aligned}
&= A.\overline{(B.\bar{C})} && \text{(de Morgan's Theorem)} \\
&= A.(\bar{B} + C) && \text{(de Morgan's Theorem)} \\
&= A.\bar{B} + A.C && \text{(Second Distributive Rule)}
\end{aligned}
$$

$A.\bar{B}$ and $A.C$ are generated separately, giving the circuit shown in fig. 13.20.

Fig. 13.20

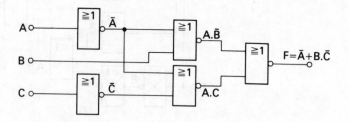

(b) For NAND gates

$$F = \bar{A} + B.\bar{C}$$

Inputs to the final NAND gate are

$$\bar{\bar{A}} = A \quad \text{and} \quad \overline{B.\bar{C}} = \bar{B} + C$$

$\bar{B} + C$ has to be generated separately, giving the circuit shown in fig. 13.21.

Fig. 13.21

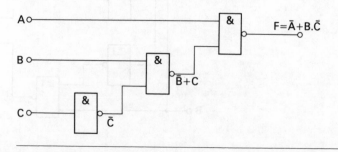

EXERCISES 13

1. Simplify the following Boolean expressions:

   (a) $F = A + B + 1$

   (b) $F = (A + B) \cdot 1$

   (c) $F = A + B \cdot 1$

2. Simplify the following Boolean expressions:

   (a) $F = A + B + 0$

   (b) $F = (A + B) \cdot 0$

   (c) $F = A + B \cdot 0$

3. Simplify the following Boolean expressions:

   (a) $F = (A + B) \cdot (A + B)$

   (b) $F = A \cdot B + B \cdot A$

   (c) $F = (A + B \cdot \bar{C}) \cdot (A + B \cdot \bar{C})$

4. Simplify the following Boolean expressions:

   (a) $F = (A + B) \cdot \overline{(A + B)}$

   (b) $F = A \cdot B + \overline{A \cdot B}$

   (c) $F = (A + \bar{B} \cdot C) + \overline{(A + \bar{B} \cdot C)}$

5. Draw a network to generate the function $F = \overline{A \cdot B + C}$

6. Draw a network to generate the function $F = \overline{A \cdot B} + A \cdot C$
   Using de Morgan's Laws, simplify this expression and hence draw the diagram of a simpler network which would produce the same result.

7. Draw a network to generate the function $F = A \cdot \bar{C} + A \cdot \bar{B} \cdot C$
   Simplify this expression and hence draw the diagram of a simpler network which produces the same result.

8. Simplify the following Boolean expression:

   $$F = A \cdot B \cdot \bar{C} + A \cdot \bar{B} \cdot C + A \cdot \bar{B} \cdot \bar{C} + A \cdot B \cdot C$$

9. Simplify the following logic functions and hence draw diagrams of circuits which will generate the functions using (a) AND, OR and NOT gates, (b) NAND gates, and (c) NOR gates.

   (i) $F = A \cdot \bar{B} \cdot \bar{C} + A \cdot \bar{B} \cdot C + \bar{A} \cdot \bar{B} \cdot \bar{C} + \bar{A} \cdot \bar{B} \cdot C$

   (ii) $F = \bar{A} \cdot B \cdot \bar{C} + \bar{A} \cdot B \cdot C + A \cdot \bar{B} \cdot C + \bar{A} \cdot \bar{B} \cdot C$

   (iii) $F = A \cdot \bar{B} \cdot \bar{C} + \bar{A} \cdot \bar{B} \cdot \bar{C} + \bar{A} \cdot B \cdot C$

   (iv) $F = A \cdot B \cdot C + \bar{A} \cdot \bar{B} \cdot \bar{C}$

   (v) $F = B \cdot \bar{C} \cdot \bar{D} + A \cdot \bar{B} \cdot D + B \cdot C \cdot \bar{D} + \bar{A} \cdot \bar{B} \cdot D$

10. Draw a circuit containing AND, OR and NOT gates to generate the function specified in the truth table.

A	B	C	Z
0	0	0	0
0	0	1	1
0	1	0	1
0	1	1	0
1	0	0	1
1	0	1	0
1	1	0	0
1	1	1	0

11. A circuit is required which will produce a logical 1 when its two inputs are identical. Indicate how such a circuit can be constructed using: (*a*) AND, OR and NOT gates; (*b*) NAND gates; and (*c*) NOR gates.

12. (*a*) The gate network shown in fig. A has three inputs *A*, *B* and *C*. Find an expression for the output *Z* and simplify this expression.

Fig. A

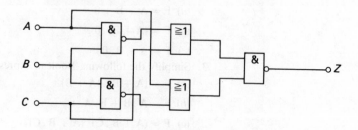

(*b*) From the simplified expression for the output *Z*, determine an equivalent network which does *not* contain NAND gates.

13. Three inputs *A*, *B* and *C* are applied to the inputs of AND, OR, NAND and NOR gates.

Give in each case the algebraic expression for the output.

Assuming that AND, OR, NAND, NOR and NOT gates are available, sketch the combinations that will realize the following:

(*a*) $A.B.\bar{C} + A.\overline{(B+C)}$

(*b*) $B.(A + \bar{B} + \bar{C}) + \overline{A + B.C}$

14. For the circuit shown in fig. B, determine the relationship between the output *Z* and the inputs *A*, *B* and *C*. Construct a truth table for the function.

Fig. B

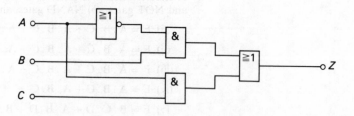

15. Simplify the following logic expressions:

(*a*) $A.(B + C) + A.B.(\bar{A} + \bar{B} + C)$

(*b*) $\bar{A}.B.C.(B.C + A.B)$

(*c*) $(\bar{A}.B + \bar{B}.A).(A + B)$

Using AND, NAND, OR, NOR or NOT gates as required, develop circuits to generate each of the functions (*a*), (*b*) and (*c*) above.

16. The gate network, shown in fig. C, has three inputs *A*, *B* and *C*. Find an expression for the output *Z* and simplify this expression.

Fig. C

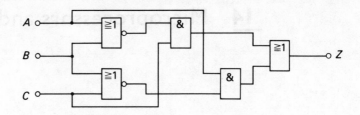

From the simplified expression, draw a simpler network that would produce the same output.

Draw the truth table for the simplified network.

17. Explain with the aid of truth tables the functions of 2-input AND, OR, NAND and NOR gates. Give the circuit symbol for each.

The gate network shown in fig. D has three inputs, A, B and C.

Fig. D

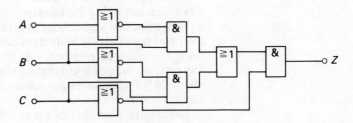

Find an expression for the output Z, and obtain a truth table to show all possible states of the network. From the truth table suggest a simpler network with the same output.

14 Microprocessors and Programs

14.1 Microprocessors

The microprocessor is an advanced electronic device which has arisen out of logic integrated circuits. The rate of development has arguably been greater than with any other electrical device. Its introduction commenced with the microelectronic developments for integrated circuits, but it has reached the point of creating circuitry with a density of about 20 000 transistors per square centimetre of the semiconductor slice.

Logic integrated circuits are capable of providing a range of functions within a fairly well defined sphere of operation for each IC; by comparison, the microprocessor seems almost to be without limit in its operation, although this is really a slight overstatement. Its greatest advantage is the vast range of functions to which a single microprocessor can be applied. In particular, the microprocessor can be applied to three principal ranges of operation:

(*a*) control (anything from a washing machine program to an oil refinery);
(*b*) calculation (anything from a pocket calculator to a simple computer);
(*c*) administration (anything from a list of names and addresses to commercial control).

The term microprocessor has become somewhat ambiguous in its development. The device is specifically a small slice of silicon on which large-scale integrated circuits have been created, the slice being mounted in a 20-pin or 40-pin DIL unit. However, many people would associate the term with the microprocessor computer, which is sold in large numbers through many stores. The name of this device has been generally curtailed to 'microprocessor', which is inappropriate to this text. In this chapter, the microprocessor is the DIL unit and not the application. In particular, it is important not to associate a keyboard with the term 'microprocessor'.

The microprocessor requires to interact with other electrical devices and cannot operate on its own. It is therefore only part of a system, in much the same manner that a transistor is only part of an amplifier.

14.2 Microprocessor operation

The microprocessor consists of thousands of electronic switching circuits. As with any logic operation, each circuit can either be ON or OFF. It follows that the operation that is effected by a microprocessor must be a binary one. Instructions fed into a microprocessor are in BInary digiTS known as *bits*. The form of the instruction must be a series of ones and zeros which relate to circuits being closed or open, i.e. on or off. Typically, a circuit which is on would supply a signal of about 5 V and a circuit which is off would supply a signal of 0 V.

A typical instruction could be 01010110. This is an eight-bit instruction and is the standard form used for the first generation of microprocessors (later generations have increased to 16 or more). The eight-bit instruction is termed a *byte*. Note that for the instructions given above, the first digit appears superfluous, but all eight bits must be given to complete an instruction, even to the point of 00000001 or 00000000.

The byte is fed into a microprocessor either by a sequence of pulses, which is known as asynchronous action, or all pulses at one time, which is known as synchronous action. The former is a serial operation while the latter is a parallel operation. The rate at which bits are fed into a microprocessor is determined by a clock which typically operates at 1 MHz. Thus in serial operation, 1 000 000 bits can be handled every second, which is equivalent to 125 000 instructions or bytes per second.

In order to make sense of a sequence of input information, the microprocessor has to be able to obey a number of instructions. It also requires a memory in which input information can be stored. The memory requires an addressing system so that it is known where the information is stored and, equally, from whence it can be recalled.

The input information comes in two distinct forms: instructions and data. Both are expressed in binary form but the microprocessor must be capable of distinguishing between them.

The instructions relate to a particular section of the memory arrangement. This section is termed the Read Only Memory, or ROM. This has a set of instructions manufactured into it and they cannot be changed. Also, when the microprocessor is switched off, the instructions remain permanently in the ROM and are called non-volatile. Thus if a microprocessor is given the instructions to place a byte of data, the instruction code introduced causes a number of instructions retained in the ROM to be recalled and the resulting effect is that the data is placed in the next available store.

Implicit in the observation is the alternative form of memory, i.e. one in which information can be temporarily stored. This form of memory is termed a Random Access

Memory or RAM. A RAM comprises a large number of stores, each of which has an address; therefore, when we insert an information byte, we require an associated address indicating the particular location in which it is to be stored. Similarly, we need to know the address should we wish to recall the information byte. Similarly, we need to know the address should we wish to recall the information byte. Unlike a ROM, the information stored in a RAM is lost when it is de-energized. The RAM is therefore said to be volatile.

The microprocessor therefore can be seen to operate on the interaction of a number of interacting processes. At the heart of these interactions is the *accumulator*. This can be considered as the section where the main activity takes place. Thus, when it is proceeding through a series of operations, the changes in the information process take place in the accumulator. Typically, a sequence of events could require two or three changes, at which stage the processor would have gone as far as it could. The result can then be stored in the RAM, clearing the accumulator ready for the next series of operations. The control of the sequence may come either from the ROM or from another section of the RAM in conjunction with the ROM.

By this stage, we have introduced a number of terms and probably confusion is setting in. However, a diagram of the microprocessor, shown in fig. 14.1, helps to relate the terms.

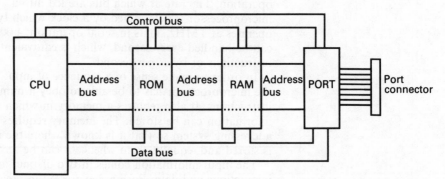

Fig. 14.1 A simple block diagram of a microprocessor system

The central processing unit is the microprocessor chip containing the accumulator. It is connected to the ROM and to the RAM by three sets of circuits, or *buses*. The address bus relays the direction of the data to be stored or recalled from memory in order that the correct storage system is used. The data, however, is transferred through the data bus. We can think of this as being like a railway where the address bus sends the information that controls the track points, thus ensuring that the train of data arrives at the appropriate destination. A subsequent changing of the points permits the following train of data to be directed to a different destination.

The diagram also shows the control bus, which carries the instructions for the organization of the sequence of operations including the commencement and termination of the sequences.

Finally, the *port* is the circuitry which connects the microprocessor system to the world around it.

In the limited content of a general book of electrical and electronic engineering, it is not possible to explain fully the operation of each of the parts of a microprocessor system. Should a reasonably detailed understanding be required, it is necessary to refer to a text relating specifically to microprocessor systems; but this introduction would not be complete without slightly expanding the simplified system shown in fig. 14.1.

In particular, fig. 14.2 has added in the clock, which provides the timing pulses. Also decoders have been added,

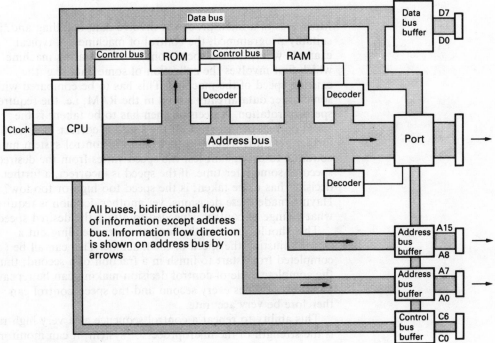

Fig. 14.2 Expanded block diagram of a microprocessor system

and require introduction. Each set of connections between the address bus and the various chips connected to it use the optimum number of address lines. A conventional 8-bit microprocessor uses 16 address lines, numbered A0 to A15. However, a typical ROM chip uses only 11 lines, i.e. A0 to A10, thus lines A11 to A15 cannot be connected.

Using the binary system, the range of information that can be carried by address lines A0 to A10 can be repeated 32 times appropriate to the range of information that could appear on lines A11 to A15, i.e. 00000, 00001, 00010, etc. The decoder uses these upper lines to control the memory chips and this is termed mapping the memory.

For a system which handles 64 K of memory (equivalent to 65 526 bytes since K = 1024), typical memory maps are shown in fig. 14.3.

Fig. 14.3 Typical memory mappings

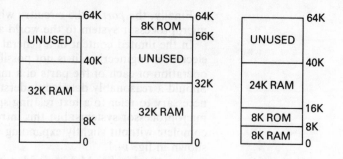

14.3 Microprocessor control

Most uses of a microprocessor involve data handling and, if suitably programmed, the control of machines. A typical example would be the speed control of a rotating machine, which first involves the collection of some data, e.g. the existing speed of the machine. This has to be compared with some other data already stored in the RAM, i.e. the required speed of rotation. A decision then has to be taken: is the speed correct or incorrect? If the speed is correct then no further action is required other than the control system must continue to be vigilant lest the speed varies from the desired speed at some later time. If the speed is incorrect, a further decision has to be taken: is the speed too high or too low? Having made these decisions, yet another decision is required: what change of control is needed to attain the desired speed?

This should result in the control system sending out a signal adjusting the speed of the machine. This can all be completed from start to finish in a fraction of a second; thus the complete cycle of control decision-making can be repeated hundreds of times every second and the speed control can therefore be very accurate.

This ability to repeat a control sequence at a very high rate is the strength of the microprocessor system; it can monitor a situation and take action within milliseconds should the need arise. Further, the decision taken can be a simple one or a highly complex one. For instance, the microprocessor might not only decide that the machine is too slow but might also make the control decision, based on the acceleration of the machine away from the reference speed desired. This makes the decision a highly sophisticated one and therefore we have to plan very carefully the program of events which we wish the microprocessor system to execute.

We also have to remember that the microprocessor can basically only make decisions of the yes/no variety. In order to obtain a decision by this means, it might be necessary to make tens or even hundreds of yes/no decisions to result in one apparently complex decision but such is the rapidity of microprocessor operation that this is still achieved very quickly. A seemingly simple stage in a program might require many instructions before it is completed, and therefore we have to appreciate that the approach to programming a microprocessor involves two steps:

(*a*) the first step is to determine the interrelated stages, each of which performs a decision-making routine;

(*b*) the second step is to write the program for each of the above stages.

Once the various stage programs have been written, they have to be combined to make the entire operational program for the microprocessor system.

A typical method of determining the required stages of a program is the drawing of a flowchart. The principal symbols used in flowcharts are shown in fig. 14.4.

Fig. 14.4 Flowchart symbols

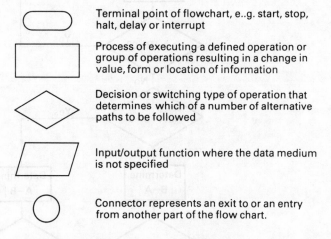

Terminal point of flowchart, e..g. start, stop, halt, delay or interrupt

Process of executing a defined operation or group of operations resulting in a change in value, form or location of information

Decision or switching type of operation that determines which of a number of alternative paths to be followed

Input/output function where the data medium is not specified

Connector represents an exit to or an entry from another part of the flow chart.

For the example of the control of the speed of a rotating machine, the flowchart is shown in fig. 14.5.

The control process need not limit itself to one action. In fact, the problem about the suggested arrangement for the machine speed control is that it is repeating too often, i.e. there is no possibility that within about a millisecond the machine will have corrected any deviation in speed. Rather it might be better to give it a chance to take some action before reassessing the need for change of speed control. This delay could be achieved by switching off the control system for a number of milliseconds before repeating the sequence.

This off time could be used for other applications, however; for instance, let us consider a number of machines which are independent of one another. The microprocessor system could consider the speed of number 1, then go on to consider number 2, and so on until, having reviewed them all and having taken action on all, it could come back to the first and start all over again. Typically, control of 40 or more machines could be kept under surveillance by one system, while at the same time other information could be obtained to promote the more effective use of the machines.

Microprocessor systems, however, are rarely used to their full capacity and it is interesting to note that again we have to consider the change of approach to engineering design which they have brought, i.e. the microprocessor is a device

Fig. 14.5 Flowchart for
machine speed control

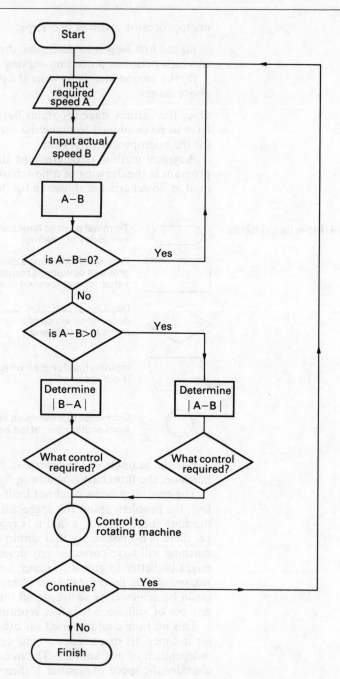

capable of an almost infinite number of functions and we
have to decide how many of these we use in any given
application. It makes little odds whether we use many of its
possibilities or few, because the initial cost of the chip
remains the same and that cost is relatively small.

Example 14.1   *Draw the flowchart of a program controlling the supply valve to a*
water tank. The valve should be energized when the water level falls to
1.0 m and should be de-energized when the water level rises to 1.5 m.

Fig. 14.6

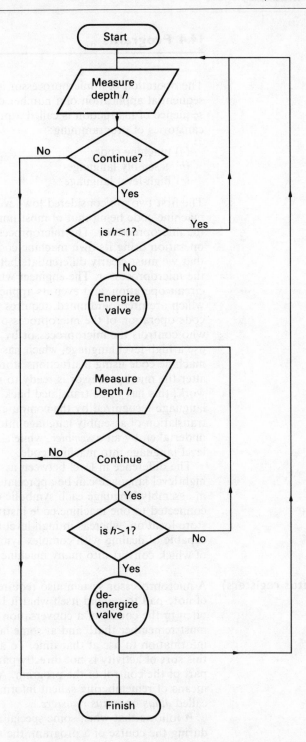

The flowchart is shown in fig. 14.6. It will be seen that there are two opportunities to discontinue the sequence, which would otherwise continue indefinitely. The arrangement is a simple one, making no allowance for component failure; e.g. what would happen if the valve jammed open or shut?

14.4 Programs

The operation of a microprocessor is effected by the sequential application of a number of instructions. Such a sequence of instruction is called a program. There are three categories of programming:

(*a*) machine code
(*b*) assembly language
(*c*) high-level language.

The first two are considered low-level programming, the machine code being that of most immediate significance to the microprocessor. The microprocessor undertakes its operation using its own machine code and it is at this point that we must clearly differentiate between the two views of the microprocessor. The engineer who is interested in the circuit operation, and even its application to control systems which are preprogrammed, requires to know the machine code operation of the microprocessor. The computer operator who controls the microprocessor by means of a keyboard will use a high-level language, which has to be translated into the machine code using instructions stored in the ROM. Equally, after the microprocessor is ready to relate the result of its work, this has to be translated back into the high-level language recognized by the computer operator. The translation of assembly language into machine code is undertaken by an *assembler*, while a *compiler* translates high-level language into machine code.

The difference in level between assembly language and high-level language can be appreciated when it is noted that in assembly language each symbolic code instruction is connected to one machine code instruction and takes up one store location, whereas in high-level languages the compiler is capable of dealing with complex symbolic instructions, each of which convert into many machine code instruction.

Flags (status registers)

A microprocessor system also requires to have its own form of note-pad to remind itself what it has done. For instance, often in the course of a conversation, we are likely to think 'I must remember that!' and at some later stage we recall the information to use at that time. To a microprocessor system this sort of activity is not directly part of the program, but is part of the control of the program. All systems have some means of remembering salient information and these are called *flags* or *status registers*.

It follows that when some special feature has been achieved during the course of a program, the microprocessor can set its flag either to 0 or to 1. At a later stage, this information can readily be made available to the control system to indicate the outcome of an instruction. For instance, let us say that the flag has been set to 1 and the program has reached the BNZ instruction (see below). This instruction considers the status of the flag: because the flag is at 1 the program continues, but had the flag been at 0 it would have turned back to an earlier stage in the program. The information held by the flag is control information and therefore it does not enter the accumulator which holds the program data. Also because it does not have to be stored in the memory, the flag status can be obtained much more quickly. Most systems incorporate a number of flags, but for this introduction to programs we shall assume there to be only one.

Simplified instruction set

For the purpose of this introduction to programming, it is necessary to select a number of simplified instructions. While the code letters used are similar to those used in a variety of applications, it is worth noting that each microprocessor system has its own instruction set with which one has to become familiar.

ADD Add memory contents to the accumulator
This instruction adds the contents of a stated memory location to the contents of the accumulator, and stores the result in the accumulator. If several successive additions are carried out, no account is taken of any carries that are generated.

AND Logical AND memory with accumulator
This instruction performs a logical AND between the contents of a stated memory location and the contents of the accumulator. The AND is between the equivalent bits of the two bytes; the result is stored in the accumulator. If there are no bit pairs which are both logic 1, the result will be zero and the zero flag will be set to logic 1.

BRK Break
This instruction stops the program. No operand is required.

BNZ Branch if not equal to zero
When this instruction is encountered, the zero flag is tested. Thus the BNZ instruction should follow an instruction which is designed to affect the zero flag. If the flag is set to logic 0, the branch is taken; otherwise the program continues to the next instruction in sequence.

BZE Branch if equal to zero
When this instruction is encountered, the zero flag is tested. Thus the BZE instruction should follow an instruction which is designed to affect the zero flag. If the flag is set to logic 1, then the branch is taken; otherwise the program continues to the next instruction in sequence.

CMP Compare memory with accumulator
This instruction compares the contents of a stated memory location with the contents of the accumulator. This is done by subtracting the memory contents from the accumulator contents. The answer to the subtraction is not stored so the data in the accumulator is not destroyed. The zero flag will be set to logic 1 if the answer is zero; it will be set to logic 0 otherwise. The CMP instruction will normally be followed by a conditional branch.

DEC Decrement memory
This instruction decrements the contents of a stated memory location by 1. If the contents of the memory become zero as a result of this operation, the zero flag will be set to logic 1.

EOR EXCLUSIVE-OR memory with accumulator
This instruction performs the EXCLUSIVE-OR between the contents of a stated memory location and the contents of the accumulator. The EXCLUSIVE-OR is between the equivalent bits of the two bytes; the result is stored in the accumulator.

JMP Jump
This instruction causes the program to jump to a stated memory location (and hence instruction) rather than proceed to the next instruction in sequence. The jump is unconditional and is always taken.

LDA Load accumulator
This instruction has two modes:
(*a*) to load the accumulator with fixed data, the data value being held in memory in the location immediately following the instruction;
(*b*) to load the accumulator with data contained in a stated memory location. The instruction is followed by the address in memory from which data is to be loaded.

ORA Logical OR memory with accumulator
This instruction performs a logical OR between the contents of a stated memory location and the contents of the accumulator. The OR is between the equivalent bits of the two bytes; the result is stored in the accumulator.

SHL Shift memory contents left
This instruction shifts the contents of the stated memory location to the left by 1 bit. The most significant bit is lost; the least significant bit is replaced by a 0.

SHR Shift memory contents right
This instruction shifts the contents of the stated memory location to the right by 1 bit. The least significant bit is lost; the most significant bit is replaced by a 0.

STA Store accumulator in memory
This instruction stores the contents of the accumulator in a stated memory location. The memory location is held in the program in the two bytes following the instruction.

SUB Subtract memory contents from the accumulator
This instruction subtracts the contents of a stated memory location from the contents of the accumulator and stores the result in the accumulator. If several successive subtractions are carried out no account is taken of borrows. If the subtraction results in a zero answer, this is indicated by the zero flag being set to logic 1.

14.5 Simple programs

The most fundamental programs are almost self-evident, but they represent an opportunity to become familiar with the terms.

Example 14.2

In a microprocessor system, a single-byte number is located at 0060 and also a single-byte number is available at the input A001. The program of this system is given below.

Explain each instruction of the following program and describe the purpose of the program.

LDA 0060
STA 0040
LDA A001
SUB 0040
STA 0041
BRK

The instructions are as follows:

LDA 0060 Load the accumulator with the number stored at location 0060 of the RAM.

STA 0040 Store this number in location 0040 of the RAM.

LDA A001 Load the accumulator with the number available at input A001, thus replacing the previous number held in the accumulator.

SUB 0040 Subtract the number held in location 0040 from the number held in the accumulator.

STA 0041 Store this new number (i.e. the difference resulting from the subtraction) in location 0041.

BRK End the program.

The program determines the difference between the input number available at A001 and the number stored in 0060, and subsequently stores the difference in 0041.

Note that there is no apparent need to shift the number held at 0060 to 0040 in order to undertake the subtraction, but this step illustrates two points. First, the loading of the accumulator replaces any previous information held there, and second, this program might be a stage in a greater program in which the information held at 0060 could later be modified for other purposes, leaving the original information still intact at 0040.

Example 14.3

The input byte to a microprocessor system is addressed at A001 and represents a hexadecimal number less than 30. The program is given below, the output being addressed A000.

Explain each instruction of the program and hence explain the function of the program.

LDA A001
STA 0041
SHL 0041
LDA 0041
STA A000
BRK

The program is modified to that shown below by the addition of three instructions. Explain these additional instructions and hence explain the function of the new program.

LDA A001
STA 0040
STA 0041
SHL 0041
SHL 0041
LDA 0041
ADD 0040
STA A000
BRK

Before giving the answer to this question, we require to consider the limitation given, i.e. a hexadecimal number less than 30 and its significance relative to the instruction SHL. The number 30 would be given in digital form as 0011 0000 and the SHL instruction moves each digit one place to the left. The application of the SHL instruction would thus give 0110 0000, which represents the hexadecimal number 60; i.e. the SHL instruction is equivalent to multiplying by 2. Similarly, the SHR instruction is equivalent to dividing by 2. In this example, we apply the SHL instruction twice, which results in 1100 0000. A further application would shift the lefthand digit outwith the 8-bit register and we would require to involve an overflow byte, which is in advance of this simple introduction to programming so we must commence with a number which ensures that, initially, the first two bits are 0.

LDA A001 Load the accumulator with the information at input A001.

STA 0041 Store the information in the accumulator in 0041.

SHL 0041 Using the information stored in 0041, shift each bit one place to the left and store this new information in 0041. Note that the accumulator operated on the information initially in 0041 and not on that in the accumulator, even though it happened to be the same. Also, after the operation, the information stored in 0041 is changed and the original formation taken from A001 is lost, unless it happens still to be available at A001.

LDA 0041 Load the accumulator with the information stored at 0041.

STA A000 Store the information held in the accumulator at output point A000.

BRK End program.

Had it been desirable to retain the original input information, this should have been stored somewhere other than 0041 as illustrated in the modified program. However, it will be noted that the previous program had the function of taking in a number from input port A001, multiplying it by 2 and making it available to output port A000.

The modified program performs as follows:

LDA A001 Load the accumulator with the information at input A001.

STA 0040 Store the information in the accumulator in 0040.

STA 0041 Store the information in the accumulator in 0041.

SHL 0041 Shift the memory contents of 0041 one digit to the left.

SHL 0041 Shift the memory content of 0041 one digit to the left (this results in the shift having been undertaken twice).

LDA 0041 Load the accumulator with the information stored in 0041.

ADD 0040 Add the information stored in 0040 to the information held in the accumulator.

STA A000   Store the information held in the accumulator at output
port A000.

BRK   End program.

In this example, we saw the need to store information twice
because, in the second instance, the stored information was modified
by the SHL instruction and was therefore effectively lost.

The program had the formation of taking in a number from input
port A001, storing it, multiplying it by 4, adding the original number
(which is equivalent in total to multiplying by 5) and making the
resultant available to output port A000.

These examples have illustrated the most simple
instructions which are generally associated with calculations
rather than with control systems. Control systems depend on
decision-making and for that we require to consider control
programming.

14.6 Control programs

Control programs depend on decisions which usually can be
thought of as yes/no decisions. However, there are terms
unknown to a logic system, that require the rephrasing of a
question as a comparison, e.g. is A = B? If the answer is yes
then the program might continue, otherwise it loops back to
an earlier step in the program. Should the decision be to loop
back, then the program will repeat until eventually the
desired answer is obtained and then the program can
proceed.

Usually the comparison question is phrased in one of two
ways. Either the accumulator should hold a number which
must be equal to zero for the program to proceed, in which
case the instruction is BNZ, or the accumulator should hold
a number other than zero for the program to proceed, in
which case the instruction is BZE. Both are used in
example 14.4.

Example 14.4   *A hoist is fitted with a protective device which operates at a height of
2.0 m. Above 2.0 m the protective device inputs logic 00 to the
controlling microprocessor system.*

*As the hoist descends toward the 2.0 m limit, the input changes to
logic 01. Should the hoist descend past the 2.0 m limit, the input
returns to logic 00. Write a program that continually checks the state
of the input signal at A001 and which outputs the byte 20 (which stops
the hoist) at A000 after the protection device input has changed from
00 to 01 back to 00.*

The program could take the following form:

LDA #01   Load the accumulator with the data 01 to ensure the
data is available for comparison.

STA 0040   Store the data in 0040.

LDA #20   Load the accumulator with the data 20 to ensure the
data is available for the protective device.

	STA 0041	Store the data in 0041.
→	LDA A001	Load the accumulator with the information at input port A001.
	AND 0040	Logical AND this memory with the accumulator information to set the zero flag required for the BZE instruction.
└	BNZ	Check the zero flag and branch back if equal to zero. (If the hoist is descending, the input is 00 and the reference is 01 hence the AND operation gives a zero-flag setting of logic 1 and the branch is taken, thus reflecting the sequence awaiting the further descent of the hoist. If the hoist reaches the 2.0 m limit the A001 changes to 01, hence the AND operation has a result of 1 and the flag is set to logic 0. If the flag is set to 0, the program moves to the next instruction.)
→	LDA A001	Load the accumulator with the information at input port A001.
	AND 0040	Logical AND this memory with the accumulator information to set the zero flag required for the BNZ instruction.
└	BZE	Check the zero flag and branch back if not equal to zero. (If the hoist is around the 2.0 m limit, then A001 gives 01, hence the AND operation has a result of 1 and the flag is set to logic 0. If the flag is set to 0 then the branch is taken, thus repeating the sequence awaiting for the action by the hoist. Should the hoist continue then A001 changes to 00, hence the AND operation has a result of 0 and the flag is set to logic 1, in which case the program moves to the next instruction.)
	LDA 0041	Load the accumulator with the data in 0041.
	STA A000	Make the accumulator data available to output port A000.
	BRK	End program.

In this program it is clear that the BZE and BNZ instructions depend on a previous logic decision, in this instance taken by the AND operation which has set the zero flag. Each instruction we program can carry a reference number, hence the BZE and BNZ instructions can include details of the instructions to which they have to branch; however, at this introductory stage an arrow is sufficient to indicate the intention.

Example 14.5
The level of liquid in a tank is measured by a float transducer which provides an input signal to a microprocessor system, the signal being proportional to the level of the liquid. It is required that the microprocessor should output a signal, stored at address 0080, to switch on an indicator lamp when the input falls to 05, and it is required that it should output a signal, stored at address 0081, to switch off the lamp when the input rises to 25. The input is addressed at A001 and the output at A000.

Write a program setting up the operating conditions and continuously monitoring the input signal. Assume that the initial level of liquid is between the limits and falling.

LDA #05	Load the accumulator with the data 05.
STA 0082	Store the data in 0082.
LDA #25	Load the accumulator with the data 25.
STA 0082	Store the data in 0082.

LDA A001	Load the accumulator with the information at input port A001.
CMP 0082	Compare the information in the accumulator with the information stored in 0082. If the bytes are identical then the zero flag is set to 1, otherwise the zero flag is set to 0.
BNZ	Check the zero flag and branch back if zero. If the flag is set to 1 (indicating that the water level is that appropriate to the input 05) then the program continues.
LDA 0080	Load the accumulator with the information at 0080.
STA A000	Make the accumulator data available to output port A000.
LDA A001	Load the accumulator with the information at input port A001.
CMP 0083	Compare the information in the accumulator with the information stored in 0083. If the bytes are identical then the zero flag is set to 1, otherwise the zero flag is set to 0.
BNZ	Check the zero flag and branch back if zero. If the flag is set to 1 (indicating the water level has risen to the input 25) then the program continues.
LDA 0081	Load the accumulator with the information at 0081.
STA A000	Make the accumulator data available to output port A000.
JMP	Return to earlier instructions, anticipating that the water level will again fall.

No indication of the manner in which the program might terminate has been given as it has been assumed that the monitoring of the water level will be an on-going process. However, the reader can write a further program to terminate the process as considered appropriate.

14.7 Data representation

There are a number of ways in which data can be represented in programmable systems. Each microprocessor machine works in binary but it is not usually convenient to write numbers as *bi*nary dig*it*s, i.e. bits.

One common method is to group bits together (simpler machines often use groups of eight) and refer to this unit as a byte.

As an example consider the 16-bit representation

$$\text{MSB} \rightarrow \underbrace{1010\,0111}_{\text{Upper byte}} \quad \underbrace{1101\,0100}_{\text{Lower byte}} \leftarrow \text{LSB}$$

where MSB is the most significant bit and LSB is the least significant bit.

Another term used in programmable systems is 'word'. A word is a general term, and the number of bits in a word varies from one system to another. Each item of data which has a specific meaning is called a word. The meaning may take one of three forms:

1. an instruction in the program;

2. an address, or memory location;

3. a numeric or logical item of data.

Number representation

The table below lists the 16 combinations of a 4-bit word:

Binary	Decimal
0000	0
0001	1
0010	2
0011	3
0100	4
0101	5
0110	6
0111	7
1000	8
1001	9
1010	10
1011	11
1100	12
1101	13
1110	14
1111	15

Binary numbers are represented in a similar way to decimal numbers except that the base, or radix, is two rather than ten. For example,

Decimal: $437 = (4 \times 10^2) + (3 \times 10^1) + (7 \times 10^0)$

Binary: $1011 = (1 \times 2^3) + (0 \times 2^2) + (1 \times 2^1) + (1 \times 2^0)$

$$= (8 + 2 + 1)_{10} = 11_{10} \qquad \text{(see table above)}$$

All binary numbers can be converted to decimal using this method, no matter how many bits in the word. For example,

$$1010\,0111\,0010 = (1 \times 2^{11}) + (0 \times 2^{10}) + (1 \times 2^9) + (0 \times 2^8)$$

$$+ (0 \times 2^7) + (1 \times 2^6) + (1 \times 2^5) + (1 \times 2^4)$$

$$+ (0 \times 2^3) + (0 \times 2^2) + (1 \times 2^1) + (0 \times 2^0)$$

$$= 2048 + 512 + 64 + 32 + 16 + 2$$

$$= 2674$$

Conversion from decimal to binary can be achieved by successive division by two and recording the remainder. If the number is even, the first remainder is zero and therefore produces the least significant bit.

For example, to convert 13_{10} to binary:

```
2 | 13
2 |  6  r1  LSB
2 |  3  r0
2 |  1  r1
  |  0  r1
```

$$13_{10} = 1101_2$$

$$= (1 \times 2^3) + (1 \times 2^2) + (0 \times 2^1) + (1 \times 2^0)$$

$$= (8 + 4 + 1)_{10} = (13)_{10}$$

Hexadecimal representation ('hex')

Binary numbers are often represented for convenience in hexadecimal code, i.e. base 16 ($= 2^4$). In the hexadecimal system 16 digits are required: the ten decimal digits, and the first six letters of the alphabet. The following table shows the comparison of binary, decimal and hexadecimal numbers:

Binary	Decimal	Hexadecimal
0000	0	0
0001	1	1
0010	2	2
0011	3	3
0100	4	4
0101	5	5
0110	6	6
0111	7	7
1000	8	8
1001	9	9
1010	10	A
1011	11	B
1100	12	C
1101	13	D
1110	14	E
1111	15	F

Using hexadecimal base, groups of four bits can be represented by a single digit:

$$(1001\,1100\,0101)_2 = 9C5_{16}$$

$$(1110\,0000\,1010)_2 = E0A_{16}$$

The decimal equivalent can be found by:

$$9C5_{16} = (9 \times 16^2) + (C \times 16^1) + (5 \times 16^0)$$
$$= (9 \times 64) + (12 \times 16) + (5 \times 1)$$
$$= (576 + 192 + 5)_{10}$$
$$= 773_{10}$$

14.8 Programming in hexadecimal representation

During the execution of each instruction, the address specified by the program counter is passed along the address bus to the memory. Since each address refers to a single register the number of bits in the program counter limits the maximum number of memory locations. For a 16-bit program counter 2^{16} or 65 536 locations can be addressed.

The program counter, however, stores 16 binary digits, usually represented in hexadecimal for convenience:

0100 1011 1001 0011   in binary

or   4   B   9   3   in hexadecimal

When using the instruction set listed in section 14.4, each

instruction requires a 4-digit hex address. For example:

LDA $0240   ($ indicates a hexadecimal number)

This causes the contents of the register at $0240 to be transferred to the accumulator.

The accumulator is an 8-bit register, which implies that each memory register is also 8-bit. Hence, a memory address can store:

> *either*: an 8-bit instruction code
> *or*: the lower half of a 16-bit address
> *or*: the upper half of a 16-bit address
> *or*: an 8-bit dataword

In a microprocessor, it is usual for the lower order of the address to precede the upper order.

The use of hexadecimal numbers is best appreciated by considering a series of instructions and their explanations as follows:

(*a*) LDA #$26
       AND $0247

If the contents of $0247 is $0F the bit-by-bit AND result will be $06:

$$0010\,0110 = \$26$$
$$0000\,1111 = \$0F$$
$$\overline{}$$
$$0000\,0110 = \$06$$

Since the result is non-zero the zero flag will be cleared.

However, if $0247 stores $91 the result is $00:

$$0010\,0110 = \$26$$
$$1001\,0001 = \$91$$
$$\overline{}$$
$$0000\,0000 = \$00$$

The result is zero and so the zero flag will be set.

(*b*) DEC $025D
       BZE $0213

When the DEC instruction is executed the contents of $025D are reduced by 1. If the result is zero the zero flag is set to 1, otherwise it will be zero.

When the BZE instruction is executed the zero flag is tested. If the flag is set to 0 (i.e. a non-zero result) the branch is *not* taken and the normal sequence is continued.

However, if the result was zero the flag is set to 1 and the branch is taken. The four hex characters ($0213) indicate the address of the first instruction in the new sequence.

(*c*) LDA #$04
       CMP $0283
       BNZ $022E

The LDA instruction causes the accumulator to be loaded with $04. The CMP instruction compares the dataword in $0283 with the contents of the accumulator and the zero flag is set to 1 if the dataword is $04, otherwise it is cleared (set to 0).

When the BNZ instruction is executed the branch is taken only if the zero flag is set to 0 (i.e. a non-zero result).

Example 14.6 *Write a program to add two numbers, stored at $0300 and $0301, and store the result at $0302.*

Location	Instruction code	Memory address	Comment
0200	LDA	0300	Get first number
0203	ADD	0301	Add second
0206	STA	0302	Store result
0209	BRK		Stop

Except for BRK, each instruction occupies three 8-bit memory locations:

0200	LDA	Load accumulator
1	00 ⎱	
2	03 ⎰	Address of data (low byte first)
3	ADD	Add to accumulator
4	01 ⎱	
5	03 ⎰	Address of data (low byte first)
6	STA	Store accumulator
7	02 ⎱	
8	03 ⎰	Memory address
9	BRK	Halt

If $0300 contains $02 and $0301 contains $06 the result will be $08 stored in $0302. (Note that $0300 and $0301 will not be changed by this program.)

The total should not exceed $FF, since this generates a carry: thus if

$$\$0300: \$D4 = 1101\,0100$$

and

$$\$0301: \$72 = 0111\,0010$$

the total is

$$\underset{\text{carry}}{1} \quad 0100\,0110$$

The carry indicates that a further byte is required.

Example 14.7 *Write a program to set the four most significant bits of the data in $0303 to zeros, leaving the remaining bits unchanged.*

Location	Code	Address	Comment
0200	LDA #	0F	Get $0F to acc. ($0F: 0000 1111)
0202	AND	0303	AND with ($0303)
0205	STA	0303	Store new data
0208	BRK		

Notice that LDA# uses only one address byte.

0200	LDA#	Load accumulator, immediate data word
0201	$0F	
0202	AND	
etc.		

If $0303 initially stores $6D:

$$\$6D: 0110\ 1101$$

AND
$$\$0F: 0000\ 1111$$
$$\overline{\hspace{3cm}}$$
$$0000\ 1101 \Rightarrow \$0D$$

i.e. 1.0 = 0 and 1.1 = 1. Hence, the lower bits are unchanged whereas the upper bits are made 0.

This is a common operation called masking.

Example 14.8   *Write a program to multiply the contents of $0309 by five and store the result at $030A. (Assume data is less than $33.)*

Location	Code	Address	Comment
0200	LDA	0309	Get number
0203	SHL	0309	Multiply number by 2
0205	SHL	0309	Multiply number by 2
0208	ADD	0309	Add to number
020B	STA	0309	Store result
020E	BRK		

The SHL instruction moves each bit one place towards the most significant bit (MSB) and therefore doubles its value:

$$\$01 \qquad 0000\ 0001$$

$$\text{SHL} \Rightarrow 0000\ 0010 = \$02$$

If data = $26 (= 38 in decimal)

$$\text{SHL} \qquad 0100\ 1100\ (= 76_{10})$$

$$\text{SHL} \qquad 1001\ 1000\ (= 152_{10})$$

$$\$26 \qquad 0010\ 0110\ (= 38_{10})$$
$$\overline{\hspace{4cm}}$$
$$1011\ 1110\ (= 190_{10})$$

Note: $0010\ 0110 = (0 \times 2^7) + (0 \times 2^6) + (1 \times 2^5) + (0 \times 2^4)$
$$+ (0 \times 2^3) + (1 \times 2^2) + (1 \times 2^1) + (0 \times 2^0)$$
$$= 0 + 0 + 32 + 0 + 0 + 4 + 2 + 0$$
$$= 38_{10}$$

EXERCISES 14

1. (*a*) Distinguish between a *register* and a *memory* in a microcomputer.

    (*b*) Describe the need for analogue-to-digital interfaces and digital-to-analogue interfaces within a microcomputer system.

2. (*a*) (i) Draw a block diagram of a bus-organized microprocessor

system illustrating the microprocessor, the memory, the input and output. (ii) Explain the function of the various buses in the above system.

(b) Distinguish between the serial and parallel transfer of digital signals, and list an example of a device employing each.

3. (a) In a microprocessor system distinguish between the following: (i) operand and instruction set; (ii) conditional and unconditional branch instructions; (iii) the two address modes for the LDA instruction LDA# xx and LDA yyyy.

(b) It is required to interface a microcomputer (via the output port) to a power circuit having an a.c. rating of 240 V, 10 A.

List *two* examples of a suitable interface between the microcomputer and the power circuit.

(c) A flowchart can be a valuable aid in programming development. List *four* of the symbols used.

4. Write a program to add two numbers, stored at 0260 and 0261, and store the result at 0262.

5. Using the instruction set provided in section 14.4, write a program which takes the number 27 from location 0030 and *converts this number to* 72 (i.e. digit interchange) which it finally stores in location 0031. The program should be accompanied by comments explaining each instruction.

6. In a microprocessor system, a single-byte number is located at 0060 and also a single-byte number is available at the input A001. The program of this system is given below.

Explain each instruction and describe the purpose of the program.

```
LDA 0060
STA 0040
LDA A001
SUB 0040
STA 0041
BRK
```

7. (a) Describe the role of the microprocessor system in a typical application indicating the origin of the data, the processing function and the destination of the output signal.

(b) A program using the instruction set in section 14.4 is given below. (i) Describe each instruction and the operation of the program. (ii) State the value stored in location 0042 after running the program.

```
        LDA   #08
        STA   0040
        LDA   #04
        STA   0041
        LDA   #00
   ┌→   ADD   0041
   │    DEC   0040
   └    BNZ
        STA   0042
        BRK
```

8. (a) Using the instruction set in section 14.4, provide the instructions for the following: (i) to load the accumulator immediately with the operand 55; (ii) to subtract the contents of memory location 0020 from the accumulator without affecting the data in the accumulator.

(b) A part of a program using the instruction set listed in section 14.4 is given below. Describe each instruction and the operation of this sequence of instructions.

```
LDA   #04
STA   0040
DEC   0040
BNZ
LDA   0040
STA   A000
```

9. A hoist is fitted with a protective device which operates at a height of 3.0 m. Above 3.0 m, the protective device inputs logic 0 to the controlling microprocessor system.

   As the hoist descends toward the 3.0-m limit, the input changes to logic 1. Should the hoist descend past the 3.0-m limit, the input returns to logic 0.

   Write a program that continually checks the state of the input signal at A001 and which outputs the byte 2F (which stops the hoist) at A000 after the protection device input has changed from 0 to 1 back to 0.

10. A microcomputer has eight switches connected to the input port A001 and eight LEDs connected to the output port A000.

    A logic 1 is applied to the input from a closed switch and an LED is lit by the application of a logic 1.

    (a) Write a program which performs the following: (i) reads the state of the switches and tests this state against a pattern of FF; (ii) if the state of the switches matches this pattern, output a pattern AA which lights alternate LEDs.

    (b) Comment on *each* instruction of the program.

11. Write a program to set the four most significant bits of the data in $0273 to zeros, leaving the remaining bits unchanged.

12. Write a program to set the four most significant bits of the data in $0284 to ones, leaving the remaining bits unaffected.

13. Write a program to invert each bit of the data stored in $0266.

14. Write a program to multiply the contents of $0266 by five and store the result in $0268 without changing the data in $0266. Assume that the data is less than 33.

15. Write a program to multiply the contents of $028A by seven and store the result in $028B without changing the data in $028A.

16. In an 8-bit dataword, the bits are numbered 0, 1, 2, 3, 4, 5, 6, 7, where the highest bit is 7. Write a program to set (make 1) bits 4, 3, 2 and 1 of $0293 and store the result in $0292.

17. Write a program to clear (make 0) the four least significant bits of $0292 and invert the remaining bits.

18. Write a program to move each bit of $0292 one place to the right and set the two most significant bits.

19. Write a program to form a dataword from the four most significant bits of $024A and the four least significant bits of $024B and store the result in $024C, leaving the data in $024A and in $024B unaffected.

20. Write a program to move the four most significant bits of $0291 to the four least significant bits of $0292, leaving $0291 unaffected and set bits 7 and 5 to logic 1.

21. Write a program which tests the most significant bit of the data in $0260 and stores data at $0900 so that: if the MSB = 1 then the data stored at $0900 is that already stores in $0700; if the MSB = 0 then the data stored at $0900 is that already stored in $0701. Write the program first using the BZE instruction and then using the BNZ instruction.

22. Write a program which sets the MSB of $0800 if both bits 7 and 6 of the data in $0250 are zeros and clears the MSB of $0800 if *either* bit is a one.

23. Write a program to store the data in $0250 at $0900 if the data in $0901 is equal to $0018 and store the data at $0251 at $0900 if the data in $0901 is equal to $0019. Write the program as a continuous loop.

15 Direct-Current Machines

15.1 General arrangement of a d.c. machine

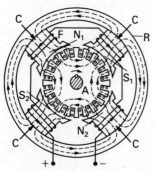

Fig. 15.1 General arrangement of a 4-pole d.c. machine

Fig. 15.1 shows the general arrangement of a four-pole d.c. generator or motor. The fixed part consists of four steel cores C, referred to as *pole cores*, attached to a steel ring R, called the *yoke*. The pole cores are usually made of steel plates riveted together and bolted to the yoke, which may be of cast steel or fabricated rolled steel. Each pole core has pole tips, partly to support the field winding and partly to increase the cross-sectional area and thus reduce the reluctance of the airgap. Each pole core carries a winding F so connected as to excite the poles alternately N and S.

The armature core A consists of steel laminations, about 0.4–0.6 mm thick, insulated from one another and assembled on the shaft in the case of small machines and on a cast-steel spider in the case of large machines. The purpose of laminating the core is to reduce the eddy-current loss (section 18.1). Slots are stamped on the periphery of the laminations, partly to accommodate and provide mechanical security to the armature winding and partly to give a shorter airgap for the magnetic flux to cross between the pole face and the armature 'teeth'. In fig. 15.1, each slot has two circular conductors, insulated from each other.

The term *conductor*, when applied to armature windings, refers to the active portion of the winding, namely that part which cuts the flux, thereby generating an e.m.f.; for example, if an armature has 40 slots and if each slot contains 8 wires, the armature is said to have 8 conductors per slot and a total of 320 conductors.

The dotted lines in fig. 15.1 represent the distribution of the *useful* magnetic flux, namely that flux which passes into the armature core and is therefore cut by the armature conductors when the armature revolves. It will be seen from fig. 15.1 that the magnetic flux which emerges from N_1 divides, half going towards S_1 and half towards S_2. Similarly, the flux emerging from N_2 divides equally between S_1 and S_2.

Suppose the armature to revolve clockwise, as shown by the curved arrow in fig. 15.1. Applying Fleming's Righthand Rule (section 2.10), we find that the e.m.f. generated in the conductors is towards the paper in those moving under the N poles and outwards from the paper in those moving under the S poles. If the airgap is of uniform length, the e.m.f. generated

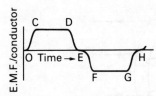

Fig. 15.2 Waveform of e.m.f.
generated in a conductor

in a conductor remains constant while it is moving under a
pole face, and then decreases rapidly to zero when the
conductor is midway between the pole tips of adjacent poles.
Fig. 15.2 shows the variation of the e.m.f. generated in a
conductor while the latter is moving through two pole
pitches, a *pole pitch* being the distance between the centres of
adjacent poles. Thus, at instant O, the conductor is midway
between the pole tips of, say, S_2 and N_1, and CD represents
the e.m.f. generated while the conductor is moving under the
pole face of N_1, the e.m.f. being assumed positive when its
direction is towards the paper in fig. 15.1. At instant E, the
conductor is midway between the pole tips of N_1 and S_1; and
portion EFGH represents the variation of the e.m.f. while the
conductor is moving through the next pole pitch. The
variation of e.m.f. during interval OH in fig. 15.2 is repeated
indefinitely, so long as the speed is maintained constant.

A d.c. generator, however, has to give a voltage that
remains constant in direction and in magnitude, and it is
therefore necessary to use a *commutator* to enable a steady or
direct voltage to be obtained from the alternating e.m.f.
generated in the rotating conductors.

Fig. 15.3 shows a longitudinal or axial section and an end
elevation of half of a relatively small commutator. It consists

Fig. 15.3 Commutator of a
d.c. machine

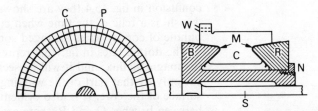

of a large number of wedge-shaped copper segments or bars
C, assembled side by side to form a ring, the segments being
insulated from one another by thin mica sheets P. The
segments are shaped as shown so that they can be clamped
securely between a V-ring B, which is part of a cast-iron bush
or sleeve, and another V-ring R which is tightened and kept
in place by a nut N. The bush is keyed to shaft S.

The copper segments are insulated from the V-rings by
collars of micanite M, namely thin mica flakes cemented
together with shellac varnish and moulded to the exact shape
of the rings. These collars project well beyond the segments
so as to reduce surface leakage of current from the
commutator to the shaft. At the end adjacent to the winding,
each segment has a milled slot to accommodate two armature
wires W which are soldered to the segment.

15.2 Ring-wound armature

The action of the commutator is most easily understood if we

Fig. 15.4 Ring-wound armature

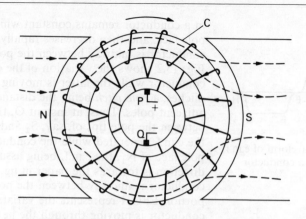

consider the earliest form of armature winding, namely the ring-wound armature shown in fig. 15.4. In this diagram, C represents a core built of sheet-iron rings or laminations, insulated from one another. The core is wound with eight coils, each consisting of two turns; and the two ends of any one coil are connected to adjacent segments of the commutator. P and Q represent two carbon brushes, namely two blocks of specially treated carbon, bearing on the commutator. Actually these brushes are pressed by springs against the outer surface of the segments, but to avoid confusion in fig. 15.4 they are shown on the inside. The term 'brush' is a relic of the time when current was collected by a bundle of copper wires arranged somewhat like a brush.

The dotted lines in fig. 15.4 represent the distribution of the magnetic flux; and it will be seen that this magnetic flux is cut only by that part of the winding that lies on the external surface of the core. It is also evident from fig. 15.4 that, between brushes Q and P, there are two paths in parallel through the armature winding and that all the conductors are divided equally between these two paths. Furthermore, as the armature rotates, this state of affairs remains unaltered.

Suppose the armature to be driven clockwise. Then, by applying the Righthand Rule of section 2.10, we find that the direction of the e.m.f.s generated in conductors under the N pole is towards the paper, and that of the e.m.f.s in the conductors under the S pole is outwards from the paper, as indicated by the arrowheads. The result is that in each of the two *parallel* paths, the same number of conductors is generating e.m.f. acting from Q towards P. The effect is similar to that obtained if equal numbers of cells (in series) were connected in two parallel circuits, as shown in fig. 15.5. In this case it is clear that P is the positive and Q the negative terminal, and that the total e.m.f. is equal to the sum of the e.m.f.s of the cells connected in series between Q and P. Also, the current in each path is a half of the total current flowing outwards at P and returning at Q. Similarly, in the armature winding, the total e.m.f. between brushes P and Q depends upon the e.m.f. per conductor and the number of conductors in series per path between Q and P; and the current in each conductor is half the total current at each brush.

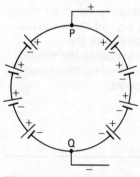

Fig. 15.5 Equivalent circuit of an armature winding

It follows from fig. 15.4 that the conductors generating e.m.f. are those which are moving opposite a pole, and that in each path the number of conductors simultaneously generating e.m.f. remains constant from instant to instant and is unaffected by the rotation of the armature. Hence, the p.d. between P and Q must also remain practically constant from instant to instant — a result made possible by the commutator.

The ring winding of fig. 15.4 has the disadvantages: (*a*) it is very expensive to wind, since each turn has to be taken round the core by hand; (*b*) only a small portion of each turn is effective in cutting magnetic flux.

An alternative method of mounting a winding is described as being 'drum' wound. It has the advantages of being considerably cheaper to produce since its conductors all lie along the outer surface of the armature core, of being mechanically stronger, and of ensuring that the largest possible area (i.e. the cross-section of the armature core) is effective in cutting the magnetic field flux. This form of winding is used in all d.c. machines but, as we shall now see, it is more difficult to understand, although it produces the same effect as the simplistic ring-wound arrangement.

Note that the term 'armature' is generally associated with the rotating part of the d.c. machine. It essentially refers only to the rotating winding into which an e.m.f. is induced, thus we have the armature winding mounted on the armature core. By usage, the term armature, however, is frequently used to describe the entire rotating arrangement, i.e. the rotor. Later this can be misleading because, in a.c. machines, the e.m.f.s are induced into the fixed windings on the stator, i.e. the yoke, in which case the armature windings are the static windings. 'Armature' therefore tends to have a rather specialized interpretation when used in respect of a d.c. machine.

15.3 Double-layer drum windings

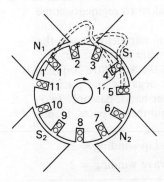

Fig. 15.6 Arrangement of a double-layer winding

Let us consider a four-pole armature with, say, 11 slots, as in fig. 15.6. In order that all the coils may be similar in shape and therefore may be wound to the correct shape before being assembled on the core, they have to be made such that if side 1 of a coil occupies the outer half of one slot, the other side 1′ occupies the inner half of another slot. This necessitates a kink in the end-connections in order that the coils may overlap one another as they are being assembled. Fig. 15.7 shows the shape of the end-connections of a single coil consisting of a number of turns, and fig. 15.8(*b*) shows how three coils, 1–1′, 2–2′ and 3–3′, are arranged in the slots so that their end-connections overlap one another, the end elevation of the end-connections of coils 1–1′ and 3–3′ being as shown in fig. 15.8(*a*). The end-connection of coil 2–2′ has

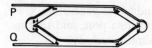

Fig. 15.7 An armature coil

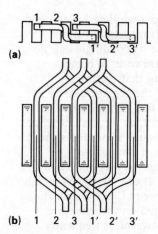

(a)

(b) 1 2 3 1' 2' 3'

Fig. 15.8 Arrangement of overlap of end-connections

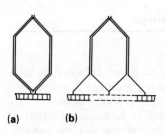

(a)　　**(b)**

Fig. 15.9 (a) Coil of a lap winding; (b) coil of a wave-winding

been omitted from fig. 15.8(a) to enable the shape of the other end-connections to be shown more clearly. In fig. 15.7, the two ends of the coil are brought out to P and Q; and as far as the connections to the commutator segments are concerned, the number of turns on each coil is of no consequence.

From fig. 15.6 it is evident that if the e.m.f.s generated in conductors 1 and 1' are to assist each other, 1' must be moving under a S pole when 1 is moving under a N pole; thus, by applying the Righthand Rule (section 2.10) to fig. 15.6 and assuming the armature to be rotated clockwise, we find that the direction of the e.m.f. generated in conductor 1 is towards the paper whereas that generated in conductor 1' is outwards from the paper. Hence, the distance between coil sides 1 and 1' must be approximately a pole pitch. With 11 slots it is impossible to make the distance between 1 and 1' exactly a pole pitch, and in fig. 15.6 one side of coil 1–1' is shown in slot 1 and the other side is in slot 4. The coil is then said to have a *coil span* of 4 − 1, namely 3. In practice, the coil span must be a whole number and is approximately equal to $\dfrac{\text{total number of slots}}{\text{total number of poles}}$.

In the example shown in fig. 15.6, a very small number of slots has for simplicity been chosen. In actual machines the number of slots per pole usually lies between 10 and 15 and the coil span is slightly less than the value given by the above expression.

Let us now return to the consideration of the 11-slot armature. The 11 coils are assembled in the slots with a coil span of 3, and we are now faced with the problem of connecting to the commutator segments the 22 ends that are projecting from the winding.

Apart from a few special windings, armature windings can be divided into two groups, depending upon the manner in which the wires are joined to the commutator, namely:

(a) lap windings,
(b) wave windings.

In lap windings the two ends of any one coil are taken to adjacent segments as in fig. 15.9(a), where a coil of two turns is shown; whereas in wave windings the two ends of each coil are bent in opposite directions and taken to segments some distance apart, as in fig. 15.9(b).*

A lap winding has as many paths in parallel between the negative and positive brushes as there are of poles; for instance, with an 8-pole lap winding, the armature conductors form eight parallel paths between the negative and positive brushes. A wave winding, on the other hand, has only two paths in parallel, irrespective of the number of poles. Hence, if a machine has p pairs of poles,

no. of parallel paths with a lap winding = $2p$

and　no. of parallel paths with a wave winding = 2.

---

* For fuller treatment of lap and wave windings, see *Principles of Electricity*, 4th edition, by A. Morley and E. Hughes.

For a given cross-sectional area of armature conductor and a given current density in the conductor, it follows that the total current from a lap winding is p times that from a wave winding. On the other hand, for a given number of armature conductors, the number of conductors in series per path in a wave winding is p times that in a lap winding. Consequently, for a given generated e.m.f. per conductor, the voltage between the negative and positive brushes with a wave winding is p times that with a lap winding. Hence it may be said that, in general, lap windings are used for low-voltage, heavy-current machines.

Example 15.1   *An eight-pole armature is wound with 480 conductors. The magnetic flux and the speed are such that the average e.m.f. generated in each conductor is 2.2 V, and each conductor is capable of carrying a full-load current of 100 A. Calculate the terminal voltage on no load, the output current on full load and the total power generated on full load when the armature is (a) lap-connected and (b) wave-connected.*

(*a*) With the armature lap-connected,

$$\left.\begin{array}{l}\text{no. of parallel paths in the}\\ \text{armature winding}\end{array}\right\} = \text{number of poles}$$

$$= 8$$

∴   no. of conductors per path $= 480/8 = 60$.

Terminal voltage on no load $=$ (e.m.f. per conductor)

$$\times \text{ (no. of conductors per path)}$$

$$= 2.2 \times 60 = 132\,\text{V}.$$

$$\text{Output current on full load} = \left(\begin{array}{c}\text{full-load current}\\ \text{per conductor}\end{array}\right)$$

$$\times \text{ (no. of parallel paths)}$$

$$= 100 \times 8 = 800\,\text{A}.$$

$$\left.\begin{array}{l}\text{Total power generated}\\ \text{on full load}\end{array}\right\} = \begin{array}{l}\text{output current}\\ \times \text{ generated e.m.f.}\end{array}$$

$$= 800 \times 132 = 105\,600\,\text{W}$$

$$= 105.6\,\text{kW}.$$

(*b*) With the armature wave-connected,

$$\text{no. of parallel paths} = 2$$

∴   no. of conductors per path $= 480/2 = 240$.

Terminal voltage on no load $= 2.2 \times 240 = 528\,\text{V}$.

Output current on full load $= 100 \times 2 = 200\,\text{A}$.

$$\left.\begin{array}{l}\text{Total power generated}\\ \text{on full load}\end{array}\right\} = 200 \times 528 = 105\,600\,\text{W}$$

$$= 105.6\,\text{kW}.$$

It will be seen from example 15.1 that the total power generated by a given machine is the same whether the armature winding is lap- or wave-connected.

15.4 Calculation of e.m.f. generated in an armature winding

When an armature is rotated through one revolution, each conductor cuts the magnetic flux emanating from all the N poles and also that entering all the S poles. Consequently,

if $\qquad \Phi = \begin{cases} \text{useful flux per pole, in webers, entering} \\ \quad \text{or leaving the armature} \end{cases}$

$\qquad p =$ number of *pairs* of poles

and $\qquad N_r =$ speed in revolutions per minute,

$$\text{time of 1 revolution} = 60/N_r \text{ seconds}$$

and time taken by a conductor to move one pole pitch

$$= \frac{60}{N_r} \cdot \frac{1}{2p} \text{ seconds}$$

∴ average rate at which conductor cuts the flux

$$= \Phi \div \left(\frac{60}{N_r} \cdot \frac{1}{2p} \right) = \frac{2\Phi N_r p}{60} \text{ webers per second}$$

and average e.m.f. generated in each conductor

$$= \frac{2\Phi N_r p}{60} \text{ volts.}$$

If $\quad Z =$ total number of armature conductors,

$$c = \begin{cases} \text{number of parallel paths through winding} \\ \quad \text{between positive and negative brushes} \end{cases}$$

$$= 2 \text{ for a wave winding}$$

and $\qquad = 2p$ for a lap winding,

∴ $\quad Z/c =$ number of conductors in series in each path.

The brushes are assumed to be in contact with segments connected to conductors in which no e.m.f. is being generated, and the e.m.f. generated in each conductor, while it is moving between positions of zero e.m.f., varies as shown by curve OCDE in fig. 15.2. The number of conductors in series in each of the parallel paths between the brushes remains practically constant; hence

$$\left. \begin{array}{l} \text{total e.m.f. between} \\ \text{brushes} \end{array} \right\} = \text{average e.m.f. per conductor} \\ \times \text{no. of conductors in series per path}$$

$$= \frac{2\Phi N_r p}{60} \times \frac{Z}{c}$$

i.e. $\qquad E = 2\frac{Z}{c} \times \frac{N_r p}{60} \times \Phi \text{ volts.} \qquad (15.1)$

Example 15.2   *A four-pole wave-connected armature has 51 slots with 12 conductors per slot and is driven at 900 r/min. If the useful flux per pole is 25 mWb, calculate the value of the generated e.m.f.*

Total number of conductors = $Z = 51 \times 12 = 612$; $c = 2$; $p = 2$; $N = 900$ r/min; $\Phi = 0.025$ weber.

Using expression (15.1), we have:

$$E = 2 \times \frac{612}{2} \times \frac{900 \times 2}{60} \times 0.025$$

$$= 459 \text{ volts}.$$

Example 15.3   *An eight-pole lap-connected armature, driven at 350 r/min, is required to generate 260 V. The useful flux per pole is about 0.05 Wb. If the armature has 120 slots, calculate a suitable number of conductors per slot.*

For an eight-pole lap winding, $c = 8$.

Hence, $$260 = 2 \times \frac{Z}{8} \times \frac{350 \times 4}{60} \times 0.05$$

$\therefore$ $Z = 890$ (approximately)

and number of conductors per slot = $890/120 = 7.4$ (approx.).

This value must be an even number; hence 8 conductors per slot would be suitable.

Since this arrangement involves a total of $8 \times 120 = 960$ conductors, and since a flux of 0.05 weber per pole with 890 conductors gave 260 V, then with 960 conductors, the same e.m.f. is generated with a flux of $0.05 \times (890/960) = 0.0464$ Wb/pole.

15.5 Armature reaction

By armature reaction is meant the effect of armature ampere-turns upon the value and the distribution of the magnetic flux entering and leaving the armature core.

Let us, for simplicity, consider a two-pole machine having an armature with eight slots and two conductors per slot, as shown in fig. 15.10. The curved lines between the conductors and the commutator segments represent the front end-connections of the armature winding and those on the outside of the armature represent the back end-connections. The armature winding — like all modern d.c. windings — is of the double-layer type, the end-connections of the outer layer being represented by full lines and those of the inner layer by dotted lines.

Brushes A and B are placed so that they are making contact with conductors which are moving midway between the poles and have therefore no e.m.f. induced in them. If the armature moves anticlockwise, the direction of the e.m.f.s generated in the various conductors is opposite to that of the currents, which are indicated in fig. 15.11(*b*) by the dots and crosses.

Fig. 15.10 A two-pole
armature winding

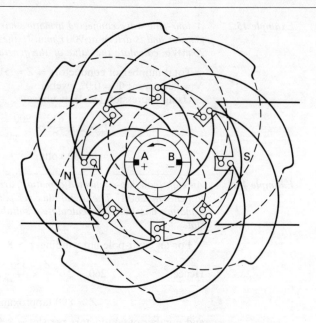

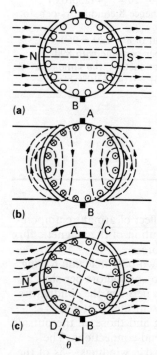

(a)

(b)

(c)

Fig. 15.11 Flux distribution
due to (*a*) field current
alone, (*b*) armature current
alone, (*c*) field and armature
currents of a d.c. motor

In diagrams where the end-connections are omitted, it is
usual to show the brushes midway between the poles, as in
fig. 15.11.

In general, an armature has ten to fifteen slots per pole, so
that the conductors are more uniformly distributed around
the armature core than is suggested by fig. 15.10; and for
simplicity we may omit the slots and consider the conductors
uniformly distributed as in fig. 15.11(*a*). The latter shows the
distribution of flux when there is no armature current, the
flux in the gap being practically radial and uniformly
distributed.

Fig. 15.11(*b*) shows the distribution of the flux set up by
current flowing through the armature winding in the direction
that it will actually flow when the machine is loaded as a
motor. It will be seen that at the centre of the armature core
and in the pole shoes the direction of this flux is at right-
angles to that due to the field winding; hence the reason why
the flux due to the armature current is termed *cross flux*.

Fig. 15.11(*c*) shows the resultant distribution of the flux
due to the combination of the fluxes in fig. 15.11(*a*) and (*b*);
thus over the trailing* halves of the pole-faces the cross flux is
in opposition to the main flux, thereby reducing the flux
density, whereas over the leading halves the two fluxes are in
the same direction, so that the flux density is strengthened.
Apart from the effect of magnetic saturation, the increase of
flux over one half of the pole face is the same as the decrease
over the other half, and the total flux per pole remains
practically unaltered. Hence, in a motor, the effect of
armature reaction is to twist or distort the flux against the
direction of rotation.

---

*The pole tip which is first met during revolution by a point on
the armature or stator surface is known as the *leading pole tip* and
the other as the *trailing pole tip*.

One important consequence of this distortion of the flux is that the magnetic neutral axis is shifted through an angle θ from AB to CD; in other words, with the machine on no load and the flux distribution of fig. 15.11(a), conductors are moving parallel to the magnetic flux and therefore generating no e.m.f. when they are passing axis AB. When the machine is loaded as a motor and the flux distorted as in fig. 15.11(c), conductors are moving parallel to the flux and generating no e.m.f. when they are passing axis CD.

An alternative and in some respects a better method of representing the effect of armature current is to draw a developed diagram of the armature conductors and poles, as in fig. 15.12(a). The direction of the current in the conductors is indicated by the dots and crosses.

Fig. 15.12 Distribution of armature m.m.f.

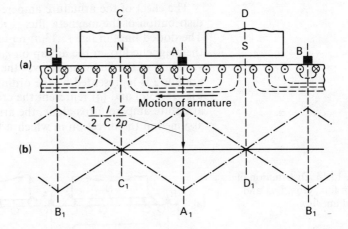

In an actual armature, the two conductors forming one turn are situated approximately a pole pitch apart, as in fig. 15.10, but *as far as the magnetic effect of the currents in the armature conductors is concerned*, the end-connections could be arranged as shown by the dotted lines in fig. 15.12(a). From the latter, it will be seen that the conductors situated between the vertical axes CC_1 and DD_1 act as if they formed concentric coils producing a magnetomotive force having its maximum value along axis AA_1. Similarly, the currents in the conductors to the left of axis CC_1 and to the right of DD_1 produce a magnetomotive force that is a maximum along axis BB_1. Since the conductors are assumed to be distributed uniformly around the armature periphery, the distribution of the m.m.f. is represented by the chain-dotted line* in fig. 15.12(b). These lines pass through zero at points C_1 and D_1 midway between the brushes.

If $I = total$ armature current, in amperes,

        $Z =$ number of armature conductors,

---

* The m.m.f. is assumed to be positive when the direction of the flux produced by it is from the airgap into the armature core (fig. 15.13).

$$c = \text{number of parallel paths}$$

and $p = \text{number of pairs of poles,}$

$$\text{current per conductor} = I/c$$

and $\text{conductors per pole} = Z/(2p)$

$$\therefore \quad \text{ampere-conductors per pole} = \frac{I}{c} \cdot \frac{Z}{2p}.$$

Since two armature conductors constitute one turn,

$$\text{ampere-turns per pole} = \frac{1}{2} \cdot \frac{I}{c} \cdot \frac{Z}{2p} \tag{15.2}$$

This expression represents the armature m.m.f. at each brush axis.

The effect of the armature ampere-turns upon the distribution of the magnetic flux is represented in fig. 15.13. The dotted lines in fig. 15.13(a) represent the distribution of the magnetic flux in the airgap on *no load*. The corresponding variation of the flux density over the periphery of the armature is represented by the ordinates of fig. 15.13(b). Fig. 15.13(c) and (d) represent the cross flux due to the armature ampere-turns alone, the armature current being assumed in the direction in which it flows when the machine

Fig. 15.13 Distribution of main flux, cross flux and resultant flux

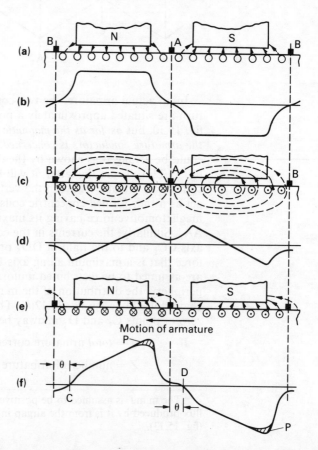

is loaded as a generator. It will be seen that the flux density in the gap increases from zero at the centre of the pole face to a maximum at the pole tips and then decreases rapidly owing to the increasing length of the path of the fringing flux, until it is a minimum midway between the poles.

Fig. 15.13(e) represents the machine operating as a motor, and the distribution of the flux density around the armature core is approximately the resultant of the graphs of fig. 15.13(b) and (d), and is represented in fig. 15.13(f). The effect of magnetic saturation would be to reduce the flux density at the leading pole tips, as indicated by the shaded areas P, and thereby to reduce the total flux per pole.

It will also be seen from fig. 15.13(f) that the points of zero flux density, and therefore of zero generated e.m.f. in the armature conductors, have been shifted through an angle θ against the direction of rotation to points C and D.

If the machine had been operated as a generator instead of as a motor, the field patterns illustrated in fig. 15.13 would still apply provided that the direction of the armature were reversed. The result of this observation is that the effect of magnetic saturation would be to reduce the flux density at the trailing pole tips, as indicated by the shaded areas P, and again thereby to reduce the total flux per pole.

It would also be seen from fig. 15.13(f) that, with the armature rotating in the opposite (clockwise) direction, the points of zero flux density, and therefore of zero generated e.m.f. in the armature conductors, have been shifted through an angle θ in the direction of rotation to points C and D.

15.6 Armature reaction in a d.c. motor

The direction of the armature current in a d.c. motor is *opposite* to that of the generated e.m.f., whereas in a generator the current is in the *same* direction as the generated e.m.f. It follows that in a d.c. motor the flux is distorted backwards; and the brushes have to be shifted backwards if they are to be on the magnetic neutral axis when the machine is loaded. A backward shift in a motor gives rise to de-magnetizing ampere-turns, and the reduction of flux tends to cause an increase of speed; in fact, this method — commutation permitting — may be used to compensate for the effect of the IR drop in the armature, thereby maintaining the speed of a shunt motor practically constant at all loads.

15.7 Compensating winding

It has already been explained in section 15.5 that the effect of armature reaction is to alter the distribution of the flux density under the pole face, as shown in fig. 15.13. If the load fluctuates rapidly over a wide range, the sudden shifting backwards and forwards of the flux may induce a sufficiently high e.m.f.* in an armature coil to produce an arc across the top of the mica sheet separating the commutator segments to which the coil is connected. Such an arc may then extend to adjacent segments until it ultimately results in a flashover between the brushes. Hence, machines subject to sudden fluctuation of load may have to be fitted with a *compensating winding*. The latter is placed in slots in the pole shoes and connected in series with the armature winding in such a way that the ampere-turns of the compensating winding are equal and opposite to those due to the armature conductors that are opposite the pole face, as shown in fig. 15.14(*a*) and (*b*).

The combined effect of currents in the armature and compensating windings of a d.c. generator is to skew the flux in the airgap as indicated in fig. 15.14(*c*), but to leave the density of the flux unaltered. The pull exerted on the

Fig. 15.14 Arrangement and connection of a compensating winding

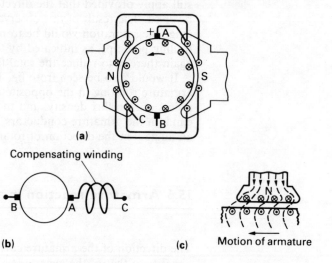

armature teeth by the skewed flux has a tangential component tending to oppose the rotation of the core. In the case of a motor, the flux is skewed in the opposite direction, thereby exerting on the teeth a pull having a tangential component responsible for the rotation of the armature.

From expression (15.2),

$$\text{armature ampere-turns per pole} = \frac{1}{2} \cdot \frac{I}{c} \cdot \frac{Z}{2p}.$$

Hence, for a uniform flux density under the pole face,

$$\left.\begin{array}{l}\text{ampere-turns per pole for} \\ \text{compensating winding}\end{array}\right\} = \frac{1}{2} \cdot \frac{I}{c} \cdot \frac{Z}{2p} \cdot \frac{\text{pole arc}}{\text{pole pitch}} \quad (15.3)$$

$$\simeq 0.7 \times \text{armature ampere-turns per pole}$$

---

*This e.m.f. is a maximum in coils that are approximately midway between the brushes.

Owing to its high cost, a compensating winding is fitted only to relatively large machines subject to sudden fluctuations of load, e.g. rolling-mill motors.

15.8 Commutation

The e.m.f. generated in a conductor of a d.c. armature is an alternating e.m.f. and the current in a conductor is in one direction when the conductor is moving under a N-pole and in the reverse direction when it is moving under a S-pole. This reversal of current in a coil has to take place while the two commutator segments to which the coil is connected are being short-circuited by a brush, and the process is termed *commutation*. The duration of this short circuit is usually about $\frac{1}{500}$ second. The reversal of, say, 100 A in an inductive circuit in such a short time is likely to present difficulty and might cause considerable sparking at the brushes.

For simplicity in considering the variation of current in the short-circuited coil we can represent the coils and the commutator segments as in fig. 15.15, where the two ends of any one coil are connected to adjacent segments, as in a lap winding.

Fig. 15.15 Portion of armature winding

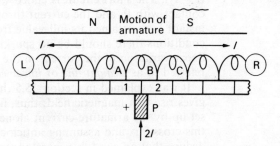

If the current per conductor is I and if the armature is moving from right to left, then — assuming the brush to be positive — coil C is carrying current from right to left (R to L), whereas coil A is carrying current from L to R. We shall therefore examine the variation of current in coil B which is connected to segments 1 and 2.

The current in coil B remains at its full value from R to L until segment 2 begins to make contact with brush P, as in fig. 15.16(a). As the area of contact with segment 2 increases,

Fig. 15.16 Coil B near the beginning and the end of commutation

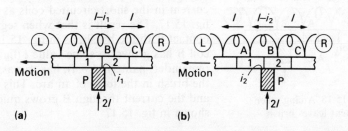

(a)   (b)

current i_1 flowing to the brush via segment 2 increases and current $(I - i_1)$ through coil B decreases. If the current distribution between segments 1 and 2 were determined by the areas of contact only, the current through coil B would decrease linearly, as shown by line M in fig. 15.17. It follows that when the brush is making equal areas of contact with segments 1 and 2, current through B would be zero; and further movement of the armature would cause the current through B to grow in the reverse direction.

Fig. 15.17 Variation of current in the short-circuited coil

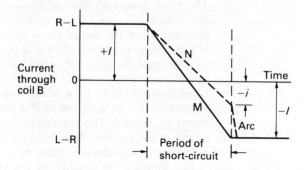

Fig. 15.16(b) represents the position of the coils near the end of the period of short circuit. The current from segment 1 to P is then i_2 and that flowing from left to right through B is $(I - i_2)$. The short circuit is ended when segment 1 breaks contact with P, and the current through coil B should by that instant have attained its full value from L to R. Under these conditions there should be no sparking at the brush, and this linear variation of the current in the short-circuited coil is referred to as *straight line* or *linear commutation*.

It was explained in section 15.5 that the armature current gives rise to a magnetic field; thus, fig. 15.11(b) shows the flux set up by the armature current alone. From the direction of this cross-flux and assuming anticlockwise rotation, we can deduce that an e.m.f. is generated towards the paper in a conductor moving in the vicinity of brush A, namely in the direction in which current was flowing in the conductor before the latter was short-circuited by the brush. The same conclusion may be derived from a consideration of the resultant distribution of flux given in fig. 15.13(e), where a conductor moving in the region of brush A is generating an e.m.f. in the same direction as that generated when the conductor was moving under the preceding main pole. This generated e.m.f. — often referred to as the *reactance voltage* — is responsible for delaying the reversal of the current in the short-circuited coils as shown by curve N in fig. 15.17. The result is that when segment 1 is due to break contact with the brush, as in fig. 15.18, the current through coil B has grown to some value i (fig. 15.17); and the remainder, namely $(I - i)$, has to pass between segment 1 and the brush in the form of an arc. This arc is rapidly drawn out and the current through B grows quickly from i to I, as shown in fig. 15.17.

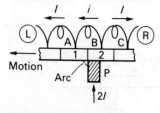

Fig. 15.18 Arcing when segment leaves brush

It is this reactance voltage that is mainly responsible for sparking at the brushes of d.c. machines, and most methods of reducing sparking are directed towards the reduction or neutralization of the reactance voltage.

15.9 Methods of improving commutation

(a) By increasing the brush-contact resistance

Let us consider the conditions shown in fig. 15.16(*b*), where segment 1 is about to break contact with the brush; and suppose that the resistance of contact with segment 1 is, say, 20 times that with segment 2. Let us also, for simplicity, neglect any reactance voltage generated in coil B.

If r = resistance of contact between segment 2 and brush,

then $20r$ = resistance of contact between segment 1 and brush.

Let R = resistance of coil B.

From fig. 15.16(*b*) it will be seen that:

p.d. between 1 and P = p.d. across B + p.d. between 2 and P

i.e. $$i_2 \times 20r = (I - i_2)R + (2I - i_2)r$$

$$\therefore \quad i_2 = I \cdot \frac{R + 2r}{R + 21r}.$$

Let us consider the effect of varying the value of r relative to that of R; thus if $r = R$, then:

$$i_2 = 0.136I.$$

If $r = 0.1R$, $\qquad i_2 = 0.39I$,

and if $r = 0.01R$, $\qquad i_2 = 0.841I$.

It will therefore be evident that if the brush contact resistance is made very low, for example by using copper gauze brushes, a large current is still flowing from the leading segment to the brush even when the area of contact has decreased to a very small value. This means that the current density becomes very high and an arc is easily formed as the segment leaves the brush. By the use of carbon brushes the contact resistance is considerably increased and commutation greatly improved. Various grades of carbon brushes possessing different contact resistances are manufactured; and the most suitable grade for a particular machine is very largely a matter of experiment.

(b) By shifting the brushes forward in a generator and backward in a motor

If the brushes were moved to the magnetic neutral zones C and D in fig. 15.16(*f*), the short-circuited coils would not be generating any e.m.f. and the reversal of the current in a short-circuited coil would then be determined by the relative resistances of that coil and of the areas of contact between the brush and the segments, as already explained for method (*a*). In general, however, it is desirable to move the brushes a

Fig. 15.19 Brush ahead of
magnetic neutral axis in a
generator

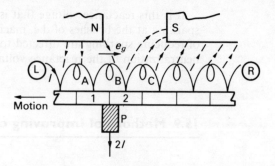

little further forward than the magnetic neutral zone in order
that the short-circuited coil may be cutting the fringing flux of
the main pole immediately in front, as shown in fig. 15.19.
The e.m.f., e_g, generated in B is then of assistance in hastening
or forcing the reversal of current in that coil.

The disadvantage of this method is that, for best
commutation, the brushes have to be shifted for every
variation of load. Also, the larger the forward shift of the
brushes of a generator or the backward shift for a motor, the
greater are the demagnetizing ampere-turns, and this method
is consequently rarely found in practice.

**(c) By using
commutating poles**

Instead of moving the brushes beyond the magnetic neutral
zone so as to generate in the short-circuited coil an e.m.f.
which assists the reversal of the current, we can obtain the
same effect by inserting auxiliary poles between the main
poles. These auxiliary poles are termed *commutating poles*,
compoles or *interpoles*, and their polarity must be the same as
that of the main pole immediately behind, as in fig. 15.20.

Fig. 15.20 Commutating
pole

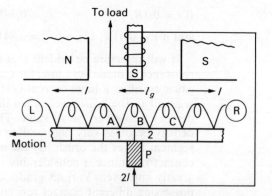

Again, if the machine were a generator, the interpole polarity
must be the same as that of the main pole immediately ahead,
as shown by reversing the motion in fig. 15.20. The brushes
are fixed on the geometric neutral axis, and the exciting
winding of the commutating poles is connected in series with
the armature winding in order that the flux set up in these
poles may be proportional to the current to be commutated.

The e.m.f., e_g, generated in that part of the short-circuited
coil that is opposite the face of the compole must be sufficient
to neutralize the e.m.f. generated by the cross-flux in the

remainder of the coil and to provide the surplus e.m.f. necessary to hasten or force the reversal of the current. Since the value of e_g should be proportional to the armature current and since the flux in the compoles is proportional to the armature current, it follows that if the value of e_g is satisfactory on full load, it is automatically satisfactory at all loads between no load and full load.

Fig. 15.21 shows the connections of the compole and shunt windings for a two-pole motor driven anticlockwise. It will be

Fig. 15.21 Connections of a shunt motor with compoles

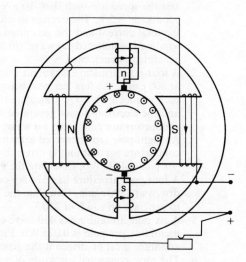

seen that the compole and armature ampere-turns are in opposition to one another; consequently the number of ampere-turns on the compoles must be sufficient to neutralize the armature ampere-turns and to provide the extra ampere-turns required to send the necessary flux through the magnetic circuit of the compoles. In practice, the number of compole ampere-turns per pole is about 1.2–1.3 times the number of armature ampere-turns per pole.

Summary of important formulae

$$E = \frac{2Z}{c} \cdot \frac{N_r p}{60} \cdot \Phi \text{ volts} \quad (15.1)$$

where

$$c = 2 \text{ for wave winding}$$

$$= 2p \text{ for lap winding}$$

and

$$p = \text{no. of } pairs \text{ of poles.}$$

$$\left.\begin{array}{r}\text{Distorting ampere-turns} \\ \text{per pole}\end{array}\right\} = \frac{1}{2} \cdot \frac{I}{c} \cdot \frac{Z}{2p}\left(1 - \frac{4\theta}{360}\right) \quad (15.2)$$

where θ = brush lead in *electrical* degrees

$$\left.\begin{array}{l}\text{Ampere-turns per pole for}\\\text{compensating winding}\end{array}\right\} = \frac{1}{2} \cdot \frac{I}{c} \cdot \frac{Z}{2p} \cdot \frac{\text{arc}}{\text{pitch}} \qquad (15.3)$$

EXERCISES 15

1. A six-pole armature is wound with 498 conductors. The flux and the speed are such that the average e.m.f. generated in each conductor is 2 V. The current in each conductor is 120 A. Find the total current and the generated e.m.f. of the armature if the winding is connected (*a*) wave, (*b*) lap. Also find the total power generated in each case.

2. A four-pole armature is wound with 564 conductors and driven at 800 r/min, the flux per pole being 20 mWb. The current in each conductor is 60 A. Calculate the total current, the e.m.f. and the electrical power generated in the armature if the conductors are connected (*a*) wave, (*b*) lap.

3. An eight-pole lap-connected armature has 96 slots with 6 conductors per slot and is driven at 500 r/min. The useful flux per pole is 0.09 Wb. Calculate the generated e.m.f.

4. A four-pole armature has 624 lap-connected conductors and is driven at 1200 r/min. Calculate the useful flux per pole required to generate an e.m.f. of 250 V.

5. A six-pole armature has 410 wave-connected conductors. The useful flux per pole is 0.025 Wb. Find the speed at which the armature must be driven if the generated e.m.f. is to be 485 V.

6. The wave-connected armature of a four-pole d.c. generator is required to generate an e.m.f. of 520 V when driven at 660 r/min. Calculate the flux per pole required if the armature has 144 slots with 2 coil slides per slot, each coil consisting of 3 turns.

7. The armature of a four-pole d.c. generator has 47 slots, each containing 6 conductors. The armature winding is wave-connected, and the flux per pole is 25 mWb. At what speed must the machine be driven to generate an e.m.f. of 250 V?

8. Develop from first principles an expression for the e.m.f. of a d.c. generator.

    Calculate the e.m.f. developed in the armature of a two-pole d.c. generator, whose armature has 280 conductors and is revolving at 1000 r/min. The flux per pole is 0.03 Wb.

    (N.C.T.E.C., O.1)

9. Draw a neat labelled diagram of the cross-section of a four-pole d.c. shunt-connected generator. What are the essential functions of the field coils, armature, commutator and brushes?

    The e.m.f. generated by a four-pole d.c. generator is 400 V when the armature is driven at 1000 r/min. Calculate the flux per pole if the wave-wound armature has 39 slots with 16 conductors per slot.   (U.E.I., O.1)

10. Explain the function of the commutator in a d.c. machine.

    A six-pole d.c. generator has a lap-connected armature with 480 conductors. The resistance of the armature circuit is 0.02 Ω. With an output current of 500 A from the armature, the terminal voltage is 230 V when the machine is driven at 900 r/min. Calculate the useful flux per pole and derive the expression employed.   (App. El., L.U.)

11. A 300-kW, 500-V, eight-pole d.c. generator has 768 armature conductors, lap-connected. Calculate the number of

demagnetizing and cross ampere-turns per pole when the brushes are given a lead of 5 electrical degrees from the geometric neutral. Neglect the effect of the shunt current.

12. Write a short essay describing the effects of (*a*) armature reaction and (*b*) poor commutation on the performance of a d.c. machine. Indicate in your answer how the effects of armature reaction may be reduced and how commutation may be improved.   (E.M.E.U., O.2)

13. (*a*) Explain how armature reaction occurs in a d.c. generator and the effect it has on the flux distribution of the machine.

    (*b*) Explain the difficulties of commutation in a d.c. generator. What methods are used to overcome these difficulties?   (W.J.E.C., O.2)

14. A four-pole motor has a wave-connected armature with 888 conductors. The brushes are displaced backwards through 5 angular degrees from the geometrical neutral. If the total armature current is 90 A, calculate: (*a*) the cross and the back ampere-turns per pole, and (*b*) the additional field current to neutralize this demagnetization, if the field winding has 1200 turns per pole.

15. Calculate the number of turns per pole required for the commutating poles of the d.c. generator referred to in Q. 11, assuming the compole ampere-turns per pole to be about 1.3 times the armature ampere-turns per pole and the brushes to be in the geometric neutral.

16. An eight-pole generator has a lap-connected armature with 640 conductors. The ratio of pole arc per pole pitch is 0.7. Calculate the ampere-turns per pole of a compensating winding to give uniform airgap density when the total armature current is 900 A.

17. Define the temperature coefficient of resistance.

    A four-pole machine has a wave-wound armature with 576 conductors. Each conductor has a cross-sectional area of 5 mm² and a mean length of 800 mm. Assuming the resistivity for copper to be $0.0173\,\mu\Omega \cdot m$ at 20°C and the temperature coefficient of resistance to be 0.004/°C at 0°C, calculate the resistance of the armature winding at its working temperature of 50°C.   (U.E.I., O.1)

16   Direct-Current Motors

16.1 Armature and field connections

The general arrangement of the brush and field connections of a four-pole machine is shown in fig. 16.1. The four brushes B make contact with the commutator. The positive brushes are connected to the positive terminal A and the negative brushes to the negative terminal A_1. From fig. 15.10 it will be seen that the brushes are situated approximately in line with the centres of the poles. This position enables them to make contact with conductors in which little or no e.m.f. is being generated since these conductors are then moving between the poles.

Fig. 16.1 Armature and field connections

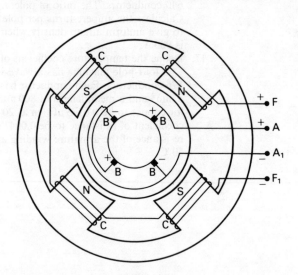

The four exciting or field coils C are usually joined in series and the ends are brought out to terminals F and F_1. These coils must be so connected as to produce N and S poles alternately. The arrowheads in fig. 16.1 indicate the direction of the field current when F is positive.

In general, we may divide the methods used for connecting the field and armature windings into the following groups:

(a) *Separately-excited machines* — the field winding being connected to a source of supply other than the armature of its own machine.

(*b*) *Self-excited machines*, which may be subdivided into:

    (i) *Shunt-wound machines* — the field winding being connected across the armature terminals.

    (ii) *Series-wound machines* — the field winding being connected in series with the armature winding.

    (iii) *Compound-wound machines* — a combination of shunt and series windings.

Before we discuss the above systems in greater detail, let us consider the relationship between the magnetic flux and the exciting ampere-turns of a machine on no load. From fig. 15.1 it will be seen that the ampere-turns of one field coil have to maintain the flux through one airgap, a pole core, part of the yoke, one set of armature teeth and part of the armature core. The number of ampere-turns required for the *airgap* is directly proportional to the flux and is represented by the straight line OA in fig. 16.2. For low values of the flux, the number of ampere-turns required to send the flux through the ferromagnetic portion of the magnetic circuit is very small; but when the flux exceeds a certain value, some parts — especially the teeth — begin to get saturated and the number of ampere-turns increases far more rapidly than the flux, as shown by curve B. Hence, if DE represents the number of ampere-turns per pole required to maintain flux OD across the airgap and if DF represents the number of ampere-turns per pole to send this flux through the ferromagnetic portion of the magnetic circuit, then:

total ampere-turns per pole to produce flux OD

$$= DE + DF = DG.$$

By repeating this procedure for various values of the flux, we can derive the *magnetization curve* C representing the relationship between the useful magnetic flux per pole and the total ampere-turns per pole.

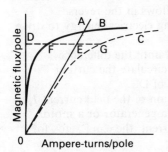

Fig. 16.2 Magnetization curve of a machine

16.2 A d.c. machine as generator or motor

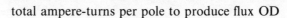

There is no difference of construction between a d.c. generator and a d.c. motor. In fact, the only difference is that in a generator the generated e.m.f. is greater than the terminal voltage, whereas in a motor the generated e.m.f. is less than the terminal voltage. For instance, suppose a shunt generator D (fig. 16.3) to be driven by an engine and connected through a centre-zero ammeter A to a battery B. If the field regulator R is adjusted until the reading on A is zero, the e.m.f., E_D, generated in D is then exactly equal to the e.m.f., E_B, of the battery. If R is now reduced, the e.m.f. generated in D exceeds that of B, and the excess e.m.f. is available to circulate a current I_D through the resistance of the armature circuit, the battery and the connecting conductors. Since I_D is

Fig. 16.3 Shunt-wound machine as generator or motor

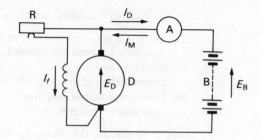

in the same direction as E_D, machine D is a generator of electrical energy.

Next, suppose the supply of steam or oil to the engine driving D to be cut off. The speed of the set falls, and as E_D decreases, I_D becomes less, until, when $E_D = E_B$, there is no circulating current. But E_D continues to decrease and becomes less than E_B, so that a current I_M flows in the reverse direction. Hence B is now supplying electrical energy to drive D as an electric motor.

The speed of D continues to fall until the difference between E_D and E_B is sufficient to circulate the current necessary to maintain the rotation of D.

It will be noticed that the direction of the field current I_f is the same whether D is running as a generator or a motor.

The relationship between the current, the e.m.f., etc., for machine D may be expressed thus:

If

$$E = \text{e.m.f. generated in armature,}$$

$$V = \text{terminal voltage,}$$

$$R_a = \text{resistance of armature circuit,}$$

and

$$I_a = \text{armature current,}$$

then, when D is operating as a generator,

$$E = V + I_a R_a. \tag{16.1}$$

When the machine is operating as a motor, the e.m.f., E, is less than the applied voltage V, and the direction of the current I_a is the reverse of that when the machine is acting as a generator; hence

$$E = V - I_a R_a$$

or

$$V = E + I_a R_a. \tag{16.2}$$

Since the e.m.f. generated in the armature of a motor is in opposition to the applied voltage, it is sometimes referred to as a *back e.m.f.*

Example 16.1   *The armature of a d.c. machine has a resistance of 0.1 Ω and is connected to a 230-V supply. Calculate the generated e.m.f. when it is running* (a) *as a generator giving 80 A and* (b) *as a motor taking 60 A.*

(*a*) Voltage drop due to armature resistance = $80 \times 0.1 = 8$ V.

From (16.1),   generated e.m.f. $= 230 + 8 = 238$ V.

(*b*) Voltage drop due to armature resistance = $20 \times 0.1 = 6$ V.

From (16.2),   generated e.m.f. $= 230 - 6 = 224$ V.

16.3 Speed of a motor

In section 15.4 it was shown that the relationship between the generated e.m.f., speed, flux, etc., is represented by:

$$E = 2\frac{Z}{c} \cdot \frac{N_r p}{60} \cdot \Phi \qquad (15.1)$$

For a given machine, Z, c and p are fixed; and in such a case we can write:

$$E = kN_r\Phi$$

where

$$k = 2\frac{Z}{c} \cdot \frac{p}{60}.$$

Substituting for E in expression (16.2) we have:

$$V = kN_r\Phi + I_a R_a$$

$$\therefore \qquad N_r = \frac{V - I_a R_a}{k\Phi}. \qquad (16.3)$$

The value of $I_a R_a$ is usually less than 5 per cent of the terminal voltage V, so that:

$$N_r \simeq \frac{V}{k\Phi}. \qquad (16.4)$$

In words, this expression means that the speed of an electric motor is approximately proportional to the voltage applied to the armature and inversely proportional to the flux; and all methods of controlling the speed involve the use of either or both of these relationships.

Example 16.2

A four-pole motor is fed at 440 V and takes an armature current of 50 A. The resistance of the armature circuit is 0.28 Ω. The armature winding is wave-connected with 888 conductors and the useful flux per pole is 0.023 Wb. Calculate the speed.

From expression (16.2) we have:

$$440 = \text{generated e.m.f.} + 50 \times 0.28$$

$$\therefore \qquad \text{generated e.m.f.} = 440 - 14 = 426 \text{ V}.$$

Substituting in the e.m.f. equation (15.1), we have:

$$426 = 2 \times \frac{888}{2} \times \frac{N_r \times 2}{60} \times 0.023$$

$$\therefore \qquad N_r = 626 \text{ r/min}.$$

Example 16.3

A motor runs at 900 r/min off a 460-V supply. Calculate the approximate speed when the machine is connected across a 200-V supply. Assume the new flux to be 0.7 of the original flux.

If Φ be the original flux, then from expression (16.4):

$$900 = \frac{460}{k\Phi}$$

$$\therefore \qquad k\Phi = 0.511$$

$$\text{and} \quad \text{new speed} = \frac{\text{new voltage}}{k \times \text{original flux} \times 0.7} \text{ (approximately)}$$

$$= \frac{200}{0.511 \times 0.7} = 559 \text{ r/min}.$$

16.4 Torque of an electric motor

If we start with equation (16.2) and multiply each term by I_a, namely the total armature current, we have:

$$VI_a = EI_a + I_a^2 R_a.$$

But VI_a represents the total electrical power supplied to the armature, and $I_a^2 R_a$ represents the loss due to the resistance of the armature circuit. The difference between these two quantities, namely EI_a, therefore represents the mechanical power developed by the armature. All of this mechanical power is not available externally, since some of it is absorbed as friction loss at the bearings and at the brushes and some is wasted as hysteresis loss (section 3.14) and in circulating eddy currents in the ferromagnetic core (section 18.1).

If T is the torque,* in newton metres, exerted on the armature to develop the mechanical power just referred to, and if N_r is the speed in revolutions per minute, then from expression (1.7),

mechanical power developed $= 2\pi T N_r / 60$ watts.

Hence $\qquad 2\pi T N_r / 60 = EI_a \qquad\qquad\qquad (16.5)$

$$= 2\frac{Z}{c} \cdot \frac{N_r p}{60} \cdot \Phi \cdot I_a$$

$$\therefore \qquad\qquad\qquad T = 0.318 \frac{I_a}{c} \cdot Zp\Phi \text{ newton metres†}$$

$$(16.6)$$

For a given machine, Z, c and p are fixed; in which case:

$$T \propto I_a \times \Phi \qquad\qquad\qquad (16.7)$$

Or, in words, the torque of a given d.c. motor is proportional to the product of the armature current and the flux per pole.

---

* In many textbooks, the value of the torque is derived from expression (2.1). This method gives the correct result; but it should be realized that with a slotted armature, flux density in the slots is extremely low, so that there is practically no force on the conductors. Practically the whole of the torque is exerted on the teeth.

† Expression (16.6) also represents the torque to be applied to the armature of a d.c. *generator* to generate the output power of and the I^2R loss in the *armature winding*. Additional torque has to be applied to supply the iron loss in the armature core and the friction and windage losses of the generator.

Example 16.4   *A d.c. motor takes an armature current of 110 A at 480 V. The resistance of the armature circuit is 0.2 Ω. The machine has six poles and the armature is lap-connected with 864 conductors. The flux per pole is 0.05 Wb. Calculate (a) the speed and (b) the gross torque developed by the armature.*

(*a*)   Generated e.m.f. $= 480 - (110 \times 0.2)$

$$= 458 \text{ V}.$$

Since the armature winding is lap-connected, $c = 6$.

Substituting in expression (15.1), we have:

$$458 = 2 \times \frac{864}{6} \times \frac{N_r \times 3}{60} \times 0.05$$

$$\therefore \qquad N_r = 636 \text{ r/min}.$$

(*b*) Mechanical power developed by armature $= 110 \times 458$

$$= 50\,380 \text{ W}.$$

Substituting in expression (16.5) we have:

$$2\pi T \times (636/60) = 50\,380$$

$$T = 756 \text{ N} \cdot \text{m}.$$

Alternatively, using expression (16.6) we have:

$$T = 0.318 \times (110/6) \times 864 \times 3 \times 0.05$$

$$= 756 \text{ N} \cdot \text{m}.$$

Example 16.5   *The torque required to drive a d.c. generator at 15 r/s is 2 kN·m. The core, friction and windage losses in the machine are 8 kW. Calculate the power generated in the armature winding.*

Driving torque $= 2 \text{ kN} \cdot \text{m} = 2000 \text{ N} \cdot \text{m}.$

From expression (1.7), power required to drive the generator

$$= 2\pi \times 2000 \, [\text{N} \cdot \text{m}] \times 15 \, [\text{r/s}]$$

$$= 188\,400 \text{ W} = 188.4 \text{ kW}.$$

Since core, friction and windage losses $= 8 \text{ kW}$

$$\therefore \qquad \text{power generated in armature winding} = 188.4 - 8$$

$$= 180.4 \text{ kW}.$$

16.5 Starting resistor

If the armature of example 16.2 were stationary and then switched directly across a 440-V supply, there would be no generated e.m.f. and the current would tend to grow to $440/0.28 = 1572$ A. Such a current, in addition to subjecting the armature to a severe mechanical shock, would blow the fuses, thereby disconnecting the supply from the motor. It is therefore necessary (except with very small motors) to connect a variable resistor in series with the armature, the resistance being reduced as the armature accelerates. Such an arrangement is termed a *starter*. If the starting current in the above example is to be limited to, say, 80 A, the total resistance of the starter and armature must be $440/80 = 5.5\,\Omega$, so that the resistance of the starter alone must be $(5.5 - 0.28) = 5.22\,\Omega$.

Fig. 16.4 shows a starting resistor R subdivided between four contact-studs S and connected to a shunt-wound motor.

Fig. 16.4 Shunt-wound motor with starter

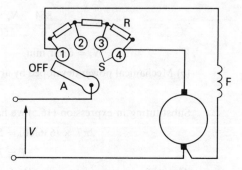

One end of the shunt winding is joined to stud 1; consequently when arm A is moved from 'Off' to that stud, the full voltage is applied to the shunt winding and the whole of R is in series with the armature. The armature current instantly grows to a value I_1 (fig. 16.5) where:

$$I_1 = \frac{\text{supply voltage } V}{\text{resistance of (armature + starter)}}. \tag{16.8}$$

Since the torque is proportional to (armature current × flux), it follows that the maximum torque is immediately available to accelerate the armature.

As the armature accelerates, its e.m.f. grows and the armature current decreases as indicated by curve *ab*. When the current has fallen to some pre-arranged value I_2, arm A is moved over to stud 2, thereby cutting out sufficient resistance to allow the current to rise once more to I_1. This operation is repeated until A is on stud 4 and the whole of the starting resistor is cut out of the armature circuit. The motor continues to accelerate and the current to decrease until it settles down at some value I (fig. 16.5) such that the torque due to this current is just sufficient to enable the motor to cope with its load.

It is evident from fig. 16.4 that when A is on stud 4 the whole of R is in the field circuit. The effect upon the field current, however, is almost negligible. Thus, for the example considered at the beginning of this section, the shunt current of such a machine would not exceed 2 A, so that the resistance of shunt winding F would be at least 440/2, namely $220\,\Omega$. Consequently the addition of $5.22\,\Omega$ means a decrease of only 2.4 per cent in the field current.

If the field winding were connected to *a* as in fig. 16.6, then at the instant of starting, the voltage across the field winding would be very small, namely that across the armature

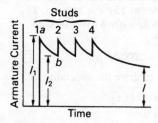

Fig. 16.5 Variation of starting current

Fig. 16.6 Wrong methods of connecting the shunt winding

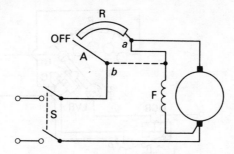

winding. In the above example, for instance, this p.d. is only $80 \times 0.28 = 22.4$ V. Consequently the torque would be extremely small and the machine would probably refuse to start. On the other hand, if the field winding were connected to b, as shown dotted in fig. 16.6, it would be directly across the supply; but if the motor were stopped by moving the starter arm back to 'Off', the field would still remain excited. If such a field current were switched off by opening switch S quickly, the sudden collapse of the flux would induce a very high e.m.f. in F and might result in a breakdown of the insulation.

With the connections shown in fig. 16.4, it is obvious that the armature winding, the starting resistor and the field winding form a closed circuit. Consequently when A returns to 'Off', the kinetic energy of the machine and its load maintains rotation for an appreciable period, and during most of that time the machine continues to excite as a shunt generator. The field current therefore decreases comparatively slowly and there is no risk of an excessive e.m.f. being induced.

16.6 Protective devices on starters

It is very desirable to provide the starter with protective devices to enable the starter arm to return to 'Off',

(a) when the supply fails, thus preventing the armature being directly across the mains when this voltage is restored, and

(b) when the motor becomes overloaded or develops a fault causing the machine to take an excessive current.

Fig. 16.7 shows a starter fitted with such features. The low-volt release LVR consists of a coil wound on a U-shaped core. The starter arm carries a steel plate B, so that when it is in the 'On' position, as shown, the core of LVR is magnetized by the field current and holds B against the tendency of a spiral spring G to return A to 'Off'. Should the supply fail, the motor stops, LVR is de-energized and the residual magnetism in it is not sufficient to hold A against the counterclockwise torque of G.

Fig. 16.7 Starter with no-volt and overload releases

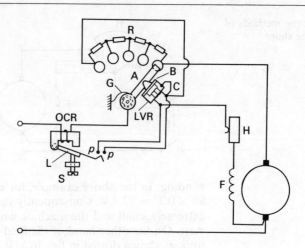

If a connection C be made between the core of LVR and the supply side of the coil, the shunt current flows from A via B and C, and the starting resistor is thereby short circuited.

The coil of the *overcurrent release* OCR is wound on a steel core and connected in series with the motor. A steel plate L, pivoted at one end, carries an extension which, when lifted, connects together two pins p, p. When the current taken by the motor exceeds a certain value, the magnetic pull on L is sufficient to lift it, thereby short circuiting the coil of LVR and releasing the starter arm A. The critical value of the overload current is dependent upon the length of airgap between L and the core of OCR and is controlled by a screw adjustment S.

16.7 Speed characteristics of electric motors

With very few exceptions, d.c. motors are shunt-, series- or compound-wound. The connections of a shunt motor have already been given in figs. 16.4 and 16.7; and figs. 16.8 and 16.9 show the connections for series and compound motors respectively, the starter in each case being shown in the 'On' position. (Students should always make a practice of including starters in diagrams of motor connections.) In compound motors, the series and shunt windings almost invariably assist each other, as indicated in fig. 16.9.

The speed characteristic of a motor usually represents the variation of speed with input current or input power, and its shape can be easily derived from expression (16.3), namely:

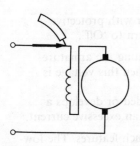

Fig. 16.8 Series-wound motor

$$N_r = \frac{V - I_a R_a}{k\Phi}.$$

In shunt motors, the flux Φ is only slightly affected by the armature current and the value of $I_a R_a$ at full load rarely exceeds 5 per cent of V, so that the variation of speed with

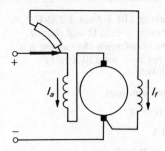

Fig. 16.9 Compound-wound motor

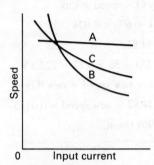

Fig. 16.10 Speed characteristics of shunt, series and compound motors

input current may be represented by curve A in fig. 16.10. Hence shunt motors are suitable where the speed has to remain approximately constant over a wide range of load.

In series motors, the flux increases at first in proportion to the current and then less rapidly owing to magnetic saturation (fig. 16.2). Also R_a in the above expression now includes the resistance of the field winding. Hence the speed is roughly inversely proportional to the current, as indicated by curve B in fig. 16.10. It will be seen that if the load falls to a very small value, the speed may become dangerously high. A series motor should therefore not be employed when there is any such risk; for instance, it should never be belt-coupled to its load except in very small machines such as vacuum cleaners.

Since the compound motor has a combination of shunt and series excitations, its characteristic (curve C in fig. 16.10) is intermediate between those of the shunt and series motors, the exact shape depending upon the values of the shunt and series ampere-turns.

16.8 Torque characteristics of electric motors

In section 16.4 it was shown that for a given motor:

$$\text{torque} \propto \text{armature current} \times \text{flux per pole}.$$

Since the flux in a shunt motor is practically independent of the armature current:

$$\therefore \quad \text{torque of a shunt motor} \propto \text{armature current}$$

and is represented by the straight line A in fig. 16.11.

In a series motor the flux is approximately proportional to the current up to full load, so that:

$$\text{torque of a series motor} \propto (\text{armature current})^2, \text{approx}.$$

Above full load, magnetic saturation becomes more marked and the torque does not increase so rapidly.

Curves A, B and C in fig. 16.11 show the relative shapes of torque curves for shunt, series and compound motors having the same full-load torque OQ with the same full-load armature current OP, the exact shape of curve C depending upon the relative value of the shunt and series ampere-turns at full load.

From fig. 16.11 it is evident that for a given current below the full-load value the shunt motor exerts the largest torque; but for a given current above that value the series motor exerts the largest torque.

The maximum permissible current at starting is usually about 1.5 times the full-load current. Consequently where a large starting torque is required, such as for hoists, cranes, electric trains, etc., the series motor is the most suitable machine.

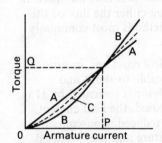

Fig. 16.11 Torque characteristics of shunt, series and compound motors

Example 16.6   *A series motor runs at 600 r/min when taking 110 A from a 230-V supply. The resistance of the armature circuit is 0.12 Ω and that of the series winding is 0.03 Ω. Calculate the speed when the current has fallen to 50 A, assuming the useful flux per pole for 110 A to be 0.024 Wb and that for 50 A to be 0.0155 Wb.*

Total resistance of armature and series windings

$$= 0.12 + 0.03 = 0.15 \, \Omega,$$

∴ e.m.f. generated when current is 110 A

$$= 230 - 110 \times 0.15 = 213.5 \, \text{V}.$$

In section 16.3 it was shown that for a given machine:

generated e.m.f. = a constant (say k) × speed × flux.

Hence with 110 A, $213.5 = k \times 600 \times 0.024$

∴ $k = 14.82.$

With 50 A,   generated e.m.f. $= 230 - 50 \times 0.12 = 222.5 \, \text{V}.$

But   the new e.m.f. generated $= k \times$ new speed × new flux

∴ $222.5 = 14.82 \times$ new speed $\times 0.0155$

∴   speed for 50 A $= 969 \, \text{r/min}.$

16.9 Speed control of d.c. motors

It has already been explained in section 16.3 that the speed of a d.c. motor can be altered by varying either the flux or the armature voltage or both; and the methods most commonly employed are:

(*a*) A variable resistor, termed a *field regulator*, in series with the shunt winding — only applicable to shunt and compound motors. Such a field regulator is indicated by H in fig. 16.7. When the resistance is increased, the field current, the flux and the generated e.m.f. are reduced. Consequently more current flows through the armature and the increased torque enables the armature to accelerate until the generated e.m.f. is again nearly equal to the applied voltage (see example 16.7).

With this method it is possible to increase the speed to three or four times that at full excitation, but it is not possible to reduce the speed below that value. Also, with any given setting of the regulator, the speed remains approximately constant between no load and full load.

(*b*) A resistor, termed a *controlled*, in series with the armature. The electrical connections for a controller are exactly the same as for a starter, the only difference being that in a controller the resistor elements are designed to carry the armature current indefinitely, whereas in a starter they can only do so for a comparatively short time without getting excessively hot.

For a given armature current, the larger the controller resistance in circuit, the smaller is the p.d. across the armature and the lower, in consequence, is the speed.

This system has several disadvantages: (i) the relatively high cost of the controller; (ii) much of the input energy may be dissipated in the controller and the overall efficiency of the motor considerably reduced thereby; (iii) the speed may vary greatly with variation of load due to the change in the p.d. across the controller causing a corresponding change in the p.d. across the motor; thus, if the supply voltage is 240 V, and if the current decreases so that the p.d. across the controller falls from, say, 100 to 40 V, then the p.d. across the motor increases from 140 V to 200 V.

The principal advantage of the system is that speeds from zero upwards are easily obtainable, and the method is chiefly used for controlling the speed of cranes, hoists, trains, etc., where the motors are frequently started and stopped and where efficiency is of secondary importance.

(*c*) Exciting the field winding off a constant-voltage system and supplying the armature from a separate generator, as shown in fig. 16.12. M represents the motor whose speed is to

Fig. 16.12 Ward–Leonard system of speed control

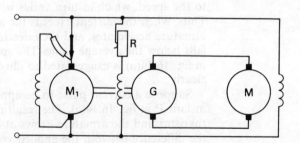

be controlled; M_1G is a motor-generator set consisting of a shunt motor coupled to a separately-excited generator. The voltage applied to the armature of M can be varied between zero and a maximum by means of R. If provision is made for reversing the excitation of G, the speed of M can then be varied from a maximum in one direction to the same maximum in the reverse direction.

This method is often referred to as the *Ward–Leonard* system and is used for controlling the speed of motors driving colliery winders, rolling mills, etc.

(*d*) When an a.c. supply is available, the voltage applied to the armature can be controlled by thyristors, the operation of which is explained in section 25.2. Briefly, the thyristor is a solid-state rectifier which is normally non-conducting in the forward and reverse directions. It is provided with an extra electrode, termed the *gate*, so arranged that when a pulse of current is introduced into the gate circuit, the thyristor is 'fired', i.e. it conducts in the forward direction. Once it is

Fig. 16.13 Thyristor system
of speed control

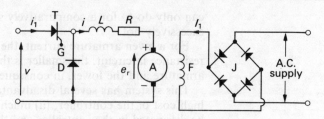

fired, the thyristor continues to conduct until the current falls
below the holding value.

Fig. 16.13 shows a simple arrangement for controlling a
d.c. motor from a single-phase supply. Field winding F is
separately excited via bridge-connected rectifiers J
(section 10.11) and armature A is supplied via thyristor T.
Gate G of the thyristor is connected to a firing circuit which
supplies a current pulse once every cycle. The arrangement of
the firing circuit is not shown as it is too involved for
inclusion in this diagram. In fig. 16.13, R and L represent the
resistance and inductance respectively of the armature
winding and an external inductor that may be inserted to
increase the inductance of the circuit. A diode D is connected
across the armature and the inductor.

The sine wave in fig. 16.14(a) represents the supply voltage
and the wavy line MN represents the e.m.f., e_r, generated by
the rotation of the armature. The value of e_r is proportional
to the speed, which in turn varies with the armature current.
Thus, when the current exceeds the average value, the
armature accelerates, and then decelerates when the current
falls below the average value. The speed fluctuation indicated
in fig. 16.14(a) is exaggerated to illustrate the effect more
clearly.

Suppose a current pulse to be applied to gate G at
instant P in fig. 16.14(a). The resulting current through the
thyristor and the armature grows at a rate depending upon
the difference between the applied voltage, v, and the
rotational e.m.f., e_r, and upon the ratio L/R for the circuit.
Thus at any instant:

$$v = e_r + iR + L \cdot di/dt$$
$$\therefore \quad i = (v - e_r - L \cdot di/dt)/R \tag{16.9}$$

Fig. 16.14 Waveforms for
fig. 16.13

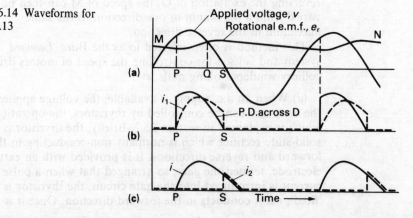

While the thyristor is conducting, the p.d. across the armature and inductor, and therefore across diode D, varies as shown by the full-line waveform in fig. 16.14(b).

At instant Q, $e_r = v$, so that $i = -(L/R) \cdot di/dt$. The current is now decreasing so that di/dt is negative; hence the current is still positive and is therefore continuing to exert a *driving* torque on the armature.

At instant S, the supply voltage is reversing its direction. A reverse current from the supply flows through D, thereby making the cathode of the thyristor *positive* relative to its anode. Consequently, the current i_1 through the thyristor falls to zero, as indicated by the dotted line in fig. 16.14(b), so that the thyristor reverts to its non-conducting state.

Current is now confined to the closed circuit formed by the armature and diode D. Since the p.d. across D is practically zero, equation (16.9) can now be written:

$$i = -(e_r - L \cdot di/dt)/R.$$

Current i is now decreasing at sufficiently high rate for the value of $L \cdot di/dt$ to exceed e_r. Since the value of di/dt is negative, the direction of the current is unaltered. Hence the current still continues to exert a driving torque on the armature, the energy supplied to the load being recovered partly from that stored in the inductance of the circuit while the current was growing and partly from that stored as kinetic energy in the motor and load during acceleration.

The dotted waveform in fig. 16.14(c) represents current i_2 through diode D. The latter is often referred to as a free-wheeling diode since it carries current when the thyristor *ceases* to conduct.

The later the instant of firing the thyristor, the smaller is the average voltage applied to the armature and the lower the speed in order that the motor may take the same average current from the supply to enable it to maintain the *same* load torque. Thus the speed of the motor can be controlled over a wide range.

An increase of motor load causes the speed to fall, thereby allowing a larger current pulse to flow during the conducting period. The fluctuation of current can be reduced by (a) using two thyristors to give full-wave rectification when the supply is single-phase and (b) using three or six thyristors when the supply is three-phase.

An important application of the thyristor is the speed control of series motors in battery-driven vehicles. The principle of operation is that pulses of the battery voltage are applied to the motor, and the *average* value of the voltage across the motor is controlled by varying the ratio of the 'on' and 'off' durations of the pulses. Thus, if the 'on' period is t_1 and the 'off' period is t_2,

average motor voltage = battery voltage $\times t_1/(t_1 - t_2)$.

Fig. 16.15 shows the essential features of this method of speed control. The series motor M is connected in series with thyristor T across the battery. A free-wheeling diode D is connected in parallel with the motor, and a switch S is used

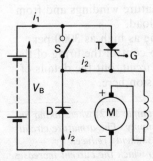

Fig. 16.15 Chopper speed control of a series motor

to short-circuit the thyristor at the end of each 'on' period.

With S open, the thyristor is fired by a current pulse applied to gate G, causing a current i_1 to flow through T and M, as shown in fig. 16.16. After an interval t_1, switch S is closed for sufficient time to allow the thyristor current to fall to zero, thereby enabling the thyristor to revert to its non-conducting state.

After S is opened, the current through M decreases at a rate such that the e.m.f. induced in the inductance of the field and armature windings exceeds the rotational e.m.f. by an amount sufficient to circulate a current i_2 around the closed circuit formed by the motor and diode D. After an interval t_2, the operation is repeated, as indicated in fig. 16.16.

Fig. 16.16 Waveforms for
fig. 16.15

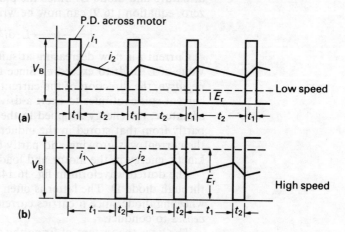

For a low speed, the ratio t_1/t_2 is small, as shown in fig. 16.16(*a*). A higher speed is obtained by increasing t_1 and reducing t_2, as in fig. 16.16(*b*). The dotted horizontal lines in fig. 16.16 represent the average values of the rotational e.m.f. E_r generated in the armature. The average value of the current is determined by the torque requirement, exactly as described in earlier sections.

The battery is supplying power only during the 'on' intervals t_1. During the 'off' intervals t_2, current t_2 is exerting a *driving* torque on the armature, the energy supplied to the mechanical load during these intervals being derived from the magnetic fields of the series and armature windings and from the kinetic energy of the motor and load.

The frequency of the pulses may be as high as 3000 per second, and the arrangement used to control the firing of the thyristor is referred to as a *chopper circuit*, but the details of this circuit are too complex for inclusion here.

Example 16.7   *A shunt motor is running at 626 r/min (example 16.2) when taking an armature current of 50 A from a 440-V supply. The armature circuit has a resistance of 0.28 Ω. If the flux is suddenly reduced by 5 per cent, find: (a) the maximum value to which the current increases momentarily and the ratio of the corresponding torque to the initial torque; (b) the ultimate steady value of the armature current, assuming* *the torque due to the load to remain unaltered.*

(a) From example 16.2:

initial e.m.f. generated $= 400 - 50 \times 0.28 = 426$ V.

Immediately after the flux is reduced 5 per cent, i.e. before the speed has begun to increase:

new e.m.f. generated $= 426 \times 0.95 = 404.7$ V

$\therefore$ corresponding voltage drop due to armature resistance

$$= 440 - 404.7 = 35.3 \text{ V}$$

and corresponding armature current

$$= 35.3/0.28 = 126 \text{ A}.$$

From expression (16.7):

torque of a given machine $\propto$ armature current $\times$ flux

$\therefore$ $\dfrac{\text{new torque}}{\text{initial torque}} = \dfrac{\text{new current}}{\text{initial current}} \times \dfrac{\text{new flux}}{\text{initial flux}}$

$$= \frac{126}{50} \times 0.95 = 2.394.$$

Hence the sudden reduction of 5 per cent in the flux is accompanied by more than a twofold increase of torque; this is the reason why the motor accelerated.

(b) After the speed and current have attained steady values, the torque will have decreased to the original value, so that:

new current $\times$ new flux $=$ original current $\times$ original flux

$\therefore$ new armature current $= 50 \times \dfrac{1}{0.95} = 52.6$ A.

Example 16.8   *A shunt motor runs on no load at 700 r/min off a 440-V supply. The resistance of the shunt circuit is 240 Ω. Calculate the additional resistance required in the shunt circuit to raise the no-load speed to 1000 r/min. The following table gives the relationship between the flux and the shunt current:*

Shunt current (A)	0.5	0.75	1.0	1.25	1.5	1.75	2.0
Flux per pole (mWb)	6.0	8.0	9.4	10.2	10.8	11.2	11.5

When the motor is on no load, the voltage drop due to the armature resistance is negligible.
Initial shunt current $= 440/240 = 1.83$ A.
From a graph representing the data given in the above table, corresponding flux per pole $= 11.3$ mWb.
Since the speed is inversely proportional to the flux,

flux per pole at 1000 r/min $= 11.3 \times 700/1000$

$$= 7.91 \text{ mWb}.$$

From the graph it is found that the shunt current to produce a flux of 7.91 mWb/pole is 0.74 A,

$\therefore$ corresponding resistance of shunt circuit $= 440/0.74$

$$= 595 \,\Omega,$$

and $\left.\begin{array}{l}\text{additional resistance required}\\ \text{in shunt circuit}\end{array}\right\} = 595 - 240 = 335 \,\Omega.$

Example 16.9   *A shunt motor, supplied at 230 V, runs at 900 r/min when the armature current is 30 A. The resistance of the armature circuit is 0.4 Ω. Calculate the resistance required in series with the armature to reduce the speed to 600 r/min, assuming that the armature current is then 20 A.*

Initial e.m.f. generated = $230 - 30 \times 0.4 = 218$ V.

Since the excitation remains constant, the generated e.m.f. is proportional to the speed.

$\therefore$ e.m.f. generated at 600 r/min = $218 \times 600/900 = 145.3$ V

Hence

$\left.\begin{array}{l}\text{voltage drop due to total}\\ \text{resistance of armature circuit}\end{array}\right\} = 230 - 145.3 = 84.7$ V,

and $\left.\begin{array}{l}\text{total resistance of}\\ \text{armature circuit}\end{array}\right\} = 84.7/20 = 4.235\ \Omega$

$\therefore$ $\left.\begin{array}{l}\text{additional resistance required}\\ \text{in armature circuit}\end{array}\right\} = 4.235 - 0.4 = 3.835\ \Omega.$

Summary of important formulae

For a generator, $E = V + I_a R_a$ (16.1)

For a motor, $V = E + I_a R_a$ (16.2)

For a given motor, $N_r = \dfrac{V - I_a R_a}{k\Phi} \simeq \dfrac{V}{k\Phi}$ (16.3)

$T = 0.381 \dfrac{I_a}{c} \cdot Zp\Phi$ newton metres   (16.6)

For a given motor, $T \propto I_a \Phi$ (16.7)

EXERCISES 16

1. A shunt machine has armature and field resistances of 0.04 Ω and 100 Ω respectively. When connected to a 460-V d.c. supply and driven as a generator at 600 r/min, it delivers 50 kW. Calculate its speed when running as a motor and taking 50 kW from the same supply.

   Show that the direction of rotation of the machine as a generator and as a motor under these conditions is unchanged.

2. Obtain from first principles an expression for the e.m.f. of a 2-pole d.c. machine, defining the symbols used.

   A 100-kW, 500-V, 750-r/min, d.c. shunt generator, connected to constant-voltage bus-bars, has field and armature resistances of 100 Ω and 0.1 Ω respectively. If the prime-mover fails, and the machine continues to run, taking 50 A from the bus-bars, calculate its speed. Neglect brush-drop and armature reaction effects.   (S.A.N.C., O.2)

3. Derive the e.m.f. equation for a d.c. machine.

A d.c. shunt motor has an armature resistance of $0.5\,\Omega$ and is connected to a 200-V supply. If the armature current taken by the motor is 20 A, what is the e.m.f. generated by the armature?

What is the effect of: (*a*) inserting a resistor in the field circuit; (*b*) inserting a resistor in the armature circuit if the armature current is maintained at 20 A? (E.M.E.U., O.1)

4. Explain clearly the effect of the back e.m.f. of a shunt motor. What precautions must be taken when starting a shunt motor?

A four-pole d.c. motor is connected to a 500-V d.c. supply and takes an armature current of 80 A. The resistance of the armature circuit is $0.4\,\Omega$. The armature is wave wound with 522 conductors and the useful flux per pole is 0.025 Wb. Calculate: (*a*) the back e.m.f. of the motor; (*b*) the speed of the motor; (*c*) the torque in newton metres developed by the armature. (U.E.I., O.1)

5. A shunt machine is running as a motor off a 500-V system, taking an armature current of 50 A. If the field current is suddenly increased so as to increase the flux by 20 per cent, calculate the value of the current that would momentarily be fed back into the mains. Neglect the shunt current and assume the resistance of the armature circuit to be $0.5\,\Omega$.

6. A shunt motor is running off a 220-V supply taking an armature current of 15 A, the resistance of the armature circuit being $0.8\,\Omega$. Calculate the value of the generated e.m.f.

If the flux were suddenly reduced by 10 per cent, to what value would the armature current increase momentarily?

7. A six-pole d.c. motor has a wave-connected armature with 87 slots, each containing six conductors. The flux per pole is 20 mWb and the armature has a resistance of $0.13\,\Omega$. Calculate the speed when the motor is running off a 240-V supply and taking an armature current of 80 A. Calculate also the torque, in newton metres, developed by the armature.

8. A four-pole, 460-V shunt motor has its armature wave-wound with 888 conductors. The useful flux per pole is 0.02 Wb and the resistance of the armature circuit is $0.7\,\Omega$. If the armature current is 40 A, calculate: (*a*) the speed, and (*b*) the torque in newton metres. (U.E.I., O.1)

9. A four-pole motor has its armature lap-wound with 1040 conductors and runs at 1000 r/min when taking an armature current of 50 A from a 250-V d.c. supply. The resistance of the armature circuit is $0.2\,\Omega$. Calculate: (*a*) the useful flux per pole; (*b*) the torque developed by the armature in newton metres. (U.E.I., O.1)

10. A d.c. shunt generator delivers 5kW at 250 V when driven at 1500 r/min. The shunt circuit resistance is $250\,\Omega$ and the armature circuit resistance is $0.4\,\Omega$. The core, friction and windage losses are 250 W. Determine the torque (in newton metres) required to drive the machine at the above load. (W.J.E.C., O.2)

11. Explain why a d.c. shunt-wound motor needs a starter on constant-voltage mains. A shunt-wound motor has a field resistance of $350\,\Omega$ and an armature resistance of $0.2\,\Omega$ and runs off a 250-V supply. The armature current is 55 A and the motor speed is 1000 r/min. Assuming a straight-line magnetization curve, calculate: (*a*) the additional resistance required in the field circuit to increase the speed to 1100 r/min for the same armature current, and (*b*) the speed with the original field current and an armature current of 100 A. (E.M.E.U., O.1)

12. Explain the necessity for using a starter with a d.c. motor.

A 240-V d.c. shunt motor has an armature of resistance of 0.2 Ω. Calculate: (*a*) the value of resistance which must be introduced into the armature circuit to limit the starting current to 40 A; (*b*) the e.m.f. generated when the motor is running at a constant speed with this additional resistance in circuit and with an armature current of 30 A.   (N.C.T.E.C., O.1)

13. A separately-excited d.c. motor is fed from a 460-V d.c. supply. The speed is to be controlled by field weakening. When the machine was driven as a separately-excited generator at 750 r/min, the following open-circuit curve was obtained:

Generated e.m.f. (V)	10	172	300	360	385	395
Field current (A)	0	1	2	3	4	5

Plot a curve showing the no-load speed against field current. Assume negligible losses.   (N.C.T.E.C., O.2)

14. A long-shunt, compound-wound motor connected to a 250-V supply has a full-load output of 15 kW, and at this load is found to have an efficiency of 82 per cent. The armature has a resistance of 0.2 Ω and the series field winding (which is designed to change the flux per pole by 0.002 Wb on full load) has a resistance of 0.3 Ω. The shunt field, which produces a flux per pole of 0.02 Wb, has a resistance of 250 Ω. When the machine runs on no load, its speed is found to be 500 r/min and the current taken from the supply is 2.5 A.

Calculate the speed on full load if the machine is cumulatively compounded.

The ferromagnetic parts of the machine operate on the linear part of their magnetization characteristics throughout and the effect of temperature on resistance may be neglected.
(E.M.E.U., O.2)

15. A series d.c. motor is fed from a 400-V d.c. supply. Calculate and plot the variation of motor speed (rad/s) against armature current, given the following information:

(*a*) When the machine is driven as a generator and separately excited, the following open-circuit curve was obtained at a speed of 100 rad/s:

Field current (A)	0	5	15	25	35	40	50
Open-circuit voltage (V)	0	87	245	335	375	385	395

(*b*) Armature resistance = 1.6 Ω; field resistance = 0.9 Ω.

Assume negligible brush voltage drop.   (N.C.T.E.C., O.2)

16. A d.c. shunt motor takes an armature current of 20 A from a 230-V supply. Resistance of the armature circuit is 0.5 Ω. Calculate the resistance required in series with the armature to halve the speed if: (*a*) the load torque is constant; (*b*) the load torque is proportional to the square of the speed.

17. Calculate the torque, in newton metres, developed by a d.c. motor having an armature resistance of 0.25 Ω and running at 750 r/min when taking an armature current of 60 A from a 480-V supply.

18. A six-pole, lap-wound, 220-V, shunt-excited d.c. machine takes an armature current of 2.5 A when unloaded at 950 r/min. When loaded, it takes an armature current of 54 A from the supply and runs at 950 r/min. The resistance of the armature circuit is 0.18 Ω and there are 1044 armature conductors.

For the loaded condition, calculate: (i) the generated e.m.f.; (ii) the useful flux per pole; (iii) the useful torque developed by the machine in newton metres.   (U.E.I., O.2)

19. A d.c. shunt motor runs at 900 r/min from a 480-V supply when

taking an armature current of 25 A. Calculate the speed at which it will run from a 240-V supply when taking an armature current of 15 A. The resistance of the armature circuit is 0.8 Ω. Assume the flux per pole at 240 V to have decreased to 75 per cent of its value at 480 V.

20. A shunt motor, connected across a 440-V supply, takes an armature current of 20 A and runs at 500 r/min. The armature circuit has a resistance of 0.6 Ω. If the magnetic flux is reduced by 30 per cent and the torque developed by the armature increases by 40 per cent, what are the values of the armature current and of the speed? (App. El., L.U.)

21. A d.c. series motor connected across a 460-V supply runs at 500 r/min when the current is 40 A. The total resistance of the armature and field circuits is 0.6 Ω. Calculate the torque on the armature in newton metres. (App. El., L.U.)

22. A d.c. series motor has the following magnetization data:

Field current (A)	5	10	15	20	25
E.M.F. (V) at 800 r/min	180	356	488	548	570

The armature resistance is 1.4 Ω and the field circuit resistance 0.6 Ω. Plot the speed/current curve when the motor is connected across a 480-V supply.

23. The output power of a shunt motor running off a 240-V supply is 16 kW. Calculate the value of the starter resistance necessary to limit the starting current to 1.5 times the full-load current if the full-load efficiency of the motor is 88 per cent and the resistance of the armature circuit is 0.2 Ω. Neglect the shunt current.

    Also, calculate the generated e.m.f. of the motor when the current has fallen to the full-load value, assuming that the whole of the starter resistance is still in circuit.

24. A d.c. shunt motor runs off a 230-V supply and has an armature resistance of 0.3 Ω. Calculate: (a) the resistance required in series with the armature to limit the armature current to 75 A at starting, and (b) the value of the generated e.m.f. when the armature current has fallen to 50 A with this value of resistance still in circuit.

25. Assuming that the e.m.f. generated by a d.c. machine is $E = k\Phi\omega$ volts, deduce the expression for the generated torque as $T = k\Phi I_a$ newton metres, where Φ = flux per pole in webers, ω = speed of armature in radians per second, I_a = armature current in amperes and k = a constant.

    A d.c. shunt motor runs at 70 rad/s from a 460-V supply and takes an armature current of 100 A. Resistors of 2.52 Ω and 1.64 Ω are connected in series with the armature and in parallel with the armature respectively. The field excitation remains direct from the 460-V supply, but the generated torque is halved. Determine the new running speed. Neglect brush voltage drop. The armature resistance is 0.4 Ω.

    (N.C.T.E.C., O.2)

26. A d.c. series motor, having armature and field resistances of 0.06 Ω and 0.04 Ω respectively, was tested by driving it at 2000 r/min and measuring the open-circuit voltage across the armature terminals, the field being supplied from a separate source. One of the readings taken was: field current, 350 A, armature p.d., 1560 V. From the above information, obtain a point on the speed/current characteristic, and one on the torque/current characteristic for normal operation at 750 V and at 350 A. Take the torque due to rotational loss as 50 N·m. Assume that brush drop and field weakening due to armature reaction can be neglected. (S.A.N.C., O.2)

27. A shunt motor runs on no load at 800 r/min off a 240-V supply with no external resistor in the field circuit. The armature current is 2 A. Calculate the resistance required in series with the shunt winding so that the motor may run at 950 r/min when taking an armature current of 30 A. Shunt winding resistance = 160 Ω; armature resistance = 0.4 Ω. It may be assumed that the flux is proportional to the field current.

28. The resistance of the armature circuit of a 250-V shunt motor is 0.3 Ω and its full-load speed is 1000 r/min. Calculate the resistance required in series with the armature to reduce the speed with full-load torque to 800 r/min, the full-load armature current being 50 A. If the load torque is then halved, at what speed will the motor run? Neglect the effect of armature reaction.

29. A d.c. series motor, connected to a 440-V supply, runs at 600 r/min when taking a current of 50 A. Calculate the value of a resistor which, when inserted in series with the motor, will reduce the speed to 400 r/min, the gross torque being then half its previous value. Resistance of motor = 0.2 Ω. Assume the flux to be proportional to the field current.

30. A series motor runs at 900 r/min when taking 30 A at 230 V. The total resistance of the armature and field circuits is 0.8 Ω. Calculate the values of the additional resistance required in series with the machine to reduce the speed to 500 r/min if the gross torque is: (*a*) constant; (*b*) proportional to the speed; (*c*) proportional to the square of the speed. Assume the magnetic circuit to be unsaturated.

17 Direct-Current Generators

17.1 Armature and field connections

Consequent upon the development of high-quality semiconductor devices such as diodes and thyristors, the demand for d.c. generators has declined rapidly. It follows that although there are the same methods used for connecting the field and armature windings as are found in d.c. motors, some now merit little attention and we need only consider the following groups:

(*a*) Separately-excited machines — the field winding being connected to a source of supply other than the armature of its own machine.

(*b*) Self-excited machines — only the shunt-wound variant remains important, in which the field winding is connected across the armature terminals.

(*c*) Self-excited machines — the compound-wound machines have characteristics of passing interest.

17.2 Separately excited generator

Fig. 17.1 shows a simple method of representing the armature and field windings, A and A_1 being the armature terminals and F and F_1 the field terminals. The field winding is connected in series with a resistor R and an ammeter A to a battery or another generator.

Fig. 17.1 Separately excited generator

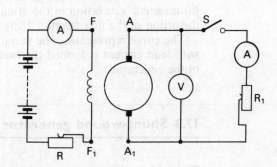

Suppose the armature to be driven at a *constant* speed on no load (i.e. with switch S open) and the exciting current to be increased from zero up to the maximum permissible value and then reduced to zero, the generated e.m.f. being noted for various values of the magnetizing current. It is found that the relationship between the two quantities is represented by curves P and Q in fig. 17.2, P being for increasing values of the excitation and Q for decreasing values. The difference between the curves is due to hysteresis (section 3.11), and OR represents the e.m.f. generated by the residual magnetism in the poles. If the test is repeated, the e.m.f. curve is found to follow the dotted line and then merge into curve P. These curves are termed the *internal* or *open-circuit characteristics* of the machine.

From expression (15.1), it is evident that for a given speed, the e.m.f. generated in a given machine is proportional to the flux per pole; and for a given field winding, the number of ampere-turns per pole is proportional to the magnetizing current. Hence the shape of the open-circuit characteristic P in fig. 17.2 is the same as that of the magnetization curve C in fig. 16.2; and curves P and Q in fig. 17.2 indicate how the flux varies with increasing and decreasing values of the magnetizing current.

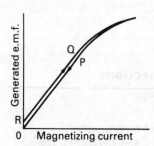

Fig. 17.2 Open-circuit characteristic

Let us next consider the effect of load upon the terminal voltage. This can be determined experimentally by closing switch S (fig. 17.1), varying the resistance of resistor R_1 and noting the terminal voltage for each load current. Curve M in fig. 17.3 is typical of the relationship for such a machine. The decrease in the terminal voltage with increase of load is mainly due to the resistance drop in the armature circuit; thus, if the load current is 100 A and the resistance of the armature circuit is 0.08 Ω, the voltage drop in the armature circuit is $100 \times 0.08 = 8$ V. Consequently, if the generated e.m.f. be 235 V, the terminal p.d. $= 235 - 8 = 227$ V. In general, if E = generated e.m.f., I_a = armature current, R_a = resistance of armature circuit and V = terminal p.d., then:

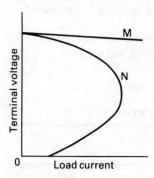

Fig. 17.3 Load characteristics of separately excited and shunt generators

$$V = E - I_a R_a \tag{17.1}$$

Another factor that may cause a decrease in the terminal voltage is the decrease of flux and therefore of the generated e.m.f. due to (i) the demagnetizing ampere-turns of the armature if the brushes are given a forward shift and (ii) magnetic saturation in the armature teeth due to distortion of the flux (section 15.5).

The curve representing the variation of terminal voltage with load current is termed the *load* or *external characteristic* of the generator.

17.3 Shunt-wound generator

The field winding is connected in series with a field regulating

Fig. 17.4 Shunt-wound
generator

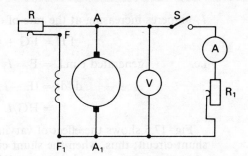

resistor R across the armature terminals as shown in fig. 17.4,
and is therefore in parallel or 'shunt' with the load. The
power absorbed by the shunt circuit is limited to about 2–3
per cent of the rated output of the generator by winding the
field coils with a large number of turns of comparatively thin
wire.

A shunt generator will excite only if the poles have some
residual magnetism and the resistance of the shunt circuit is
less than some critical value, the actual value depending upon
the machine and upon the speed at which the armature is
driven. Suppose curve P in fig. 17.5 to represent the open-
circuit characteristic of a shunt generator with *increasing*
excitation; then, for a shunt current OA, the e.m.f. is AB and

$$\text{corresponding resistance of shunt circuit} = \frac{\text{terminal voltage}}{\text{shunt current}}$$

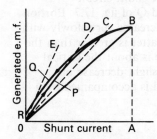

Fig. 17.5 Variation of e.m.f.
with shunt circuit resistance

$$= AB/OA = \tan BOA$$

$$= \text{slope of OB}.$$

When the machine is running at its rated speed with the
shunt circuit disconnected, the residual magnetism generates
an e.m.f. OR. Immediately the shunt circuit is closed, this
e.m.f., OR, circulates a current OC (fig. 17.5) through the
shunt winding; but the magnetic flux produced by this
current generates an e.m.f. CD, which in turn circulates a
larger shunt current, etc., until the e.m.f. reaches the value
AB. During this continuous process of building up the
excitation, the e.m.f. generated by a given shunt current is
greater than that required to send that current through the
resistance of the shunt circuit. For instance, when the shunt
current is OF (fig. 17.6), the generated e.m.f. is FH, but the
voltage required to send this current through the resistance of
the shunt circuit is FG. The difference between FH and FG,
namely GH, is the voltage required to neutralize the e.m.f.
induced in the field winding* due to the growth of the field
current; thus if R_f and L are the resistance and inductance
respectively of the shunt circuit, then for a field current

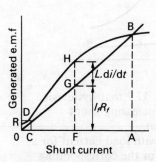

Fig. 17.6 Growth of the
excitation of a shunt
generator

---

* The resistance and inductance of the armature winding are being
neglected, since they are usually very small compared with the
corresponding values of the shunt circuit.

I_f amperes increasing at the rate of di/dt amperes per second,

$$FH = FG + GH$$

i.e.   generated e.m.f. $= E = I_f R_f + L \cdot di/dt$

$\therefore$ $di/dt = (E - I_f R_f)/L$

                    $= HG/L$ in fig. 17.6

Fig. 17.5 shows the effect of varying the resistance of the shunt circuit; thus, when the shunt circuit is broken, the resistance is infinite and the e.m.f. OR is due to the residual magnetism. When the shunt circuit is closed and its resistance reduced to tan EOA, the e.m.f. is only slightly larger than OR; but when it is reduced to tan DOA, the value of the e.m.f. is extremely sensitive to variation of the resistance — a very slight reduction of the shunt resistance is accompanied by a relatively large increase of e.m.f. Since tan DOA is the value of the shunt circuit resistance at which the excitation increases rapidly, this value is known as the *critical resistance of the shunt circuit*.

By drawing lines such as OE, OD, OC and OB corresponding to various values of the shunt circuit resistance, we can derive curves P and Q of fig. 17.7. Portion *ab* of curve P shows that the e.m.f. increases very slowly with decrease of resistance so long as the latter is greater than the critical value; but when the resistance is about the critical value, the e.m.f. is very unstable — a slight decrease of resistance or a slight increase of speed is accompanied by a large increase of e.m.f.

Fig. 17.7 Variation of e.m.f. with shunt resistance

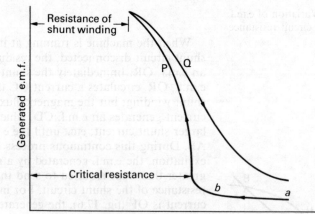

It will be seen from curve Q in fig. 17.7 that hysteresis causes the variation of voltage with increasing shunt resistance to be comparatively gradual.

The variation of terminal voltage with load current for a shunt generator is greater than that for the corresponding separately-excited generator and is represented by curve N in fig. 17.3. Thus, when the load on a shunt generator is increased, the decrease in the terminal voltage, due partly to the increased $I_a R_a$ drop in the armature winding and partly to armature reaction, is accompanied by a decrease in the shunt current. Consequently, the flux and the generated e.m.f. are reduced, thereby causing a further reduction in the terminal voltage.

The load characteristic of a shunt generator can be derived from the open-circuit characteristic by the construction shown in fig. 17.8, on the assumption that the effect of armature reaction (section 15.5) is negligible. Curve Q represents the

Fig. 17.8 Derivation of the load characteristics of a shunt generator

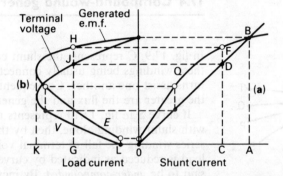

variation of generated e.m.f. with *decreasing* excitation, and AB represents the terminal voltage of the generator on no load when the resistance of the shunt circuit is adjusted to tan BOA. For this shunt resistance, the relationship between the terminal voltage and the shunt current is represented by the straight line OB. Hence, when the machine is loaded so that the terminal voltage is CD, the shunt current is OC and the generated e.m.f. is CF; and from expression (17.1) we have

$$CD = CF - I_a R_a$$

$$\therefore \qquad I_a = \frac{CF - CD}{R_a} = \frac{DF}{R_a} \qquad (17.2)$$

where $\qquad I_a$ = armature current

and $\qquad R_a$ = resistance of armature circuit.

Corresponding load current = armature current − shunt curren

$$= DF/R_a - OC$$

$$= OG \text{ in fig. 17.8}(b).$$

It follows from expression (17.2) that the armature current is proportional to the vertical intercept between curve Q and straight line OB. Hence we can derive curves E and V in fig. 17.8(b) representing the relationships between the load current and the generated e.m.f. and terminal voltage respectively; thus, on the vertical line at G corresponding to load current OG, points H and J are the horizontal projections of F and D respectively. It will be seen that as the resistance of the load is reduced, the load current increases to a maximum value OK and then decreases to OL when the

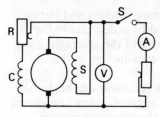

Fig. 17.9 Compound-wound generator

terminals are short-circuited, OL being the current due to the e.m.f. generated by the residual magnetism.

The shunt machine is the type of d.c. generator most frequently employed, but the load current must be limited to a value that is well below the maximum value, thereby avoiding excessive variation of the terminal voltage.

17.4 Compound-wound generator

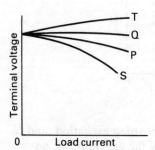

Fig. 17.10 Load characteristics of shunt-wound and compound-wound generators

In fig. 17.9, C represents the shunt coils and S the series coils, these windings being usually connected* so that their ampere-turns assist one another. Consequently the larger the load, the greater are the flux and the generated e.m.f.

If curve S in fig. 17.10 represents the load characteristic with shunt winding alone, then by the addition of a small series winding the fall of terminal voltage with increase of load is reduced as indicated by curve P. Such a machine is said to be *under-compounded*. By increasing the number of series turns we can arrange for the machine to maintain its terminal voltage (curve Q) practically constant between no load and full load, in which case the machine is said to be *level-compounded*. If the number of series turns be increased still further, the terminal voltage increases with increase of load — as represented by curve T. The machine is then said to be *over-compounded*.

Example 17.1   *A shunt generator is to be converted into a level-compounded generator by the addition of a series field winding. From a test on the machine with shunt excitation only, it is found that the shunt current is 4.1 A to give 440 V on no load and 5.8 A to give the same voltage when the machine is supplying its full load of 200 A. The shunt winding has 1200 turns per pole. Find the number of series turns required per pole.*

Ampere-turns per pole required on no load

$$= 4.1 \times 1200 = 4920 \text{ At.}$$

Ampere-turns per pole required on full load

$$= 5.8 \times 1200 = 6960 \text{ At.}$$

Hence ampere-turns per pole to be provided by the series winding

$$= 6960 - 4920 = 2040 \text{ At,}$$

∴   number of series turns per pole

$$= 2040/200 = 10.$$

---

* In practice, it is of little consequence whether the shunt winding is connected 'long-shunt' as in fig. 17.9 or 'short-shunt', i.e. directly across the armature terminals, since the shunt current is very small compared with the full-load current and the number of series turns is very small compared with the number of shunt turns.

The shunt current is so small compared with the full-load current that it is of little consequence whether it is included in or omitted from the calculation.

EXERCISES 17

1. Sketch neat graphs to show the internal or open-circuit characteristics of a separately-excited d.c. generator. Why is a field regulator necessary with such a machine?

   A four-pole d.c. generator gives 410 V on open circuit when driven at 900 r/min. Calculate the flux per pole if the wave-connected armature winding has 39 slots with 16 conductors per slot. (U.E.I.)

2. A six-pole d.c. generator having a lap-connected armature winding is required to give a terminal voltage of 240 V when supplying an armature current of 400 A. The armature has 84 slots and is driven at 700 r/min. The resistance of the armature circuit is 0.03 Ω and the useful flux per pole is about 0.03 Wb. Calculate the number of conductors per slot and the actual value of the useful flux per pole.

3. Explain why the terminal voltage of a d.c. shunt-excited generator falls as the current supplied by the machine is increased.

   The resistance of the field circuit of a shunt-excited d.c. generator is 200 Ω. When the output of the generator is 100 kW, the terminal voltage is 500 V and the generated e.m.f. 525 V. Calculate: (a) the armature resistance, and (b) the value of the generated e.m.f. when the output is 60 kW, if the terminal voltage is then 520 V. (E.M.E.U.)

4. A short-shunt compound generator has armature, shunt-field and series-field resistances of 0.8 Ω, 45 Ω and 0.6 Ω respectively and supplies a load of 5 kW at 250 V. Calculate the e.m.f. generated in the armature.

5. The following table gives the open-circuit voltages for different field currents of a shunt generator driven at a constant speed:

Terminal voltage (V)	120	240	334	400	444	470
Field current (A)	0.5	1.0	1.5	2.0	2.5	3.0

   Plot a graph showing the variation of generated e.m.f. with exciting current and from this graph derive the value of the generated e.m.f. when the shunt circuit has a resistance of (a) 160 Ω, (b) 210 Ω and (c) 300 Ω. Also find the value of the critical resistance of the shunt circuit.

6. The following table gives open-circuit voltages for different values of field current for a d.c. generator:

Open-circuit voltage (V)	120	240	334	400	444	470
Field current (A)	0.5	1.0	1.5	2.0	2.5	3.0

   Express these values graphically and determine therefrom the generated e.m.f. when the shunt field circuit has a resistance of 200 Ω. Find also the critical resistance of the shunt field circuit. The effects of armature resistance and brush volt drop may be neglected. (E.M.E.U., O.1)

7. What conditions must be fulfilled for the self-excitation of a d.c. shunt generator?

   The open-circuit characteristic of a shunt-excited d.c. machine, run at 1200 r/min, is:

Terminal voltage (V)	47	85	103	114	122	127	135	141
Field current (A)	0.2	0.4	0.6	0.8	1.0	1.2	1.6	2.0

The field coil resistance is 55 Ω. Determine: (i) the value of the field-regulating resistance to enable the machine to generate 120 V on open circuit when run at 1200 r/min; (ii) the value of the open-circuit voltage when the regulator is set to 20 Ω and the speed is reduced to 800 r/min.   (U.E.I., O.2)

8. The open-circuit characteristic of a certain d.c. generator, when running at 500 r/min, is given below:

Field current (A)	1	4	6	8	10
Generated e.m.f. (V)	71	135	179	202	214

The generator is now shunt-connected and run at 600 r/min, the resistance of the field circuit being 25 Ω. What e.m.f. will it generate?

If the generator is placed on load and the terminal p.d. becomes 230 V, what is the load current? The armature resistance is 0.2 Ω and armature reaction is to be neglected.

(App. El., L.U.)

9. Calculate the number of series turns per pole required on a compound generator to enable it to maintain the voltage constant at 460 V between no load and full load of 100 kW. Without any series winding, it is found that the shunt current has to be 2 A on no load and 2.65 A on full load to maintain the voltage constant at 460 V. Number of turns per pole on the shunt winding = 2000.

If the series coils were wound with 8 turns per pole and had a total resistance of 0.01 Ω, what value of diverter resistance would be required to give level compounding?

18 Efficiency of D.C. Machines

18.1 Losses in generators and motors

The losses in d.c. machines can be classified thus:

i. Armature losses

(a) I^2R loss in armature winding. The resistance of an armature winding can be measured by the voltmeter–ammeter method. If the resistance measurement is made at room temperature, the resistance at normal working temperature should be calculated. Thus, if the resistance be R_1 at room temperature of, say, 15°C and if 50°C be the temperature *rise* of the winding after the machine has been operating on full load for 3 or 4 hours, then from expression (1.16), we have:

$$\text{resistance at } 65°C = R_1 \times \frac{1 + (0.004\,26 \times 65)}{1 + (0.004\,26 \times 15)} = 1.2R_1.$$

(b) Core loss in the armature core due to hysteresis and eddy currents. Hysteresis loss has been discussed in section 3.14 and is dependent upon the quality of the steel. It is proportional to the frequency and is approximately proportional to the square of the flux. The eddy-current loss is due to circulating currents set up in the steel laminations. Had the core been of solid steel, as shown in fig. 18.1(*a*) for a two-pole machine, then if the armature were rotated, e.m.f.s would be generated in the core in exactly the same way as they are generated in conductors placed on the armature, and these e.m.f.s would circulate currents — known as *eddy currents* — in the core as shown dotted in fig. 18.1(*a*), the rotation being assumed clockwise when the armature is viewed from the righthand side of the machine. Owing to the very low resistance of the core, these eddy currents would be considerable and would cause a large loss of power in and excessive heating of the armature.

If the core is made of laminations insulated from one another, the eddy currents are confined to their respective sheets, as shown dotted in fig. 18.1(*b*), and the eddy-current loss is thereby reduced. Thus, if the core is split up into five laminations, the e.m.f. per lamination is only a fifth of that generated in the solid core. Also, the cross-sectional area per path is reduced to about a fifth, so that the resistance per path is roughly five times that of the solid core. Consequently the current per path is about $\frac{1}{25}$ of that in the solid core.

Fig. 18.1 Eddy currents

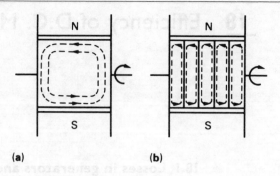

(a)   **(b)**

Hence:

$$\frac{I^2R \text{ loss per lamination}}{I^2R \text{ loss in solid core}} = \left(\frac{1}{25}\right)^2 \times 5 = \frac{1}{125} \text{ (approx.)}$$

Since there are five laminations,

$$\frac{\text{total eddy-current loss in laminated core}}{\text{total eddy-current loss in solid core}} = \frac{5}{125} = \left(\frac{1}{5}\right)^2.$$

It follows that the eddy-current loss is approximately proportional to the square of the thickness of the laminations. Hence the eddy-current loss can be reduced to any desired value, but if the thickness of the laminations is made less than about 0.4 mm, the reduction in the loss does not justify the extra cost of construction. Eddy-current loss can also be reduced considerably by the use of silicon-iron alloy — usually about 4 per cent of silicon — due to the resistivity of this alloy being much higher than that of ordinary steel.

Since the e.m.f.s induced in the core are proportional to the frequency and the flux, therefore the eddy-current loss is proportional to (frequency × flux)^2.

ii. Commutator losses

(*a*) Loss due to the contact resistance between the brushes and the segments. This loss is dependent upon the quality of the brushes. For carbon brushes, the p.d. between a brush and the commutator, over a wide range of current, is usually about 1 volt per positive set of brushes and 1 volt per negative set, so that the total contact-resistance loss, in watts, is approximately 2 × total armature current.

(*b*) Loss due to friction between the brushes and the commutator. This loss depends upon the total brush pressure, the coefficient of friction and the peripheral speed of the commutator.

iii. Excitation losses

(*a*) Loss in the shunt circuit (if any) equal to the product of the shunt current and the terminal voltage. In shunt generators this loss increases a little between no load and full load, since the shunt current has to be increased to maintain the terminal voltage constant; but in shunt and compound motors, it remains approximately constant.

(*b*) Losses in series, compole and compensating windings (if any). These losses are proportional to the square of the armature current.

iv. Bearing friction and windage losses	The bearing friction loss is roughly proportional to the speed, but the windage loss, namely the power absorbed in setting up circulating currents of air, is proportional to the cube of the speed. The windage loss is very small unless the machine is fitted with a cooling fan.
v. Stray load loss	It was shown in section 15.5 that the effect of armature reaction is to distort the flux, the flux densities at certain points of the armature being increased; consequently the core loss is also increased. This stray loss is usually neglected as it is difficult to estimate its value.

18.2 Efficiency of a d.c. motor

If R_a = total resistance of armature circuit (including the brush-contact resistance and the resistance of series and compole windings, if any),

I = input current,

I_s = shunt current

and I_a = armature current = $I - I_s$,

then   total loss in armature circuit = $I_a^2 R_a$.

If V = terminal voltage,

loss in shunt circuit = $I_s V$.

This includes the loss in the shunt regulating resistor.
If C = sum of core, friction and windage losses,

$$\text{total losses} = I_a^2 R_a + I_s V + C$$

$$\text{Input power} = IV$$

$\therefore$ output power = $IV - I_a^2 R_a - I_s V - C$

and   efficiency $\eta = \dfrac{IV - I_a^2 R_a - I_s V - C}{IV}$ (18.1)

18.3 Approximate condition for maximum efficiency

Let us assume (*a*) that the shunt current is negligible compared with the armature current at load corresponding to maximum efficiency and (*b*) that the shunt, core and friction losses are independent of the load; then from expression (18.1),

$$\text{efficiency of a motor, } \eta = \frac{IV - I^2 R_\text{a} - I_\text{s} V - C}{IV}$$

$$= \frac{V - I R_\text{a} - \dfrac{1}{I}(I_\text{s} V + C)}{V} \qquad (18.2)$$

This efficiency is a maximum when the numerator of (18.2) is a maximum, namely when:

$$\frac{\text{d}}{\text{d}I} \left\{ V - I R_\text{a} - \frac{1}{I}(I_\text{s} V + C) \right\} = 0$$

i.e.

$$R_\text{a} - \frac{1}{I^2}(I_\text{s} V + C) = 0 \qquad (18.3)$$

$$\therefore \qquad I^2 R_\text{a} = I_\text{s} V + C \qquad (18.4)$$

The condition for the numerator in expression (18.2) to be a maximum, and therefore the efficiency to be a maximum, is that the lefthand side of expression (18.3), when differentiated with respect to I, should be positive; thus,

$$\frac{\text{d}}{\text{d}I} \{ R_\text{a} - I^{-2}(I_\text{s} V + C) \} = 2 I^{-3} (I_\text{s} V + C)$$

Since this quantity is positive, it follows that expression (18.4) represents the condition for maximum efficiency; i.e. the efficiency of the motor is a maximum when the load is such that the variable loss is equal to the constant loss.

Precisely the same conclusion can be derived for a generator.

18.4 Determination of efficiency

By direct measurement of input and output powers

In the case of small machines the output power can be measured by some form of mechanical brake as that shown in fig. 18.2, where a belt (or ropes) on an air- or water-cooled pulley has one end attached via a spring balance S to the floor and has a mass of m kilograms suspended at the other end.

Weight of the suspended mass $= W$

$$\simeq 9.81m \text{ newtons.}$$

If   reading on spring balance $= S$ newtons,

$\therefore$ net pull due to friction $= (W - S)$ newtons.

If $r = $ effective radius of brake, in metres

and $N = $ speed of pulley, in revolutions per minute,

torque due to brake friction $= (W - S)r$ newton metres

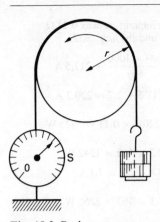

Fig. 18.2 Brake test on a motor

and $\qquad$ output power $= 2\pi(W - S)rN/60$ watts (18.5)

If $\quad V =$ supply voltage, in volts

and $\quad I =$ current taken by motor, in amperes,

$\qquad$ input power $= IV$ watts

and $\qquad$ efficiency $= \dfrac{2\pi(W - S)rN}{60 \times IV} = \dfrac{2\pi(9.81m - S)rN}{60 \times IV}$ (18.6)

The size of machine that can be tested by this method is limited by the difficulty of dissipating the heat generated at the brake.

The principal methods of testing larger machines are the Swinburne's and Hopkinson's Methods. The main advantages of the Swinburne's Method are that the power required is small compared with the machine's rating and the test data enable the efficiency to be determined at any load but the method makes no allowance for stray losses and there is no check on the performance. These disadvantages are overcome by the Hopkinson's Method which has only the drawback that it requires to test two identical machines simultaneously. These methods are too specialized for further consideration in this text.

Example 18.1

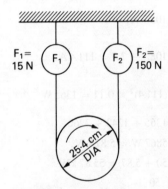

Fig. 18.3

In a load test on a 240-V, d.c. series motor, the following data were recorded at one value of load:

terminal voltage	238 V
line current	6.90 A
belt brake	See fig. 18.3
speed	755 r/min

Evaluate: (i) the torque, (ii) the brake power, (iii) the efficiency. The resistance between terminals is 2.0 Ω, and it has been found for this machine that the windage and friction losses are equal to $(1.8 \times 10^{-4})N^2$ watts, where N is the speed in revolutions per minute. Estimate the core loss for the load.

Torque $T = (F_2 - F_1)r = (150 - 15) \times \dfrac{25.4 \times 10^{-2}}{2}$

$\qquad = 17.15\ \text{N} \cdot \text{m}$

Brake power $P_o = T\omega_r = 17.15 \times \dfrac{2\pi \times 755}{60} = 1355.5\ \text{W}$

Input power $P_i = VI = 238 \times 6.9 = 1642.2\ \text{W}$

Losses $= P_i - P_o = 1642.2 - 1355.5 = 286.7\ \text{W}$

$\qquad = I^2R \text{ loss} + \text{windage and friction loss}$

$\qquad\quad + \text{core loss}$

$\qquad = 6.9^2 \times 2 + (1.8 \times 10^{-4}) \times 755^2 + \text{core loss}$

$\qquad = 95.2 + 102.6 + \text{core loss}$

$\therefore \quad \text{core loss} = 286.7 - 95.2 - 102.6 = 88.9\ \text{W}.$

Example 18.2 *A 100-kW, 460-V shunt generator was run as a motor on no load at its rated voltage and speed. The total current taken was 9.8 A, including a shunt current of 2.7 A. The resistance of the armature circuit*

(including compoles) at normal working temperature was $0.11\,\Omega$.
Calculate the efficiencies at (a) *full load and* (b) *half load.*

(a)   Output current at full load $= \dfrac{100 \times 1000}{460} = 217.5\,\text{A}$,

$\therefore$ armature current at full load $= 217.5 + 2.7 = 220.2\,\text{A}$.

$\left.\begin{array}{l}I^2R \text{ loss in armature} \\ \text{circuit at full load}\end{array}\right\} = (220)^2 \times 0.11 = 5325\,\text{W}$.

Loss in shunt circuit $= 2.7 \times 460 = 1242\,\text{W}$.

Armature current on no load $= 9.8 - 2.7 = 7.1\,\text{A}$,

$\left.\begin{array}{l}\text{input power to armature} \\ \text{on no load}\end{array}\right\} = 7.1 \times 460 = 3265\,\text{W}$.

But $\left.\begin{array}{l}\text{loss in armature} \\ \text{circuit on no load}\end{array}\right\} = (7.1)^2 \times 0.11 = 5.5\,\text{W}$.

This loss is less than 0.2 per cent of the input power and could therefore have been neglected.

Hence, core, friction and windage losses $= 3260\,\text{W}$,

and   total losses at full load $= 5325 + 1242 + 3260$

$= 9827\,\text{W} = 9.83\,\text{kW}$.

Input power at full load $= 100 + 9.83 = 109.8\,\text{kW}$

and   efficiency at full load $= \dfrac{100}{109.8} = 0.911$ per unit*

$= 91.1$ per cent.

(b) $\left.\begin{array}{l}\text{Output current} \\ \text{at half load}\end{array}\right\} = 108.7\,\text{A}$,

$\therefore$ $\left.\begin{array}{l}\text{armature current} \\ \text{at half load}\end{array}\right\} = 108.7 + 2.7 = 111.4\,\text{A}$

and $\left.\begin{array}{l}I^2R \text{ loss in armature} \\ \text{circuit at half load}\end{array}\right\} = (111.4)^2 \times 0.11 = 1365\,\text{W}$

$\therefore$ total losses at half load $= 1365 + 1242 + 3260$

$= 5867\,\text{W} = 5.87\,\text{kW}$.

Input power at half load $= 50 + 5.87 = 55.87\,\text{kW}$

and   efficiency at half load $= \dfrac{50}{55.87} = 0.895$ per unit

$= 89.5$ per cent.

---

* Per-unit value $= \dfrac{\text{actual value (in any unit)}}{\text{basic or rated value (in same unit)}}$

Thus, if the rated voltage of a system is 200 V, a voltage of 190 V is $190/200 = 0.95$ p.u.; or if an ammeter is reading 5.2 A when the correct value is 5 A, the error is $(5.2 - 5)/5 = 0.04$ p.u. It will be obvious that the percentage value is 100 times the per-unit value.

Summary of important formulae

$$\text{Efficiency of motor} = \frac{IV - I_a^2 R_a - I_s V - C}{IV} \qquad (18.2)$$

For maximum efficiency,

$$\text{variable loss} = \text{constant loss}$$

For the brake shown in fig. 18.2,

$$\text{efficiency} = \frac{2\pi(9.81m - S)rN}{60 \times IV} \qquad (18.6)$$

EXERCISES 18

1. The armature of a four-pole d.c. motor has an armature eddy-current loss of 500 W when driven at a given speed and field excitation. If the speed is increased by 15 per cent and the flux is increased by 10 per cent, calculate the new value of the eddy-current loss.

2. A d.c. shunt motor has an output of 8 kW when running at 750 r/min off a 480-V supply. The resistance of the armature circuit is 1.2 Ω and that of the shunt circuit is 800 Ω. The efficiency at that load is 83 per cent. Determine: (*a*) the no-load armature current; (*b*) the speed when the motor takes 12 A; and (*c*) the armature current when the gross torque is 60 N · m. Assume the flux to remain constant.

3. In a brake test on a d.c. motor, the effective load on the brake pulley was 265 N, the effective diameter of the pulley 650 mm and the speed 720 r/min. The motor took 35 A at 220 V. Calculate the output power, in kilowatts, and the efficiency at this load.

4. In a test on a d.c. motor with the brake shown in fig. 18.2, the mass suspended at one end of the belt was 30 kg and the reading on the spring balance was 65 N. The effective diameter of the brake wheel was 400 mm and the speed was 960 r/min. The input to the motor was 23 A at 240 V. Calculate: (*a*) the output power and (*b*) the efficiency.

5. A d.c. shunt machine has an armature resistance of 0.5 Ω and a field-circuit resistance of 750 Ω. When run under test as a motor, with no mechanical load, and with 500 V applied to the terminals, the line current was 3 A. Allowing for a drop of 2 V at the brushes, estimate the efficiency of the machine when it operates as a generator with an output of 20 kW at 500 V, the field-circuit resistance remaining unchanged. State the assumptions made. (S.A.N.C., O.2)

6. Enumerate the various losses which take place in a shunt motor when it is running on no load.

    A 230-V shunt motor, running on no load and at normal speed, takes an armature current of 2.5 A from 230-V mains. The field-circuit resistance is 230 Ω and the armature-circuit resistance is 0.3 Ω. Calculate the motor output, in kilowatts, and the efficiency when the total current taken from the mains is 35 A.

    If the motor is used as a 230-V shunt generator, find the efficiency and input power for an output current of 35 A. (E.M.E.U.)

19 Two- and Three-Phase Circuits

19.1 Disadvantages of the single-phase system

The earliest application of alternating current was for heating the filaments of electric lamps. For this purpose the single-phase system was perfectly satisfactory. Some years later, a.c. motors were developed, and it was found that for this application the single-phase system was not very satisfactory. For instance, the single-phase induction motor — the type most commonly employed — was not self-starting unless it was fitted with an auxiliary winding. By using two separate windings with currents differing in phase by a quarter of a cycle or three windings with currents differing in phase by a third of a cycle, it was found that the induction motor was self-starting and had better efficiency and power factor than the corresponding single-phase machine.

The system utilizing two windings is referred to as a *two-phase system* and that utilizing three windings is referred to as a *three-phase system*. We shall now consider these two systems in detail.

19.2 Generation of two-phase e.m.f.s

In fig. 19.1, AA_1 and BB_1 represent two similar loops fixed to each other at an angle of $90°$ and driven anticlockwise in the magnetic field due to poles NS. For the position shown in fig. 19.1, the e.m.f. generated in AA_1 is zero and that in BB_1 is a maximum. The direction of the e.m.f. generated in BB_1 is indicated by the arrowheads. It follows that while loop AA_1 is moving through half a revolution from the position shown in fig. 19.1, the direction of the e.m.f. generated in AA_1 is from the 'start' terminal towards the 'finish' terminal. Let us regard this direction as positive; consequently, the e.m.f. generated in AA_1 can be represented by curve E_A in fig. 19.2.

Since the loops are rotating anticlockwise, it is evident from fig. 19.1 that the e.m.f. generated in side B has the same magnitude as that in side A but lags by $90°$ and is therefore represented by curve E_B.

If the two phases are kept electrically isolated from each

Fig. 19.1 Generation of two-phase e.m.f.s

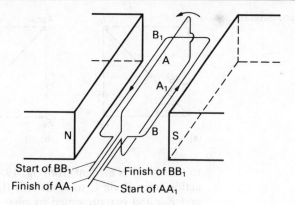

Fig. 19.2 Waveforms of two-phase e.m.f.s

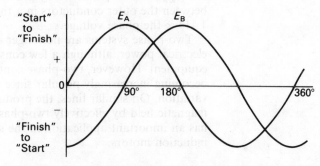

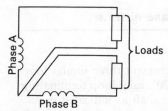

Fig. 19.3 Two-phase four-wire system

other, we have the *two-phase, four-wire system* shown in fig. 19.3. The volume of copper in the conductors between the generator and the load can be reduced by joining the phases at C and using a common return wire CC_1, as shown in fig. 19.4; thus we have the *two-phase, three-wire system*. If the load is balanced, i.e. if the phase currents I_A and I_B are equal and differ in phase by a quarter of a cycle, as in fig. 19.5(*a*), the current I_C in the *common* or *neutral wire* is $\sqrt{2} \times$ the phase current in each *outer conductor*. Hence, if *a* is the cross-sectional area of each conductor of the four-wire system, the total cross-sectional area of the line conductors for this system is $4a$; whereas, for the same current density, the total cross-sectional area for the three-wire system is $(a + a = 1.414a)$, namely $3.414a$.

The phasors $E_{CA}*$ and E_{CB} in fig. 19.5(*b*) represent the r.m.s. values of the e.m.f.s in phases A and B respectively, the e.m.f. being assumed positive when its direction is from 'start'

Fig. 19.4 Two-phase, three-wire system

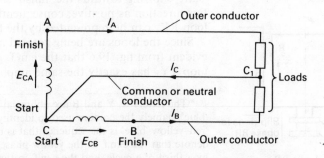

* See note on double subscripts, section 1.19.

Fig. 19.5 Phasor diagram
for fig. 19.4

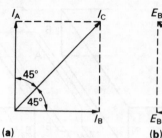
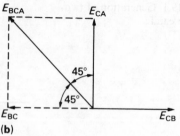

(a)   **(b)**

to 'finish'. It is evident from fig. 19.4 that the resultant e.m.f.
acting from B via C towards A is the phasor difference of E_{CA}
and E_{CB} and is represented by phasor E_{BCA}. Hence the voltage
between the outer conductors in a three-wire system is
1.414 × the phase voltage.

Two-phase systems are no longer used to distribute
electrical power, although a few consumers still use two-phase
equipment. However, two-phase control systems are
becoming increasingly popular since only one phase requires
variation. On similar lines, the production of a rotating
magnetic field by effectively two-phase currents (section 22.1)
has an important application in the starting of single-phase
induction motors.

19.3 Generation of three-phase e.m.f.s

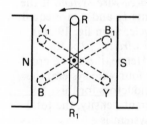

Fig. 19.6 Generation of
three-phase e.m.f.s

In fig. 19.6, RR_1, YY_1 and BB_1* represent three similar loops
fixed to one another at angles of 120°, each loop terminating
in a pair of slip-rings carried on the shaft as indicated in
fig. 19.7. We shall refer to the slip-rings connected to sides R,
Y and B as the 'finishes' of the respective phases and those
connected to R_1, Y_1 and B_1 as the 'starts'.

Suppose the three coils to be rotated anticlockwise at a
uniform speed in the magnetic field due to poles NS. The
e.m.f. generated in loop RR_1 is zero for the position shown in
fig. 19.6. When the loop has moved through 90° to the
position shown in fig. 19.7, the generated e.m.f. is at its
maximum value, its direction round the loop being from the
'start' slip-ring towards the 'finish' slip-ring. Let us regard
this direction as positive; consequently the e.m.f. induced in
loop RR_1 can be represented by the full-line curve of fig. 19.8.

Since the loops are being rotated anticlockwise, it is
evident from fig. 19.6 that the e.m.f. generated in side Y of
loop YY_1 has exactly the same amplitude as that generated in

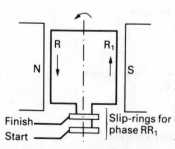

Fig. 19.7 Loop RR_1 at
instant of maximum e.m.f.

* The letters R, Y and B are abbreviations of 'red', 'yellow' and
'blue', namely the colours used to identify the three phases. Also,
'red–yellow–blue' is the sequence that is universally adopted to
denote that the e.m.f. in the yellow phase lags that in the red phase
by a third of a cycle, and the e.m.f. in the blue phase lags that in the
yellow phase by another third of a cycle.

Fig. 19.8 Waveforms of three-phase e.m.f.s

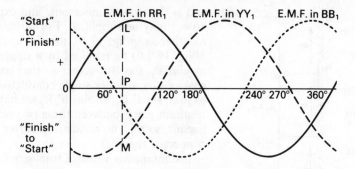

side R, but lags by 120° (or 1/3 cycle). Similarly, the e.m.f. generated in side B of loop BB_1 is equal to but lags that in side Y by 120°. Hence the e.m.f.s generated in loops RR_1, YY_1 and BB_1 are represented by the three equally spaced curves of fig. 19.8, the e.m.f.s being assumed positive when their directions round the loops are from 'start' to 'finish' of their respective loops.

If the instantaneous value of the e.m.f. generated in phase RR_1 is represented by $e_R = E_m \sin \theta$,

then   instantaneous e.m.f. in $YY_1 = e_Y = E_m \sin (\theta - 120°)$

and   instantaneous e.m.f. in $BB_1 = e_B = E_m \sin (\theta - 240°)$.

19.4 Delta connection of three-phase windings

The three phases of fig. 19.6 can, for convenience, be represented as in fig. 19.9, where the phases are shown isolated from one another. L_1, L_2 and L_3 represent loads

Fig. 19.9 Three-phase windings with six line conductors

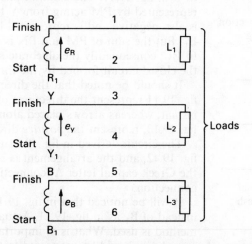

connected across the respective phases. Since we have assumed the e.m.f.s to be positive when acting from 'start' to 'finish', they can be represented by the arrows e_R, e_Y and e_B in fig. 19.9. This arrangement necessitates six line conductors

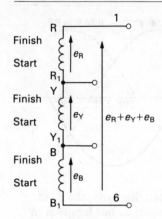

Fig. 19.10 Resultant e.m.f.
in a mesh-connected winding

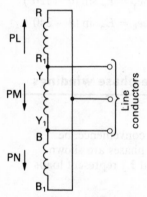

Fig. 19.11 Mesh connection
of three-phase winding

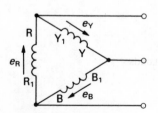

Fig. 19.12 Conventional
representation of a mesh-
connected winding

and is therefore cumbersome and expensive, so let us consider how it may be simplified. For instance, let us join R_1 and Y together as in fig. 19.10, thereby enabling conductors 2 and 3 of fig. 19.9 to be replaced by a single conductor. Similarly, let us join Y_1 and B together so that conductors 4 and 5 may be replaced by another single conductor. Before we can proceed to join 'start' B_1 to 'finish' R, we have to prove that the resultant e.m.f. between these two points is zero at every instant, so that no circulating current is set up when they are connected together.

Instantaneous value of total e.m.f. acting from B_1 to R

$$= e_R + e_Y + e_B$$

$$= E_m\{\sin\theta + \sin(\theta - 120°) + \sin(\theta - 240°)\}$$

$$= E_m(\sin\theta + \sin\theta \cdot \cos 120° - \cos\theta \cdot \sin 120°$$
$$+ \sin\theta \cdot \cos 240° - \cos\theta \cdot \sin 240°)$$

$$= E_m(\sin\theta - 0.5\sin\theta - 0.866\cos\theta$$
$$- 0.5\sin\theta + 0.866\cos\theta)$$

$$= 0.$$

Since this condition holds for every instant, it follows that R and B_1 can be joined together, as in fig. 19.11, without any circulating current being set up around the circuit. The three line conductors are joined to the junctions thus formed.

It might be helpful at this stage to consider the actual values and directions of the e.m.f.s at a particular instant. For instance, at instant P in fig. 19.8, the e.m.f. generated in phase R is positive and is represented by PL acting from R_1 to R in fig. 19.11. The e.m.f. in phase Y is negative and is represented by PM acting from Y to Y_1, and that in phase B is also negative and is represented by PN acting from B to B_1. But the sum of PM and PN is exactly equal numerically to PL; consequently the algebraic sum of the e.m.f.s round the closed circuit formed by the three windings is zero.

It should be noted that the directions of the arrows in fig. 19.11 represent the directions of the e.m.f. at a *particular instant*, whereas arrows placed alongside symbol *e*, as in fig. 19.10, represent the *positive* directions of the e.m.f.s.

The circuit derived in fig. 19.11 is usually drawn as in fig. 19.12, and the arrangement is referred to as *delta* (from the Greek capital letter Δ) connection, also known as a mesh connection.

It will be noticed that in fig. 19.12, R is connected to Y_1 instead of B_1 as in fig. 19.11. Actually, it is immaterial which method is used. What is of importance is that the 'start' of one phase should be connected to the 'finish' of another phase, so that the arrows representing the positive directions of the e.m.f.s point in the same direction round the mesh formed by the three windings.

19.5 Star connection of three-phase windings

Let us go back to fig. 19.9 and join together the three 'starts', R_1, Y_1 and B_1 at N, as in fig. 19.13, so that the three conductors 2, 4 and 6 of fig. 19.9 can be replaced by the single conductor NM of fig. 19.13.

Fig. 19.13 Star connection of three-phase winding

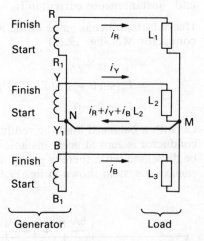

Since the generated e.m.f. has been assumed positive when acting from 'start' to 'finish', the current in each phase must also be regarded as positive when flowing in that direction, as represented by the arrows in fig. 19.13. If i_R, i_Y and i_B be the instantaneous values of the currents in the three phases, the instantaneous value of the current in the common wire MN is $(i_R + i_Y + i_B)$, having its positive direction from M to N.

This arrangement is referred to as a *four-wire star-connected* system and is more conveniently represented as in fig. 19.14; and junction N is referred to as the *star* or *neutral point*. Three-phase motors are connected to the line conductors R, Y and B, whereas lamps, heaters, etc., are usually connected between the line and neutral conductors, as indicated by L_1, L_2 and L_3, the total load being distributed

Fig. 19.14 Four-wire star-connected system

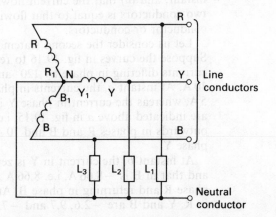

as equally as possible between the three lines. If these three loads are exactly alike, the phase currents have the same peak value, I_m, and differ in phase by 120°. Hence if the instantaneous value of the current in load L_1 be represented by:

$$i_1 = I_m \sin \theta$$

instantaneous current in $L_2 = i_2 = I_m \sin (\theta - 120°)$

and   instantaneous current in $L_3 = i_3 = I_m \sin (\theta - 240°)$.

Hence instantaneous value of the resultant current in neutral conductor MN (fig. 19.13)

$$= i_1 + i_2 + i_3$$
$$= I_m\{\sin \theta + \sin (\theta - 120°) + \sin (\theta - 240°)\}$$
$$= I_m \times 0 = 0,$$

i.e. with a balanced load the resultant current in the neutral conductor is zero at *every* instant; hence this conductor can be dispensed with, thereby giving us the *three-wire star-connected* system shown in fig. 19.15.

Fig. 19.15 Three-wire star connected system with balanced load

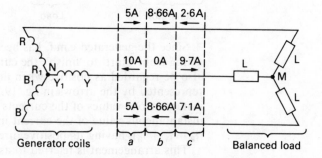

When we are considering the distribution of current in a three-wire three-phase system it is helpful to bear in mind: (*a*) that arrows such as those of fig. 19.13, placed alongside *symbols*, indicate the direction of the current when it is assumed to be *positive* and not the direction at a particular instant; and (*b*) that the current flowing outwards in one or two conductors is equal to that flowing back in the remaining conductor or conductors.

Let us consider the second statement in greater detail. Suppose the curves in fig. 19.16 to represent the three currents differing in phase by 120° and having a peak value of 10 A. At instant *a*, the currents in phases R and B are each 5 A, whereas the current in phase Y is − 10 A. These values are indicated above *a* in fig. 19.15, i.e. 5 A are flowing outwards in phases R and B and 10 A are returning in phase Y.

At instant *b*, the current in Y is zero, that in R is 8.66 A and that in B is −8.66 A, i.e. 8.66 A are flowing outwards in phase R and returning in phase B. At instant *c*, the currents in R, Y and B are −2.6, 9.7 and −7.1 A respectively;

Fig. 19.16 Waveforms of current in a balanced three-phase system

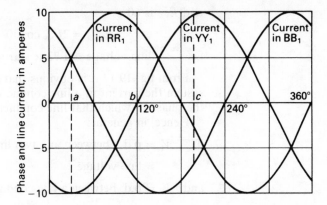

i.e. 9.7 A flow outwards in phase Y and return via phases R (2.6 A) and B (7.1 A).

It will be seen that the distribution of currents between the three lines is continually changing; but at every instant the algebraic sum of the currents is zero.

19.6 Relationships between line and phase voltages and currents in a star-connected system

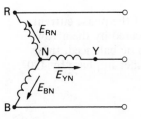

Fig. 19.17 Star-connected generator

Let us again assume the e.m.f. in each phase to be positive when acting from the neutral point outwards, so that the r.m.s. values of the e.m.f.s generated in the three phases can be represented by E_{NR}, E_{NY} and E_{NB} in figs. 19.17 and 19.18.* The value of the e.m.f. acting from Y via N to R is the phasor *difference* of E_{NR} and E_{NY}. Hence E_{YN} is drawn equal and opposite to E_{NY} and added to E_{NR}, giving E_{YNR} as the e.m.f. acting from Y to R via N. Note that the *three* subscript letters YNR are necessary to indicate unambiguously the *positive* direction of this e.m.f.

Having decided on YNR as the positive direction of the line e.m.f. between Y and R, we must adhere to the same sequence for the e.m.f.s between the other lines, i.e. the sequence must be YNR, RNB and BNY. E_{RNB} is obtained by subtracting E_{NR} from E_{NB} and E_{BNY} is obtained by subtracting E_{NB} from E_{NY} as shown in fig. 19.18. From the symmetry of this diagram it is evident that the line voltages are equal and are spaced 120° apart. Further, since the sides of all the parallelograms are of equal length, the diagonals bisect one another at right-angles. Also, they bisect the angles of their respective parallelograms; and, since the angle between E_{NR}

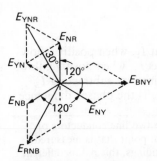

Fig. 19.18 Phasor diagram for fig. 19.17

* When the relationships between line and phase quantities are being derived for either the star- or the delta-connected system, it is essential to relate the phasor diagram to a circuit diagram and to indicate on each phase the direction in which the voltage or current is assumed to be positive. A phasor diagram by itself is meaningless.

and E_{YN} is $60°$,

$$\therefore \qquad E_{YNR} = 2E_{NR} \cos 30° = \sqrt{3}E_{NR}$$

i.e.   line voltage $= 1.73 \times$ star (or phase) voltage.

From fig. 19.17 it is obvious that in a star-connected system the current in a line conductor is the same as that in the phase to which that line conductor is connected.

Hence, in general:

if $V_L = $ p.d.* between any two line conductors

$= $ line voltage

and $V_P = $ p.d. between a line conductor and the neutral point

$= $ star voltage (or voltage to neutral)

and if I_L and $I_P = $ line and phase currents respectively, then for a star-connected system,

$$V_L = 1.73 \, V_P \tag{19.1}$$

and

$$I_L = I_P \tag{19.2}$$

19.7 Relationships between line and phase voltages and currents in a delta-connected system with a balanced load

Let I_1, I_2 and I_3 be the r.m.s. values of the phase currents having their positive directions as indicated by the arrows in fig. 19.19. Since the load is assumed to be balanced, these currents are equal in magnitude and differ in phase by $120°$, as shown in fig. 19.20.

Fig. 19.19 Delta-connected system with balanced load

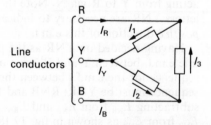

From fig. 19.19 it will be seen that I_1, when positive, flows away from line conductor R, whereas I_3, when positive, flows towards it. Consequently, I_R is obtained by subtracting I_3 from I_1, as in fig. 19.20. Similarly, I_Y is the phasor difference

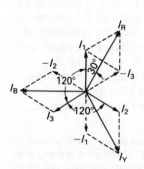

Fig. 19.20 Phasor diagram for fig. 19.19

---

* In practice, it is the voltage between two line conductors or between a line conductor and the neutral point that is measured. Owing to the impedance drop in the windings, this p.d. is different from the corresponding e.m.f. generated in the winding, except when the generator is on open circuit; hence, in general, it is preferable to work with the potential difference, V, rather than with the e.m.f., E.

of I_2 and I_1, and I_B is the phasor difference of I_3 and I_2. From fig. 19.20 it is evident that the line currents are equal in magnitude and differ in phase by 120°. Also,

$$I_R = 2I_1 \cos 30° = \sqrt{3}I_1.$$

Hence for a delta-connected system with a balanced load,

line current = 1.73 × phase current

i.e.

$$I_L = 1.73\, I_P \qquad (19.3)$$

From fig. 19.19 it is obvious that in a delta-connected system, the line and the phase voltages are the same, i.e.

$$V_L = V_P \qquad (19.4)$$

Example 19.1   *In a three-phase four-wire system the line voltage is 415 V and non-inductive loads of 10 kW, 8 kW and 5 kW are connected between the three line conductors and the neutral as in fig. 19.21. Calculate: (a) the current in each line, and (b) the current in the neutral conductor.*

Fig. 19.21 Circuit diagram for example 19.1

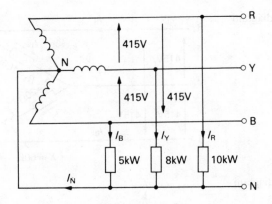

(a)   Voltage to neutral $= \dfrac{\text{line voltage}}{1.73} = \dfrac{415}{1.73} = 240 \text{ V}.$

If I_R, I_Y and I_B be the currents taken by the 10-kW, 8-kW and 5-kW loads respectively,

$$I_R = 10 \times 1000/240 = 41.67 \text{ A},$$
$$I_Y = 8 \times 1000/240 = 33.33 \text{ A}$$

and

$$I_B = 5 \times 1000/240 = 20.83 \text{ A}.$$

These currents are represented by the respective phasors in fig. 19.22.

(b) The current in the neutral is the phasor sum of the three line currents. In general, the most convenient method of adding such quantities is to calculate the resultant horizontal and vertical components thus:

horizontal component $= I_H = I_Y \cos 30° - I_B \cos 30°$
$$= 0.866(33.33 - 20.83) = 10.83 \text{ A}.$$

and   vertical component $= I_V = I_R - I_Y \cos 60° - I_B \cos 60°$
$$= 41.67 - 0.5(33.33 - 20.83) = 14.59 \text{ A}.$$

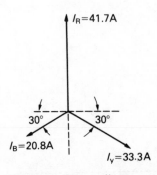

Fig. 19.22 Phasor diagram for fig. 19.21

These components are represented in fig. 19.23.

$$\therefore \quad \text{current in neutral} = I_N = \sqrt{\{(10.83)^2 + (14.59)^2\}}$$
$$= 18.2 \text{ A}.$$

Example 19.2 *A delta-connected load is arranged as in fig. 19.24. Calculate: (a) the phase currents, and (b) the line currents. The supply voltage* is 415 V at 50 Hz.*

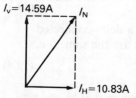

$I_v = 14.59\text{A}$ I_N

$I_H = 10.83\text{A}$

Fig. 19.23 Vertical and horizontal components of I_N

(*a*) Since the phase sequence is R, Y, B, the voltage having its positive direction from R to Y leads 120° on that having its positive direction from Y to B, i.e. V_{RY} is 120° in front of V_{YB} (see section 1.19).

Similarly, V_{YB} is 120° in front of V_{BR}. Hence the phasors representing the line (and phase) voltages are as shown in fig. 19.25.

If I_1, I_2, I_3 be the phase currents in loads RY, YB and BR respectively:

$$I_1 = 415/100 = 4.15 \text{ A}, \quad \text{in phase with } V_{RY}.$$

$$I_2 = \frac{415}{\sqrt{(20^2 + 60^2)}} = \frac{415}{63.25} = 6.56 \text{ A}.$$

Fig. 19.24 Circuit diagram for example 19.2

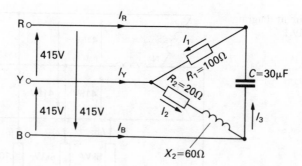

Fig. 19.25 Phasor diagram for fig. 19.24

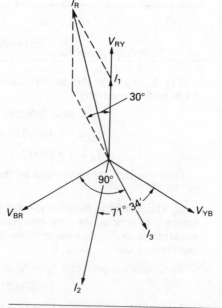

---

* The voltage given for a three-phase system is always the line voltage unless it is stated otherwise.

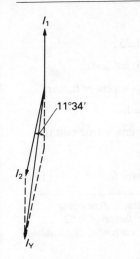

11°34′

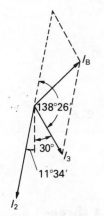

Fig. 19.26 Phasor diagram for deriving I_Y

Fig. 19.27 Phasor diagram for deriving I_B

I_2 lags V_{YB} by an angle ϕ_2 such that

$$\phi_2 = \tan^{-1} 60/20 = 71° 34'.$$

Also

$$I_3 = 2 \times 3.14 \times 50 \times 30 \times 10^{-6} \times 415$$
$$= 3.91 \, \text{A}, \quad \text{leading } V_{BR} \text{ by } 90°.$$

(b) If the current I_R in line conductor R be assumed to be positive when flowing towards the load, the phasor representing this current is obtained by subtracting I_3 from I_1, as in fig. 19.25.

$$\therefore \quad I_R^2 = (4.15)^2 + (3.91)^2 + 2 \times 4.15 \times 3.91 \cos 30° = 60.53$$

$$\therefore \quad I_R = 7.78 \, \text{A}.$$

Current in line conductor Y is obtained by subtracting I_1 from I_2, as shown separately in fig. 19.26.

But angle between I_2 and I_1 reversed $= \phi_2 - 60° = 71° 34' - 60°$

$$= 11° 34'.$$

$$\therefore \quad I_Y^2 = (4.15)^2 + (6.56)^2 + 2 \times 4.15 \times 6.56 \times \cos 11° 34' = 113.6$$

$$\therefore \quad I_Y = 10.66 \, \text{A}.$$

Similarly, the current in line conductor B is obtained by subtracting I_2 from I_3, as shown in fig. 19.27.

Angle between I_3 and I_2 reversed $= 180° - 30° - 11° 34'$

$$= 138° 26'.$$

$$\therefore \quad I_B^2 = (6.56)^2 + (3.91)^2 + 2 \times 6.56 \times 3.91 \times \cos 138° 26'$$

$$= 19.94$$

$$\therefore \quad I_B = 4.47 \, \text{A}.$$

This problem could be solved graphically, but in that case it would be necessary to draw the phasors to a large scale to ensure reasonable accuracy.

19.8 Power in a three-phase system with a balanced load

If $\quad I_P = $ r.m.s. value of the current in each phase

and $\quad V_P = $ r.m.s. value of the p.d. across each phase,

active power per phase $= I_P V_P \times$ power factor

and total active power $= 3 I_P V_P \times$ power factor (19.5)

If I_L and V_L be the r.m.s. values of the line current and voltage respectively, then for a *star-connected system*,

$$V_P = V_L/1.73 \quad \text{and} \quad I_P = I_L.$$

Substituting for I_P and V_P in (19.5), we have:

total active power in watts $= 1.73 \, I_L V_L \times$ power factor.

For a *delta-connected system,*

$$V_P = V_L \quad \text{and} \quad I_P = I_L/1.73,$$

Again, substituting for I_P and V_P in (19.5), we have:

total active power in watts = $1.73\, I_L V_L$ × power factor.

Hence it follows that for any balanced load,

active power in watts = 1.73 × line current × line voltage

× power factor

$$= 1.73\, I_L V_L \times \text{power factor} \tag{19.6}$$

Example 19.3
A three-phase motor operating off a 415-V system is developing 20 kW at an efficiency of 0.87 p.u. and a power factor of 0.82. Calculate: (a) the line current and (b) the phase current if the windings are delta-connected.

(a) Since

$$\text{efficiency} = \frac{\text{output power in watts}}{\text{input power in watts}}$$

$$= \frac{\text{output power in watts}}{1.73\, I_L V_L \times \text{p.f.}}$$

$$\therefore \qquad 0.87 = \frac{20 \times 1000}{1.73 \times I_L \times 415 \times 0.82}$$

and

$$\text{line current} = I_L = 39\ \text{A}.$$

(b) For a delta-connected winding,

$$\text{phase current} = \text{line current}/1.73 = 39/1.73 = 22.6\ \text{A}.$$

19.9 Measurement of active power in a three-phase three-wire system

Case (a). Star-connected balanced load, with neutral point accessible

If a wattmeter W be connected with its current coil in one line and the voltage circuit between that line and the neutral point, as shown in fig. 19.28, the reading on the wattmeter gives the power per phase:

$$\therefore \qquad \text{total active power} = 3 \times \text{wattmeter reading}.$$

Case (b). Balanced or unbalanced load, star- or delta-connected. The two-wattmeter method

Suppose the three loads L_1, L_2 and L_3 to be connected in star, as in fig. 19.29. The current coils of the two wattmeters are connected in any two lines, say the 'red' and 'blue' lines, and the voltage circuits are connected between these lines and the third line.

Suppose v_{RN}, v_{YN} and v_{BN} to be the instantaneous values of the p.d.s across the loads, these p.d.s being assumed positive when the respective line conductors are positive in relation to the neutral point. Also, suppose i_R, i_Y and i_B to be the corresponding instantaneous values of the line (and phase)

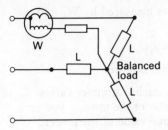

Fig. 19.28 Measurement of active power in a star-connected balanced load

currents.

$\therefore$ instantaneous power in load $L_1 = i_R v_{RN}$

instantaneous power in load $L_2 = i_Y v_{YN}$

and instantaneous power in load $L_3 = i_B v_{BN}$

$\therefore$ total instantaneous power $= i_R v_{RN} + i_Y v_{YN} + i_B v_{BN}$.

From fig. 19.29 it is seen that:

instantaneous current through current coil of $W_1 = i_R$

and $\left.\begin{array}{c}\text{instantaneous p.d. across}\\ \text{voltage circuit of } W_1\end{array}\right\} = v_{RN} - v_{YN}$

$\therefore$ instantaneous power measured by $W_1 = i_R(v_{RN} - v_{YN})$.

Fig. 19.29 Measurement of power by two wattmeters

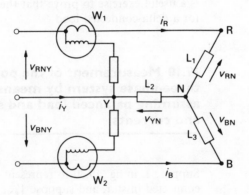

Similarly,

instantaneous current through current coil of $W_2 = i_B$

and $\left.\begin{array}{c}\text{instantaneous p.d.* across}\\ \text{voltage circuit of } W_2\end{array}\right\} = v_{BN} - v_{YN}$

$\therefore$ instantaneous power measured by $W_2 = i_B(v_{BN} - v_{YN})$.

Hence the sum of the instantaneous powers of W_1 and W_2

$$= i_R(v_{RN} - v_{YN}) + i_B(v_{BN} - v_{YN})$$
$$= i_R v_{RN} + i_B v_{BN} - (i_R + i_B)v_{YN}.$$

From Kirchhoff's First Law (section 1.14), the algebraic sum of the instantaneous currents at N is zero,

i.e. $i_R + i_Y + i_B = 0$

$\therefore$ $i_R + i_B = -i_Y$

---

* It is important to note that this p.d. is not $v_{YN} - v_{BN}$. This is due to the fact that a wattmeter reads positively when the currents in the current and voltage coils are *both* flowing from the junction of these coils or *both* towards that junction; and since the positive direction of the current in the current coil of W_2 has already been taken as that of the arrowhead alongside i_B in fig. 19.29, it follows that the current in the voltage circuit of W_2 is positive when flowing from the 'blue' to the 'yellow' line.

so that sum of instantaneous powers measured by W_1 and W_2

$$= i_R v_{RN} + i_B v_{BN} + i_Y v_{YN}$$

$$= \text{total instantaneous power.}$$

Actually, the power measured by each wattmeter varies from instant to instant, but the inertia of the moving system causes the pointer to read the average value of the power. Hence the sum of the wattmeter readings gives the average value of the total power absorbed by the three phases, i.e. the active power.

Since the above proof does not assume a balanced load or sinusoidal waveforms, it follows that the sum of the two wattmeter readings gives the total power under all conditions. The above proof was derived for a star-connected load, and it is a useful exercise to prove that the same conclusion holds for a delta-connected load.

19.10 Measurement of the power factor of a three-phase system by means of two wattmeters, assuming balanced load and sinusoidal voltages and currents

Suppose L in fig. 19.30 to represent three similar loads connected in star, and suppose V_{RN}, V_{YN} and V_{BN} to be the r.m.s. values of the phase voltages and I_R, I_Y and I_B to be the r.m.s. values of the currents. Since these voltages and currents

Fig. 19.30 Measurement of active power and power factor by two wattmeters

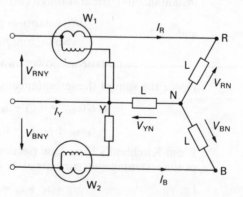

are assumed sinusoidal, they can be represented by phasors, as in fig. 19.31, the currents being assumed to lag the corresponding phase voltages by an angle ϕ.

Current through current coil of $W_1 = I_R$.

P.D. across voltage circuit of W_1 = phasor difference of

$$V_{RN} \text{ and } V_{YN}$$

$$= V_{RNY} \text{ (see section 1.19).}$$

Fig. 19.31 Phasor diagram
for fig. 19.30

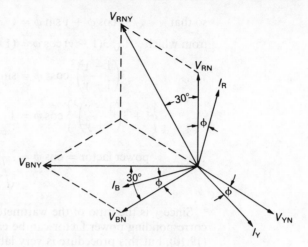

Phase difference between I_R and $V_{RNY} = 30° + \phi$

$\therefore$ reading on $W_1 = P_1 = I_R V_{RNY} \cos(30° + \phi)$.

Current through current coil of $W_2 = I_B$.

P.D. across voltage circuit of W_2 = phasor difference of

$$V_{BN} \text{ and } V_{YN}$$

$$= V_{BNY}.$$

Phase difference between I_B and $V_{BNY} = 30° - \phi$

$\therefore$ reading on $W_2 = P_2 = I_B V_{BNY} \cos(30° - \phi)$.

Since the load is balanced,

$$I_R = I_Y = I_B = \text{(say)}\, I_L, \quad \text{numerically}$$

and $\qquad V_{RNY} = V_{BNY} = \text{(say)}\, V_L, \quad \text{numerically}.$

Hence $\qquad P_1 = I_L V_L \cos(30° + \phi) \qquad\qquad (19.7)$

and $\qquad P_2 = I_L V_L \cos(30° - \phi) \qquad\qquad (19.8)$

$\therefore \quad P_1 + P_2 = I_L V_L \{\cos(30° + \phi) + \cos(30° - \phi)\}$

$$= I_L V_L (\cos 30° \cdot \cos \phi - \sin 30° \cdot \sin \phi$$

$$+ \cos 30° \cdot \cos \phi + \sin 30° \cdot \sin \phi)$$

$$= 1.73\, I_L V_L \cos \phi \qquad\qquad (19.9)$$

namely the expression deduced in section 19.8 for the total
active power in a balanced three-phase system. This is an
alternative method of proving that the sum of the two
wattmeter readings gives the total active power, but it should
be noted that this proof assumed a balanced load and
sinusoidal voltages and currents.

Dividing (19.7) by (19.8), we have:

$$\frac{P_1}{P_2} = \frac{\cos(30° + \phi)}{\cos(30° - \phi)} = \text{(say)}\, y$$

$$\therefore \quad y = \frac{(\sqrt{3}/2)\cos\phi - (1/2)\sin\phi}{(\sqrt{3}/2)\cos\phi + (1/2)\sin\phi}$$

so that $\qquad \sqrt{3}y\cos\phi + y\sin\phi = \sqrt{3}\cos\phi - \sin\phi$

from which $\qquad \sqrt{3}(1-y)\cos\phi = (1+y)\sin\phi$

$$\therefore \qquad 3\left(\frac{1-y}{1+y}\right)^2 \cos^2\phi = \sin^2\phi = 1 - \cos^2\phi$$

$$\left\{1 + 3\left(\frac{1-y}{1+y}\right)^2\right\}\cos\phi = 1$$

$$\therefore \qquad \text{power factor} = \cos\phi = \frac{1}{\sqrt{\left\{1 + 3\left(\dfrac{1-y}{1+y}\right)^2\right\}}} \qquad (19.10)$$

Since y is the ratio of the wattmeter readings, the corresponding power factor can be calculated from expression (19.10), but this procedure is very laborious. A more convenient method is to draw a graph of the power factor for various ratios of P_1/P_2; and in order that these ratios may lie between $+1$ and -1, as in fig. 19.32, it is always the practice to take P_1 as the smaller of the two readings. By adopting this practice, it is possible to derive reasonably accurate values of power factor from the graph.

Fig. 19.32 Relationship between power factor and ratio of wattmeter readings

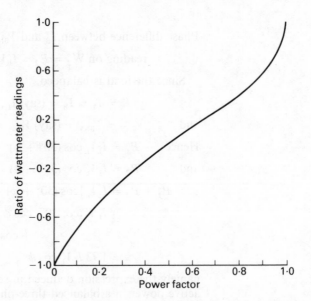

When the power factor of the load is 0.5 lagging, ϕ is $60°$; and from (19.7), the reading on $W_1 = I_L V_L \cos 90° = 0$. When the power factor is less than 0.5 lagging, ϕ is greater than $60°$ and $(30° + \phi)$ is therefore greater than $90°$. Hence the reading on W_1 is negative. To measure this active power it is necessary to reverse the connections to either the current or the voltage coil, but the reading thus obtained must be taken as negative when the total active power and the ratio of the wattmeter readings are being calculated.

An alternative method of deriving the power factor is as follows:

From (19.7), (19.8) and (19.9), $P_2 - P_1 = I_L V_L \sin \phi$

and $\qquad \tan \phi = \dfrac{\sin \phi}{\cos \phi} = 1.73\left(\dfrac{P_2 - P_1}{P_2 + P_1}\right)$ (19.11)

Hence, ϕ and $\cos \phi$ can be determined with the aid of trigonometrical tables.

Example 19.4   *The input power to a three-phase motor was measured by the two-wattmeter method. The readings were 5.2 kW and − 1.7 kW, and the line voltage was 415 V. Calculate: (a) the total active power; (b) the power factor; and (c) the line current.*

(*a*) Total power $= 5.2 - 1.7 = 3.5 \, \text{kW}$.

(*b*) Ratio of wattmeter readings $= -1.7/5.2 = -0.327$.

From fig. 19.32, power factor $= 0.28$.
Or alternatively, from (19.11),

$$\tan \phi = 1.73\left\{\frac{5.2 - (-1.7)}{5.2 + (-1.7)}\right\} = 3.41$$

$$\therefore \qquad \phi = 73° \, 39'$$

and $\qquad$ power factor $= \cos \phi = 0.281$.

From the data it is impossible to state whether the power factor is lagging or leading.

(*c*) From (19.6), $\quad 3500 = 1.73 \times I_L \times 415 \times 0.281$

$$\therefore \qquad I_L = 17.3 \, \text{A}.$$

Summary of important formulae

For a star-connected system,

$$V_L = 1.73 \, V_P \tag{19.1}$$

and $\qquad\qquad\qquad I_L = I_P$ (19.2)

For delta-connected system with balanced load,

$$V_L = V_P \tag{19.4}$$

and $\qquad\qquad\qquad I_L = 1.73 \, I_P$ (19.3)

For star- or delta-connected system with balanced load,

total active power $= 1.73 \times$ line current $\times$ line voltage

$\times$ power factor

$= 1.73 \, I_L V_L \times$ power factor (19.6)

If P_1 and P_2 be the readings obtained with the two-wattmeter method on a three-wire, three-phase system,

total active power $= P_1 + P_2$ under all conditions,

and for a balanced load and sinusoidal waveforms,

$$\text{power factor} = \frac{1}{\sqrt{\left\{1 + 3\left(\dfrac{P_2 - P_1}{P_2 + P_1}\right)^2\right\}}} \tag{19.10}$$

and

$$\tan\phi = 1.73\left(\frac{P_2 - P_1}{P_2 + P_1}\right) \tag{19.11}$$

EXERCISES 19

1. In a two-phase, three-wire system, the phase voltage is 3000 V. Calculate the voltage between the outers.

   If the current in each phase is 50 A and the two currents are in quadrature, what is the value of the current in the common wire?

2. The currents in the two phases of a two-phase, three-wire system are 40 A and 70 A and are in phase with their respective voltages. Calculate the current in the neutral conductor.

3. The two-phase generator of fig. 19.4 has a terminal voltage of 230 V per phase and the frequency is 50 Hz. A coil having a resistance of 20 Ω and an inductance of 0.1 H is connected across phase A, and a 70-μF capacitor is connected across phase B. Assuming the voltage of phase B to lag that of phase A by a quarter of a cycle, calculate the current in each outer conductor and that in the neutral conductor. What is the voltage between the outer conductors?

4. Deduce the relationship between the phase and the line voltages of a three-phase star-connected generator.

   If the phase voltage of a three-phase star-connected generator be 200 V, what will be the line voltages: (a) when the phases are correctly connected; (b) when the connections to one of the phases are reversed?   (App. El., L.U.)

5. Show with the aid of a phasor diagram that for both star-connected and delta-connected balanced loads, the total active power is given by $\sqrt{3}VI\cos\phi$, where V and I are the line values of voltage and current respectively and ϕ is the angle between phase values of voltage and current.

   A balanced three-phase load consists of three coils, each of resistance 4 Ω and inductance 0.02 H. Determine the total active power when the coils are (a) star-connected, (b) delta-connected to a 415-V, three-phase, 50-Hz supply.   (E.M.E.U., O.2)

6. Derive, for both star- and delta-connected systems, an expression for the total power input for a balanced three-phase load in terms of line voltage, line current and power factor.

   The star-connected secondary of a transformer supplies a delta-connected motor taking a power of 90 kW at a lagging power factor of 0.9. If the voltage between lines is 600 V, calculate the current in the transformer winding and in the motor winding. Draw circuit and phasor diagrams, properly labelled, showing all voltages and currents in the transformer secondary and the motor.   (U.L.C.I., O.2)

7. A three-phase delta-connected load, each phase of which has an inductive reactance of 40 Ω and a resistance of 25 Ω, is fed from the secondary of a three-phase star-connected transformer, which has a phase voltage of 240 V. Draw the circuit diagram of the system and calculate: (a) the current in each phase of the load; (b) the p.d. across each phase of the load; (c) the current

in the transformer secondary windings; (*d*) the total active power taken from the supply and its power factor.

(N.C.T.E.C., O.2)

8. Three similar coils, connected in star, take a total power of 1.5 kW, at a power factor of 0.2, from a three-phase, 415-V, 50-Hz supply. Calculate: (*a*) the resistance and inductance of each coil, and (*b*) the line currents if one of the coils is short-circuited.

9. (*a*) Three 20-μF capacitors are star-connected across a 415-V, 50-Hz, three-phase, three-wire supply. Calculate the current in each line.

   (*b*) If one of the capacitors is short-circuited, calculate the line currents.

   (*c*) If one of the capacitors is open-circuited, calculate (i) the line currents and (ii) the p.d. across each of the other two capacitors.

10. (*a*) Explain the advantages of three-phase supply for distribution purposes.

   (*b*) Assuming the relationship between the line and phase values of currents and voltages, show that the active power input to a three-phase balanced load is $\sqrt{3}VI\cos\phi$, where V and I are line quantities.

   (*c*) Three similar inductors, each of resistance 10 Ω and inductance 0.019 H, are delta-connected to a three-phase, 415-V, 50-Hz sinusoidal supply. Calculate: (i) the value of the line current; (ii) the power factor; (iii) the active power input to the circuit.

(W.J.E.C., O.2)

11. A three-phase, 415-V, star-connected motor has an output of 50 kW, with an efficiency of 90 per cent and a power factor of 0.85. Calculate the line current. Sketch a phasor diagram showing the voltages and currents.

   If the motor windings were connected in mesh, what would be the correct voltage of a three-phase supply suitable for the motor?

12. Discuss the importance of power-factor correction in a.c. systems.

   A 415-V, 50-Hz, three-phase distribution system supplies a 20-kVA, three-phase induction motor load at a power factor of 0.8 lagging, and a star-connected set of impedances, each having a resistance of 10 Ω and an inductive reactance of 8 Ω. Calculate the capacitance of delta-connected capacitors required to improve the overall power factor to 0.95 lagging.

(E.M.E.U., O.2)

13. State the advantages to be gained by raising the power factor of industrial loads.

   A 415-V, 50-Hz three-phase motor takes a line current of 15.0 A when operating at a lagging power factor of 0.65. When a capacitor bank is connected across the motor terminals, the line current is reduced to 11.5 A. Calculate the rating (in kvar) and the capacitance per phase of the capacitor bank for: (*a*) star connection; (*b*) delta connection.

   Find also the new overall power factor.

(S.A.N.C., O.2)

14. Three coils, each having a resistance of 20 Ω and a reactance of 15 Ω, are connected in star to a 415-V, three-phase, 50-Hz supply. Calculate: (i) the line current; (ii) the active power supplied; (iii) the power factor.

   If three capacitors, each of the same capacitance, are connected in delta to the same supply so as to form a parallel network with the above coils, calculate: (i) the capacitance of each capacitor to obtain a resultant power factor of

0.95 lagging; (ii) the line current taken by the combined circuits. Draw to scale a phasor diagram showing: (*a*) the line voltage; (*b*) the voltage across each of the coils; (*c*) the line current taken by the combined circuits; (*d*) the current in each capacitor; (*e*) the current in each coil. (App. El., L.U.)

15. Derive the numerical relationship between the line and phase currents for a balanced three-phase delta-connected load.

   Three coils are connected in delta to a three-phase, three-wire, 415-V, 50-Hz supply and take a line current of 5 A at 0.8 power factor lagging. Calculate the resistance and inductance of the coils.

   If the coils are star-connected to the same supply, calculate the line current and the total power.

   Calculate the line currents if one coil becomes open-circuited when the coils are connected in star. (E.M.E.U.)

16. The load connected to a three-phase supply comprises three similar coils connected in star. The line currents are 25 A and the apparent and active power inputs are 20 kVA and 11 kVA respectively. Find the line and phase voltages, active power input and the resistance and reactance of each coil.

   If the coils are now connected in delta to the same three-phase supply, calculate the line currents and the active power taken. (U.L.C.I.)

17. Non-reactive loads of 10, 6 and 4 kW are connected between the neutral and the red, yellow and blue phases respectively of a three-phase, four-wire system. The line voltage is 415 V. Find the current in each line conductor and in the neutral.
(App. El., L.U.)

18. Explain the advantage of connecting the low-voltage winding of distribution transformers in star.

   A factory has the following load with power factor of 0.9 lagging in each phase. Red phase 40 A, yellow phase 50 A and blue phase 60 A. If the supply is 415-V, three-phase, four-wire, calculate the current in the neutral and the total active power. Draw a phasor diagram for phase and line quantities. Assume that, relative to the current in the red phase, the current in the yellow phase lags by 120° and that in the blue phase leads by 120°.

19. A three-phase, 415-V system has the following load connected in mesh: between the red and yellow lines, a non-reactive resistor of 100 Ω; between the yellow and blue lines, a coil having a reactance of 60 Ω and negligible resistance; between the blue and red lines, a loss-free capacitor having a reactance of 130 Ω. Calculate: (*a*) the phase currents; (*b*) the line currents. Assume the phase sequence to be R–Y, Y–B and B–R. Also, draw the complete phasor diagram.

20. The phase currents in a delta-connected three-phase load are as follows: between the red and yellow lines, 30 A at p.f. 0.707 leading; between the yellow and blue lines, 20 A at unity p.f.; between the blue and red lines, 25 A at p.f. 0.866 lagging. Calculate the line currents and draw the complete phasor diagram.

21. (*a*) If, in a laboratory test, you were required to measure the total power taken by a three-phase balanced load, show how to do this, using two wattmeters. Explain the principle of the method.

   Draw the phasor diagram for the balanced-load case with a lagging power factor and use this to explain why the two wattmeter readings differ.

   (*b*) The load taken by a three-phase induction motor was

measured by the two-wattmeter method, and the readings were 860 W and 240 W. What is the active power taken by the motor and at what power factor is it working? (W.J.E.C., O.2)

22. With the aid of a circuit diagram, show that two wattmeters can be connected to read the total power in a three-phase, three-wire system.

    Two wattmeters connected to read the total power in a three-phase system supplying a balanced load read 10.5 kW and −2.5 kW respectively. Calculate the total active power.

    Drawing suitable phasor diagrams, explain the significance of: (a) equal wattmeter readings, and (b) a zero reading on one wattmeter. (U.L.C.I., O.2)

23. Two wattmeters are used to measure power in a three-phase, three-wire network. Show by means of connection and complexor (phasor) diagrams that the sum of the wattmeter readings will measure the total active power.

    Two such wattmeters read 120 W and 50 W when connected to measure the active power taken by a balanced three-phase load. Find the power factor of the load.

    If one wattmeter tends to read in the reverse direction, explain what changes may have occurred in the circuit. (S.A.N.C., O.2)

24. Each branch of a three-phase star-connected load consists of a coil of resistance 4.2 Ω and reactance 5.6 Ω. The load is supplied at a line voltage of 415 V, 50 Hz. The total active power supplied to the load is measured by the two-wattmeter method. Draw a circuit diagram of the wattmeter connections and calculate their separate readings.

    Derive any formula used in your calculations. (S.A.N.C., O.2)

25. Three non-reactive loads are connected in delta across a three-phase, three-wire, 415-V supply in the following way: (i) 10 kW across R and Y lines; (ii) 6 kW across Y and B lines; (iii) 4 kW across B and R lines.

    Draw a phasor diagram showing the three line voltages and the load currents and determine: (a) the current in the B line and its phase relationship to the line voltage V_{BR}; (b) the reading of a wattmeter whose current coils are connected in the B line and whose voltage circuit is connected across the B and R lines. The phase rotation is RYB.

    Where would a second wattmeter be connected for the two-wattmeter method and what would be its reading?

26. Two wattmeters connected to measure the input to a balanced three-phase circuit indicate 2500 W and 500 W respectively. Find the power factor of the circuit: (a) when both readings are positive, and (b) when the latter reading is obtained after reversing the connections to the current-coil of one instrument. Draw the phasor and connection diagrams.

27. With the aid of a phasor diagram show that the active power and power factor of a balanced three-phase load can be measured by two wattmeters.

    For a certain load, one wattmeter indicated 20 kW and the other 5 kW after the voltage circuit of this wattmeter had been reversed. Calculate the active power and the power factor of the load. (App. El., L.U.)

28. The current coil of a wattmeter is connected in the red line of a three-phase system. The voltage circuit can be connected between the red line and either the yellow line or the blue line by means of a two-way switch. Assuming the load to be balanced, show with the aid of a phasor diagram that the sum

of the wattmeter indications obtained with the voltage circuit connected to the yellow and the blue lines respectively gives the total active power.

29. A single wattmeter is used to measure the total active power taken by a 415-V, three-phase induction motor. When the output power of the motor is 15 kW, the efficiency is 88 per cent and the power factor is 0.84 lagging. The current coil of the wattmeter is connected in the yellow line. With the aid of a phasor diagram, calculate the wattmeter indication when the voltage circuit is connected between the yellow line and (a) the red line, (b) the blue line. Show that the sum of the two wattmeter indications gives the total active power taken by the motor. Assume the phase sequence to be R–Y–B.

30. A wattmeter has its current coil connected in the yellow line, and its voltage circuit is connected between the red and blue lines. The line voltage is 415 V and the balanced load takes a line current of 30 A at a power factor of 0.7 lagging. Draw circuit and phasor diagrams and derive an expression for the reading on the wattmeter in terms of the line voltage and current and of the phase difference between the phase voltage and current. Calculate the value of the wattmeter indication.

20 Transformers

20.1 Introductory

One of the main advantages of a.c. transmission and distribution is the ease with which an alternating voltage can be increased or reduced. For instance, the general practice is to generate at voltages of about 11–22 kV, then step up by means of transformers to higher voltages, for the transmission lines. At suitable points other transformers are installed to step the voltage down to values suitable for motors, lamps, heaters, etc. A medium-size transformer has a full-load efficiency of about 97–98 per cent, so that the loss at each point of transformation is very small. Also, since there are no moving parts, the amount of supervision required is practically negligible.

20.2 Principle of action of a transformer

Fig. 20.1 shows the general arrangement of a transformer. A steel core C consists of laminated sheets, about 0.35 mm thick, insulated from one another by thin layers of paper or varnish or by spraying the laminations with a mixture of flour, chalk and water which, when dried, adheres to the metal. The purpose of laminating the core is to reduce the loss due to eddy currents induced by the alternating magnetic flux (section 18.1(b)). The vertical portions of the core are referred to as the *limbs* and the top and bottom portions are

Fig. 20.1 A transformer

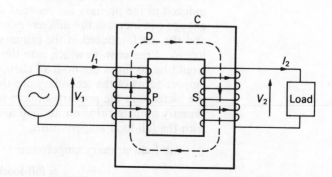

the *yokes*. Coils P and S are wound on the limbs. Coil P is connected to the supply and is therefore termed the *primary*; coil S is connected to the load and is termed the *secondary*.

An alternating voltage applied to P circulates an alternating current through P and this current produces an alternating flux in the steel core, the mean path of this flux being represented by the dotted line D. If the whole of the flux produced by P passes through S, the e.m.f. induced in each turn is the same for P and S. Hence, if N_1 and N_2 be the number of turns on P and S respectively,

$$\frac{\text{total e.m.f. induced in S}}{\text{total e.m.f. induced in P}} = \frac{N_2 \times \text{e.m.f. per turn}}{N_1 \times \text{e.m.f. per turn}} = \frac{N_2}{N_1}$$

When the secondary is on open circuit, its terminal voltage is the same as the induced e.m.f. The primary current is then very small, so that the applied voltage V_1 is practically equal and opposite to the e.m.f. induced in P. Hence:

$$\frac{V_2}{V_1} \simeq \frac{N_2}{N_1} \tag{20.1}$$

Since the full-load efficiency of a transformer is nearly 100 per cent,

$I_1 V_1 \times$ primary power factor $\simeq I_2 V_2 \times$ secondary power factor.

But the primary and secondary power factors at full load are nearly equal,

$$\therefore \qquad \frac{I_1}{I_2} \simeq \frac{V_2}{V_1} \tag{20.2}$$

An alternative and more illuminating method of deriving the relationship between the primary and secondary currents is based upon a comparison of the primary and secondary ampere-turns. When the secondary is on open circuit, the primary current is such that the primary ampere-turns are just sufficient to produce the flux necessary to induce an e.m.f. that is practically equal and opposite to the applied voltage. This magnetizing current is usually about 3–5 per cent of the full-load primary current.

When a load is connected across the secondary terminals, the secondary current — by Lenz's Law — produces a demagnetizing effect. Consequently the flux and the e.m.f. induced in the primary are reduced slightly. But this small change can increase the difference between the applied voltage and the e.m.f. induced in the primary from, say, 0.05 per cent to, say, 1 per cent, in which case the new primary current would be 20 times the no-load current. The demagnetizing ampere-turns of the secondary are thus *nearly* neutralized by the increase in the primary ampere-turns; and since the primary ampere-turns on no load are very small compared with the full-load ampere-turns,

$\therefore$ full-load primary ampere-turns

$\simeq$ full-load secondary ampere-turns,

i.e. $\qquad\qquad I_1N_1 \simeq I_2N_2$

so that $\qquad\qquad \dfrac{I_1}{I_2} \simeq \dfrac{N_2}{N_1} \simeq \dfrac{V_2}{V_1}$ $\qquad\qquad$ (20.3)

It will be seen that the magnetic flux forms the connecting link between the primary and secondary circuits and that any variation of the secondary current is accompanied by a small variation of the flux and therefore of the e.m.f. induced in the primary, thereby enabling the primary current to vary approximately proportionally to the secondary current.

This balance of primary and secondary ampere-turns is an important relationship wherever transformer action occurs.

20.3 E.M.F. equation of a transformer

Suppose the maximum value of the flux to be Φ_m webers and the frequency to be f hertz. From fig. 20.2 it is seen that the flux has to change from $+\Phi_m$ to $-\Phi_m$ in half a cycle, namely in $1/2f$ seconds.

$\therefore \quad$ average rate of change of flux $= 2\Phi_m \div \tfrac{1}{2}f$

$\qquad\qquad\qquad\qquad\qquad\qquad = 4f\Phi_m$ webers per second

and $\qquad$ average e.m.f. induced per turn $= 4f\Phi_m$ volts.

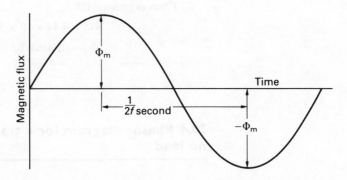

Fig. 20.2 Waveform of flux variation

But for a sinusoidal wave the r.m.s. or effective value is 1.11 times the average value,

$\therefore \quad$ r.m.s. value of e.m.f. induced per turn $= 1.11 \times 4f\Phi_m$.

Hence, r.m.s. value of e.m.f. induced in primary

$\qquad\qquad\qquad\qquad = E_1 = 4.44N_1 f\Phi_m$ volts $\qquad\qquad$ (20.4)

and r.m.s. value of e.m.f. induced in secondary

$\qquad\qquad\qquad\qquad = E_2 = 4.44N_2 f\Phi_m$ volts. $\qquad\qquad$ (20.5)

An alternative method of deriving these formulae is as follows:

If ϕ = instantaneous value of flux in webers

$$= \Phi_m \sin 2\pi ft$$

$\therefore$ instantaneous value of induced e.m.f. per turn

$$= -d\phi/dt \text{ volts}$$

$$= -2\pi f \Phi_m \times \cos 2\pi ft \text{ volts}$$

$$= 2\pi f \Phi_m \times \sin (2\pi ft - \pi/2) \qquad (20.6)$$

$\therefore$ maximum value of induced e.m.f. per turn $= 2\pi f \Phi_m$ volts

and r.m.s. value of induced e.m.f. per turn

$$= 0.707 \times 2\pi f \Phi_m = 4.44 f \Phi_m \text{ volts.}$$

Hence r.m.s. value of primary e.m.f. $= E_1 = 4.44 N_1 f \Phi_m$ volts

and r.m.s. value of secondary e.m.f. $= E_2 = 4.44 N_2 f \Phi_m$ volts.

Example 20.1   *A 250-kVA, 11000-V/415-V, 50-Hz single-phase transformer has 80 turns on the secondary. Calculate: (a) the approximate values of the primary and secondary currents; (b) the approximate number of primary turns; and (c) the maximum value of the flux.*

(a) Full-load primary current $\simeq \dfrac{250 \times 1000}{11\,000} = 22.7 \text{ A}$,

and full-load secondary current $= \dfrac{250 \times 1000}{415} = 602 \text{ A}$.

(b)   No. of primary turns $\simeq \dfrac{80 \times 11\,000}{415} = 2121$

(c) From expression (20.5),

$$415 = 4.44 \times 80 \times 50 \times \Phi_m$$

$\therefore$ $\Phi_m = 23.4 \text{ mWb}$.

20.4 Phasor diagram for a transformer on no load

It is most convenient to commence the phasor diagram with the phasor representing the quantity that is common to the two windings, namely the flux Φ. This phasor can be made any convenient length and may be regarded merely as a reference phasor, relative to which other phasors have to be drawn.

In the preceding section, expression (20.6) shows that the e.m.f. induced by a sinusoidal flux lags the flux by a quarter of a cycle. Consequently the e.m.f. E_1 induced in the primary winding is represented by a phasor drawn 90° behind Φ, as in fig. 20.3.

The e.m.f. E_2 also lags the flux by 90°, but the effect which this produces at the terminals of the transformer depends on

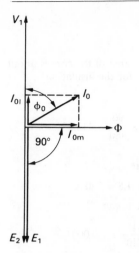

V_1

I_{01} ϕ_0 I_0

$90°$ I_{0m} Φ

E_2 E_1

Fig. 20.3 Phasor diagram for transformer on no load

the manner in which the secondary winding is constructed and, more importantly, the manner in which the ends of the winding are connected to the transformer terminals. In practice, the normal procedure ensures that V_2 is in phase with V_1 and only a very few transformers depart from this arrangement.

However, if V_2 and V_1 were drawn in phase with one another on fig. 20.3, the diagram would become cluttered and therefore, for convenience, it is usual to show E_1 and E_2 in phase with one another, thus ensuring that V_2 appears in the opposite quadrant of the phasor diagram from V_1. This gives the appearance that the voltages are in antiphase and it should be remembered that the manner of drawing is for convenience only and that the voltages are in fact in phase.

The values of E_2 and E_1 are proportional to the number of turns on the secondary and primary windings, since practically the whole of the flux set up by the primary winding is linked with the secondary winding when the latter is on open circuit. Another matter of convenience in drawing the transformer phasor diagrams in this chapter is that it has been assumed that N_2 and N_1 are equal so that $E_2 = E_1$, as shown in fig. 20.3.

Since the difference between the value of the applied voltage V_1 and that of the induced e.m.f. E_1 is only about 0.05 per cent when the transformer is on no load, the phasor representing V_1 can be drawn equal and opposite to that representing E_1.

The no-load current, I_0, taken by the primary consists of two components: (*a*) a reactive or magnetizing component,* I_{0m}, producing the flux and therefore in phase with the latter, and (*b*) an active or power component, I_{0l}, supplying the hysteresis and eddy-current losses in the core and the negligible I^2R loss in the primary winding. Component I_{0l} is in phase with the applied voltage: i.e. $I_{0l}V_1 = $ core loss. This component is usually very small compared with I_{0m}, so that the no-load power factor is very low. From fig. 20.3, it will be seen that:

$$\text{no-load current} = I_0 = \sqrt{(I_{0l}^2 + I_{0m}^2)} \qquad (20.7)$$

and
$$\text{power factor on no load} = \cos \phi_0 = I_{0l}/I_0 \qquad (20.8)$$

Example 20.2 *A single-phase transformer has 480 turns on the primary and 90 turns on the secondary. The mean length of the flux path in the core is 1.8 m and the joints are equivalent to an airgap of 0.1 mm. If the maximum value of the flux density is to be 1.1 T when a p.d. of 2200 V at 50 Hz is applied to the primary, calculate: (a) the cross-sectional area of the core; (b) the secondary voltage on no load; (c) the primary current and power factor on no load. Assume the value of the magnetic field strength for 1.1 T in the core to be 400 A/m, the corresponding core loss to be 1.7 W/kg at 50 Hz and the density of the core to be 7800 kg/m³.*

(a) From (20.4), $2200 = 4.44 \times 480 \times 50 \times \Phi_m$

$\therefore$ $\Phi_m = 0.0206$ Wb

---

* The waveform of this component is discussed in section 20.22.

$$\left.\begin{array}{r}\text{and} \quad \text{cross sectional} \\ \text{area of core}\end{array}\right\} = \frac{0.0206}{1.1} = 0.0187\,\text{m}^2.$$

This is the net area of the core; the gross area of the core is about 10 per cent greater than this value to allow for the insulation between the laminations.

(b) Secondary voltage on no load $= 2200 \times \dfrac{90}{480} = 412.5\,\text{V}.$

(c) Total magnetomotive force for the core

$$= 400 \times 1.8 = 720\,\text{A}$$

and magnetomotive force for the equivalent airgap

$$= \frac{1.1}{4\pi \times 10^{-7}} \times 0.0001$$

$$= 87.5\,\text{A}$$

∴ total m.m.f. to produce the maximum flux density

$$= 720 + 87.5 = 807.5\,\text{A}$$

∴ maximum value of magnetizing current

$$= \frac{807.5}{480} = 1.682\,\text{A}.$$

Assuming the current to be sinusoidal,

r.m.s. value of magnetizing current

$$= I_{0m} = 0.707 \times 1.682$$

$$= 1.19\,\text{A}.$$

Volume of core $= 1.8 \times 0.0187$

$$= 0.0337\,\text{m}^3$$

∴ mass of core $= 0.0337 \times 7800$

$$= 263\,\text{kg}$$

and core loss $= 263 \times 1.7 = 447\,\text{W}$

∴ core-loss component of current $= I_{0l} \dfrac{447}{2200}$

$$= 0.203\,\text{A}.$$

From (20.7), no-load current $= I_0$

$$= \sqrt{\{(1.19)^2 + 0.203)^2\}}$$

$$= 1.21\,\text{A},$$

and from (20.8), power factor on no load

$$= \frac{0.203}{1.21}$$

$$= 0.168\,\text{lagging}.$$

20.5 Phasor diagram for a loaded transformer, assuming the voltage drop in the windings to be negligible

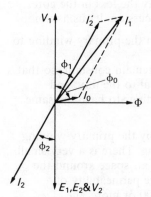

Fig. 20.4 Phasor diagram for a loaded transformer having negligible voltage drop in windings

With this assumption, it follows that the secondary terminal voltage V_2 is the same as the e.m.f. E_2 induced in the secondary, and the primary applied voltage V_1 is equal and opposite in phase to the e.m.f. E_1 induced in the primary winding. Also, if we again assume equal number of turns on the primary and secondary windings, then $E_1 = E_2$.

Let us consider the general case of a load having a lagging power factor $\cos \phi_2$; hence the phasor representing the secondary current I_2 lags V_2 by an angle ϕ_2, as shown in fig. 20.4. Phasor $I_{2'}$ represents the component of the primary current to neutralize the demagnetizing effect of the secondary current and is drawn equal and opposite to I_2. $I_{2'}$ is described as 'I_2 referred'. I_0 is the no-load current of the transformer, already discussed in section 20.4. The phasor sum of $I_{2'}$ and I_0 gives the total current I_1 taken from the supply, and the power factor on the primary side is $\cos \phi_1$, where ϕ_1 is the phase difference between V_1 and I_1.

In fig. 20.4, the phasor representing I_0 has, for clearness, been shown far larger relative to the other current phasors than it is in an actual transformer.

Example 20.3

A single-phase transformer has 1000 turns on the primary and 200 turns on the secondary. The no-load current is 3 A at a power factor 0.2 lagging. Calculate the primary current and power factor when the secondary current is 280 A at a power factor of 0.8 lagging. Assume the voltage drop in the windings to be negligible.

If $I_{2'}$ represents the component of the primary current to neutralize the demagnetizing effect of the secondary current, the ampere-turns due to $I_{2'}$ must be equal and opposite to those due to I_2, i.e.

$$I_{2'} \times 1000 = 280 \times 200$$

$$\therefore \qquad I_{2'} = 56 \, \text{A}.$$

$$\cos \phi_2 = 0.8, \quad \therefore \quad \sin \phi_2 = 0.6$$

and $\qquad \cos \phi_0 = 0.2, \quad \therefore \quad \sin \phi_0 = 0.98.$

From fig. 20.4 it will be seen that:

$$I_1 \cos \phi_1 = I_{2'} \cos \phi_2 + I_0 \cos \phi_0$$
$$= (56 \times 0.8) + (3 \times 0.2) = 45.4 \, \text{A}$$

and $\qquad I_1 \sin \phi_1 = I_{2'} \sin \phi_2 + I_0 \sin \phi_0$
$$= (56 \times 0.6) + (3 \times 0.98) = 36.54 \, \text{A}.$$

Hence, $\qquad I_1^2 = (45.4)^2 + (36.45)^2 = 3398$

so that $\qquad I_1 = 58.3 \, \text{A}.$

Also, $\qquad \tan \phi_1 = 36.54/45.4 = 0.805$

so that $\qquad \phi_1 = 38° \, 50'.$

Hence   primary power factor $= \cos \phi_1 = \cos 38° \, 50'$
$$= 0.78 \, \text{lagging}.$$

20.6 Useful and leakage fluxes in a transformer

When the secondary winding of a transformer is on open circuit, the current taken by the primary winding is responsible for setting up the magnetic flux and providing a very small power component to supply the loss in the core. To simplify matters in the present discussion, let us assume:

(a) the core loss and the I^2R loss in the primary winding to be negligible;

(b) the permeability of the core to remain constant, so that the magnetizing current is proportional to the flux; and

(c) the primary and secondary windings to have the same number of turns, i.e. $N_1 = N_2$.

Fig. 20.5 shows all the flux set up by the primary winding passing through the secondary winding. There is a very small amount of flux returning through the air space around the primary winding, but since the relative permeability of transformer core is of the order of 1000 or more, the reluctance of the air path is 1000 times that of the parallel path through the limb carrying the secondary winding. Consequently the flux passing through the air space is negligible compared with that through the secondary. It follows that the e.m.f.s induced in the primary and secondary windings are equal and that the primary applied voltage, V_1, is equal and opposite to the e.m.f., E_1, induced in the primary, as shown in figs. 20.6 and 20.7.

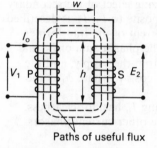

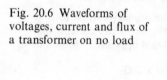

Paths of useful flux

Fig. 20.5 Transformer on no load

Fig. 20.6 Waveforms of voltages, current and flux of a transformer on no load

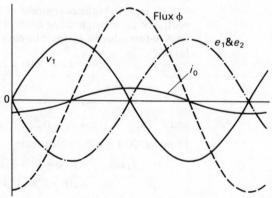

Next, let us assume a load having a power factor such that the secondary current is in phase with E_2. As already explained in section 20.2, the primary current, I_1, must now have two components:

(a) I_{0m} to maintain the useful flux, the maximum value of which remains constant within about 2 per cent between no load and full load; and

(b) a component, $I_{2'}$, to neutralize the demagnetizing effect of the secondary current, as shown in figs. 20.8(a) and 20.9.

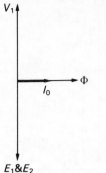

Fig. 20.7 Phasor diagram
for fig. 20.6

At instant A in fig. 20.8(*a*), the magnetizing current is *zero*, but I_2 and $I_{2'}$ are at their *maximum* values; and if the direction of the current in primary winding P is such as to produce flux upwards in the lefthand limb of fig. 20.10, the secondary current must be in such a direction as to produce flux upwards in the righthand limb, and the flux of each limb has to return through air. Since the flux of each limb is linked only with the winding by which it is produced, it is referred to as *leakage* flux and is responsible for inducing an e.m.f. of self inductance in the winding with which it is linked. The reluctance of the paths of the leakage flux, Φ_L, is almost entirely due to the long air paths and is therefore practically constant. Consequently the value of the leakage flux is proportional to the load current, whereas the value of the useful flux remains almost independent of the load. The reluctance of the paths of the leakage flux is very high, so

Fig. 20.8 Waveforms of
induced e.m.f.s, currents and
fluxes in a transformer on
load

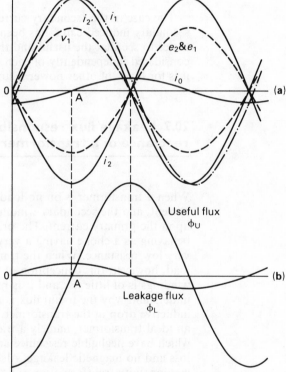

that the value of this flux is relatively small even on full load when the values of $I_{2'}$ and I_2 are about 20–30 times the magnetizing current I_{0m}.

From the above discussion it follows that the actual flux in a transformer can be regarded as being due to the two components shown in fig. 20.8(*b*), namely:

(*a*) the useful flux, Φ_U, linked with both windings and remaining practically constant in value at all loads; and
(*b*) the leakage flux, Φ_L, half of which is linked with the

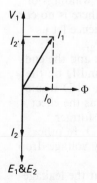

Fig. 20.9 Phasor diagram
for fig. 20.8

Fig. 20.10 Paths of leakage
flux

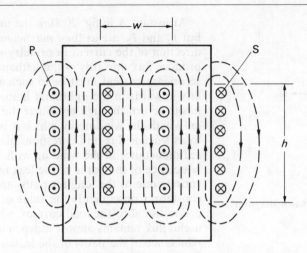

primary winding and half with the secondary, and its value is
proportional to the load.

The case of the secondary current in phase with the
secondary induced e.m.f. has been considered because it is
easier to see that the useful and the leakage fluxes can be
considered independently of each other for this condition than
it is for loads of other power factor.

20.7 Leakage flux responsible for the inductive reactance of a transformer

When a transformer is on no load there is no secondary
current, and the secondary winding has not the slightest effect
upon the primary current. The primary winding is then
behaving as a choke having a very high inductance and a
very low resistance. When the transformer is supplying a
load, however, this conception of the inductance of the
windings is of little use and it is rather difficult at first to
understand why the useful flux is not responsible for any
inductive drop in the transformer. So let us first of all assume
an ideal transformer, namely a transformer the windings of
which have negligible resistance and in which there is no core
loss and no magnetic leakage. Also for convenience let us
assume unity transformation ratio, i.e. $N_1 = N_2$.

If such a transformer were enclosed in a box and the ends
of the windings brought to terminals A, B, C and D on the
lid, as shown in fig. 20.11, the p.d. between C and D would
be equal to that between A and B; and, as far as the effect
upon the output voltage is concerned, the transformer
behaves as if A were connected to C and B to D. In other
words, the useful flux is not responsible for any voltage drop
in a transformer.

In the preceding section it was explained that the leakage
flux is proportional to the primary and secondary currents
and that its effect is to induce e.m.f.s of self induction in the

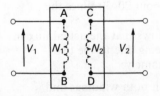

Fig. 20.11 Ideal transformer
enclosed in a box

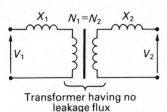

Fig. 20.12 Transformer with leakage reactances

Transformer having no leakage flux

windings. Consequently the effect of leakage flux can be considered as equivalent to inductive reactors X_1 and X_2 connected in series with a transformer having no leakage flux, as shown in fig. 20.12, these reactors being such that the flux-linkages produced by the primary current through X_1 are equal to those due to the leakage flux linked with the primary winding, and the flux-linkages produced by the secondary current through X_2 are equal to those due to the leakage flux linked with the secondary winding of the actual transformer. The straight line drawn between the primary and secondary windings in fig. 20.12 is the symbol used to indicate that the transformer has a ferromagnetic core.

20.8 Methods of reducing leakage flux

The leakage flux can be practically eliminated by winding the primary and secondary, one over the other, uniformly around a laminated ferromagnetic ring of uniform cross-section. But such an arrangement is not commercially practicable except in very small sizes, owing to the cost of threading a large number of turns through the ring.

The principal methods used in practice are:

(*a*) Making the transformer 'window' long and narrow.
(*b*) Arranging the primary and secondary windings concentrically (see fig. 20.13).

Fig. 20.13 Concentric windings

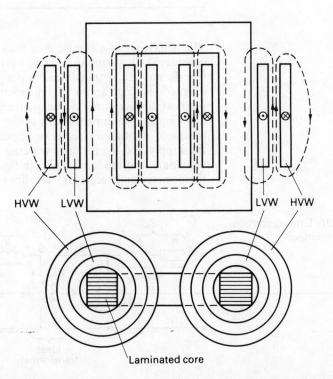

HVW   LVW   LVW   HVW

Laminated core

(c) Sandwiching the primary and secondary windings (see fig. 20.14).

Fig. 20.14 Sandwiched windings

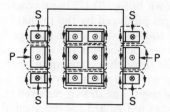

(d) Using shell-type construction (see fig. 20.15).

Fig. 20.15 Shell-type construction

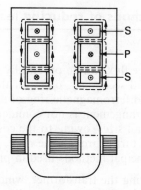

20.9 Equivalent circuit of a transformer

The behaviour of a transformer may be conveniently considered by assuming it to be equivalent to an ideal transformer, i.e. a transformer having no losses and no magnetic leakage and a ferromagnetic core of infinite permeability requiring no magnetizing current, and then allowing for the imperfections of the actual transformer by means of additional circuits or impedances inserted between the supply and the primary winding and between the secondary and the load. Thus, in fig. 20.16, P and S represent the primary and secondary windings of the ideal transformer.

Fig. 20.16 Equivalent circuit of a transformer

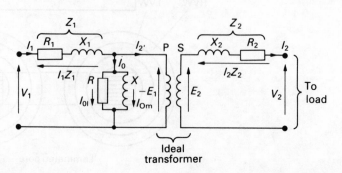

R_1 and R_2 are resistances equal to the resistances of the primary and secondary windings of the actual transformer. Similarly, inductive reactances X_1 and X_2 represent the reactances of the windings due to leakage flux in the actual transformer, as already explained in section 20.7.

The inductive reactor X is such that it takes a reactive current equal to the magnetizing current I_{0m} of the actual transformer. The core losses due to hysteresis and eddy currents are allowed for by a resistor R of such value that it takes a current I_{0l} equal to the core-loss component of the primary current, i.e. $I_{0l}^2 R$ is equal to the core loss of the actual transformer. The resultant of I_{0m} and I_{0l} is I_0, namely the current which the transformer takes on no load. The phasor diagram for the equivalent circuit on no load is exactly the same as that given in fig. 20.3.

20.10 Phasor diagram for a transformer on load

For convenience let us assume an equal number of turns on the primary and secondary windings, so that $E_1 = E_2$. Both E_1 and E_2 lag the flux by $90°$, as shown in fig. 20.17, and $-E_1$ represents the voltage across the primary of the ideal transformer (the voltage is equal in magnitude but opposite to the e.m.f., hence the negative sign).

Let us also assume the general case of a load having a lagging power factor; consequently, in fig. 20.17, I_2 has been drawn lagging E_2 by about $45°$. Then:

Fig. 20.17 Phasor diagram for a transformer on load

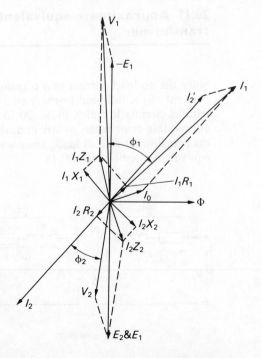

I_2R_2 = voltage drop due to secondary resistance,

I_2X_2 = voltage drop due to secondary leakage reactance

and I_2Z_2 = voltage drop due to secondary impedance.

The secondary terminal voltage V_2 is the phasor difference of E_2 and I_2Z_2; in other words, V_2 must be such that the phasor sum of V_2 and I_2Z_2 is E_2, and the derivation of the phasor representing V_2 is evident from fig. 20.17. The power factor of the load is $\cos\phi_2$, where ϕ_2 is the phase difference between V_2 and I_2. $I_{2'}$ represents the component of the primary current to neutralize the demagnetizing effect of the secondary current and is drawn equal and opposite to I_2. I_0 is the no-load current of the transformer (section 20.4). The phasor sum of $I_{2'}$ and I_0 gives the total current I_1 taken from the supply.

I_1R_1 = voltage drop due to primary resistance,

I_1X_1 = voltage drop due to primary leakage reactance,

I_1Z_1 = voltage drop due to primary impedance

and V_1 = phasor sum of $-E_1$

and I_1Z_1 = supply voltage.

If ϕ_1 is the phase difference between V_1 and I_1, then $\cos\phi_1$ is the power factor on the primary side of the transformer. In fig. 20.17 the phasors representing the no-load current and the primary and secondary voltage drops are, for clearness, shown far larger relative to the other phasors than they are in an actual transformer.

20.11 Approximate equivalent circuit of a transformer

Since the no-load current of a transformer is only about 3–5 per cent of the full-load primary current, we can omit the parallel circuits R and X in fig. 20.16 without introducing an appreciable error when we are considering the behaviour of the transformer on full load. Thus we have the simpler equivalent circuit of fig. 20.18.

Fig. 20.18 Approximate equivalent circuit of a transformer

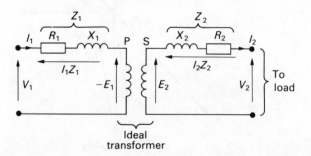

Ideal transformer

20.12 Simplification of the approximate equivalent circuit of a transformer

We can replace the resistance R_2 of the secondary of fig. 20.18 by inserting additional resistance $R_{2'}$ in the primary circuit such that the power absorbed in $R_{2'}$ when carrying the primary current is equal to that in R_2 due to the secondary current.

i.e.
$$I_1^2 R_{2'} = I_2^2 R_2$$
$$\therefore \qquad R_{2'} = R_2(I_2/I_1)^2 \simeq R_2(V_1/V_2)^2.$$

Hence if R_e be a single resistance in the primary circuit equivalent to the primary and secondary resistances of the actual transformer, then:

$$R_e = R_1 + R_{2'} = R_1 + R_2(V_1/V_2)^2 \qquad (20.9)$$

Similarly, since the inductance of a coil is proportional to the square of the number of turns, the secondary leakage reactance X_2 can be replaced by an equivalent reactance X in the primary circuit, such that:

$$X_{2'} = X_2(N_1/N_2)^2 \simeq X_2(V_1/V_2)^2.$$

If X_e be the single reactance in the primary circuit equivalent to X_1 and X_2 of the actual transformer,

$$X_e = X_1 + X_{2'} = X_1 + X_2(V_1/V_2)^2 \qquad (20.10)$$

If Z_e be the equivalent impedance of the primary and secondary windings referred to the primary circuit,

$$Z_e = \sqrt{(R_e^2 + X_e^2)} \qquad (20.11)$$

If ϕ_e be the phase difference between I_1 and $I_1 Z_e$, then

$$R_e = Z_e \cos \phi_e \quad \text{and} \quad X_e = Z_e \sin \phi_e.$$

The simplified equivalent circuit of the transformer is given in fig. 20.19, and fig. 20.20(*a*) is the corresponding phasor diagram.

Fig. 20.19 Simplified equivalent circuit of a transformer

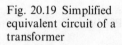
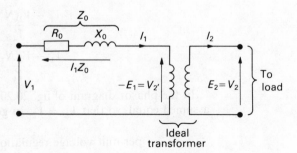

20.13 Voltage regulation of a transformer

The voltage regulation of a transformer is defined as the

Fig. 20.20 Phasor diagram
for fig. 20.19

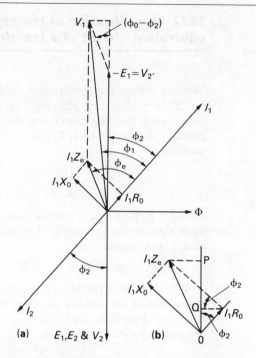

(a) E_1, E_2 & V_2 (b)

variation of the secondary voltage between no load and full
load, expressed as either a per-unit or a percentage of the *no-
load* voltage, the primary voltage being assumed constant, i.e.

$$\left.\begin{array}{r}\text{voltage}\\\text{regulation}\end{array}\right\} = \frac{\text{no-load voltage} - \text{full-load voltage}}{\text{no-load voltage}} \quad (20.12)$$

If V_1 = primary applied voltage,

secondary voltage on no load = $V_1 \times N_2/N_1$,

since the voltage drop in the primary winding due to the no-
load current is negligible.

If V_2 = secondary terminal voltage on full load,

$$\text{voltage regulation} = \frac{V_1(N_2/N_1) - V_2}{V_1(N_2/N_1)}$$

$$= \frac{V_1 - V_2(N_1/N_2)}{V_1} \text{ per unit}$$

$$= \frac{V_1 - V_2(N_1/N_2)}{V_1} \times 100 \text{ per cent}$$

In the phasor diagram of fig. 20.20, N_1 and N_2 were
assumed equal, so that $V_{2'} = V_2$. In general, $V_{2'} = V_2(N_1/N_2)$,

$$\therefore \qquad \text{per-unit voltage regulation} = \frac{V_1 - V_{2'}}{V_1} \quad (20.13)$$

In fig. 20.20(*a*), let us draw a perpendicular from V_1 to
meet the extension of $V_{2'}$ at A; then:

$$V_1^2 = (V_{2'} + V_{2'}\text{A})^2 + (V_1\text{A})^2$$

$$= \{V_{2'} + I_1 Z_e \cos(\phi_e - \phi_2)\}^2 + \{I_1 Z_e \sin(\phi_e - \phi_2)\}^2$$

In actual practice, $I_1 Z_e \sin(\phi_e - \phi_2)$ is very small compared with $V_{2'}$, so that:

$$V_1 \simeq V_{2'} + I_1 Z_e \cos(\phi_e - \phi_2).$$

Hence,

$$\text{per-unit voltage regulation} = \frac{V_1 - V_{2'}}{V_1}$$

$$= \frac{I_1 Z_e \cos(\phi_e - \phi_2)}{V_1} \quad (20.14)$$

Since $\quad Z_e \cos(\phi_e - \phi_2) = Z_e(\cos\phi_e \cdot \cos\phi_2 + \sin\phi_e \cdot \sin\phi_2)$

$$= R_e \cos\phi_2 + X_e \sin\phi_2$$

$$\therefore \quad \left.\begin{array}{r}\text{per-unit voltage} \\ \text{regulation}\end{array}\right\} = \frac{I_1(R_e \cos\phi_2 + X_e \sin\phi_2)}{V_1} \quad (20.15)$$

This expression can also be derived by projecting $I_1 R_e$ and $I_1 Z_e$ on to OA, as shown enlarged in fig. 20.20(b), from which it follows that:

$$V_{2'}\text{A in fig. } 20.20(a) = \text{OP in fig. } 20.20(b)$$

$$= \text{OQ} + \text{QP}$$

$$= I_1 R_e \cos\phi_2 + I_1 X_e \sin\phi_2$$

$$\therefore \text{ per-unit voltage regulation} = \frac{V_1 - V_{2'}}{V_1} \simeq \frac{V_{2'}\text{A}}{V_1} = \frac{\text{OP}}{V_1}$$

$$= \frac{I_1(R_e \cos\phi_2 + X_e \sin\phi_2)}{V_1}.$$

The above expressions have been derived on the assumption that the power factor is lagging. Should the power factor be leading, the angle in expression (20.14) would be $(\phi_e + \phi_2)$ and the term in brackets in expression (20.15) would be $(R_e \cos\phi_2 - X_e \sin\phi_2)$.

Example 20.4   *A 100-kVA transformer has 400 turns on the primary and 80 turns on the secondary. The primary and secondary resistances are 0.3 Ω and 0.01 Ω respectively, and the corresponding leakage reactances are 1.1 Ω and 0.035 Ω respectively. The supply voltage is 2200 V. Calculate:* (a) *the equivalent impedance referred to the primary circuit, and* (b) *the voltage regulation and the secondary terminal voltage for full load having a power factor of* (i) *0.8 lagging and* (ii) *0.8 leading.*

(a) From (20.9), equivalent resistance referred to primary

$$= R_e = 0.3 + 0.01(400/80)^2$$

$$= 0.55\,\Omega.$$

From (20.10), equivalent leakage reactance referred to primary

$$= X_e = 1.1 + 0.035(400/80)^2$$

$$= 1.975\,\Omega.$$

From (20.11), equivalent impedance referred to primary

$$= Z_e = \sqrt{\{(0.55)^2 + (1.975)^2\}}$$

$$= 2.05\,\Omega.$$

(*b*) (i) Since $\cos \phi_2 = 0.8$, $\therefore \sin \phi_2 = 0.6$.

$$\text{Full-load primary current} \simeq \frac{100 \times 1000}{2200} = 45.45 \text{ A}.$$

Substituting in (20.15), we have:

$$\left.\begin{array}{l}\text{voltage regulation for power}\\ \text{factor 0.8 lagging}\end{array}\right\} = \frac{45.45(0.55 \times 0.8 + 1.975 \times 0.6)}{2200}$$

$$= 0.0336 \text{ per unit}$$

$$= 3.36 \text{ per cent}.$$

Secondary terminal voltage on no load $= 2200 \times 80/400 = 440$ V

$\therefore$ decrease of secondary terminal voltage between no load and full load

$$= 440 \times 0.0336 = 14.8 \text{ } V$$

$\therefore$ secondary terminal voltage on full load

$$= 440 - 14.8 = 425.2 \text{ V}.$$

(ii) Voltage regulation for power factor 0.8 leading

$$= \frac{45.45(0.55 \times 0.8 - 1.975 \times 0.6)}{2200} = -0.0154 \text{ per unit}$$

$$= -1.54 \text{ per cent}.$$

Increase of secondary terminal voltage between no load and full load

$$= 440 \times 0.0154 = 6.78 \text{ V}$$

$\therefore$ secondary terminal voltage on full load

$$= 440 + 6.78 = 446.8 \text{ V}.$$

Example 20.5　*Calculate the per-unit and the percentage resistance and leakage reactance drops of the transformer referred to in example 20.4.*

Per-unit resistance drop of a transformer

$$= \frac{\left(\begin{array}{c}\text{full-load primary}\\ \text{current}\end{array}\right) \times \left(\begin{array}{c}\text{equivalent resistance}\\ \text{referred to primary circuit}\end{array}\right)}{\text{primary voltage}}$$

$$= \frac{\left(\begin{array}{c}\text{full-load secondary}\\ \text{current}\end{array}\right) \times \left(\begin{array}{c}\text{equivalent resistance}\\ \text{referred to secondary circuit}\end{array}\right)}{\text{secondary voltage on no load}}$$

Thus, for example 20.4,

$$\text{full-load primary current} \simeq 45.45 \text{ A},$$

and equivalent resistance referred to primary circuit

$$= 0.55 \text{ } \Omega$$

$\therefore$

$$\text{resistance drop} = \frac{45.45 \times 0.55}{2200} = 0.0114 \text{ per unit}$$

$$= 1.14 \text{ per cent}.$$

Alternatively, full-load secondary current

$$\simeq 45.45 \times 400/80$$

$$= 227.2 \text{ A},$$

and equivalent resistance referred to secondary circuit

$$= 0.01 + 0.3\left(\frac{80}{400}\right)^2 = 0.022\,\Omega.$$

Secondary voltage on no load = 440 V,

$$\therefore \qquad \text{resistance drop} = \frac{227.2 \times 0.022}{440} = 0.0114 \text{ per unit}$$

$$= 1.14 \text{ per cent}.$$

Similarly, leakage reactance drop of a transformer

$$= \frac{\left(\begin{array}{c}\text{full-load primary}\\ \text{current}\end{array}\right) \times \left(\begin{array}{c}\text{equivalent leakage reactance}\\ \text{referred to primary circuit}\end{array}\right)}{\text{primary voltage}}$$

$$= \frac{45.45 \times 1.975}{2200} = 0.0408 \text{ per unit} = 4.08 \text{ per cent}.$$

It is usual to refer to the per-unit or the percentage resistance and leakage reactance drops on full load as merely the per-unit or the percentage resistance and leakage reactance of the transformer; thus, the above transformer has a per-unit resistance and leakage reactance of 0.0114 and 0.0408 respectively or a percentage resistance and leakage reactance of 1.14 and 4.08 respectively.

20.14 Efficiency of a transformer

The losses which occur in a transformer on load can be divided into two groups:

(*a*) I^2R losses in primary and secondary windings, namely $I_1^2R_1 + I_2^2R_2$;

(*b*) core losses due to hysteresis and eddy currents. The factors determining these losses have already been discussed in sections 3.14 and 18.1(*b*).

Since the maximum value of the flux in a normal transformer does not vary by more than about 2 per cent between no load and full load, it is usual to assume the core loss constant at all loads.

Hence, if P_c = total core loss,

$$\text{total losses in transformer} = P_c + I_1^2R_1 + I_2^2R_2$$

and   efficiency $= \dfrac{\text{output power}}{\text{input power}} = \dfrac{\text{output power}}{\text{output power} + \text{losses}}$

$$= \frac{I_2V_2 \times \text{power factor}}{I_2V_2 \times \text{p.f.} + P_c + I_1^2R_1 + I_2^2R_2}. \qquad (20.16)$$

Greater accuracy is possible by expressing the efficiency thus:

$$\text{efficiency} = \frac{\text{output power}}{\text{input power}} = \frac{\text{input power} - \text{losses}}{\text{input power}}$$

$$= 1 - \frac{\text{losses}}{\text{input power}} \qquad (20.17)$$

Example 20.6   *The primary and secondary windings of a 500-kVA transformer have resistances of 0.42 Ω and 0.0019 Ω respectively. The primary and secondary voltages are 11 000 V and 415 V respectively and the core loss is 2.9 kW. Calculate the efficiency on* (a) *full load and* (b) *half load, assuming the power factor of the load to be 0.8.*

(a) Full-load secondary current $= \dfrac{500 \times 1000}{415} = 1205$ A

and   full-load primary current $\simeq \dfrac{500 \times 1000}{11\,000} = 45.5$ A

∴   secondary copper loss on full load

$$= (1205)^2 \times 0.0019 = 2760 \text{ W}$$

and primary copper loss on full load

$$= (45.5)^2 \times 0.42 = 870 \text{ W}$$

∴   total I^2R loss on full load $= 3630$ W $= 3.63$ kW

and   total loss on full load $= 3.63 + 2.9 = 6.53$ kW.

Output power on full load $= 500 \times 0.8 = 400$ kW

∴   input power on full load $= 400 + 6.53 = 406.53$ kW.

From (20.17), efficiency on full load

$$= \left(1 - \frac{6.53}{406.53}\right) = 0.9839 \text{ per unit}$$

$$= 98.39 \text{ per cent}.$$

(b) Since the I^2R loss varies as the square of the current,

∴   total I^2R loss on half load $= 3.63 \times (0.5)^2 = 0.91$ kW

and   total loss on half load $= 0.91 + 2.9 = 3.81$ kW

∴   efficiency on half load $= \left(1 - \dfrac{3.81}{203.81}\right) = 0.9813$ per unit

$$= 98.13 \text{ per cent}.$$

20.15 Condition for maximum efficiency of a transformer

If R_{2e} be the equivalent resistance of the primary and secondary windings referred to the *secondary* circuit,

$$R_{2e} = R_1(N_2/N_1)^2 + R_2$$

$$= \text{a constant for a given transformer.}$$

Hence for any load current I_2,

$$\text{total } I^2R \text{ loss} = I_2^2 R_{2e}$$

and
$$\text{efficiency} = \frac{I_2 V_2 \times \text{p.f.}}{I_2 V_2 \times \text{p.f.} + P_c + I_2^2 R_{2e}}$$

$$= \frac{V_2 \times \text{p.f.}}{V_2 \times \text{p.f.} + P_c/I_2 + I_2 R_{2e}} \quad (20.18)$$

For a normal transformer, V_2 is approximately constant; hence for a load of given power factor, the efficiency is a maximum when the denominator of (20.18) is a minimum,

i.e. when
$$\frac{d}{dI_2}(V_2 \times \text{p.f.} + P_c/I_2 R_{2e}) = 0$$

$$\therefore \qquad\qquad -P_c/I_2^2 + R_{2e} = 0$$

or
$$\qquad\qquad I_2^2 R_{2e} = P_c \quad (20.19)$$

To check that this condition gives the minimum and not the maximum value of the denominator in expression (20.18), $(-P_c/I_2^2 + R_{2e})$ should be differentiated with respect to I_2, thus:

$$\frac{d}{dI_2}(-P_c/I_2^2 + R_{2e}) = 2P_c/I_2^3$$

Since this quantity is positive, expression (20.19) is the condition for the minimum value of the denominator of (20.18) and therefore the maximum value of the efficiency. Hence the efficiency is a maximum when the variable I^2R loss is equal to the constant core loss.

Example 20.7 *Find the output at which the efficiency of the transformer of example 20.6 is a maximum and calculate its value, assuming the power factor of the load to be 0.8.*

With the full-load output of 500 kVA, the total I^2R loss is 3.63 kW.

Let n = fraction of full-load apparent power (in kVA) at which the efficiency is a maximum.

$$\text{Corresponding total } I^2R \text{ loss} = n^2 \times 3.63 \text{ kW}$$

Hence, from (20.19),
$$\qquad\qquad n^2 \times 3.63 = 2.9$$

$$\therefore \qquad\qquad n = 0.894,$$

and
$$\qquad \text{output at maximum efficiency} = 0.894 \times 500$$

$$= 447 \text{ kVA}.$$

It will be noted that the value of the apparent power at which the efficiency is a maximum is independent of the power factor of the load.

Since the I^2R and core losses are equal when the efficiency is a maximum,

$$\therefore \qquad\qquad \text{total loss} = 2 \times 2.9 = 5.8 \text{ kW}$$

$$\text{Output power} = 447 \times 0.8 = 357.6 \text{ kW}$$

$$\therefore \quad \text{maximum efficiency} = \left(1 - \frac{5.8}{357.6 + 5.8}\right) = 0.984 \text{ per unit}$$

$$= 98.4 \text{ per cent.}$$

20.16 Open-circuit and short-circuit tests on a transformer

These two tests enable the efficiency and the voltage regulation to be calculated without actually loading the transformer and with an accuracy far higher than is possible by direct measurement of input and output powers and voltages. Also, the power required to carry out these tests is very small compared with the full-load output of the transformer.

Open-circuit test   The transformer is connected as in fig. 20.21 to a supply at the rated voltage and frequency, namely the voltage and frequency given on the nameplate. The ratio of the voltmeter readings, V_1/V_2, gives the ratio of the number of turns. Ammeter A gives the no-load current, and its reading is a check on the magnetic quality of the ferromagnetic core and joints. The primary current on no load is usually less than 5 per cent of the full-load current, so that the I^2R loss on no load is less than 1/400 of the primary I^2R loss on full load and is therefore negligible compared with the core loss. Hence the wattmeter reading can be taken as the core loss of the transformer.

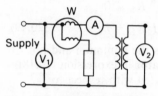

Fig. 20.21 Open-circuit test on a transformer

Short-circuit test   The secondary is short-circuited through a suitable ammeter A_2, as shown in fig. 20.22, and a *low* voltage is applied to the primary circuit. This voltage should, if possible, be adjusted to circulate full-load currents in the primary and secondary circuits. Assuming this to be the case, the I^2R loss in the windings is the same as that on full load. On the other hand, the core loss is negligibly small, since the applied voltage and therefore the flux are only about one-twentieth to one-thirtieth of the rated voltage and flux, and the core loss is approximately proportional to the square of the flux. Hence the power registered on wattmeter W can be taken as the I^2R loss in the windings.

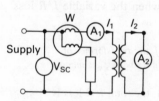

Fig. 20.22 Short-circuit test on a transformer

20.17 Calculation of efficiency from the open-circuit and short-circuit tests

If P_{oc} = input power in watts on the open-circuit test,

= core loss,

and P_{sc} = input power in watts on the short-circuit test with full-load currents,

= total I^2R loss on full load,

then   total loss on full load = $P_{oc} + P_{sc}$

and efficiency on full load

$$= \frac{\text{full-load } S \times \text{p.f.}}{(\text{full-load } S \times \text{p.f.}) + P_{oc} + P_{sc}} \quad (20.20)$$

where S is the apparent power (in VA). Also, for any load equal to $n \times$ full load,

$$\text{corresponding total loss} = P_{oc} + n^2 P_{sc}$$

and corresponding efficiency

$$= \frac{n \times \text{full-load } S \times \text{p.f.}}{n \times \text{full-load } S \times \text{p.f.} + P_{oc} + n^2 P_{sc}} \quad (20.21)$$

20.18 Calculation of the voltage regulation from the short-circuit test

Since the secondary voltage is zero, the whole of the applied voltage on the short-circuit test is absorbed in sending currents through the impedances of the primary and secondary windings; and since ϕ_e in fig. 20.20 is the phase angle between the primary current and the voltage drop due to the equivalent impedance referred to the primary circuit,

$$\cos \phi_e = \text{power factor on short-circuit test},$$

$$= \frac{P_{sc}}{I_1 V_{sc}}.$$

If V_{sc} be the value of the primary applied voltage on the s.c. test when *full-load* currents are flowing in the primary and secondary windings, then from expression (20.14),

$$\text{per-unit voltage regulation} = \frac{V_{sc} \cos (\phi_e - \phi_2)}{V_1} \quad (20.22)$$

Example 20.8   *The following results were obtained on a 50-kVA transformer: O.C. test — primary voltage, 3300 V; secondary voltage, 415 V; primary power, 430 W. S.C. test – primary voltage, 124 V; primary current, 15.3 A; primary power, 525 W; secondary current, full-load value. Calculate: (a) the efficiencies at full load and at half load for 0.7 power factor; (b) the voltage regulations for power factor 0.7 (i) lagging, (ii) leading; and (c) the secondary terminal voltages corresponding to (i) and (ii).*

(a)   Core loss = 430 W,

loss on full load = 525 W

∴   total loss on full load = 955 W = 0.955 kW

and   efficiency on full load $= \dfrac{50 \times 0.7}{(50 \times 0.7) + 0.955}$

$$= \left(1 - \frac{0.955}{35.95}\right) = 0.9734 \text{ per unit}$$

$$= 97.34 \text{ per cent}.$$

$$\text{loss on half load} = 525 \times (0.5)^2 = 131\,\text{W},$$

$$\therefore \quad \text{total loss on half load} = 430 + 131 = 561\,\text{W} = 0.561\,\text{kW}$$

$$\text{and efficiency on half load} = \frac{25 \times 0.7}{(25 \times 0.7) + 0.561}$$

$$= \left(1 - \frac{0.561}{18.06}\right) = 0.969 \text{ per unit}$$

$$= 96.9 \text{ per cent.}$$

$$(b) \qquad\qquad \cos\phi_e = \frac{525}{124 \times 15.3} = 0.2765$$

$$\therefore \qquad\qquad \phi_e = 73°\,57'$$

For $\cos\phi_2 = 0.7$, $\qquad \phi_2 = 45°\,34'.$

From expression (20.22), for power factor 0.7 lagging,

$$\text{voltage regulation} = \frac{124 \cos(73°\,57' - 45°\,34')}{3300}$$

$$= 0.033 \text{ per unit} = 3.3 \text{ per cent.}$$

For power factor 0.7 leading,

$$\text{voltage regulation} = \frac{124 \cos(73°\,57' + 45°\,34')}{3300}$$

$$= -0.0185 \text{ per unit}$$

$$= -1.85 \text{ per cent.}$$

(c) Secondary voltage on open circuit = 415 V

$\therefore$ secondary voltage on full load, p.f. 0.7 lagging

$$= 415(1 - 0.033) = 401.3\,\text{V}$$

and secondary voltage on full load, p.f. 0.7 leading

$$= 415(1 + 0.0185) = 422.7\,\text{V}.$$

20.19 Three-phase core-type transformers

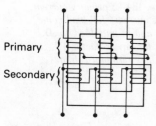

Primary

Secondary

Fig. 20.23 Three-phase core-type transformer

Modern large transformers are usually of the three-phase core-type shown in fig. 20.23. Three similar limbs are connected by top and bottom yokes, each limb having primary and secondary windings, arranged concentrically. In fig. 20.23 the primary is shown star-connected and the secondary mesh-connected. Actually, the windings may be connected star–delta, delta–star, star–star or delta–delta, depending upon the conditions under which the transformer is to be used.

Example 20.9   *A three-phase transformer has 420 turns on the primary and 36 turns on the secondary winding. The supply voltage is 3300 V. Find the secondary line voltage on no load when the windings are connected (a) star–delta and (b) delta–star.*

(a) Primary phase voltage = 3300/1.73 = 1908 V,

∴ secondary phase voltage = 1908 × 36/420 = 163.5 V

= secondary line voltage.

(b) Primary phase voltage = 3300 V,

∴ secondary phase voltage = 3300 × 36/420 = 283 V,

∴ secondary line voltage = 283 × 1.73 = 490 V.

20.20 Auto-transformers

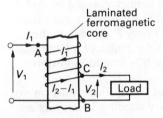

Fig. 20.24 An auto-transformer

An auto-transformer is a transformer having a part of its winding common to the primary and secondary circuits; thus, in fig. 20.24, winding AB has a tapping at C, the load being connected across CB and the supply voltage applied across AB.

I_1 and I_2 = primary and secondary currents respectively,

N_1 = no. of turns between A and B,

N_2 = no. of turns between C and B

and n = ratio of the *smaller* voltage to the *larger* voltage.

Neglecting the losses, the leakage reactance and the magnetizing current, we have for fig. 20.24,

$$n = \frac{V_2}{V_1} = \frac{I_1}{I_2} = \frac{N_2}{N_1}.$$

The nearer the ratio of transformation is to unity, the greater is the economy of conductor material. Also, for the same current density in the windings and the same peak values of the flux and of the flux density, the I^2R loss in the auto-transformer is lower and the efficiency higher than in the two-winding transformer.

Auto-transformers are mainly used for (a) interconnecting systems that are operating at roughly the same voltage and (b) starting cage-type induction motors (section 24.11). Should an auto-transformer be used to supply a low-voltage system from a high-voltage system, it is essential to earth the common connection, for example, B in fig. 20.24, otherwise there is a risk of serious shock. In general, however, an auto-transformer should not be used for interconnecting high-voltage and low-voltage systems.

20.21 Current transformers

It is difficult to construct ammeters and the current coils of wattmeters, energy (kWh) meters and relays to carry alternating currents greater than about 100 A. Furthermore, if the voltage of the system exceeds 500 V, it is dangerous to connect such instruments directly to the high-voltage. These difficulties are overcome by using current transformers.

Fig. 20.25 shows an ammeter A supplied through a current transformer. The ammeter is usually arranged to give full-scale deflection with 5 A, and the ratio of the primary to secondary turns must be such that full-scale ammeter reading is obtained with full-load current in the primary. Thus, if the primary has 4 turns and the full-load primary current is 50 A, the full-load primary ampere-turns are 200; consequently, to circulate 5 A in the secondary, the number of secondary turns must be 200/5, namely 40.

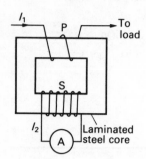

Fig. 20.25 A current transformer

If the number of primary turns were reduced to *one* and the secondary winding had 40 turns, the primary current to give full-scale reading of 5 A on the ammeter would be 200 A. Current transformers having a single-turn primary are usually constructed as shown in fig. 20.26, where P represents the primary conductor passing through the centre of a laminated steel ring C. The secondary winding S is wound uniformly around the ring.

The secondary circuit of a current transformer must on no account be opened while the primary winding is carrying a current, since all the primary ampere-turns would then be available to produce flux. The core loss due to the high flux density would cause excessive heating of the core and windings, and a dangerously high e.m.f. might be induced in the secondary winding. Hence if it is desired to remove the ammeter from the secondary circuit, the secondary winding must first be short circuited. This will not be accompanied by an excessive secondary current, since the latter is proportional to the primary current; and since the primary winding is in *series* with the load, the primary current is determined by the value of the load and not by that of the secondary current.

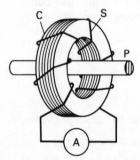

Fig. 20.26 A bar-primary current transformer

20.22 Waveform of the magnetizing current of a transformer

In fig. 20.3, the phasor for the no-load current of a transformer is shown leading the magnetic flux. A student may ask: is the flux not a maximum when this current is a maximum? The answer is that they are at their maximum values at the same instant (assuming the eddy-current loss to be negligible); but if the applied voltage is sinusoidal, then the magnetizing current of a ferromagnetic-core transformer is not sinusoidal, and a non-sinusoidal quantity cannot be represented by a phasor.

Suppose the relationship between the flux and the magnetizing current for the ferromagnetic core to be represented by the hysteresis loop in fig. 20.27(a). Also, let us assume that the waveform of the flux is sinusoidal as shown

Fig. 20.27 Waveform of magnetizing current

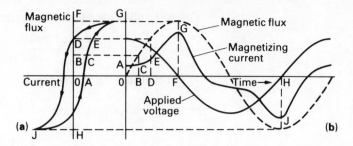

by the dotted curve in fig. 20.27(*b*). It was shown in section 20.3 that when the flux is sinusoidal, the e.m.f. induced in the primary is also sinusoidal and lags the flux by a quarter of a cycle. Hence the voltage applied to the primary must be sinusoidal and leads the flux by a quarter of a cycle as shown in fig. 20.27(*b*).

At instant O in fig. 20.27(*b*), the flux is zero and the magnetizing current is OA. At instant B, the flux is OB in fig. 20.27(*a*), and the current is BC. When the flux is at its maximum value OF, the current is also at its maximum value FG. Thus, by projecting from fig. 20.27(*a*) to the flux curve in fig. 20.27(*b*) and erecting ordinates representing the corresponding values of the current, the waveform of the magnetizing current can be derived. It can be shown that this waveform contains a sinusoidal component having the same frequency as the supply voltage and referred to as the *fundamental* component, together with other sinusoidal components having frequencies that are odd multiples of the supply frequency. The fundamental current component lags the applied voltage by an angle ϕ that is a little less than 90°, and

hysteresis loss = r.m.s. fundamental current component

$\times$ r.m.s. voltage $\times \cos \phi$.

Summary of important formulae

$$\frac{V_2}{V_1} \simeq \frac{N_2}{N_1} \simeq \frac{I_1}{I_2} \qquad (20.2)$$

$$E_1 = 4.44 N_1 f \Phi_m \qquad (20.4)$$

$$E_2 = 4.44 N_2 f \Phi_m \qquad (20.5)$$

Equivalent resistance referred to primary

$$= R_e = R_1 + R_2 (V_1/V_2)^2 \qquad (20.9)$$

Equivalent leakage reactance referred to primary

$$= X_e = X_1 + X_2 (V_1/V_2)^2 \qquad (20.10)$$

Equivalent impedance referred to primary

$$= Z_e = \sqrt{(R_e^2 + X_e^2)} \qquad (20.11)$$

Per-unit voltage regulation of a transformer

$$= \frac{\text{no-load voltage} - \text{full-load voltage}}{\text{no-load voltage}} \quad (20.12)$$

For load having power factor $= \cos \phi_2$ lagging,

$$\text{per-unit voltage regulation} = \frac{I_1 Z_e \cos(\phi_e - \phi_2)}{V_1} \quad (20.14)$$

$$= \frac{I_1(R_e \cos \phi_2 + X_e \sin \phi_2)}{V_1} \quad (20.15)$$

and

$$\text{per-unit efficiency} = \frac{I_2 V_2 \times \text{power factor}}{(I_2 V_2 \times \text{p.f.}) + P_c + I_1^2 R_1 + I_2^2 R_2} \quad (20.16)$$

$$= \left(1 - \frac{\text{losses}}{\text{input power}}\right) \quad (20.17)$$

Condition for maximum efficiency:

$$I^2 R \text{ loss} = \text{core loss} \quad (20.19)$$

From open-circuit and short-circuit tests:

Per-unit efficiency at $n \times$ full load

$$= \frac{n \times \text{full-load } S \times \text{p.f.}}{n \times \text{full-load } S \times \text{p.f.} + P_{oc} + n^2 P_{sc}} \quad (20.21)$$

and $\text{per-unit voltage regulation} = \dfrac{V_{sc} \cos(\phi_e - \phi_2)}{V_1} \quad (20.22)$

EXERCISES 20

1. The design requirements of an 11 000-V/415-V, 50-Hz, single-phase, core-type transformer are: approximate e.m.f./turn, 15 V; maximum flux density, 1.5 T. Find a suitable number of primary and secondary turns, and the net cross-sectional area of the core.   (S.A.N.C., O.2)

2. The primary winding of a single-phase transformer is connected to a 240-V, 50-Hz supply. The secondary winding has 1500 turns. If the maximum value of the core flux is 0.002 07 Wb, determine: (*a*) the number of turns on the primary winding; (*b*) the secondary induced voltage; (*c*) the net cross-sectional core area if the flux density has a maximum value of 0.465 T.
                                                         (E.M.E.U., O.2)

3. An 11 000-V/240-V, 50-Hz, single-phase, core-type transformer has a core section 25 cm × 25 cm. Allowing for a space factor of 0.9, find a suitable number of primary and secondary turns, if the flux density is not to exceed 1.2 T.   (S.A.N.C., O.2)

4. A single-phase 50-Hz transformer has 80 turns on the primary winding and 400 turns on the secondary winding. The net cross-sectional area of the core is 200 cm². If the primary winding is connected to a 240-V, 50-Hz supply, determine: (*a*) the e.m.f. induced in the secondary winding; (*b*) the maximum value of the flux density in the core.   (N.C.T.E.C., O.2)

5. A 50-kVA single-phase transformer has a turns ratio of 300/20. The primary winding is connected to a 2200-V, 50-Hz supply.

Calculate: (*a*) the secondary voltage on no load; (*d*) the approximate values of the primary and secondary currents on full load; (*c*) the maximum value of the flux.

6. A 200-kVA, 3300-V/240-V, 50-Hz, single-phase transformer has 80 turns on the secondary winding. Assuming an ideal transformer, calculate: (*a*) the primary and secondary currents on full load; (*b*) the maximum value of the flux; (*c*) the number of primary turns.

7. The following data applies to a single-phase transformer:

peak flux density in the core   = 1.41 T,
net core area   = 0.01 m^2,
current density in conductors   = 2.5 MA/m^2,
conductor diameter   = 2.0 mm,
primary supply (assume sinusoidal) = 240 V, 50 Hz.

Calculate the rating (in kVA) of the transformer and the number of turns on the primary winding. (U.L.C.I., O.2)

8. A transformer for a radio receiver has a 240-V, 50-Hz primary and three secondary windings as follows: a 1000-V winding with a centre tapping, a 4-V winding with a centre tapping and a 6.3-V winding. The net cross-sectional area of the core is 14 cm^2. Calculate the number of turns on each winding if the maximum flux density is not to exceed 1 T.

Note. Since the number of turns on a winding must be an integer, it is best to calculate the number of turns on the low-voltage winding first. In this question, the 4-V winding must have an even number of turns since it has a centre tapping.

9. The primary of a certain transformer takes 1 A at a power factor of 0.4 when connected across a 240-V, 50-Hz supply and the secondary is on open circuit. The number of turns on the primary is twice that on the secondary. A load taking 50 A at a lagging power factor of 0.8 is now connected across the secondary.

Sketch, and explain briefly, the phasor diagram for this condition, neglecting voltage drops in the transformer. What is now the value of the primary current? (App. El., L.U.)

10. A 4:1 ratio step-down transformer takes 1 A at 0.15 power factor on no load. Determine the primary current and power factor when the transformer is supplying a load of 25 A at 0.8 power factor lag. Ignore internal voltage drops.

(W.J.E.C., O.2)

11. A 3300-V/240-V, single-phase transformer, on no load, takes 2 A at power factor 0.25. Determine graphically, or otherwise, the primary current and power factor when the transformer is supplying a load of 60 A at power factor 0.9 leading.

(App. El., L.U.)

12. A three-phase transformer has its primary winding delta-connected and its secondary winding star-connected. The number of turns per phase on the primary is 4 times that on the secondary, and the secondary line voltage is 415 V. A balanced load of 20 kW, at power factor 0.8, is connected across the secondary terminals. Assuming an ideal transformer, calculate the primary voltage and the phase and line currents on the secondary and primary sides. Sketch a circuit diagram and indicate the values of the voltages and currents on the diagram. (App. El., L.U.)

13. A 50-Hz, three-phase, core-type transformer is connected star–delta and has a line voltage ratio of 11 000/415 V. The cross-section of the core is square with a circumscribing circle of 0.6 m diameter. If the maximum flux density is about 1.2 T,

calculate the number of turns per phase on the low-voltage and on the high-voltage windings. Assume the insulation to occupy 10 per cent of the gross core area.

14. If three transformers, each with a turns ratio of 12:1, are connected star–delta and the primary line voltage is 11 000 V, what is the value of the secondary no-load voltage?

   If the transformers are reconnected delta–star with the same primary voltage, what is the value of the secondary line voltage?

15. A 415-V, three-phase supply is connected through a three-phase loss-free transformer of 1:1 ratio, which has its primary connected in mesh and secondary in star, to a load comprising three 20-Ω resistors connected in mesh. Calculate the currents in the transformer windings, in the resistors and in the lines to the supply and the load. Find also the total power supplied and the power dissipated by each resistor.

16. The no-load current of a transformer is 5.0 A at 0.3 power factor when supplied at 240 V, 50 Hz. The number of turns on the primary winding is 200. Calculate: (i) the maximum value of the flux in the core; (ii) the core loss; (iii) the magnetizing current. (N.C.T.E.C.)

17. Calculate: (*a*) the number of turns required for an inductor to absorb 240 V on a 50-Hz circuit; (*b*) the length of airgap required if the coil is to take a magnetizing current of 3 A (r.m.s.); and (*c*) the phase difference between the current and the terminal voltage. Mean length of steel path, 500 mm; maximum flux density in core, 1 T; sectional area of core, 3000 mm²; maximum magnetic field strength for the steel, 250 A/m; core loss, 1.7 W/kg; density of core, 7800 kg/m³. Neglect the resistance of the winding and any magnetic leakage and fringing. Assume the current waveform to be sinusoidal.

18. Calculate the no-load current and power factor for the following 60-Hz transformer: mean length of core path, 700 mm; maximum flux density, 1.1 T; maximum magnetic field strength for the core, 300 A/m; core loss, 2.4 W/kg. All the joints may be assumed equivalent to a single airgap of 0.2 mm. Number of primary turns, 120; primary voltage, 2.0 V. Neglect the resistance of the primary winding and assume the current waveform to be sinusoidal.

19. The ratio of turns of a single-phase transformer is 8, the resistances of the primary and secondary windings are 0.85 Ω and 0.012 Ω respectively, and the leakage reactances of these windings are 4.8 Ω and 0.07 Ω respectively. Determine the voltage to be applied to the primary to obtain a current of 150 A in the secondary when the secondary terminals are short-circuited. Ignore the magnetizing current. (App. El., L.U.)

20. A single-phase transformer operates from a 240-V supply. It has an equivalent resistance of 0.1 Ω and an equivalent leakage reactance of 0.5 Ω referred to the primary. The secondary is connected to a coil having a resistance of 200 Ω and a reactance of 100 Ω. Calculate the secondary terminal voltage. The secondary winding has four times as many turns as the primary.

21. A 10-kVA single-phase transformer, for 2000 V/400 V at no load, has resistances and leakage reactances as follows. *Primary winding*: resistance, 5.5 Ω; reactance, 12 Ω. *Secondary winding*: resistance, 0.2 Ω; reactance, 0.45 Ω. Determine the approximate value of the secondary voltage at full load, 0.8 power factor (lagging), when the primary supply voltage is 2000 V. (App. El., L.U.)

22. Calculate the voltage regulation at 0.8 lagging power factor for a transformer which has an equivalent resistance of 2 per cent

and an equivalent leakage reactance of 4 per cent.

23. A 75-kVA transformer, rated at 11 000 V/240 V on no load, requires 310 V across the primary to circulate full-load currents on short-circuit, the power absorbed being 1.6 kW. Determine (a) the percentage voltage regulation and (b) the full-load secondary terminal voltage for power factors of (i) unity, (ii) 0.8 lagging and (iii) 0.8 leading.

    If the input power to the transformer on no load is 0.9 kW, calculate the per-unit efficiency at full load and at half load for power factor 0.8 and find the load (in kVA) at which the efficiency is a maximum.

24. The primary and secondary windings of a 30-kVA, 11 000/240-V transformer have resistances of 10 Ω and 0.016 Ω respectively. The total reactance of the transformer referred to the primary is 23 Ω. Calculate the percentage regulation of the transformer when supplying full-load current at a power factor of 0.8 lagging. (W.J.E.C., O.2)

25. A 50-kVA, 6360-V/240-V transformer is tested on open and short circuit to obtain its efficiency, the results of the test being as follows:

    Open circuit: primary voltage, 6360 V; primary current, 1 A; power input, 2 kW.

    Short circuit: voltage across primary winding, 180 V; current in secondary winding, 175 A; power input, 2 kW.

    Find the efficiency of the transformer when supplying full load at a power factor of 0.8 lagging and draw a phasor diagram (neglecting impedance drops) for this condition.

    (E.M.E.U., O.2)

26. A 240-V/400-V single-phase transformer absorbs 35 W when its primary winding is connected to a 240-V, 50-Hz supply, the secondary being on open circuit.

    When the primary is short-circuited and a 10-V, 50-Hz supply is connected to the secondary winding, the power absorbed is 48 W when the current has the full-load value of 15 A.

    Estimate the efficiency of the transformer at half load, 0.8 power factor lagging. (U.E.I., O.2)

27. A 1-kVA transformer has a core loss of 15 W and a full-load I^2R loss of 20 W. Calculate the full-load efficiency, assuming the power factor to be 0.9.

    An ammeter is scaled to read 5 A, but it is to be used with a current transformer to read 15 A. Draw a diagram of connections for this, giving terminal markings and currents. (N.C.T.E.C., O.2)

28. Discuss fully the energy losses in single-phase transformers. Such a transformer working at unity power factor has an efficiency of 90 per cent at both one-half load and at the full load of 500 W. Determine the efficiency at 75 per cent of full load. (App. El., L.U.)

29. A single-phase transformer is rated at 10 kVA, 240 V/100 V. When the secondary terminals are open-circuited and the primary winding is supplied at normal voltage (240 V), the current input is 2.6 A at a power factor of 0.3. When the secondary terminals are short-circuited, a voltage of 18 V applied to the primary causes the full-load current (100 A) to flow in the secondary, the power input to the primary being 240 W. Calculate: (a) the efficiency of the transformer at full load, unity power factor; (b) the load at which maximum efficiency occurs; (c) the value of the maximum efficiency. (App. El., L.U.)

30. A 400-kVA transformer has a core loss of 2 kW and the maximum efficiency at 0.8 power factor occurs when the load is 240 kW. Calculate: (*a*) the maximum efficiency at unity power factor, and (*b*) the efficiency on full load at 0.71 power factor.
(W.J.E.C., O.2)

31. A 40-kVA transformer has a core loss of 450 W and a full-load I^2R loss of 850 W. If the power factor of the load is 0.8, calculate: (*a*) the full-load efficiency; (*b*) the maximum efficiency; and (*c*) the load at which maximum efficiency occurs.
(App. El., L.U.)

32. Each of two transformers, A and B, has an output of 40 kVA. The core losses in A and B are 500 and 250 W respectively, and the full-load I^2R losses are 500 and 750 W respectively. Tabulate the losses and efficiencies at quarter, half and full load for a power factor of 0.8. For each transformer, find the load at which the efficiency is a maximum.

33. If the transformers referred to in Q. 32 be used for supplying a lighting load (unity power factor) and have their primaries permanently connected to the supply system, compare the all-day efficiencies of the two transformers, assuming the output to be 4 hours at full load, 8 hours at half load and the remaining hours on no load. What would be the saving effected in 12 weeks (7 days per week) with the better transformer if the charge for electrical energy be 5.2 p/kW · h?

   Note.
   $$\text{All-day efficiency} = \frac{\text{output energy in kW} \cdot \text{h in 24 hours}}{\text{input energy in kW} \cdot \text{h in 24 hours}}.$$

34. A 100-kVA lighting transformer has a full-load loss of 3 kW, the losses being equally divided between the core and I^2R. During one day the transformer operates on full load for 3 hours, on half load for 4 hours, the output being negligible for the remainder of the day. Calculate the all-day efficiency.

35. What is meant by the term *magnetic hysteresis*? Explain why, when steel is subjected to alternating magnetization, energy losses occur due to both hysteresis and eddy currents. The core loss in a transformer core at normal flux density was measured at frequencies of 30 and 50 Hz, the results being 30 W and 54 W respectively. Calculate: (*a*) the hysteresis loss; (*b*) the eddy-current loss at 50 Hz.
(App. El., L.U.)

   Note. For a given specimen, with maximum flux density B_m, hysteresis loss $= k_h f B_m^x$ and eddy current loss $= k_e f^2 B_m^2$ (section 9.1), where k_h, k_e and x are constants.
   $$\text{total loss} = P = k_h f B_m^x + k_e f^2 B_m^2$$
   For a given B_m, $P = k_h' f + k_e' f^2$

   Hence, if the total core loss for a given maximum flux density is known at two frequencies, k_h' and k_e' can be calculated and the hysteresis and eddy-current components of the core loss determined at any desired frequency.

36. Explain why the ferrous magnetic circuits subject to alternating magnetism are usually laminated and give examples of typical core construction.
   The total core loss in a 415-V, 50-Hz single-phase transformer is 2400 W. When a 240-V, 25-Hz supply is applied, the total core loss is 800 W. Calculate the eddy-current and hysteresis losses at normal voltage and frequency.
(E.M.E.U., O.2)

37. The core loss in a certain transformer is 80 W at 25 Hz and

204 W at 60 Hz, the maximum flux density being the same. Calculate the total core loss at 100 Hz at the same maximum flux density.

38. The following table gives the relationship between the flux and the magnetizing current in the primary winding of a transformer:

Current (A)	0	0.25	0.5	0.75	1.0	1.25	1.5
Flux (mWb)	− 1.76	− 1.48	−0.8	1.0	1.56	1.88	2.08
Current (A)	1.75	2.0	1.5	1.0	0.5	0	
Flux (mWb)	2.17	2.24	2.2	2.14	2.04	1.76	

Plot the above values and derive the waveform of the magnetizing current of the transformer, assuming the waveform of the flux to be sinusoidal and to have a peak value of 2.24 mWb.

21 Introduction to the Generalized Theory of Electrical Machines

21.1 Introductory

Electrical machines are, in general, used to convert mechanical energy into electrical energy, as in electric generators, or electrical energy into mechanical energy, as in electric motors. Most electrical machines consist of an outer stationary member and an inner rotating member. The stationary and rotating members consist of steel cores, separated by an airgap, and form a magnetic circuit in which magnetic flux is produced by currents flowing through windings situated on the two members.

21.2 Windings

One member of an electrical machine has a winding, referred to as a *field winding*, the function of which is to produce a magnetic flux. The other member has a winding or group of coils, termed an *armature winding*, in which an e.m.f. is generated by the movement of this winding relative to the magnetic flux produced by the field winding.

In d.c. machines the field winding is stationary and the armature winding is located on a rotating steel core. In a.c. synchronous generators and motors, the field winding is usually on the rotor and the armature winding is stationary, for reasons given in section 22.1.

The types of windings used on electrical machines can be grouped thus:

(a) *Concentrated or coil windings* wound around the salient poles of d.c. machines and relatively slow-speed a.c. synchronous generators and motors, as shown in figs. 3.9, 15.1 and 22.2.

(b) *Phase or distributed windings* distributed in slots located on the inner surface of a stator, as shown in figs. 22.4 and 22.5, or in slots located on the outer surface of a cylindrical rotor, as in fig. 22.3. The conductors of this type of winding are, in the majority of machines, connected in separate circuits: for instance, figs. 22.7 and 22.9 show how the conductors forming the phase windings of a three-phase stator may be connected. In these diagrams, each group of conductors occupies a phase band of two slots per pole, distributed in regular sequence over the various pole pitches.

(c) *Commutator windings* in which the coils are connected to commutator segments as shown in figs. 15.4 and 15.10. This type of winding must be wound on the rotating member of a machine in order that the commutator may act as a switch which automatically reverses the direction of the current in a coil as the latter passes brushes bearing on the commutator surface, as explained in section 15.8.

Commutator windings are used on the armatures of (i) d.c. machines, (ii) single-phase series motors (as in vacuum cleaners) and (iii) three-phase variable-speed commutator motors.

21.3 Electromechanical energy conversion

In the process of converting mechanical energy into electrical energy, or vice-versa, some of the input energy is converted into heat owing to I^2R, core and friction losses in the machine. There is also an increase or a decrease of energy stored in the magnetic field of the machine if the magnetic flux is increased or decreased. If a change of load on an electric motor is accompanied by a change of speed, some of the input energy is converted into kinetic energy during the period that the machine and its load are being accelerated. On the other hand, if the speed is decreasing, the motor and its load give up some of their kinetic energy, thereby reducing the value of the input energy during the deceleration period.

A machine is said to operate under *steady-state* conditions if its load, speed and excitation remain constant. Under these conditions,

input power = output power + power dissipated as heat.

The losses in a loaded d.c. machine have already been discussed in section 18.1, and a table indicating what happens to the electrical power supplied to an induction motor is given on p. 541. The following table shows what happens to the mechanical power supplied to a loaded a.c. generator having an exciter carried on an extension of its shaft:

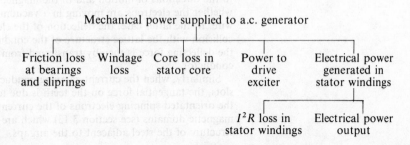

Mechanical power supplied to a.c. generator

Friction loss at bearings and sliprings	Windage loss	Core loss in stator core	Power to drive exciter	Electrical power generated in stator windings
				I^2R loss in stator windings / Electrical power output

21.4 Production of torque

It is explained in sections 15.5, 23.1, 23.6 and 24.1 that in all generators and motors, the torque produced by the magnetic flux on the rotating member is due to the distortion or skewing of that flux in the airgap between the rotating and stationary members. An equal and opposite torque is exerted on the stationary member and is transmitted through the frame of the machine to its foundations.

If a conductor carrying a current I amperes is located on the *surface* of a smooth armature core of length l metres, and if the flux density in the airgap is B teslas, the force on the conductor* is BlI newtons in a tangential direction (see section 2.8). If the distance between the conductor and the axis of rotation is r metres, the torque on the conductor is $BlIr$ newton metres; and the total torque on the armature is the algebraic sum of the torques on the conductors.

With such an arrangement it would be necessary to insert numerous wedges radially into the armature core to prevent the winding moving relative to the core, i.e. the torque would be transmitted to the core via these wedges. Such a necessity is avoided in actual machines by embedding the conductors in slots, as shown in fig. 15.10 for a rotor and in fig. 22.1 for a stator. The flux density in the slot is very low compared with that in the airgap, so that the torque exerted on the conductors is almost negligible. Practically the whole of the torque is exerted on the teeth. The value of this torque, however, is the same as that which would be produced by the current-carrying conductors if they were located on the surface of the core.

The fact that the conductors are embedded in slots and are therefore situated in regions of low flux density does not affect the magnitude of the e.m.f. generated in the conductors. Thus, when a slot containing a conductor moves towards the right in fig. 21.1, the conductor must pass through the *whole* of the flux entering the armature from the N pole when it moves under the pole face from position A to position B. The conductor therefore generates the same e.m.f. as it would have done had it been located on the surface of the core.

One can imagine the lines of magnetic flux to move

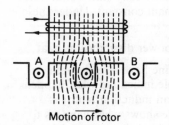

Motion of rotor

Fig. 21.1 Generation of e.m.f.

* When electrons are moving at right-angles to a magnetic field, there is a force acting on the electrons in a direction perpendicular to the directions of motion and of the magnetic field, irrespective of whether the electrons are moving in a vacuum or along a conducting wire. In the latter case, the deflection of the electrons is halted by collision with the lattice structure of the conducting material; and the deflecting force is thereby transferred from the electrons to the conductor.

Similarly, when the current-carrying conductors are embedded in slots, the tangential force on the teeth is due to the force exerted on the orientated spinning electrons of the current rings located in the magnetic domains (see section 3.13) which are held rigidly in the structure of the steel adjacent to the airgaps.

comparatively slowly across the teeth where the density is high, and then sweep rapidly across the slots where the density is very low. In this way, the *rate* at which the conductor cuts the flux is the same irrespective of whether the conductor is located on the core surface or embedded in a slot.

21.5 Relationship between torque and dM/dθ

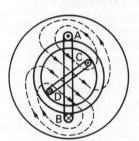

Fig. 21.2 Resultant distribution of magnetic flux

Let us consider cylindrical stator and rotor steel cores, the stator having coil AB and the rotor having coil CD wound as shown in fig. 21.2. For simplicity, each coil is indicated as a single turn. Suppose the planes of the coils to be displaced by an angle θ radians relative to each other and the coils to carry currents in the directions indicated by the dots and crosses.

The resultant magnetic field is roughly as shown by the dotted lines. (There is practically no flux in the airgaps between A and C and between B and D when the currents in the two coils are equal.) It will be seen that the flux in the airgap is skewed in such a direction as to exert an anticlockwise torque on the rotor and an equal clockwise torque on the stator. Suppose this torque to be T newton metres.

If the friction at the bearings is negligible, we can turn the rotor clockwise by the application of an external torque T equal in magnitude to the anticlockwise torque exerted by the flux on the rotor. If the rotor is thereby turned through an angle $d\theta$ radians in a clockwise direction, as shown in fig. 21.3,

$$\text{mechanical work done} = T \cdot d\theta \text{ joules.}$$

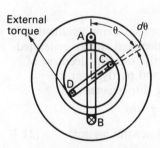

Fig. 21.3 Displacement of rotor

If L_1 and $L_2 = \begin{cases} \text{self inductances of coils} \\ \quad \text{AB and CD respectively,} \end{cases}$

$M = \begin{cases} \text{mutual inductance between} \\ \quad \text{coils AB and CD} \end{cases}$

and I_1 and I_2 = currents in coils AB and CD respectively,

then from expression (4.27),

total energy stored in the magnetic field

$$= \tfrac{1}{2}L_1 I_1^2 + \tfrac{1}{2}L_2 I_2^2 + M I_1 I_2 \text{ joules.}$$

The values of the self inductances of coils AB and CD are independent of the position of the rotor and are therefore constant. On the other hand, the value of M decreases from a maximum positive value to zero as the displacement θ is increased from zero to $\pi/2$ radians. If dM henrys is the change of mutual inductance when coil CD is turned clockwise through $d\theta$ radians, and if the currents in coils AB and CD are maintained constant at I_1 and I_2 respectively, change of energy stored in the magnetic field $= I_1 I_2 \cdot dM$ joules.

Since the value of M has been reduced by the increased displacement $d\theta$, dM is negative. Consequently the expression $I_1 I_2 \cdot dM$ represents a *decrease* of energy stored in the magnetic field.

The effect of reducing the mutual inductance is to reduce the number of flux-linkages with each coil and thus to induce an e.m.f. in each coil.

If dt = time taken to produce displacement $d\theta$,

e.m.f. induced in coil AB = $-I_2 \cdot dM/dt$ volts

and   e.m.f. induced in coil CD = $-I_1 \cdot dM/dt$ volts.

By Lenz's Law (section 2.10), the e.m.f. induced in coil AB tries to prevent the reduction in the number of flux-linkages caused by the *increased* displacement $d\theta$. Consequently the direction of this induced e.m.f. is the same as that of the current I_1. This means that the voltage applied to coil AB has to be *reduced* by $I_2 \cdot dM/dt$ volts in order to maintain the current in coil AB *constant* at I_1 during the time that the rotor is being turned through $d\theta$ radians.

Hence, electrical energy *generated* in coil AB due to this induced e.m.f.

$$= I_1 \times (I_2 \cdot dM/dt) \times dt$$
$$= I_1 I_2 \cdot dM \text{ joules.}$$

Similarly, electrical energy *generated* in coil CD

$$= I_2 \times (I_1 \cdot dM/dt) \times dt$$
$$= I_1 I_2 \cdot dM \text{ joules.}$$

Hence, total electrical energy *generated* in coils AB and CD

$$= 2I_1 I_2 \cdot dM \text{ joules.}$$

Since the friction loss at the bearings is being assumed negligible, it follows that the total electrical energy generated in coils AB and CD must come partly from the decrease of energy stored in the magnetic field and partly from the mechanical work done in producing displacement $d\theta$,

i.e. $2I_1 I_2 \cdot dM = I_1 I_2 \cdot dM + T \cdot d\theta$

$\therefore$ $T = I_1 I_2 \cdot dM/d\theta$ newton metres*   (21.1)

Figure 21.4 shows coils AB and CD in four different positions relative to each other. The dots and crosses represent the directions of the currents in the coils.

Figure 21.4(*b*) shows the coils in the same positions as in fig. 21.2, and it has already been explained that for this condition, the torque exerted on the rotor by the magnetic

---

* Expression (21.1) can be used to calculate the torque on the moving coil of the electrodynamic (or dynamometer) instrument described in section 26.8. The value of M is determined experimentally for various positions of the moving coil; and from a graph representing the values of M for different values of θ, the values of $dM/d\theta$ can be derived for various positions of the pointer over the whole of the scale.

Fig. 21.4 Direction of torque
for different relative
positions of coils

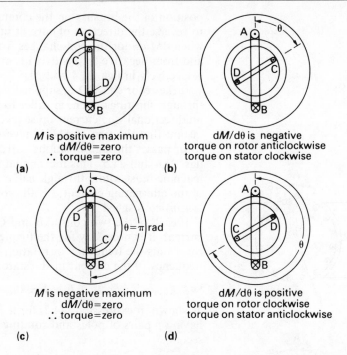

M is positive maximum
dM/dθ=zero
∴ torque=zero

(a)

dM/dθ is negative
torque on rotor anticlockwise
torque on stator clockwise

(b)

M is negative maximum
dM/dθ=zero
∴ torque=zero

(c)

dM/dθ is positive
torque on rotor clockwise
torque on stator anticlockwise

(d)

flux is anticlockwise. As the displacement θ of the rotor from
the position shown in fig. 21.4(a) is increased from zero up to
$\pi/2$ radians, the mutual inductance decreases from a positive
maximum to zero, as represented in fig. 21.5(a). Further
displacement of the rotor through $\pi/2$ radians causes the
mutual inductance to vary between zero and a negative
maximum. During this displacement of π radians, $dM/d\theta$ is
negative and constant in magnitude; hence the torque on the
rotor remains constant in an anticlockwise direction, as
shown in fig. 21.5(b). It is assumed that the reluctance of the
steel cores is negligible compared with that of the airgap, so
that the flux density in the airgap due to current in coil AB
alone would be uniform.

While the displacement is varied between π and 2π radians,
$dM/d\theta$ is positive and constant in magnitude. Consequently
the torque on the rotor remains constant in a clockwise
direction.

In a machine with d.c. excitation on the stationary member
and a commutator winding on the rotating member, the

Fig. 21.5 Variation of
mutual inductance and
torque with rotor
displacement

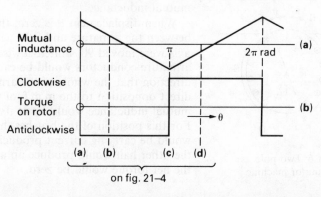

Mutual
inductance

π   2π rad   **(a)**

Clockwise

Torque
on rotor   θ

Anticlockwise

(a) (b)   (c) (d)

on fig. 21–4

position of the brushes on the commutator must be such as to reverse the direction of current in each coil at the instant when $dM/d\theta$ for the coil changes from positive to negative and from negative to positive, i.e. at instants (a) and (c) respectively in figs. 21.4 and 21.5.

If the rotor coil CD is supplied with alternating current through slip-rings, then, in order to maintain the torque unidirectional, the current must reverse its direction at the instant that the sign of $dM/d\theta$ reverses, namely each time the rotor passes through positions corresponding to (a) and (c) in figs. 21.4 and 21.5. This means that the rotor of a 2-pole machine must rotate through half a revolution in half a cycle of the alternating current; i.e. the rotor must rotate at *synchronous* speed.

If each of the two coils, AB and CD, carries an alternating current, it is necessary for the frequency of the current in the rotor plus the frequency of rotation to be equal to the frequency of the current in the stator coil,

i.e. $$f_s = f_r + (n_r \times p)$$

as shown in expression (24.2) for a 3-phase induction motor having p pairs of poles and rotating at n_r revolutions per second.

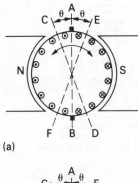

(a)

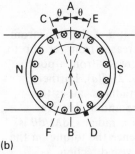

(b)

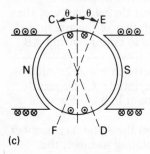

(c)

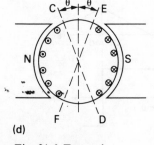

(d)

Fig. 21.6 Two-pole commutator machine

21.6 Application to commutator machines

Let us consider the *two-pole machine* of fig. 21.6(a). The axis of the poles is referred to as the *direct* or *d* axis; and axis AB, displaced 90° from the *d*-axis, is termed the *quadrature* or *q* axis.

Fig. 21.6(b) shows that when the brushes are moved forward through an angle θ from the *q*-axis, the conductors in an angle 4θ shown in fig. 21.6(c) carry current in such a direction as to produce a flux along the *d*-axis. This flux is linked with the field winding and is therefore responsible for *mutual inductance* between the armature and field windings. The current in the remaining conductors (fig. 21.6(d)) produces flux along the *q*-axis. This flux is not linked with the field winding and therefore contributes nothing towards the mutual inductance.

When displacement θ is zero, the mutual inductance between the armature and field windings is zero. If the brush axis were moved 90° anticlockwise in fig. 21.6(c), all the armature conductors would be carrying current in such a direction that the whole of the armature m.m.f. would be in direct opposition to the m.m.f. of the field winding. Hence the mutual inductance would be at its negative maximum value. For this position of the brushes, half the armature conductors would be carrying current producing a clockwise torque and the other half would produce an anticlockwise torque, so that the net torque would be zero.

If the brushes were moved 90° clockwise from the *q*-axis, the whole of the armature m.m.f. would be acting in the same direction as the field m.m.f. and the mutual inductance would be at its positive maximum value. For this brush position, half the conductors would be exerting a clockwise torque and the other half an anticlockwise torque, so that the net torque would again be zero.

With the brushes on the *q*-axis, i.e. axis AB in fig. 15.11, all the conductors moving in the magnetic field are carrying current in such a direction as to exert a clockwise torque. Hence the torque is a maximum when the brush displacement θ from the *q*-axis is zero.

It is found experimentally (see *The Unified Theory of Electrical Machines* by C. V. Jones, p. 340) that the relationship between the mutual inductance and the brush displacement is practically sinusoidal, i.e. for a brush displacement θ from the *q*-axis.

$$\text{mutual inductance} = M \sin \theta$$

$$\therefore \qquad dM/d\theta = M \cos \theta.$$

In a d.c. machine, the brushes are normally on or very near the *q*-axis, so that $\theta = 0°$ and $dM/d\theta = M$ henrys per radian.

As far as the present discussion is concerned, the d.c. machine could be totally enclosed, with two pairs of wires brought out to external terminals, namely armature terminals AA_1 and field terminals FF_1, as shown in fig. 17.1. If the armature and field currents are represented by I_a and I_f respectively, we have from expression (21.1):

$$\text{torque} = T = I_a I_f \, dM/d\theta = I_a I_f M \text{ newton metres} \qquad (21.2)$$

If $\qquad E_r$ = rotational e.m.f. generated in the armature

and $\qquad \omega_r$ = angular speed in radians per second = $2\pi n_r$

where $\qquad n_r$ = speed in revolutions per second,

then, if the machine is operating as a motor (see section 16.3),

$$\text{mechanical power developed} = E_r I_a.$$

From expression (16.5),

$$E_r I_a = \omega_r T = \omega_r I_a I_f M$$

$$\therefore \qquad M = \frac{E_r}{\omega_r I_f} = \frac{E_r}{2\pi n_r I_f} \qquad (21.3)$$

The value of *M* for a given machine can therefore be determined experimentally from the open-circuit characteristic (fig. 17.4) obtained by running the machine as a generator at a known speed; and the torque for given values of I_a and I_f can then be calculated from expression (21.2).

21.7 Application to three-phase synchronous machines

In order to consider the application of the generalized theory of machines when applied to synchronous machines, we need first of all to have some understanding of the construction of such machines. This is covered in Chapters 22 and 23 so that the analysis of the 3-phase synchronous machine in terms of the generalized theory is deferred to section 23.11.

It is interesting to note that it was the performance of such machines in response to fault and other transient conditions which first caused the development of the generalized theory, and subsequently this has played a major role in the development of automated systems.

EXERCISES 21

1. An electrodynamic milliammeter has a deflection of 50° when a direct current of 20 mA flows through it. The inductance of the instrument changes at the rate of 1.2 mH/degree over the working range. Calculate the control torque, in micronewton metres per degree, exerted by the hairsprings. (Note that $dM/d\theta$ must be expressed in henrys per *radian*).

2. The hairsprings of an electrodynamic wattmeter exert a control torque of $0.6\,\mu\text{N}\cdot\text{m}$/degree. The voltage circuit has a resistance of $6000\,\Omega$ and negligible reactance. The mutual inductance between the fixed and moving coils is $2\,\mu\text{H}$/degree over the working range. The voltage circuit is connected across a 240-V a.c. supply and the current in the current coil is 5 A, lagging the voltage by 30°. Calculate the deflection.

3. For the machine referred to in Q. 13, Exercises 16, p. 412, calculate the value of M when the machine is taking an armature current of 60 A and a field current of 4 A. If the speed is 600 r/min, what is the value of the mechanical power developed?

4. A universal series motor exerts a gross torque of $0.4\,\text{N}\cdot\text{m}$ when taking a direct current of 0.6 A. Calculate the value of M. If the total resistance of the armature and field windings is $50\,\Omega$ and the speed is 3000 r/min, what is the value of the applied voltage?

5. The motor of Q. 4 is connected across a 240-V, 50-Hz supply and is exerting a torque of $0.4\,\text{N}\cdot\text{m}$. The total self inductance of the field and armature windings is 0.6 H and the core and friction losses are negligible. Determine (*a*) the current, (*b*) the rotational e.m.f., (*c*) the speed, (*d*) the output power, (*e*) the power factor and (*f*) the efficiency. Note that $V^2 = (IR + E_r)^2 + (IX)^2$, where R and X represent the total resistance and reactance respectively and E_r is the rotational e.m.f. in phase with the current.

22 A.C. Synchronous Machine Windings

22.1 General arrangement of synchronous machines

Let us first consider why synchronous machines, whether motors or generators, are usually constructed with stationary armature windings and rotating poles. Suppose we have a 20-MVA, 11-kV, three-phase synchronous machine; then from expression (19.6),

$$20 \times 10^6 = 1.73 \times I_{\mathrm{L}} \times 11 \times 10^3$$

$$\therefore \qquad \text{line current} = I_{\mathrm{L}} = 1050 \, \text{A}.$$

Hence, if the machine was constructed with stationary poles and a rotating three-phase winding, three slip-rings would be required, each capable of dealing with 1050 A, and the insulation of each ring together with that of the brushgear would be subjected to a working voltage of 11/1.73, namely 6.35 kV. Further, it is usual to connect synchronous generator windings in star and to join the star-point through a suitable resistor to a metal plate embedded in the ground so as to make good electrical contact with earth; consequently a fourth slip-ring would be required.

By using a stationary a.c. winding and a rotating field system, only two slip-rings are necessary and these have to deal with only the exciting current. Assuming the power required for exciting the poles of the above machine to be 150 kW and the voltage to be 400 V,

$$\text{exciting current} = 150 \times 1000/400 = 375 \, \text{A}.$$

In other words, the two slip-rings and brushgear would have to deal with only 375 A and be insulated for merely 400 V. Hence, by using a stationary a.c. winding and rotating poles, the construction is considerably simplified and the slip-ring losses are reduced.

Further advantages of this arrangement are: (a) the extra space available for the a.c. winding makes it possible to use more insulation and to enable operating voltages of up to 33 kV; (b) with the simpler and more robust mechanical construction of the rotor, a high speed is possible, so that a greater output is obtainable from a machine of given dimensions.

The slots in the laminated stator core of a synchronous machine are usually semi-enclosed, as shown in fig. 22.1, so

Fig. 22.1 Portion of stator of a.c. synchronous machine

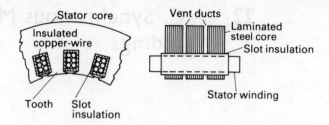

as to distribute the magnetic flux as uniformly as possible in the airgap, thereby minimizing the ripple that would appear in the e.m.f. waveform if open slots were used.

In section 6.2 it was explained that if a synchronous machine has p pairs of poles and the speed is n revolutions per second,

$$\text{frequency} = f = np \qquad (22.1)$$

Hence for a 50-Hz supply, a two-pole synchronous machine must operate at 3000 r/min, a four-pole synchronous machine at 1500 r/min, etc.

22.2 Types of rotor construction

Synchronous machines can be divided into two categories: (a) those with salient or projecting poles; and (b) those with cylindrical rotors.

The salient-pole construction is used in comparatively small machines and machines driven at a relatively low speed. For instance, if a 50-Hz synchronous machine is to operate at, say, 375 r/min, then, from expression (22.1), the machine must have 16 poles; and to accommodate all these poles, the synchronous machine must have a comparatively large diameter. Since the output of a machine is roughly proportional to its volume, such a synchronous machine would have a small axial length. Fig. 22.2 shows one arrangement of salient-pole construction. The poles are made

Fig. 22.2 Portion of a salient-pole rotor

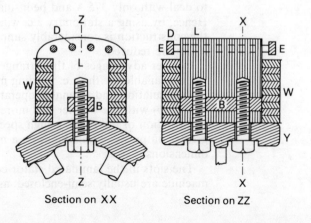

Section on XX   Section on ZZ

of fairly thick steel laminations, L, riveted together and bolted
to a steel yoke wheel Y, a bar of mild steel B being inserted
to improve the mechanical strength. The exciting winding W
is usually an insulated copper strip wound on edge, the coil
being held firmly between the pole tips and the yoke wheel.
The pole tips are well rounded so as to make the flux
distribution around the periphery/nearer a sine wave and thus
improve the waveform of the generated e.m.f. Copper rods D,
short-circuited at each end by copper bars E, are usually
inserted in the pole shoes.

Most synchronous machines are essentially high-speed
machines. The centrifugal force on a high-speed rotor is
enormous: for instance, a mass of 1 kg on the outside of a
rotor of 1 m diameter, rotating at a speed of 3000 r/min, has a
centrifugal force ($= mv^2/r$) of about 50 kN acting upon it. To
withstand such a force the rotor is usually made of a solid
steel forging with longitudinal slots cut as indicated in
fig. 22.3, which shows a two-pole rotor with 8 slots and 2
conductors per slot. In an actual rotor there are more slots
and far more conductors per slot; and the winding is in the
form of insulated copper strip held securely in position by
phosphor-bronze wedges. The regions forming the centres of
the poles are usually left unslotted. The horizontal dotted
lines joining the conductors in fig. 22.3 represent the end-
connections. If the rotor current has the direction represented
by the dots and crosses in fig. 22.3, the flux distribution is
indicated by the light dotted lines. In addition to its
mechanical robustness, this cylindrical construction has the
advantage that the flux distribution around the periphery is
nearer a sine wave than is the case with the salient-pole
machine. Consequently, a better e.m.f. waveform is obtained.

Fig. 22.3 A cylindrical rotor

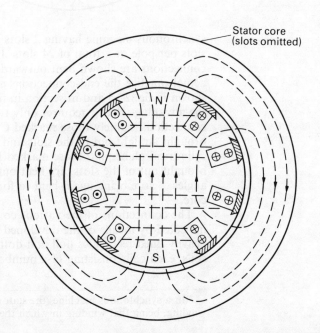

Stator core
(slots omitted)

22.3 Stator windings

In this book we shall consider only two types of three-phase winding, namely (*a*) single-layer winding and (*b*) double-layer winding; and of these types we shall consider only the simplest forms.*

Single-layer winding   The main difficulty with single-layer windings is to arrange the end-connections so that they do not obstruct one another. Fig. 22.4 shows one of the most common methods of arranging these end-connections for a four-pole, three-phase

Fig. 22.4 End-connections of a three-phase single-layer winding

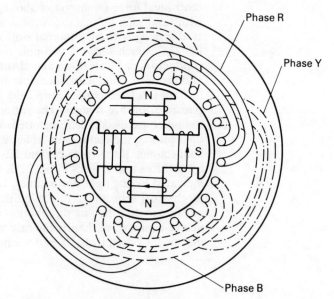

Phase R

Phase Y

Phase B

synchronous machine having 2 slots per pole per phase, i.e. 6 slots per pole or a total of 24 slots. In fig. 22.4, all the end-connections are shown bent outwards for clearness; but in actual practice the end-connections are usually shaped as shown in fig. 22.5 and in section in fig. 22.6. This method has the advantage that it requires only two shapes of end-connections, namely those marked C in fig. 22.6, which are brought straight out of the slots and bent so as to lie on a cylindrical plane, and those marked D. The latter, after being brought out of the slots, are bent outwards roughly at right-angles, before being again bent to form an arch alongside the core.

The connections of the various coils are more easily indicated by means of the developed diagram of fig. 22.7. The heavily lined rectangles (full and dotted lines) represent the coils, each coil consisting of a number of turns; and the thin

---

* In a synchronous machine, the stator winding is the armature winding, being that winding in which the operating e.m.f. is induced.

Fig. 22.5 End-connections of
a three-phase single-layer
winding

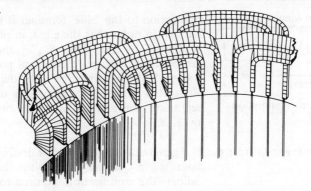

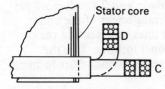

Fig. 22.6 Sectional view of
end-connections

lines — other than those representing the poles — indicate the
connections between the various coils. The width of the pole
face has been made two-thirds of the pole pitch, a pole pitch
being the distance between the centres of adjacent poles. The
poles in fig. 22.7 are assumed to be *behind* the winding and
moving towards the right. From the righthand rule
(section 2.10) — bearing in mind that the thumb represents
the direction of *motion of the conductor relative to the flux*,
namely towards the left in fig. 22.7 — the e.m.f.s in the

Fig. 22.7 Three-phase single-
layer winding

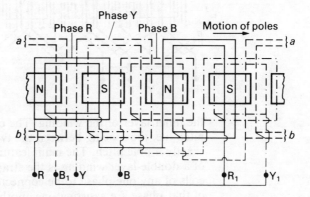

conductors opposite the poles are as indicated by the
arrowheads. The connections between the groups of coils
forming any one phase must be such that all the e.m.f.s are
assisting one another.

Since the stator has 6 slots per pole and since the rotation
of the poles through one pole pitch corresponds to half a
cycle of the e.m.f. wave or 180 electrical degrees, it follows
that the spacing between two adjacent slots corresponds to
180°/6, namely 30 electrical degrees. Hence, if the wire
forming the beginning of the coil occupying the first slot is
taken to the 'red' terminal R, the connection to the 'yellow'
terminal Y must be a conductor from a slot 4 slot-pitches
ahead, namely from the 5th slot, since this allows the e.m.f. in
phase Y to lag the e.m.f. in phase R by 120°. Similarly, the

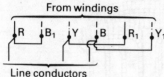

Fig. 22.8 Delta-connection of windings

connection to the 'blue' terminal B must be taken from the 9th slot in order that the e.m.f. in phase B may lag the e.m.f. in phase Y by 120°. Ends R_1, Y_1 and B_1 of the three phases can be joined to form the neutral point of a star-connected system. If the windings are to be delta-connected, end R_1 of phase R is joined to the beginning of Y, end Y_1 to the beginning of B and end B_1 to the beginning of R, as shown in fig. 22.8.

Double-layer winding

The general arrangement of a double-layer winding has been described in Chapter 15, and it is therefore only necessary to indicate the modification required to give a three-phase supply.

Let us consider a four-pole three-phase synchronous machine with 2 slots per pole per phase and 2 conductors per slot. Fig. 22.9 shows the simplest arrangement of the end-connections of one phase, the thick lines representing the conductors (and their end-connections) forming, say, the outer layer and the thin lines representing conductors forming

Fig. 22.9 One phase of a three-phase double-layer winding

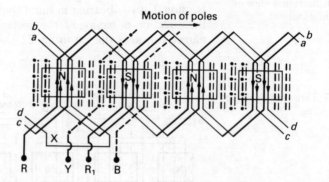

the inner layer of the winding. The coils are assumed full-pitch, i.e. the spacing between the two sides of each turn is exactly a pole pitch. The main feature of the end-connections of a double-layer winding is the strap X, which enables the coils of any one phase to be connected so that all the e.m.f.s of that phase are assisting one another.

Since there are 6 slots per pole, the phase difference between the e.m.f.s of adjacent slots is 180°/6, namely 30 electrical degrees; and since there is a phase difference of 120° between the e.m.f.s of phases R and Y, there must be 4 slot pitches between the first conductor of phase R and that of phase Y. Similarly, there must be 4 slot pitches between the first conductors of phases Y and B. Hence, if the outer conductor of the third slot is connected to terminal R, the corresponding conductor in the seventh slot, in the direction of rotation of the poles, is connected to terminal Y and that in the eleventh slot to terminal B.

In general, the single-layer winding is employed where the machine has a large number of conductors per slot, whereas the double-layer winding is more convenient when the number of conductors per slot does not exceed 8.

22.4 Expression for the e.m.f. of a stator winding

Let Z = no. of conductors in series per *phase*,

Φ = useful flux per pole, in webers,

p = no. of pairs of poles

and n_1 = speed in revolutions per second.

Magnetic flux cutting a conductor in 1 revolution = $\Phi \times 2p$

Magnetic flux cutting a conductor in 1 second = $2\Phi p \times n_1$

∴ average e.m.f. generated in 1 conductor = $2\Phi p n_1$ volts

If the stator of a three-phase machine had only 3 slots per pole, i.e. 1 slot/p/ph,* and if the coils were full-pitch, the e.m.f.s generated in all the conductors of one phase would be in phase with one another and could therefore be added arithmetically. Hence, for a winding concentrated in 1 slot/p/ph,

average e.m.f. per phase = $Z \times 2\Phi p n_1$ volts.

If the e.m.f. wave be assumed to be sinusoidal,

$$\left.\begin{array}{l}\text{r.m.s. value of e.m.f. per}\\ \text{phase for 1 slot/p/ph}\end{array}\right\} = 1.11 \times 2Z \times n_1 p \times \Phi$$

$$= 2.22 Z f \Phi \qquad (22.2)$$

A winding concentrated in 1 slot per pole per phase would have two disadvantages:

(*a*) The size of such slots and the number of conductors per slot would be so great that it would be difficult to prevent the insulation on the conductors in the centre of the slot becoming overheated, since most of the heat generated in the slots has to flow radially outwards to the steel core.

(*b*) The waveform of the e.m.f. would be similar to that of the flux distribution around the inner periphery of the stator; and, in general, this would not be sinusoidal.

22.5 Effect of distributing the winding: Distribution factor

By distributing the winding in 2 or more slots per pole per phase, the number of conductors per slot is reduced, thereby reducing the temperature rise in the centre of the slot. Also, the e.m.f. waveform is improved; for instance, consider a three-phase winding distributed in 2 slots per pole per phase as in fig. 22.4. With a salient-pole synchronous machine

---

* '/p/ph' is an abbreviation for 'per pole per phase'.

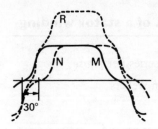

Fig. 22.10 E.M.F.
waveforms of a salient-pole

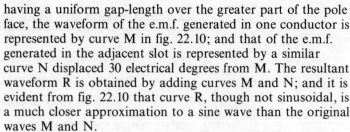

Fig. 22.11 E.M.F.s in a
three-phase synchronous
machine with 2 slots/p/ph

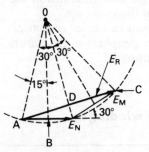

Fig. 22.12 E.M.F.s in a
three-phase synchronous
machine with 2 slots/p/ph

having a uniform gap-length over the greater part of the pole face, the waveform of the e.m.f. generated in one conductor is represented by curve M in fig. 22.10; and that of the e.m.f. generated in the adjacent slot is represented by a similar curve N displaced 30 electrical degrees from M. The resultant waveform R is obtained by adding curves M and N; and it is evident from fig. 22.10 that curve R, though not sinusoidal, is a much closer approximation to a sine wave than the original waves M and N.

Let us consider how the magnitude of the e.m.f. is affected when the winding is distributed in 2 slots per pole per phase; and for simplicity it will be assumed that the e.m.f. generated in each conductor is sinusoidal. The e.m.f.s, E_M and E_N, induced in coils occupying two adjacent slots M and N of any one phase are equal in magnitude but differ in phase by 30°. The resultant e.m.f. is the phasor sum of E_M and E_N and is represented by the diagonal E_R in fig. 22.11. Since the diagonals of the parallelogram bisect each other at right-angles,

$$\text{resultant e.m.f.} = E_R = 2E_M \cos 15° = 2 \times 0.966E_M$$
$$= 1.932E_M.$$

Had the same number of conductors been concentrated in 1 slot/p/ph, the resultant e.m.f. would have been $2E_M$;

$$\text{hence} \quad \frac{\text{e.m.f. with winding in 2 slots/p/ph}}{\text{e.m.f. with winding in 1 slot/p/ph}} = \frac{1.932E_M}{2E_M}$$
$$= 0.966.$$

The ratio $\dfrac{\text{e.m.f. with a distributed winding}}{\text{e.m.f. with a concentrated winding}}$ is termed the *distribution*, *breadth* or *spread factor* of the winding and may be represented by the symbol k_d.

An alternative method — a method that can be used to derive a general expression for k_d — of deriving the value of the distribution factor for a three-phase winding distributed in 2 slots/p/ph is to draw the phasors representing E_M and E_N end-on as in fig. 22.12. Perpendiculars drawn at mid-points of the phasors meet at a point O, which is the centre of the circumscribing circle shown dotted. The phasors are chords of this circle, and each phasor subtends an angle of 30° at the centre, namely an angle equal to the phase difference between the phasors.

The resultant e.m.f., E_R, is represented by chord AC of the circle. From fig. 22.12, it is evident that:

$$E_R = 2AD = 2OA \sin 30°.$$

Also, from fig. 22.12, it can be seen that:

$$E_M = E_N = 2AB = 2OA \sin 15°$$

Hence, for a three-phase winding distributed in 2 slots/p/ph,

$$\text{distribution factor} = k_d = \frac{E_R}{2E_M} = \frac{2OA \sin 30°}{2 \times 2OA \sin 15°}$$

$$= \frac{\sin 30°}{2 \sin 15°} = 0.966.$$

In general,

if $\quad m$ = no. of slots per pole per phase,

$$\alpha = \begin{cases} \text{phase difference between e.m.f.s generated in} \\ \text{conductors occupying adjacent slots,} \end{cases}$$

and $\quad OA = \begin{cases} \text{radius of the circumscribing circle for the} \\ \text{polygon of phasors, drawn as in fig. 22.12,} \end{cases}$

angle subtended at centre of circumscribing circle by *each* phasor = α

$\therefore \qquad$ arithmetic sum of e.m.f.s = $m \times$ e.m.f. per coil

$$= m \times 2OA \sin \alpha/2.$$

Angle subtended at centre of circumscribing circle by phasor representing the resultant e.m.f. = $m\alpha$,

$\therefore \qquad$ resultant e.m.f. = $E_R = 2OA \sin m\alpha/2$,

$\therefore \qquad$ distribution factor = $k_d = \dfrac{\text{e.m.f. with distributed winding}}{\text{e.m.f. with concentrated winding}}$

$$= \frac{\text{phasor sum of e.m.f.s}}{\text{arithmetic sum of e.m.f.s}}$$

$$= \frac{2OA \sin m\alpha/2}{m \times 2OA \sin \alpha/2}$$

$$= \frac{\sin m\alpha/2}{m \sin \alpha/2} \qquad (22.3)$$

Example 22.1 *Calculate the value of the distribution factor for a three-phase winding of a 4-pole synchronous machine having 36 slots.*

No. of slots/pole = 36/4 = 9,

$\therefore \qquad$ no. of slots/p/ph = m = 9/3 = 3.

$$\alpha = 180°/9 = 20°,$$

$\therefore \qquad m\alpha/2 = 3 \times 20°/2 = 30°.$

Substituting in expression (22.3), we have:

$$\text{distribution factor} = k_d = \frac{\sin 30°}{3 \sin 10°} = 0.960.$$

Example 22.2 *Calculate the distribution factor for a single-phase synchronous motor having 6 slots per pole (a) when all the slots are wound and (b) when only four adjacent slots per pole are wound, the remaining slots being unwound.*

(*a*) When all the slots are wound,

$$m = 6; \quad \alpha = 180°/6 = 30°; \quad \text{and} \quad m\alpha/2 = 6 \times 30°/2 = 90°;$$

$$\therefore \qquad k_d = \frac{\sin 90°}{6 \sin 15°} = 0.644.$$

(b) When only 4 adjacent slots per pole are wound,

$$m = 4; \quad \alpha = 30°; \quad \text{and} \quad m\alpha/2 = 4 \times 30°/2 = 60°;$$

$$\therefore \quad k_d = \frac{\sin 60°}{4 \sin 15°} = 0.837.$$

It follows from the above values of k_d that if the number of conductors per slot remained the same:

$$\frac{\text{e.m.f. with all slots wound}}{\text{e.m.f. with 4 adjacent slots/pole wound}} = \frac{6 \times 0.644}{4 \times 0.837} = 1.15.$$

This means that an increase of 50 per cent in the number of conductors gives only 15 per cent increase of e.m.f. This is the reason why it is customary to wind only about two-thirds of the slots in a single-phase stator winding.

22.6 Short-pitch winding: Pitch factor

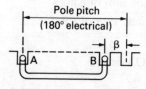

Fig. 22.13 Short-pitch coil

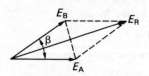

Fig. 22.14 E.M.F.s in the coil-sides of a short-pitch coil

The waveform of the resultant e.m.f. induced in a synchronous machine can be improved by making the coil pitch less than a pole pitch, as in fig. 22.13. This practice is only possible with the two-layer type of winding shown in fig. 22.9.

With a full-pitch coil, the e.m.f.s generated in the two sides are in phase with each other as far as the resultant e.m.f. of the coil is concerned. When the coil is short-pitch by an angle β electrical degrees, as shown in fig. 22.13, the e.m.f.s generated in coil sides A and B differ in phase by an angle β, and can be represented by the phasors E_A and E_B respectively in fig. 22.14. Since the diagonals of the parallelogram drawn on E_A and E_B bisect at right-angles,

$$\text{resultant e.m.f.} = E_R = 2E_A \cos \beta/2,$$

$$\therefore \quad \textit{pitch or coil-span factor} = k_p = \frac{\text{e.m.f. with short-pitch coil}}{\text{e.m.f. with full-pitch coil}}$$

$$= \frac{2E_A \cos \beta/2}{2E_A} = \cos \beta/2 \qquad (22.4)$$

For a full-pitch winding, $k_p = 1.0$.

Example 22.3   *A synchronous generator has 9 slots per pole. If each coil spans 8 slot pitches, what is the value of the pitch factor?*

Since the winding pitch is one slot pitch less than the pole pitch,

$$\beta = 180°/9 = 20°.$$

Substituting in expression (22.4), we have:

$$\text{pitch factor} = k_p = \cos 10° = 0.985.$$

22.7 General expression for the e.m.f. of a synchronous machine

If k_d = distribution factor of the winding

and k_p = pitch factor of the winding,

then from expression (22.2),

$$\text{r.m.s. value of e.m.f./phase} = 2.22 k_d k_p Z f \Phi \qquad (22.5)$$

Example 22.4 *A three-phase star-connected synchronous generator on open circuit is required to generate a line voltage of 3600 V, 50 Hz, when driven at 500 r/min. The stator has 3 slots/p/ph and 10 conductors per slot. Calculate: (a) the number of poles and (b) the useful flux per pole. Assume all the conductors per phase to be connected in series and the coils to be full-pitch.*

(*a*) From expression (22.1), $50 = 500 \times p/60$

∴   no. of poles = $2p = 12$.

(*b*)   No. of slots per phase = $3 \times 12 = 36$

∴   no. of conductors per phase = $36 \times 10 = 360$

$$\text{e.m.f. per phase} = \frac{3600}{1.73} = 2080 \text{ V}.$$

From example 22.1, $k_d = 0.96$ for 3 slots/p/ph.

For full-pitch coils, $k_p = 1.0$.

Substituting in expression (22.5), we have:

$$2080 = 2.22 \times 0.96 \times 1.0 \times 360 \times 50 \times \Phi$$

∴ $\Phi = 0.0543 \text{ Wb}.$

22.8 Production of rotating magnetic flux by two-phase currents

Let us consider a two-pole, two-phase winding having, for simplicity, only one slot per pole per phase, as shown in fig. 22.15, where AA_1 and BB_1 represent the coils of the two

Fig. 22.15 Distribution of magnetic flux due to two-phase currents

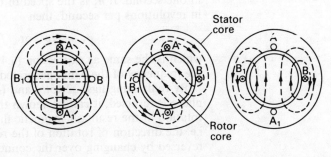

(a)   (b)   (c)

phases. Let us assume the currents to be positive when they are flowing towards the paper in conductors A and B and outwards in conductors A_1 and B_1. If the currents in the two phases are represented by curves A and B in fig. 22.16 then at instant a the current in phase A is positive and at its maximum value, and is represented by the cross and dot on conductors A and A_1 respectively in fig. 22.15(a). The current in conductors B and B_1 is zero at that instant; consequently the distribution of the magnetic flux is represented by the thin dotted lines in fig. 22.15(a).

At instant b in fig. 22.16, the current in each phase is positive and is 0.707 of the maximum value; and the

Fig. 22.16 Two-phase currents

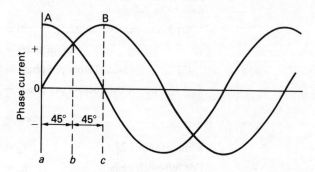

distribution of the resultant flux due to these currents is shown in fig. 22.15(b). It will be seen that the axis of the resultant flux has turned clockwise through 45° from that of fig. 22.15(a). At instant c in fig. 22.16, the current in phase A has decreased to zero; hence the distribution of the flux is as shown in fig. 22.15(c) and its axis has rotated through another 45°.

It will be seen that the resultant magnetic flux rotates through a quarter of a pole pitch in an eighth of a cycle, and therefore through two pole pitches in one cycle. If the stator is wound with p pairs of poles, the flux rotates through $1/p$ revolution in one cycle, and therefore through f/p revolutions in one second. If n_1 is the speed of the resultant magnetic flux in revolutions per second, then

$$n_1 = f/p \quad \text{or} \quad f = n_1 p$$

which is the same as expression (6.4) derived in section 6.2 for the frequency of the e.m.f. generated in an a.c. generator.

By re-drawing figs. 22.15(b) and (c) with the current in, say, phase B reversed, it can be shown that the direction of rotation of the resultant magnetic flux is then anticlockwise, i.e. the direction of rotation of the resultant flux can be reversed by changing over the connections to *one* of the phases.

22.9 Mathematical derivation of the magnitude and speed of the resultant magnetic flux due to two-phase currents

In an actual machine the windings are distributed in two or more slots per pole per phase, and the distribution of the flux due to one phase depends upon the spread of the coils and the saturation of the teeth. For simplicity, we shall represent the airgap as a horizontal plane (fig. 22.17) and assume that the space distribution of the *flux density* produced by the current in each phase is sinusoidal over a pole pitch. Thus, fig. 22.17(*a*) shows the distribution of the flux density over two pole pitches when the current in phase A is a maximum and that in phase B is zero, namely the values at instant *a* in fig. 22.16. B_m represents the maximum value of the flux density at that instant.

At instant *b* in fig. 22.16, the current in each phase is 0.707 of the maximum value, so that the distributions of the flux densities are represented by the full-line curves in fig. 22.17(*b*). Assuming the magnetic circuit to be unsaturated, we can add together the flux densities due to the two phases and thus derive the dotted curve representing the distribution of the

Fig. 22.17 Distribution of flux densities due to two-phase currents

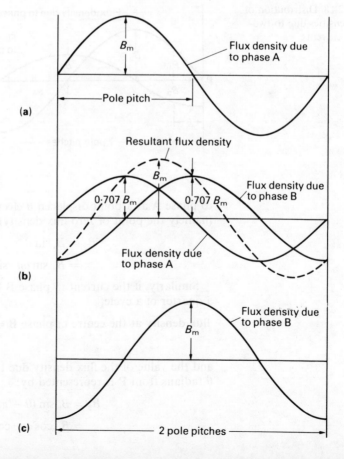

resultant flux density. At the point of maximum resultant density, the value of the flux density due to each phase is $0.707 \times 0.707 B_m$, namely $0.5 B_m$; hence it follows that the maximum value of the resultant density is B_m. Also, it is seen that the point of maximum resultant density has shifted a quarter of a pole pitch in an eighth of a cycle.

Fig. 22.17(*a*) shows the flux density distribution at instant *c* in fig. 22.16. Again, it will be seen that the maximum value is B_m and that its position has shifted another quarter of a pole pitch in an eighth of a cycle. Hence, for the instants considered, the peak value of the resultant flux density remains constant at B_m, and its position moves a quarter of a pole pitch during each eighth of a cycle and therefore through two pole pitches in one cycle. It will now be shown that the peak value of the resultant flux density is constant at *every* instant and that its position moves at a *uniform* speed.

Let B_m be again the maximum value of the flux density due to one phase. Since the value of the flux density at a given point varies sinusoidally with *time*, its maximum value, *t* seconds after it has passed through zero from negative to positive values, is represented for phase A by $B_a = B_m \sin \omega t$, as shown in fig. 22.18. Hence the value of the flux density due

Fig. 22.18 Distribution of flux densities due to two-phase currents

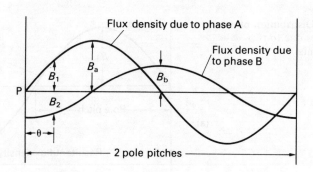

to phase A at a point displaced θ electrical radians from P (namely the point of zero flux density) is represented by:

$$B_1 = B_a \sin \theta$$

$$= B_m \sin \omega t \cdot \sin \theta.$$

Similarly, if the current in phase B lags that in phase A by a quarter of a cycle,

flux density at the centre of phase B $= B_b = B_m \sin (\omega t - \pi/2)$

$$= -B_m \cos \omega t,$$

and the value of the flux density due to phase B at a point θ radians from P is represented by:

$$B_2 = B_b \sin (\theta - \pi/2)$$

$$= B_m \cos \omega t \cdot \cos \theta.$$

Assuming the magnetic circuit to be unsaturated,

resultant flux density at a point θ radians from P $\left.\right\}$ $= B_1 + B_2$

$$= B_m(\sin \omega t \cdot \sin \theta + \cos \omega t \cdot \cos \theta)$$
$$= B_m \cos (\omega t - \theta)$$

and is constant at B_m when $(\omega t - \theta) = 0$, i.e. when

$$\theta = \omega t = 2\pi f t.$$

For a value of t equal to the duration of one cycle, namely $1/f$ second,

corresponding value of $\theta = 2\pi$ radians.

Since the distribution of the resultant flux density remains sinusoidal and the peak value remains constant, it follows that the total flux over a pole pitch is constant and rotates at a uniform speed through 2π electrical radians, namely two pitches, in one cycle. If the machine has p pairs of poles, the resultant magnetic flux rotates through $1/p$ revolution in 1 cycle,

$\therefore$ speed of rotating magnetic flux

$$= n_1 = f/p \text{ revolutions per second}.$$

22.10 Production of rotating magnetic flux by three-phase currents

Let us again consider a two-pole machine and suppose the three-phase winding to have only one slot per pole per phase, as shown in fig. 22.19. The end-connections of the coils are

Fig. 22.19 Distribution of magnetic flux due to three-phase currents

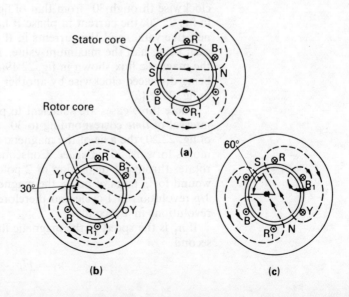

(a)

(b)

(c)

not shown, but are assumed to be similar to those already shown in fig. 22.4; thus R and R_1 represent the 'start' and the 'finish' of the 'red' phase, etc. It will be noted that R, Y and B are displaced 120 electrical degrees relative to one another. Let us also assume that the current is positive when it is flowing inwards in conductors R, Y and B, and therefore outwards in R_1, Y_1 and B_1. As far as the present discussion is concerned, the rotor core need only consist of circular steel laminations to provide a path of low reluctance for the magnetic flux.

Suppose the currents in the three phases to be represented by the curves in fig. 22.20; then at instant a the current in phase R is positive and at its maximum value, whereas in phases Y and B the currents are negative and each is half the maximum value. These currents, represented in direction by

Fig. 22.20 Three-phase currents

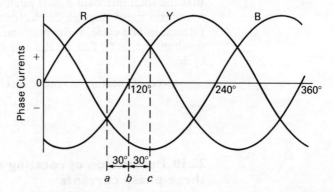

dots and crosses in fig. 22.19(a), produce the magnetic flux represented by the dotted lines. At instant b in fig. 22.20, the currents in phases R and B are each 0.866 of the maximum; and the distribution of the magnetic flux due to these currents is shown in fig. 22.19(b). It will be seen that the axis of this field is in line with coil YY_1 and therefore has turned clockwise through 30° from that of fig. 22.19(a). At instant c in fig. 22.20, the current in phase B has attained its maximum negative value, and the currents in R and Y are both positive, each being half the maximum value. These currents produce the magnetic flux shown in fig. 22.19(c), the axis of this flux being displaced clockwise by another 30° compared with that in fig. 22.19(b).

These three cases are sufficient to prove that for every interval of *time* corresponding to 30° along the horizontal axis of fig. 22.20, the axis of the magnetic flux in a two-pole stator moves forward 30° in *space*. Consequently, in 1 cycle, the flux rotates through 1 revolution or 2 pole pitches. If the stator is wound for p pairs of poles, the magnetic flux rotates through $1/p$ revolution in 1 cycle and therefore through f/p revolutions in 1 second.

If n_1 is the speed of the magnetic flux in revolutions per second,

$$n_1 = f/p \tag{22.6}$$

or
$$f = n_1 p$$

which is the same as expression (6.4) derived for a synchronous machine in section 6.2. It follows that if the stator in fig. 22.19 had the same number of poles as the synchronous generator supplying the three-phase currents, the magnetic flux in fig. 22.19 would rotate at *exactly* the same speed as the poles of the synchronous generator. It also follows that when a two-phase or a three-phase synchronous generator is supplying a balanced load, the stator currents of that synchronous generator set up a resultant magnetic flux that rotates at exactly the same speed as the poles. Hence the magnetic flux is said to rotate at *synchronous speed*, n_1.

22.11 Mathematical derivation of the magnitude and speed of the resultant flux due to three-phase currents

We shall proceed in exactly the same way as we did in section 22.9 by assuming the distribution of the flux density due to each phase to be sinusoidal over a pole pitch. Hence, the full-line curves in fig. 22.21(*a*) represent the distribution of the flux densities due to the three phases at instant *a* in fig. 22.20; and the dotted curve represents the resultant flux density on the assumption that the magnetic circuit is unsaturated. If B_m represents the maximum flux density due to the maximum current in one phase alone, it will be seen from fig. 22.21(*a*) that the contributions made to the maximum resultant flux density by phases R, Y and B are B_m, $0.25B_m$ and $0.25B_m$ respectively, and that the peak value of the resultant flux density is therefore $1.5B_m$.

Fig. 22.21 Distribution of flux densities due to three-phase currents

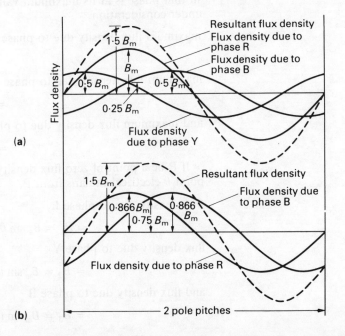

Fig. 22.21(*b*) represents the distribution of the flux densities due to phases R and B at instant *b* in fig. 22.20; and again it will be seen that the peak value of the resultant flux density is $1.5B_m$ and that its position has shifted a sixth of a pole pitch in a twelfth of a cycle.

Similarly, if the distribution of the resultant flux density were derived for instant *c* of fig. 22.20, it would be found that the peak value of the resultant flux density would again be $1.5B_m$ and that it would have moved through a further sixth of a pole pitch in a twelfth of a cycle and therefore through two pole pitches in one cycle.

It will now be shown that the peak value of the resultant flux density remains constant at *every* instant and that its position rotates at a *uniform* speed.

Suppose the curves in fig. 22.22 to represent the distributions of the flux densities due to the three phases at an instant *t* seconds after the flux due to phase R has passed

Fig. 22.22 Distribution of flux densities due to three-phase currents

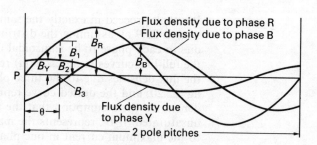

through zero from negative to positive values, and suppose B_m to be the peak density due to one phase when the current in that phase is at its maximum value. Then, at the instant under consideration,

maximum flux density due to phase R

$$= B_R = B_m \sin \omega t$$

maximum flux density due to phase Y

$$= B_Y = B_m \sin (\omega t - 2\pi/3)$$

and maximum flux density due to phase B

$$= B_B = B_m \sin (\omega t - 4\pi/3).$$

If P be a point of zero flux density for phase R, then for a point θ electrical radians from P,

flux density due to phase R

$$= B_1 = B_R \sin \theta = B_m \sin \omega t \cdot \sin \theta$$

flux density due to phase Y

$$= B_2 = B_m \sin (\omega t - 2\pi/3) \cdot \sin (\theta - 2\pi/3)$$

and flux density due to phase B

$$= B_3 = B_m \sin (\omega t - 4\pi/3) \cdot \sin (\theta - 4\pi/3)$$

If the magnetic circuit is unsaturated, the resultant flux density at a point θ electrical radians from **P**

$$= B_1 + B_2 + B_3$$
$$= B_m\{\sin \omega t \cdot \sin \theta + \sin (\omega t - 2\pi/3) \cdot \sin (\theta - 2\pi/3)$$
$$+ \sin (\omega t - 4\pi/3) \cdot \sin (\theta - 4\pi/3)\}$$
$$= 1.5B_m(\sin \omega t \cdot \sin \theta + \cos \omega t \cdot \cos \theta)$$
$$= 1.5B_m \cos (\omega t - \theta)$$

and is constant at $1.5B_m$ when $\theta = \omega t = 2\pi f t$.

For a value of t equal to $1/f$, namely the duration of one cycle,

$$\theta = 2\pi \text{ electrical radians,}$$

i.e. the position of the peak value of the resultant flux density rotates through two pole pitches in one cycle.

Since the distribution of the resultant flux density remains sinusoidal and the peak value remains constant, it follows that the total flux over a pole pitch is constant and rotates at a uniform speed through two pole pitches in one cycle. If the machine has p pairs of poles, the resultant magnetic flux rotates through $1/p$ revolution in one cycle,

$\therefore$ speed of rotating magnetic flux

$$= n_1 = f/p \text{ revolutions per second}$$

or $$f = n_1 p \qquad (22.7)$$

which is the same as expression (6.4) derived in section 6.2 for the frequency of the e.m.f. generated in the stator winding.

22.12 Reversal of direction of rotation of the magnetic flux produced by three-phase currents

Suppose the stator winding to be connected as in fig. 22.23(a) and that this arrangement corresponds to that already shown in fig. 22.19 and discussed in section 22.10. The resultant magnetic flux was found to rotate clockwise. Let us interchange the connections between two of the supply lines, say Y and B, and the stator windings, as shown in fig. 22.23(b). The distribution of currents at instant a in fig. 22.20 will be exactly as shown in fig. 22.19(a); but at instant b, the current in the winding that was originally the

Fig. 22.23 Reversal of direction of rotation

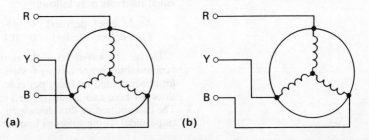

(a) (b)

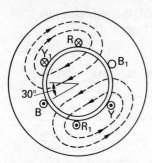

Fig. 22.24 Distribution of magnetic flux at instant *b* of fig. 22.20

'yellow' phase is now 0.866 of the maximum and the distribution of the resultant magnetic flux is as shown in fig. 22.24. From a comparison of figs. 22.19(*a*) and 22.24 it is seen that the axis of the magnetic flux is now rotating anticlockwise. The same result may be represented thus:

R		R		Y		B
$\circlearrowleft$	becomes	$\circlearrowleft$	or	$\circlearrowleft$	or	$\circlearrowleft$
B Y		Y B		B R		R Y
Original sequence		Inter-changing Y and B		Inter-changing R and Y		Inter-changing R and B

From this it will be seen that the direction of rotation of the resultant magnetic flux can be reversed by reversing the connections to any two of the three terminals of the motor. The ease with which it is possible to reverse the direction of rotation constitutes one of the advantages of three-phase motors.

Summary of important formulae

$$f = n_1 p \qquad (22.1)$$

R.M.S. value of e.m.f. per phase $= 2.22 k_d k_p Z f \Phi$ (22.5)

$$\text{Distribution factor} = \frac{\text{e.m.f. with distributed winding}}{\text{e.m.f. with concentrated winding}}$$

$$\text{Pitch factor} = \frac{\text{e.m.f. with short-pitch coil}}{\text{e.m.f. with full-pitch coil}}$$

EXERCISES 22

1. Explain why the e.m.f. generated in a conductor of an alternating-current generator is seldom sinusoidal.

   A rectangular coil of 55 turns, carried by a spindle placed at right-angles to a magnetic field of uniform density, is rotated at a constant speed. The mean area per turn is 300 cm². Calculate (*a*) the speed in order that the frequency of the generated e.m.f. may be 60 Hz, (*b*) the density of the magnetic field if the r.m.s. value of the generated e.m.f. is 10 V. (App. El., L.U.)

2. The flux density in the airgap of a synchronous machine at equal intervals is as follows:

Angle (elect. degrees)	0	15	30	45	60	75	90
Flux density (teslas)	0	0.1	0.4	0.9	1.0	1.0	1.0

   Derive the waveform of the resultant e.m.f. of a single-phase synchronous motor having 6 slots per pole when the winding is (*a*) concentrated in 1 slot per pole, (*b*) distributed in 2 adjacent slots per pole and (*c*) distributed in 4 adjacent slots per pole.

3. The distribution of flux density in a synchronous machine is trapezoidal, being uniform under the pole faces and decreasing

uniformly to zero at points midway between the poles. The ratio of pole arc to pole pitch is 0.6. The machine has 4 slots per pole. Derive curves representing the waveform of the generated e.m.f. when the winding is (a) concentrated in one slot per pole, (b) distributed in two adjacent slots per pole and (c) distributed in the four slots per pole.

4. In a synchronous motor, the flux density may be assumed uniform under the poles and zero between the poles. The ratio of pole arc to pole pitch is 0.7. Calculate the form factor of the e.m.f. generated in a full-pitch coil.

5. The field-form of a synchronous machine taken from the pole centre line, in electrical degrees, is given below, the points being joined by straight lines. Determine the form factor of the e.m.f. generated in a full-pitch coil.

Distance from pole centre, degrees:

0	20	45	60	75	90	105	120	135	

Flux density, teslas: etc.

0.7	0.7	0.6	0.15	0	0	0	−0.15	−0.6

6. A stator has two poles of arc equal to two-thirds of the pole pitch, producing a uniform radial flux of density 1 T. The length and diameter of the armature are both 0.2 m and the speed of rotation is 1500 r/min. Neglecting fringing, draw to scale the waveform of e.m.f. induced in a single fully-pitched armature coil of 10 turns. If the coil is connected through slip-rings to a resistor of a value which makes the total resistance of the circuit 4 Ω, calculate the mean torque on the coil.

   Explain with the aid of sketch how a torque is produced on a coil carrying a current in a magnetic field.   (U.L.C.I., O.2)

7. Calculate the value of the distribution factor for a three-phase synchronous motor having 12 slots per pole.

8. A full-pitched one-turn coil, the ends of which are connected to slip-rings, is wound on a cylindrical steel core and rotates in a two-pole field, the pole arc being 75 per cent of the pole pitch. Sketch, and account for, the waveform of the e.m.f. for one revolution of the coil, allowing for fringing of the flux at the pole tips.

   The stator of a three-phase, eight-pole, 750-r/min synchronous generator has 72 slots, each of which contains 10 conductors. Calculate the r.m.s. value of the e.m.f. per phase if the flux per pole is 0.1 Wb, sinusoidally distributed. Assume full-pitch coils and a winding distribution factor of 0.96.
   (App. El., L.U.)

9. Derive an expression for the terminal voltage of a synchronous generator in terms of the frequency, flux per pole and the number of conductors, discussing the assumptions that are made.

   Define the term *winding distribution factor* and show how it can be calculated. A three-phase, four-pole synchronous generator has a single-layer winding with 8 conductors per slot. The armature has a total of 36 slots. Calculate the distribution factor.

   What is the induced voltage per phase when the generator is driven at 1800 r/min, with a flux of 0.041 Wb in each pole?
   (App. El., L.U.)

10. A three-phase, star-connected synchronous generator, driven at 900 r/min is required to generate a line voltage of 460 V at 60 Hz on open circuit. The stator has 2 slots per pole per phase and 4 conductors per slot. Calculate (a) the number of poles, (b) the useful flux per pole.

11. Find the number of stator conductors per slot for a three-phase, 50-Hz synchronous generator if the winding is star-connected and has to give a line voltage of 13 kV when the machine is on open circuit. The flux per pole is about 0.15 Wb. Assume full-pitch coils and the stator to have 3 slots per pole per phase. The speed is 300 r/min.

12. The following figures refer to a three-phase generator: number of poles, 8; flux per pole, 0.1 Wb; number of slots, 96; conductors per slot, 6; speed, 750 r/min. The windings are star-connected. Calculate the line voltage on open circuit.

13. A four-pole, 50-Hz, three-phase delta-connected synchronous motor has a single-layer stator winding distributed in 36 slots, each slot containing 16 conductors. The flux per pole is 0.04 Wb. Calculate the terminal e.m.f. on open circuit.

14. A six-pole, 50-Hz, three-phase motor has 54 slots, only 36 of which are wound. The number of conductors per slot is 12 and the coils are full-pitch. The flux per pole is 0.04 Wb, sinusoidally distributed. Calculate (*a*) the distribution factor and (*b*) the r.m.s. value of the terminal voltage on open circuit.

15. A certain synchronous motor has 6 slots per pole and the coils are short-pitch by 1 slot (i.e. the coil pitch is 5 slot pitches). What is the value of the pitch factor?

16. A 50-Hz synchronous motor has a flux of 0.1 Wb/pole, sinusoidally distributed. Calculate the r.m.s. value of the e.m.f. induced in one turn which spans $\frac{3}{4}$ of a pole pitch.

17. A 10-MVA, 11-kV, 50-Hz, three-phase star-connected synchronous motor is driven at 300 r/min. The winding is housed in 360 slots and has 6 conductors per slot, the coils spanning five-sixths of a pole pitch. Calculate the sinusoidally-distributed flux/pole required to give a line voltage of 11 kV on open circuit, and the full-load current per conductor.

    (E.M.E.U., O.2)

18. A star-connected balanced three-phase load of 30 Ω resistance per phase is supplied by a 415-V, three-phase generator of efficiency 90 per cent. Calculate the power input to the generator.   (N.C.T.E.C., O.2)

19. The stator of an a.c. machine is wound for 6 poles, three-phase. If the supply frequency is 25 Hz, what is the value of the synchronous speed?

20. A stator winding supplied from a three-phase, 60-Hz system is required to produce a magnetic flux rotating at 1800 r/min. Calculate the number of poles.

21. A three-phase, two-pole motor is to have a synchronous speed of 9000 r/min. Calculate the frequency of the supply voltage.

23 Characteristics of A.C. Synchronous Machines

23.1 Armature reaction in a three-phase synchronous generator

By 'armature reaction' is meant the influence of the stator m.m.f. upon the value and the distribution of the magnetic flux in the airgaps between the poles and the stator core. It has already been explained in Chapter 22 that balanced three-phase currents in a three-phase winding produce a resultant magnetic flux of constant magnitude rotating at synchronous speed. We shall now consider the application of this principle to a synchronous generator.

Case (a). When the current and the generated e.m.f. are in phase

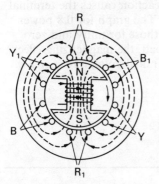

Fig. 23.1 Magnetic flux due to rotor current alone

Consider a two-pole, three-phase synchronous generator with 2 slots per pole per phase. If the machine is on open circuit there is no stator current, and the magnetic flux due to the rotor current is distributed symmetrically as shown in fig. 23.1. If the direction of rotation of the poles be clockwise, the e.m.f. generated in phase RR_1 is at its maximum and is towards the paper in conductors R and outwards in R_1.

Let us next consider the distribution of flux (fig. 23.2) due to the stator currents alone at the instant when the current in phase R is at its maximum positive value (instant a in fig. 22.20 and when the rotor (unexcited) is in the position shown in fig. 23.1. This magnetic flux rotates clockwise at synchronous speed and is therefore stationary relative to the rotor.

We can now derive the resultant magnetic flux due to the rotor and stator currents by superimposing the fluxes of figs. 23.1 and 23.2 on each other. Comparison of these figures shows that over the leading half of each pole face the two fluxes are in opposition, whereas over the trailing half of each pole face they are in the same direction. Hence the effect is to distort the magnetic flux as shown in fig. 23.3. It will be noticed that the direction of most of the lines of flux in the airgaps has been skewed and thereby lengthened. But lines of flux behave like stretched elastic cords and consequently in fig. 23.3 they exert a backward pull on the rotor; and to overcome the tangential component of this pull, the engine driving the generator has to exert a larger torque than that required on no load. Since the magnetic flux due to the stator currents rotates synchronously with the rotor, the flux distortion shown in fig. 23.3 remains the same for all positions of the rotor.

Case (b). When the current lags the generated e.m.f. by a quarter of a cycle

When the e.m.f. in phase R is at its maximum value, the poles are in the position shown in fig. 23.1. By the time the current in phase R reaches its maximum value, the poles will have moved forward through half a pole pitch to the position shown in fig. 23.4. A reference to fig. 23.2 shows that the stator m.m.f., acting alone, would send a flux from right to left through the rotor, namely in direct opposition to the flux produced by the rotor m.m.f. Hence it follows that the effect of armature reaction due to a current lagging the e.m.f. by 90° is to reduce the flux. The resultant distribution of the flux, however, is symmetrical over the two halves of the pole face, so that no torque is required to drive the rotor, apart from that to overcome losses.

Case (c). When the current leads the generated e.m.f. by a quarter of a cycle

In this case the current in phase R is a positive maximum when the N and S poles of the rotor are in the positions occupied by the S and N poles respectively in fig. 23.4. Consequently the flux due to the stator m.m.f. is now in the same direction as that due to the rotor m.m.f., so that the effect of armature reaction due to a leading current is to increase the flux.

The influence of armature reaction upon the variation of terminal voltage with load is shown in fig. 23.5, where it is assumed that the field current is maintained constant at a value giving an e.m.f. OA on open circuit. When the power factor of the load is unity, the fall in voltage with increase of load is comparatively small. With an inductive load, the demagnetizing effect of armature reaction causes the terminal voltage to fall much more rapidly. The graph for 0.8 power factor is roughly midway between those for unity and zero power factors. With a capacitive load, the magnetizing effect of armature reaction causes the terminal voltage to increase with increase of load.

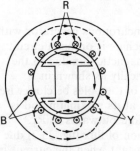

Fig. 23.2 Magnetic flux due to stator currents alone

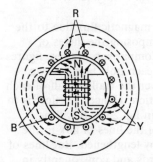

Fig. 23.3 Resultant magnetic flux for case (a)

23.2 Voltage regulation of a synchronous generator

An a.c. generator is always designed to give a certain terminal voltage when supplying its rated current at a specified power factor — usually unity or 0.8 lagging. For instance, suppose OB in fig. 23.6 to represent the full-load current and OA the rated terminal voltage of a synchronous generator. If the field current is adjusted to give the terminal voltage OA when the generator is supplying current OB at unity power factor, then when the load is removed but with the field current and speed kept unaltered, the terminal voltage rises to OC. This variation of the terminal voltage between full load and no load, expressed as a per-unit value or a percentage of the full-

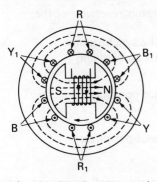

Fig. 23.4 Resultant magnetic flux for case (b)

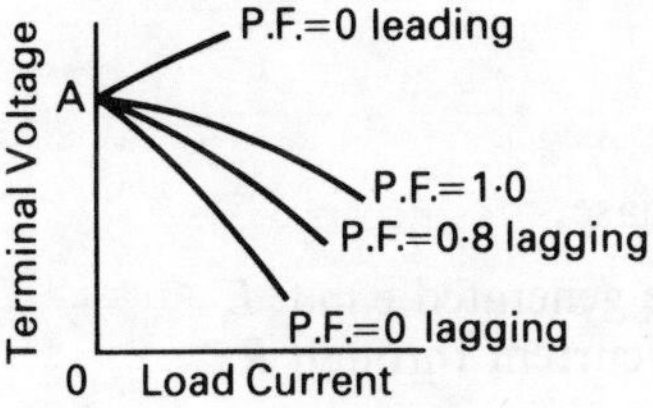

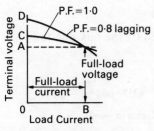

Fig. 23.5 Variation of terminal voltage with load

load voltage, is termed the per-unit or the percentage *voltage regulation* of the generator; thus:

per-unit voltage regulation

$$= \frac{\text{change of terminal voltage when full load is removed}}{\text{full-load terminal voltage}} \quad (23.1)$$

$$= \frac{AC}{OA} \text{ for unity power factor}$$

$$= \frac{AD}{OA} \text{ for p.f. of 0.8 lagging}.$$

The voltage regulation for a power factor of 0.8 lagging is usually far greater than that at unity power factor, and it is therefore important to include the power factor when stating the voltage regulation. (See example 23.1.)

23.3 Synchronous impedance

In figs. 23.3 and 23.4, the resultant flux was shown as the combination of the flux due to the stator m.m.f. alone and that due to the rotor m.m.f. alone. For the purpose of deriving the effect of load upon the terminal voltage, however, it is convenient to regard these two component fluxes as if they existed independently of each other and to consider the cylindrical-rotor type rather than the salient-pole type of generator (see section 22.2). Thus the flux due to the rotor m.m.f. may be regarded as generating an e.m.f., E, due to the rotation of the poles, this e.m.f. being a maximum in any one phase when the conductors of that phase are opposite the centres of the poles. On the other hand, the rotating magnetic field due to the stator currents can be regarded as generating an e.m.f. lagging the current by a quarter of a cycle. For instance, in fig. 22.15(a), the current in R is at its maximum value flowing towards the paper, but the e.m.f. induced in R by the rotating flux due to the stator currents is zero at that instant.

A quarter of a cycle later, this rotating flux will have turned clockwise through 90°; and since we are considering a generator having a cylindrical rotor and therefore a uniform airgap, R is then being cut at the maximum rate by flux passing from the rotor to the stator. Hence the e.m.f. induced in R at that instant is at its maximum value acting towards the paper. Since the e.m.f. induced by the rotating flux in any one phase lags the current in that phase by a quarter of a cycle, the effect is exactly similar to that of inductive reactance, i.e. the rotating magnetic flux produced by the stator currents can be regarded as being responsible for the reactance of the stator winding. Furthermore, since the rotating flux revolves synchronously with the poles, this

Fig. 23.6 Variation of terminal voltage with load

reactance is referred to as the *synchronous reactance* of the winding.

By combining the resistance with the synchronous reactance of the winding, we obtain its *synchronous impedance*. Thus,

if X_s = synchronous reactance per phase,

                          R = resistance per phase

and Z_s = synchronous impedance per phase

then $Z_s = \sqrt{(R^2 + X_s^2)}$.

In generator windings, R is usually very small compared with X_s, so that for many practical purposes, Z_s can be assumed to be the same as X_s.

The relationship between the terminal voltage V of the generator and the e.m.f. E generated by the flux due to the rotor m.m.f. alone can now be derived. Thus in fig. 23.7, S represents *one* phase of the stator winding, and R and X_s represent the resistance and synchronous reactance of that phase. If the load takes a current I at a lagging power factor $\cos\phi$, the various quantities can be represented by phasors as in fig. 23.8, where:

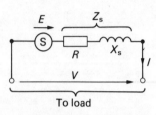

Fig. 23.7 Equivalent circuit of a synchronous generator

OI = current per phase,

OV = terminal voltage per phase,

OA = IR = component of the generated e.m.f. E absorbed in sending current through R,

OB = e.m.f. per phase induced by the rotating flux due to stator currents and lags OI by 90°,

OC = component of the generated e.m.f. E required to neutralize OB,

     = voltage drop due to synchronous reactance X_s

     = IX_s

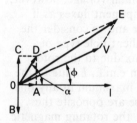

Fig. 23.8 Phasor diagram for a synchronous generator

OD = component of the generated e.m.f. absorbed in sending current through the synchronous impedance Z_s,

α = phase angle between OI and OD = $\tan^{-1} X_s/R$

and   OE = resultant of OV and OD

     = e.m.f. per phase generated by the flux due to the rotor

From fig. 23.8,

$$OE^2 = OV^2 + OD^2 + 2 \cdot OV \cdot OD \cdot \cos(\alpha - \phi)$$

i.e. $$E^2 = V^2 + (IZ_s)^2 + 2V \cdot IZ_s \cos(\alpha - \phi)$$

from which the e.m.f. E generated by the flux due to the rotor m.m.f., namely the open-circuit voltage, can be calculated. It follows that if V is the rated terminal voltage per phase of the synchronous generator and I is the full-load current per phase, the terminal voltage per phase obtained when the load is removed, the exciting current and speed remaining unaltered, is given by E; and from expression (23.1),

$$\text{per-unit voltage regulation} = \frac{E - V}{V}.$$

The synchronous impedance is important when we come to deal with the parallel operation of generators and with synchronous motors (section 23.8); but numerical calculations involving synchronous impedance are usually unsatisfactory, owing mainly to magnetic saturation in the poles and stator teeth and, in the case of salient-pole machines, to the value of Z_s varying with the p.f. of the load and with the excitation of the poles.

One method of estimating the value of the synchronous impedance is to run the generator on open circuit and measure the generated e.m.f., and then short-circuit the terminals through an ammeter and measure the short-circuit current, the exciting current and speed being kept constant. Since the e.m.f. generated on open circuit can be regarded as being responsible for circulating the short-circuit current through the synchronous impedance of the winding, the value of the synchronous impedance is given by the ratio of the open-circuit voltage per phase to the short-circuit current per phase.

Example 23.1

A three-phase, 600-MVA generator has a rated terminal voltage of 22 kV (line). The stator winding is star-connected and has a resistance of 0.014 Ω/phase and a synchronous impedance of 0.16 Ω/phase. Calculate the voltage regulation for a load having a power factor of (a) unity and (b) 0.8 lagging.

(a) Since $\quad 600 \times 10^6 = 1.73 I_L \times 22 \times 10^3$

$\therefore \qquad$ line current $= I_L = 15.7 \text{ kA} = $ phase current.

Terminal voltage per phase on full load $\quad = 22\,000/1.73 = 12\,700 \text{ V}.$

Voltage drop per phase on full load due to synchronous impedance

$$= 15\,700 \times 0.16 = 2540 \text{ V}$$

From fig. 23.8 it follows that at unity power factor,

$$\text{OV} = 12\,700 \text{ V}; \quad \text{OD} = 2540 \text{ V}; \quad \text{and} \quad \phi = 0.$$

Also, $\quad \cos \alpha = \text{OA}/\text{OD} = R/Z_s = 0.014/0.16 = 0.088$

$\therefore \qquad \text{OE}^2 = (12\,700)^2 + (2540)^2 + 2 \times 12\,700 \times 2540 \times 0.088$

$$= 1.734 \times 10^8$$

$\therefore \qquad \text{OE} = 13\,170 \text{ V},$

and $\qquad \left.\begin{array}{c} \text{voltage regulation at unity} \\ \text{power factor} \end{array}\right\} = \dfrac{13\,170 - 12\,700}{12\,700}$

$$= 0.037 \text{ per unit}$$

$$= 3.7 \text{ per cent}.$$

(b) Since the rating of the alternator is 600 MVA, the full-load current is the same whatever the power factor.

For power factor of 0.8, $\quad \phi = 36.87°$

Also $\alpha = \cos^{-1} 0.088 = 84.95°,$

so that $(\alpha - \phi) = 48.08°$ and $\cos 48.08° = 0.668.$

Hence, $OE^2 = (12\,700)^2 + (2540)^2 + 2 \times 12\,700 \times 2540 \times 0.668$

$$= 2.111 \times 10^8$$

$\therefore$ $OE = 14\,530\,V,$

and $\left.\begin{array}{l} \text{voltage regulation for power} \\ \text{factor 0.8 lagging} \end{array}\right\} = \dfrac{14\,530 - 12\,700}{12\,700}$

$$= 0.144 \text{ per unit}$$

$$= 14.4 \text{ per cent}.$$

23.4 Synchronizing of synchronous generators

For simplicity, let us assume single-phase generators; thus, in fig. 23.9, A represents a generator already connected to the busbars and B is a generator to be connected in parallel. To

Fig. 23.9 Synchronizing of generators

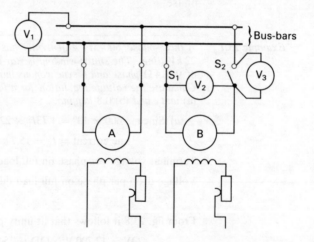

enable this to be done, the following conditions must be fulfilled:

(*a*) the frequency of B must be the same as that of A,
(*b*) the e.m.f. generated in B must be equal to the busbar voltage,
(*c*) the e.m.f. of B must be in phase with the busbar voltage.

The procedure is to start up the engine driving generator B and adjust its speed to about its rated value. The excitation of B is then adjusted so that the reading on voltmeter V_2 is the same as the busbar voltage given by voltmeter V_1. Switch S_1 is closed and voltmeter V_3 connected across switch S_2. The

Fig. 23.10 Waveforms of E_A, E_B and $(E_A - E_B)$

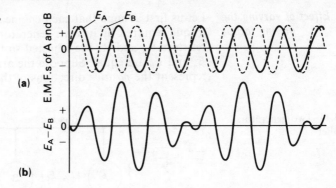

(a)

(b)

pointer of V_3 will then oscillate between zero and twice the busbar voltage at a frequency equal to the difference between the frequencies of A and B. This will be evident from fig. 23.10, where curves E_A and E_B represent the e.m.f.s of A and B respectively, the frequencies being assumed such that E_B varies through 4 cycles for every 3 cycles of E_A. The p.d. across switch S_2 is the difference between E_A and E_B and fig. 23.10(*b*) represents two 'beats' of this voltage. If the frequency of A is 50 Hz and that of B is 50.5 Hz, the beat frequency is 0.5 Hz and the pointer of V_3 makes one complete oscillation in 2 seconds.

The speed of B is adjusted until the pointer of voltmeter V_3 oscillates very slowly and switch S_2 is closed just as the pointer is reaching zero, i.e. when the e.m.f.s of A and B are in phase opposition relative to each other but in phase with each other relative to the busbars.

The above method of paralleling two generators does not indicate whether the frequency of the incoming alternator B is higher or lower than that of A. In actual practice it is customary to use an instrument, called a *synchroscope*, which shows whether the incoming machine is running too fast or too slow as well as indicating the correct moment for closing switch S_2.

23.5 Parallel operation of synchronous generators

It is not obvious why two generators continue running in synchronism after they have been paralleled; so let us consider the case of two similar single-phase generators, A and B, connected in parallel to the busbars, as shown in fig. 23.11, and let us, for simplicity, assume that there is no external load connected across the busbars. We will consider two separate cases:

(*a*) the effect of varying the torque of the driving engine, e.g. by varying the steam supply;

(*b*) the effect of varying the exciting current.

Effect of varying the driving torque

Let us first assume that each engine is exerting exactly the torque required by its own generator and that the field resistors R_A and R_B are adjusted so that the generated e.m.f.s E_A and E_B are equal. Suppose the arrows in fig. 23.11 to represent the *positive* directions of these e.m.f.s. It will be seen

Fig. 23.11 Two generators in parallel

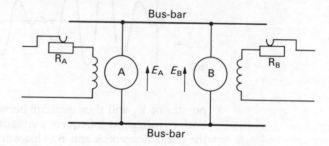

that, *relative to the busbars*, these e.m.f.s are acting in the same direction, i.e. when the e.m.f. generated in each machine is positive, each e.m.f. is making the top busbar positive in relation to the bottom busbar. But *in relation to each other*, these e.m.f.s are in opposition, i.e. if we trace the closed circuit formed by the two generators we find that the e.m.f.s oppose each other.

In the present discussion we want to find out if any current is being circulated in this closed circuit. It is therefore more convenient to consider these e.m.f.s in relation to each other rather than to the busbars. This condition is shown in fig. 23.12(a) when E_A and E_B are in exact phase opposition relative to each other; and since they are equal in magnitude, their resultant is zero and consequently no current is circulated.

Let us next assume the driving torque of B's engine to be reduced, e.g. by a reduction of its steam supply. The rotor of B falls back in relation to that of A, and fig. 23.12(b) shows the conditions when B's rotor has fallen back by an angle θ. The resultant e.m.f. in the closed circuit formed by the generator windings is represented by E_Z, and this e.m.f. circulates a current I lagging E_Z by an angle α,

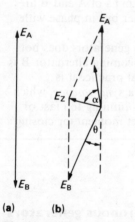

(a)   (b)

Fig. 23.12 Effect of varying the driving torque

where $I = \dfrac{E_Z}{2Z_s}$ and $\alpha = \tan^{-1} X_s/R$,

R = resistance of each generator,

X_s = synchronous reactance of each generator,

and Z_s = synchronous impedance of each generator.

Since the resistance is very small compared with the synchronous reactance, α is nearly 90°, so that the current I is almost in phase with E_A and in phase opposition to E_B. This means that A is generating and B is motoring, and the power supplied from A to B compensates for the reduction of the power supplied by B's engine. If the frequency is to be maintained constant, the driving torque of A's engine has to be increased by an amount equal to the decrease in the driving torque of B's engine.

The larger the value of θ (so long as it does not exceed about 80°), the larger is the circulating current and the greater is the power supplied from A to B. Hence machine B falls back in relation to A until the power taken from the latter exactly compensates for the reduction in the driving power of B's engine. Once this balance has been attained, B and A will run at exactly the same speed.

Effect of varying the excitation

Let us again revert to fig. 23.12(a) and assume that each engine is exerting the torque required by its generator and that the e.m.f.s, E_A and E_B, are equal. The resultant e.m.f. is zero and there is therefore no circulating current.

Suppose the exciting current of generator B to be increased so that the corresponding open-circuit e.m.f. is represented by E_B in fig. 23.13. The resultant e.m.f., E_Z ($= E_B - E_A$), circulates a current I through the synchronous impedances of the two generators; and since the machines are assumed similar, the impedance drop per machine is $\frac{1}{2}E_Z$, so that:

$$\text{terminal voltage} = E_B - \tfrac{1}{2}E_Z$$

or

$$= E_A + \tfrac{1}{2}E_Z$$

Hence one effect has been to increase the terminal voltage. Further, since angle α is nearly 90°, the circulating current I is almost in quadrature with the generated e.m.f.s, so that very little power is circulated from one machine to the other.

In general, we can therefore conclude:

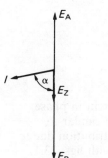

Fig. 23.13 Effect of varying the excitation

(a) the distribution of load between generators operating in parallel can be varied by varying the driving torques of the engines and only slightly by varying the exciting currents;

(b) the terminal voltage is controlled by varying the exciting currents.

23.6 Three-phase synchronous motor: Principle of action

In section 23.5 it was explained that when two generators, A and B, are in parallel, with *no load* on the busbars, a reduction in the driving torque applied to B causes the latter to fall back by some angle θ (fig. 23.12) in relation to A, so that power is supplied from A to B. Machine B is then operating as a *synchronous motor*.

It will also be seen from fig. 23.12 that the current in a synchronous motor is approximately in phase opposition to the e.m.f. generated in that machine. This effect is represented in fig. 23.14, where the rotor poles are shown in the same position relative to the three-phase winding as in fig. 23.3. The latter represented a synchronous generator with the current in phase with the generated e.m.f., whereas fig. 23.14

Fig. 23.14 Principle of
action of a three-phase
synchronous motor

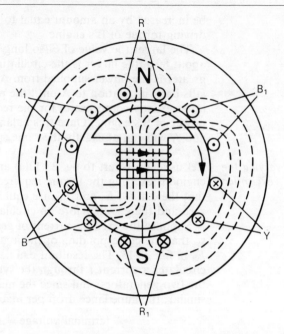

represents a synchronous motor with the current in phase
opposition to this generated e.m.f. A diagram similar to
fig. 23.2 could be drawn showing the flux distribution due to
the stator currents alone, but a comparison with figs. 23.1,
23.2 and 23.3 can be sufficient to indicate that in fig. 23.14
the effect of armature reaction is to increase the flux in the
leading half of each pole and to reduce it in the trailing half.
Consequently the flux is distorted in the direction of rotation
and the lines of flux in the gap are skewed in such a direction
as to exert a clockwise torque on the rotor. Since the
resultant magnetic flux due to the stator currents rotates at
synchronous speed, the rotor must also rotate at exactly the
same speed for the flux distribution shown in fig. 23.14 to
remain unaltered.

23.7 Effect of varying the load on a synchronous motor

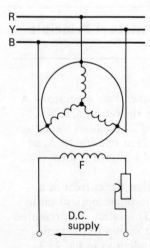

Fig. 23.15 Connections of a
three-phase synchronous
motor

Fig. 23.15 represents diagrammatically a three-phase, star-
connected synchronous motor connected to the supply mains
and excited by a field winding F in series with a variable
resistor.

Suppose V in fig. 23.16 to represent the voltage *applied* to
one phase and E the e.m.f.* *generated* in that phase. Angle θ
represents the displacement of E from exact phase opposition
to V. Let us assume that E is exactly equal to V in

* This e.m.f. is the open-circuit voltage of the machine when
driven as a generator at the same speed with the same field current.

Fig. 23.16 Phasor diagrams
for a synchronous motor

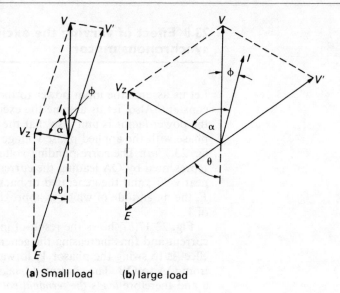

(a) Small load   **(b)** large load

magnitude. Consequently, if E had been in exact phase opposition to V, i.e. if $\theta = 0$, there would have been no stator current and therefore no power taken from the a.c. supply. The motor would have to slow down and the phase of the generated e.m.f. E would fall back in relation to the applied voltage V by some angle θ, the value of which depends upon the load.

Let us first assume that the load is small so that θ is also small, as shown in fig. 23.16(a). The resultant voltage available to send a current I through the synchronous impedance of the winding is represented by V_Z, the resultant of V and E. Alternatively, we can regard the applied voltage V as having to provide two components: (a) V' to neutralize the generated e.m.f. E, and (b) V_Z to provide the voltage drop due to the synchronous impedance of the stator winding.

If the synchronous impedance per phase is Z_s, the current is V_Z/Z_s and lags V_Z by an angle α approximating 90°, since the resistance is very small compared with the synchronous reactance. If the phase angle between I and the supply voltage V is ϕ,

$$\text{power factor of motor} = \cos \phi$$

and   power per phase taken by motor $= I V \cos \phi$.

Let us next consider the effect of an increase in load. The rotor of the motor slows down momentarily until the displacement θ is as shown in fig. 23.16(b). From the latter it is seen that V_Z and I have increased and that the motor is taking more power from the a.c. supply. The displacement θ adjusts itself automatically so that the motor takes from the supply exactly the power required by the load plus that lost in the machine.

23.8 Effect of varying the excitation of a synchronous motor

Let us assume the input power to the motor to remain constant. Also, let us assume the excitation to be such that the power factor is unity, so that the phase current I is in phase with the applied phase voltage V, as shown in fig. 23.17(a). The corresponding voltage drop per phase, V_Z, is represented by OA leading the current by an angle α which is nearly 90°, and the generated or back e.m.f. is represented by E, the magnitude of which is approximately the same as that of V.

Fig. 23.17(b) shows the result of increasing the rotor field current and thus increasing the generated e.m.f. to E_1. The effect is to swing the phasor V_Z forward through an angle ϕ_1 from OA to OB. The current, I_1, lags OB by the same angle α and therefore *leads the terminal voltage, V,* by angle ϕ_1.

Fig. 23.17 Effect of varying the excitation of a synchronous motor

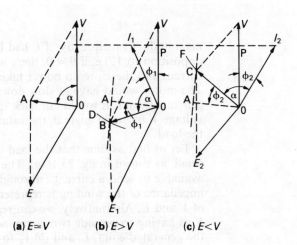

(a) $E \simeq V$ **(b)** $E > V$ **(c)** $E < V$

The projection of OI_1 on OV, namely OP, represents the active or power component of the current, and this component is being assumed constant. Considering triangles OPI_1 and OAB, we have:

$$OA = OP \times \text{synchronous impedance per phase}$$

and

$$OB = OI_1 \times \text{synchronous impedance per phase,}$$

$$\therefore \qquad OA/OB = OP/OI_1 .$$

Also, $\angle POI_1 = \angle AOB = \phi_1;$

hence the two triangles are similar, so that:

$$\angle OAB = \angle OPI_1 = 90°.$$

Consequently the locus of B is the dotted line drawn through A at right-angles to OA. Point B can therefore be determined by drawing an arc D of radius equal to E_1, using point V as centre. Phasor E_1 is then drawn parallel and equal to the line joining V and B.

The effect of reducing the excitation is shown in fig. 23.17(c). With centre V and radius equal to E_2, an arc F is drawn to cut the locus through A at point C. Phasor E_2 is then drawn parallel and equal to line VC. The current phasor OI_2 lags OC by angle α and therefore *lags the terminal voltage* by angle ϕ_2.

It follows from these phasor diagrams that the power factor of a synchronous motor can be controlled by varying the field current, as indicated in fig. 23.18, the actual range of power-factor variation being dependent upon the value of the load.

Fig. 23.18 Variation of supply current and power factor with field current for constant input power to a synchronous motor

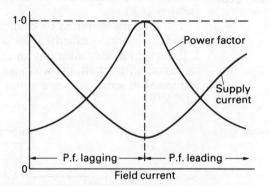

23.9 Methods of starting a three-phase synchronous motor

When the rotor of a three-phase synchronous motor is stationary, the rotating magnetic field due to the stator currents produce an alternating torque on the rotor, i.e. at one instant, the rotor is being urged clockwise, and at the next instant anticlockwise. Since the net torque is zero, a synchronous motor is not self-starting. The methods most commonly used to bring the motor up to synchronism are:

(a) By means of damping grids in the pole shoes

These grids consist of copper bars short-circuited at each end, as shown in fig. 22.2. The rotating magnetic flux induce currents in these grids and the machine accelerates like a cage-type induction motor (section 24.1).

During the starting period, the rotor field winding is usually closed through the armature of the exciter, namely a d.c. shunt generator carried on an extension of the shaft; and the current in the stator winding is limited to a permissible value by applying a reduced voltage, e.g. by a star–delta switch or an auto-transformer (section 24.11).

When the machine has reached nearly full speed as an induction motor, the rated voltage is applied to the stator winding and the exciter voltage will have built up sufficiently to magnetize the rotor poles. The resultant magnetic flux due to the stator currents is then moving past the rotor poles at a slow speed and produces a low-frequency alternating torque superimposed upon that exerted by the damping grids. Consequently, the rotor is accelerated at one instant and retarded at the next instant, and the fluctuation of speed may be sufficient to bring the rotor up to synchronous speed. Once this speed has been attained, the rotor continues to run in synchronism.

Owing to the starting torque being relatively small and the starting currents being relatively large, this method is only suitable when the load — if any — is small.

(b) By means of a wound rotor

With this method, the rotor of the synchronous motor is similar to the wound rotor of an induction motor (section 24.8).

The machine is started by connecting the slip-rings to variable resistors — exactly as for a slip-ring induction motor. If the synchronous motor has an exciter carried on an extension of the shaft, the windings of this machine can be connected in series with one of the resistors, as shown in fig. 23.19.

Fig. 23.19 Synchronous-induction motor

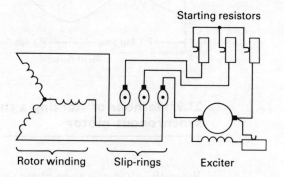

Starting resistors

Rotor winding   Slip-rings   Exciter

When the speeds exceeds a critical value, the exciter voltage builds up and the direct current from this exciter magnetizes the rotor of the synchronous motor with the same number of poles* as the stator. Consequently, by the time the starting resistors are short-circuited, the machine is running at full speed as an induction motor and is also being subjected to a low-frequency alternating torque superimposed upon the induction-motor torque, as already mentioned in section (a). The fluctuation of speed caused by this alternating torque enables the motor to be pulled into synchronism.

This type of machine is referred to as a *synchronous-induction motor* and is capable of exerting a large starting torque with a relatively low starting current.

---

* With the arrangement shown in fig. 23.19, the whole of the direct current flows through one phase of the rotor winding and half of the current flows through each of the other two phases. The distribution of the magnetic flux produced by these direct currents in a three-phase winding having 1 slot per pole per phase is exactly similar to that shown in fig. 22.5(*a*).

23.10 Advantages and disadvantages of the synchronous motor

The principal advantages of the synchronous motor are:

(1) The ease with which the power factor can be controlled. An over-excited synchronous motor having a leading power factor can be operated in parallel with induction motors having lagging power factor, thereby improving the power factor of the supply system.

Synchronous motors are sometimes run on *no load* for power-factor correction or for improving the voltage regulation of a transmission line. In such applications, the machine is referred to as a *synchronous capacitor*.

(2) The speed is constant and independent of the load. This characteristic is mainly of use when the motor is required to drive another alternator to generate a supply at a different frequency, as in frequency-changers.

The principal disadvantages are:

(1) The cost per kilowatt is generally higher than that of an induction motor.

(2) A d.c. supply is necessary for the rotor excitation. This is usually provided by a small d.c. shunt generator carried on an extension of the shaft.

(3) Some arrangement must be provided for starting and synchronizing the motor.

23.11 Application of generalized theory to 3-phase synchronous machines

Let us consider the two-pole, three-phase synchronous *generator* of fig. 23.1. When the position of the rotor winding relative to the winding of phase R is that shown in fig. 23.1, the mutual inductance between the two windings is zero. If the rotor is displaced 90° *forward* relative to phase R, as in fig. 23.4, the m.m.f.s of the two windings are in direct opposition, so that the mutual inductance is at its negative maximum value. This is the rotor position when the current lags the generated e.m.f. by 90°.

If the rotor is displaced 90° *backward* from the position shown in fig. 23.1, the direction of the rotor m.m.f. is the same as that due to phase R, so that the mutual inductance is at its positive maximum value. This is the rotor position when the current leads the generated e.m.f. by 90°.

It is found experimentally that if the rotor is displaced by an angle α* from the position of zero mutual inductance,

---

* Symbol α rather than θ is used to avoid confusion with θ when used to represent $2\pi ft$ in a.c. theory.

namely the position shown in fig. 23.1,

$$\left.\begin{array}{l}\text{mutual inductance between}\\ \text{rotor winding and phase R}\end{array}\right\} = \pm M \sin \alpha \qquad (23.2)$$

Hence $dM/d\alpha = M \cos \alpha$ (neglecting the signs).   (23.3)

When the field winding is in the position shown in fig. 23.4 relative to phase R, we can assume that the whole of the flux Φ produced by field current I_f is linked with the winding of phase R. Hence, if the number of conductors per phase is Z, the flux-linkages with phase R is $\Phi Z/2$.

From expression (4.26), it follows that for a 2-pole machine:

$$\left.\begin{array}{l}\text{maximum mutual inductance between the}\\ \text{field winding and a stator phase winding}\end{array}\right\} = M = \frac{\Phi Z}{2I_f}$$

If the machine has p pairs of poles,

$$\text{maximum mutual inductance} = M = \frac{\Phi Z p}{2I_f} \qquad (23.4)$$

From expression (22.5),

$$\text{r.m.s. value of e.m.f. per phase} = E = 2.22 k_d k_p Z f \Phi$$

Since $k_d \simeq 0.96$ for 3-phase windings and since 3-phase windings are usually full-pitch, so that $k_p = 1.0$,

$$\therefore \qquad E \simeq 2.22 Z f \Phi = 2.22 Z n_1 p \Phi$$

where n_1 = speed in revolutions per second.

Substituting for $\Phi Z p$ in expression (23.4), we have:

$$M = \frac{E}{2.22 n_1} \times \frac{1}{2I_f} = \frac{E \times 2\pi}{2.22 \omega_1} \times \frac{1}{2I_f}$$

where ω_1 = angular speed of *machine*, in radians per second.

But $2.22 = 2 \times 1.11 = 2 \times \dfrac{1/\sqrt{2}}{2/\pi} = \dfrac{\pi}{\sqrt{2}}$

Hence $M = \dfrac{E \times \pi}{2.22 \omega_1} \times \dfrac{1}{I_f} = \dfrac{\sqrt{2}E}{\omega_r I_f}$ (23.5)

The value of M can therefore be determined by running the machine on open circuit at synchronous speed and noting the terminal voltage per phase for a given field current. For a synchronous generator,

$$\text{electrical power } \textit{generated} \text{ per phase} = I_s E \cos \alpha$$

where I_s and E represent the r.m.s. values of the current and generated e.m.f. per phase respectively in the stator winding. Angle α in this expression is the same as angle EOI in fig. 23.8 and also represents the displacement of the rotor relative to phase R from the position shown in fig. 23.1, namely the position of zero mutual inductance between the field winding and phase R,

$$\left.\begin{array}{r}\text{total electrical power}\\\text{generated}\end{array}\right\} = P = 3I_s E \cos \alpha$$

$$= 3I_s \times \frac{\omega_r I_f M \cos \alpha}{\sqrt{2}}$$

and total torque $= T = P/\omega_r = (3/\sqrt{2})I_s I_f M \cos \alpha$

$$= (3/\sqrt{2})I_s I_f \times \mathrm{d}M/\mathrm{d}\alpha \qquad (23.6)$$

Expression (23.6) shows how expression (23.3) has to be modified when applied to a 3-phase synchronous machine.

When the machine is run as a synchronous motor, α is the angle between I and E in fig. 23.16, i.e. $\alpha = 180° - (\theta \pm \phi)$. The plus sign applies when the power factor is leading and the minus sign when it is lagging.

For fuller discussion of this subject, see *Higher Electrical Engineering* by Shepherd, Morton and Spence, and *Basic Electrical Engineering Science* by McKenzie Smith and Hosie.

EXERCISES 23

1. A single-phase synchronous generator has a rated output of 500 kVA at a terminal voltage of 3300 V. The stator winding has a resistance of 0.6 Ω and a synchronous reactance of 4 Ω. Calculate the percentage voltage regulation at power factor of (a) unity, (b) 0.8 lagging and (c) 0.8 leading. Sketch the phasor diagram for each case.
2. A single-phase generator having a synchronous reactance of 5.5 Ω and a resistance of 0.6 Ω, delivers a current of 100 A. Calculate the e.m.f. generated in the stator winding when the terminal voltage is 2000 V and the power factor of the load is 0.8 lagging. Sketch the phasor diagram.
3. A three-phase, star-connected, 50-Hz generator has 96 conductors per phase and a flux per pole of 0.1 Wb. The stator winding has a synchronous reactance of 5 Ω/ph and negligible resistance. The distribution factor for the stator winding is 0.96.

   Calculate the terminal voltage when three non-inductive resistors, of 10 Ω/ph, are connected in star across the terminals. Sketch the phasor diagram for one phase.
4. A 1500-kVA, 6.6-kV, three-phase, star-connected synchronous generator has a resistance of 0.5 Ω/ph and a synchronous reactance of 5 Ω/ph. Calculate the percentage change of voltage when the rated output of 1500 kVA at power factor 0.8 lagging is switched off. Assume the speed and the exciting current to remain unaltered.
5. Two single-phase generators are connected in parallel, and the excitation of each machine is such as to generate an open-circuit e.m.f. of 3500 V. The stator winding of each machine has a synchronous reactance of 30 Ω and negligible resistance. If there is a phase displacement of 40 electrical degrees between the e.m.f.s, calculate: (a) the current circulating between the two machines; (b) the terminal voltage; and (c) the power supplied from one machine to the other. Assume that there is no external load. Sketch the phasor diagram.

6. Two similar three-phase star-connected generators are connected in parallel. Each machine has a synchronous reactance of 4.5 Ω/ph and negligible resistance, and is excited to generate an e.m.f. of 1910 V/ph. The machines have a phase displacement of 30 electrical degrees relative to each other. Calculate: (a) the circulating current; (b) the terminal voltage/phase; (c) the active power supplied from one machine to the other. Sketch the phasor diagram for one phase.

7. If the two generators of Q. 6 are adjusted to be in exact phase opposition relative to each other and if the excitation of one machine is adjusted to give an open-circuit voltage of 2240 V/ph and that of the other machine adjusted to give an open-circuit voltage of 1600 V/ph, calculate: (a) the circulating current; (b) the terminal voltage.

8. A single-phase synchronous motor, having a synchronous reactance of 5 Ω and negligible resistance, is connected across a 2200-V supply. If the input power remains constant at 100 kW, calculate the generated e.m.f. when the power factor is: (a) unity; (b) 0.8 lagging; and (c) 0.8 leading. Sketch the phasor diagram for each case.

9. A three-phase, star-connected synchronous motor is connected across a 400-V supply. The stator winding has a synchronous reactance of 0.8 Ω/ph and negligible resistance. If the input power remains constant at 30 kW, calculate the generated e.m.f. per phase when the power factor is: (a) unity; (b) 0.7 lagging; and (c) 0.7 leading. Sketch, for each case, the phasor diagram showing the phase current and voltages for one phase only.

10. A 200-V single-phase synchronous motor takes a current of 10 A at unity power factor from the supply. Calculate the e.m.f. and the angle of retard if the synchronous reactance is 5 Ω and the resistance is negligible.

    If the e.m.f. is decreased by 10 per cent, calculate the current and the power factor. Assume the angle of retard to remain unaltered. (E.M.E.U.)

11. Explain why a synchronous motor will only develop a continuous torque at synchronous speed. How does a synchronous motor reach synchronous speed?

    A three-phase synchronous motor has 12 poles and operates from a 440-V, 50-Hz supply. Calculate its speed. If it takes a line current of 100 A at 0.8 power factor lead, what torque will the machine be developing? Neglect losses. (N.C.T.E.C., O.2)

12. A generator supplying 2800 kW at power factor 0.7 lagging is loaded to its full capacity, i.e. its maximum rating in kVA. If the power factor is raised to unity by means of an over-excited synchronous motor, how much more active power can the generator supply and what must be the power factor of the synchronous motor, assuming that the latter absorbs all extra power obtainable from the generator? Sketch the phasor diagram.

13. A factory takes 600 kVA at a lagging power factor of 0.6. A synchronous motor is to be installed to raise the power factor to 0.9 lagging when the motor is taking 200 kW. Calculate the corresponding apparent power (in kVA) taken by the synchronous motor and the power factor at which the motor will be operating.

14. A three-phase induction motor, taking 200 kW at 0.8 power factor lagging, is in parallel with a three-phase synchronous motor taking 250 kVA at 0.9 power factor leading. The supply voltage is 3300 V. Calculate the line current and the power factor of the total load. Sketch a phasor diagram.

15. A 3-phase, 50-Hz, star-connected generator, driven at 1000 r/min, has an open-circuit line voltage of 460 V when the field current is 16 A. The stator winding has a synchronous reactance of $2\,\Omega$/ph and negligible resistance. Calculate the value of M and therefrom determine the driving torque when the machine is supplying 50 A/ph at a p.f. of 0.8 lagging, assuming the field current and the speed to remain 16 A and 1000 r/min respectively. Neglect the core and friction losses. Note that $E^2 = (V\cos\phi)^2 + (V\sin\phi + IX)^2$ and $\cos\alpha = (V\cos\phi)/W$, where $\cos\phi$ = p.f. of load.

16. The generator of Q. 15 is run as a synchronous motor off a 3-phase, 400-V, 50-Hz supply, taking 40 A/ph at a p.f. of 0.9 leading. Using the value of M calculated in Q. 15, determine the field current.

24 Three-Phase Induction Motors

24.1 Principle of action

The stator of an induction motor is similar to that of a synchronous machine; and in the case of a machine supplied with three-phase currents, a rotating magnetic flux is produced, as already explained in section 22.3. The rotor core is laminated and the conductors often consist of uninsulated copper or aluminium bars in semi-enclosed slots, the bars being short-circuited at each end by rings or plates to which the bars are brazed or welded. In motors below about 50 kW, the aluminium rotor bars and end-rings are often cast in one operation. This type is known as the *cage* or *short-circuited* rotor. The airgap between the rotor and the stator is uniform and made as small as is mechanically possible. For simplicity, the stator slots and winding have been omitted in fig. 24.1.

Fig. 24.1 Induction motor with a cage rotor

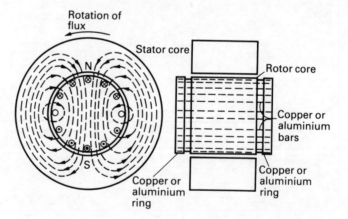

If the stator is wound for two poles, the distribution of the magnetic flux due to the stator currents at a particular instant is shown in fig. 24.1. The e.m.f. generated in a rotor conductor is a maximum in the region of maximum flux density; and if the flux be assumed to rotate anticlockwise, the directions of the e.m.f.s generated in the stationary rotor conductors can be determined by the righthand rule and are indicated by the crosses and dots in fig. 24.1. The e.m.f. generated in the rotor conductor shown in fig. 24.2 circulates a current the effect of which is to strengthen the flux density

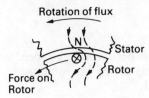

Fig. 24.2 Force on rotor

on the righthand side and weaken that on the lefthand side; i.e. the flux in the gap is distorted as indicated by the dotted lines in fig. 24.2. Consequently, a force is exerted on the rotor tending to drag it in the direction of the rotating flux.

The higher the speed of the rotor, the lower is the speed of the rotating field relative to the rotor winding and the smaller is the e.m.f. generated in the latter. Should the speed of the rotor attain the synchronous value, the rotor conductors would be stationary in relation to the rotating flux. There would therefore be no e.m.f. and no current in the rotor conductors and consequently no torque on the rotor. Hence the latter could not continue rotating at synchronous speed. As the rotor speed falls more and more below the synchronous speed, the values of the rotor e.m.f. and current and therefore of the torque continue to increase until the latter is equal to that required by the rotor losses and by any load there may be on the motor.

The speed of the rotor relative to that of the rotating flux is termed the *slip*; thus for a torque OA in fig. 24.3, the rotor speed is AC and the slip is AD, where

$$AD = AB - AC = CB.$$

For torques varying between zero and the full-load value, the slip is practically proportional to the torque. It is usual to express the slip either as a per-unit or fractional value or as a percentage of the synchronous speed; thus in fig. 24.3,

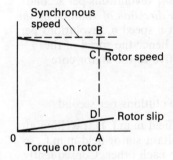

Fig. 24.3 Slip and rotor speed of an induction motor

$$\text{per-unit } slip = \frac{\text{slip in r/min}}{\text{synchronous speed in r/min}} = \frac{AD}{AB}$$

$$= \frac{\text{synchronous speed} - \text{rotor speed}}{\text{synchronous speed}}$$

$$= \frac{n_1 - n_r}{n_1} \tag{24.1}$$

and percentage slip = per unit or fractional slip × 100

$$= \frac{AD}{AB} \times 100.$$

The value of the slip at full load varies from about 6 per cent for small motors to about 2 per cent for large machines. The induction motor may therefore be regarded as practically a constant-speed machine; and the difficulty of varying its speed economically constitutes one of its main disadvantages (see section 24.10).

24.2 Frequency of rotor e.m.f. and current

It was shown in section 22.3 that for a three-phase winding with p pairs of poles supplied at a frequency of f hertz the speed of the rotating flux is given by n_1 revolutions per second, where

$$f = n_1 p$$

If n_r is the rotor speed in revolutions per second, the speed at which the rotor conductors are being cut by the rotating flux is $(n_1 - n_r)$ revolutions per second,

$$\text{frequency of rotor e.m.f.} = f_r = (n_1 - n_r)p$$

If $\quad s = $ per-unit or fractional slip $= (n_1 - n_r)/n_1$

then $\qquad\qquad\qquad\qquad n_1 - n_r = sn_1$

and $\qquad\qquad\qquad\qquad f_r = sn_1 p = sf \qquad\qquad (24.2)$

Polyphase currents in the stator winding produce a resultant magnetic field, the axis of which rotates at synchronous speed, n_1 revolutions per second, relative to the stator. Similarly the polyphase currents in the rotor winding produce a resultant magnetic field, the axis of which rotates [by expression (24.2)] at a speed sn_1 revolutions per second relative to the rotor surface, *in the direction of rotation of the rotor*. But the rotor is revolving at a speed n_r revolutions per second relative to the stator core; hence the speed of the resultant rotor magnetic field relative to the stator core

$$= sn_1 + n_r$$
$$= (n_1 - n_r) + n_r = n_1 \text{ revolutions per second},$$

i.e. the axis of the resultant rotor field m.m.f. is travelling at the same speed as that of the resultant stator field m.m.f., so that they are stationary relative to each other. Consequently the polyphase induction motor can be regarded as being equivalent to a transformer having an airgap separating the steel portions of the magnetic circuit carrying the primary and secondary windings.

Owing to this gap, the magnetizing current and the magnetic leakage for an induction motor are large compared with the corresponding values for a transformer of the same apparent power rating. Also, the friction and windage losses contribute towards making the efficiency of the induction motor less than that of the corresponding transformer. On the other hand, the stator field m.m.f. has to balance the rotor field m.m.f. and also provide the magnetizing and no-load loss components of the stator current, as in a transformer. Hence, an increase of slip due to increase of load is accompanied by an increase of the rotor currents and therefore by a corresponding increase of the stator currents.

24.3 Rotor e.m.f. and current

Let $\quad V_p = $ voltage per phase applied to stator winding,

$\qquad Z_s = $ no. of stator conductors in series per phase,

k_d = distribution factor of winding,

k_p = pitch factor of winding

= 1.0 for full-pitch coils,

and　Φ = flux per pole, i.e. total flux entering or leaving the stator over one pole pitch, namely the distance between 2 adjacent points of zero flux density.

Since the back e.m.f. generated in the stator winding is approximately equal to the applied voltage, then from expression (22.5) we have:

$$V_p \simeq 2.22 k_d k_p Z_s f \Phi \tag{24.3}$$

When the rotor is at standstill, the rotating flux Φ cuts the rotor at the same speed as it cuts the stator winding, so that the frequency of the rotor e.m.f. is then the same as the supply frequency, namely f hertz. Hence:

if　E_0 = rotor e.m.f. generated per phase* at standstill,

and　Z_r = no. of rotor conductors in series per phase,

then $E_0 = 2.22 k_d k_p Z_r f \Phi$　　　　(24.4)

Assuming the distribution and pitch factors to be the same for the stator and rotor windings, we have from (24.3) and (24.4):

$$E_0 \simeq V_p \times \frac{Z_r}{Z_s} \tag{24.5}$$

If E_r is the rotor e.m.f. generated per phase when the per-unit slip is s and the rotor frequency is $f_r = sf$,

$$E_r = 2.22 \times k_d k_p Z_r f_r \Phi$$

$$= s E_0 \tag{24.6}$$

If　R = resistance per phase of the rotor winding,

and　X_0 = leakage reactance per phase of rotor winding at standstill

= $2\pi f \times$ leakage inductance per phase of rotor winding,

then for per-unit or fractional slip s,

corresponding reactance per phase = $X_r = s X_0$

and　corresponding impedance per phase

$$= Z_r = \sqrt{\{R^2 + (sX_0)^2\}} \quad (24.7)$$

---

* In the cage rotor, the number of bars is usually a prime number, such as 47. Consequently the e.m.f.s in all the rotor bars differ from one another in phase, so that the number of phases is the same as the number of rotor bars. The rotor winding may also be of the three-phase type, star- or delta-connected, with its ends joined to 3 slip-rings. The relative advantages and disadvantages of the two types of rotor are discussed in section 24.12.

If I_0 = rotor current per phase at standstill,

and I_r = rotor current per phase at slip s,

$$I_0 = \frac{E_0}{\sqrt{(R^2 + X_0^2)}}$$

and $$I_r = \frac{E_r}{\sqrt{\{R^2 + (sX_0)^2\}}} = \frac{sE_0}{\sqrt{\{R^2 + (sX_0)^2\}}} \qquad (24.8)$$

If ϕ_r be the phase difference between E_r and I_r,

$$\tan \phi_r = X_r/R = sX_0/R \qquad (24.9)$$

and $$\cos \phi_r = \frac{R}{\sqrt{(R^2 + X_r^2)}} \qquad (24.10)$$

Example 24.1   *A three-phase induction motor is wound for 4 poles and is supplied from a 50-Hz system. Calculate: (a) the synchronous speed; (b) the speed of the rotor when the slip is 4 per cent; and (c) the rotor frequency when the speed of the rotor is 600 r/min.*

(a) From (22.1),   synchronous speed $= \dfrac{60f}{p} = \dfrac{60 \times 50}{2}$

$$= 1500 \text{ r/min}.$$

(b) From (24.1), $0.04 = \dfrac{1500 - \text{rotor speed}}{1500}$

∴   rotor speed $= 1440$ r/min.

(c) Also from (24.1),   per-unit slip $= \dfrac{1500 - 600}{1500} = 0.6$

Hence, from (24.2),   rotor frequency $= 0.6 \times 50 = 30$ Hz.

Example 24.2   *The stator winding of the motor of example 24.1 is delta-connected with 240 conductors per phase and the rotor winding is star-connected with 48 conductors per phase. The rotor winding has a resistance of 0.013 Ω/ph and a leakage reactance of 0.048 Ω/ph at standstill. The supply voltage is 415 V. Assuming the distribution factor to be 0.96 and the pitch factor to be 1.0 for each winding, calculate: (a) the flux per pole; (b) the rotor e.m.f. per phase at standstill with the rotor on open circuit; (c) the rotor e.m.f. and current per phase at 4 per cent slip; and (d) the phase difference between the rotor e.m.f. and current for a slip of (i) 4 per cent and (ii) 100 per cent. Assume the impedance of the stator winding to be negligible.*

(a) From (24.3), $415 = 2.22 \times 0.96 \times 240 \times 50 \times \Phi$

∴ $\Phi = 0.016\,23$ Wb.

(b) From (24.4), $E_0 = 2.22 \times 0.96 \times 48 \times 50 \times 0.01$

$$= 83 \text{ V}.$$

Alternatively, $E_0 = 415 \times 48/240 = 83$ V.

(c) From (24.6),

rotor e.m.f. for 4 per cent slip $= 83 \times 0.04 = 3.32$ V

From (24.7),

impedance per phase for 4 per cent slip

$$= \sqrt{\{(0.013)^2 + (0.04 \times 0.048)^2\}}$$

$$= 0.013\,14 \ \Omega,$$

$\therefore$ rotor current $= 3.32/0.013\ 14 = 252.7\ \text{A}$.

(*d*) From (24.9), it follows that for 4 per cent slip,

$$\tan \phi_r = \frac{0.04 \times 0.048}{0.013} = 0.1477$$

$$\phi_r = 8° \ 24'.$$

For 100 per cent slip, $\tan \phi_r = 0.048/0.013 = 3.692,$

$\therefore$ $\phi_r = 74° \ 51'.$

Example 24.2 shows that the slip has a considerable effect upon the phase difference between the rotor e.m.f. and current — a fact that is very important when we come to discuss the variation of torque with slip.

24.4 Relationship between the rotor *I²R* loss and the rotor slip

The following table indicates concisely what becomes of the power supplied to the stator of the induction motor:

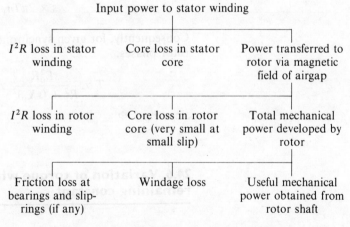

If T = torque, in newton-metres, exerted on the rotor by the rotating flux

and n_1 = synchronous speed in revolutions per second,

power transferred from stator to rotor $= 2\pi T n_1$ watts.

If n_r = rotor speed in revolutions per second,

total mechanical power developed by rotor $= 2\pi T n_r$ watts.
But from the table above it is seen that:

total $I^2 R$ loss in rotor

$\simeq$ power transferred from stator to rotor

$-$ total mechanical power developed by rotor

$$= 2\pi T(n_1 - n_r) \text{ watts}$$

$$\therefore \quad \frac{\text{total rotor } I^2R \text{ loss}}{\text{input power to rotor}} = \frac{2\pi T(n_1 - n_r)}{2\pi T n_1} = s \quad (24.11)$$

or   total rotor I^2R loss (in watts)

$$= s \times \text{input power to rotor (in watts)}$$

24.5 Factors determining the torque

If $m = $ number of rotor phases,

then, using the symbols given in section 24.3, we have:

electrical power generated in rotor $= mI_r E_r \cos \phi_r$ watts

$$= \frac{ms^2 E_0^2 R}{R^2 + (sX_0)^2}.$$

All this power is dissipated as I^2R loss in the rotor circuits.

Since   input power to rotor $= 2\pi T n_1$ watts,

hence, from (24.11), we have:

$$s \times 2\pi T n_1 = \frac{ms^2 E_0^2 R}{R^2 + (sX_0)^2}.$$

Consequently, for given synchronous speed and number of rotor phases,

$$T \propto \frac{sE_0^2 R}{R^2 + (sX_0)^2} \propto \frac{s\Phi^2 R}{R^2 + (sX_0)^2} \quad (24.12)$$

since $E_0 \propto \Phi$.

24.6 Variation of torque with slip, other factors remaining constant

If the impedance of the stator winding is assumed to be negligible, then for a given supply voltage, Φ and E_0 remain constant,

$$\therefore \quad \text{torque} \propto \frac{sR}{R^2 + (sX_0)^2} \quad (24.13)$$

The value of X_0 is usually far greater than the resistance of the rotor winding; so let us for simplicity assume $R = 1\,\Omega$ and $X_0 = 8\,\Omega$, and calculate the value of $sR/(R^2 + s^2 X_0^2)$ for various values of the slip between 1 and 0. The results are represented by curve A in fig. 24.4. It will be seen that for small values of the slip, the torque is almost directly proportional to the slip; whereas for slips between about 0.2

Fig. 24.4 Torque/slip curves for an induction motor

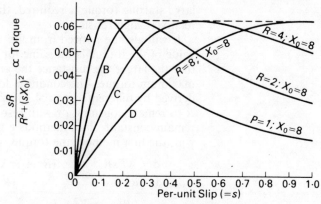

and 1, the torque is almost inversely proportional to the slip. These relationships can be easily deduced from expression (24.13). Thus, in the case of the cage rotor, R is small compared with X_0, but for values of the slip less than about 0.1 per unit, $(sX_0)^2$ is very small compared with R^2, so that:

$$\text{torque} \propto \frac{sR}{R^2} \propto \frac{s}{R} \tag{24.14}$$

i.e. the torque is directly proportional to the slip when the latter is very small.

For large values of the slip, R^2 is very small compared with $(sX_0)^2$ for the cage rotor and for the slip-ring rotor with no external resistance (section 24.8),

$$\therefore \qquad \text{torque} \propto \frac{sR}{(sX_0)^2} \propto \frac{R}{s} \tag{24.15}$$

since X_0 is constant for a given motor; i.e. the torque is inversely proportional to the slip when the latter is large. The term R has been left in the above expressions as it is referred to in the next section.

24.7 Effect of rotor resistance upon the torque/slip relationship

From expression (24.15) it is seen that when R is small compared with sX_0, the torque for a given slip is directly proportional to the value of R; whereas from expression (24.14) it follows that when R is large compared with sX_0, the torque for a given slip is inversely proportional to the value of R. The simplest method of demonstrating this effect is to repeat the calculation of $sR/(R^2 + s^2X_0^2)$ with $R = 2\,\Omega$, $R = 4\,\Omega$ and $R = 8\,\Omega$. The results are represented by curves B, C and D respectively in fig. 24.4. It will be seen that for a slip of, say, 0.05 p.u., the effect of doubling the rotor resistance is to reduce the torque by about 0.45 per unit, whereas for a slip of 1, the torque is nearly doubled when the resistance is increased from $1\,\Omega$ to $2\,\Omega$. Hence, if a large starting torque is required, the rotor must have a relatively high resistance.

It will also be noticed from fig. 24.4 that the maximum value of the torque is the same for the four values of R and that the larger the resistance the greater is the slip at maximum torque. The condition for maximum torque can be derived by differentiating (24.13) with respect to s, assuming R to remain constant, or with respect to R, assuming s to remain constant. Both methods give the same result; thus with the first method, the torque is maximum when:

$$\frac{d}{ds}\left(\frac{sR}{R^2 + s^2 X_0^2}\right) = \frac{(R^2 + s^2 X_0^2)R - sR \times 2sX_0^2}{(R^2 + s^2 X_0^2)^2} = 0$$

i.e.
$$R^2 - s^2 X_0^2 = 0$$

so that
$$sX_0 = R \qquad (24.16)$$

Hence the torque is a maximum* when the reactance is equal to the resistance. For instance, with $R = 1\,\Omega$ and $X_0 = 8\,\Omega$, maximum torque occurs when $s = 0.125$ p.u.; whereas with $R = 8\,\Omega$ and $X_0 = 8\,\Omega$, maximum torque occurs when $s = 1$, namely when the rotor is at standstill.

Substituting R for sX_0 in expression (24.13), we have:

$$\text{maximum torque} \propto \frac{sR}{2R^2}$$

$$\propto \frac{1}{2X_0}.$$

But X_0 is the leakage reactance at standstill and is a constant for a given rotor; hence the maximum torque is the same whatever the value of the rotor resistance.

24.8 Starting torque

At the instant of starting, $s = 1$, and it will be seen from fig. 24.4 that with a motor having a low-resistance rotor, such as the usual type of cage rotor, the starting torque is small compared with the maximum torque available. On the other hand, if the bars of the cage rotor were made with sufficiently high resistance to give the maximum torque at standstill, the slip for full-load torque — usually about one-third to one-half of the maximum torque — would be relatively large and the I^2R loss in the rotor winding would be high, with the result that the efficiency would be low; and if this load was maintained for an hour or two, the temperature rise would be excessive. Also the variation of speed with load would be large (see section 24.10). Hence, when a motor is required to

---

* This can be checked by differentiating expression $(R^2 - s^2 X_0^2)$ with respect to s. The result, namely $-2sX_0^2$, is negative; hence expression (24.16) gives the condition for *maximum* torque.

exert its maximum torque at starting, the usual practice is to insert extra resistance into the rotor circuit and to reduce the resistance as the motor accelerates. Such an arrangement involves a three-phase winding on the rotor, the three ends of the winding being connected via slip-rings on the shaft to external star-connected resistors R, as shown in fig. 24.5. The three arms, A, are mechanically and electrically connected together.

Fig. 24.5 Induction motor with slip-ring rotor

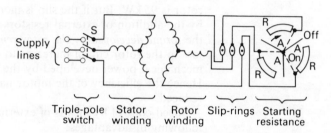

Triple-pole switch   Stator winding   Rotor winding   Slip-rings   Starting resistance

The starting procedure is to close the triple-pole switch S and then move arms A clockwise as the motor accelerates, until, at full speed, the arms are in the *ON* position shown dotted in fig. 24.5, and the starting resistors have been cut out of the rotor circuit. Large motors are often fitted with a short-circuiting and brush-lifting device which first short-circuits the three slip-rings and then lifts the brushes off the rings, thereby eliminating losses due to the brush-contact resistance and the brush friction and reducing the wear of the brushes and of the slip-rings.

24.9 Variation of torque with stator voltage, other factors remaining constant

From expression (24.12), it is seen that for given values of the slip and of the rotor resistance and reactance,

$$\text{torque on rotor} \propto \Phi^2.$$

But for a given stator winding, it follows from expression (24.3) that Φ is approximately proportional to the voltage applied to the stator winding:

$$\therefore \qquad \text{torque on rotor} \propto (\text{stator applied voltage})^2 \qquad (24.17)$$

This relationship will be referred to in section 24.11 in connection with methods of starting up motors having cage rotors.

24.10 Speed control by means of external rotor resistors

From expression (24.11), it follows that for a given input power to the rotor and therefore for a given torque exerted by the rotor, the total rotor I^2R loss is proportional to the slip. Thus, if a motor has 100 kW transferred from the stator to the rotor when the slip is 5 per cent, the total rotor I^2R loss is 5 kW and the mechanical power developed by the rotor is 95 kW. But if the slip is increased to, say, 40 per cent by the addition of external resistors in the rotor circuit and if the *torque developed by the rotor remains unaltered*, the I^2R loss in the rotor circuit increases to 40 kW and the mechanical power developed by the rotor decreases to 60 kW. Hence the efficiency of the motor has been considerably reduced.

Speed control by means of external rotor resistors has the following disadvantages:

(*a*) reduction of speed is accompanied by reduced efficiency;

(*b*) with a large resistance in the rotor circuit, the speed varies considerably with variation of torque (see fig. 24.4);

(*c*) the external rotor resistors are comparatively bulky and expensive as they may have to dissipate a good deal of power without becoming overheated.

The main advantage of this method of speed control is its simplicity.

An alternative means of speed control has been introduced in recent years, whereby an a.c. supply is changed into a direct current supply and then changed back to an a.c. supply at a different frequency from that of the original supply. This new frequency a.c. supply is injected into the rotor windings through the slip-rings and, by varying the frequency of the current in the rotor windings, it is possible to vary the speed. It should be remembered that the speed of the induction motor normally determines the frequency of the rotor currents but in this instance it is the frequency which is determining the speed.

The frequency change is effected by thyristors, which are described in Chapter 25.

Example 24.3   *The power supplied to a three-phase induction motor is 40 kW and the corresponding stator losses are 1.5 kW. Calculate: (a) the total mechanical power developed and the rotor I^2R loss when the slip is 0.04 per unit; (b) the output power of the motor if the friction and windage losses are 0.8 kW; and (c) the efficiency of the motor. Neglect the rotor core loss.*

(*a*) Input power to rotor $= 40 - 1.5 = 38.5$ kW.

From (24.11) $\dfrac{\text{rotor } I^2R \text{ loss in kW}}{38.5} = 0.04$

∴   rotor I^2R loss $= 1.54$ kW

so that mechanical power developed by the rotor

$$= 38.5 - 1.54$$
$$= 36.96 \, \text{kW}$$

(b)   Output power of motor $= 36.96 - 0.8$
$$= 36.16 \, \text{kW}$$

(c)   Efficiency of motor $= 36.16/40 = 0.904$ p.u.
$$= 90.4 \text{ per cent.}$$

Example 24.4   *If the speed of the motor of example 24.3 is reduced to 40 per cent of its synchronous speed by means of external rotor resistors, calculate: (a) the total rotor I^2R loss, and (b) the efficiency, assuming the torque and the stator losses to remain unaltered. Also, assume that the increase in the rotor core loss is equal to the reduction in the friction and windage loss.*

(a)   New slip $= (100 - 40)/100 = 0.6$ p.u.

and   input power to rotor $= 38.5 \, \text{kW}$

From (24.11),   total rotor I^2R loss $= 0.6 \times 38.5 = 23.1 \, \text{kW}$.

(b)   Total losses in rotor $= 23.1 + 0.8 = 23.9 \, \text{kW}$

∴   output power of motor $= 38.5 - 23.9 = 14.6 \, \text{kW}$

and   efficiency of motor $= 14.6/40 = 0.365$ p.u.
$$= 36.5 \text{ per cent.}$$

24.11 Starting of a three-phase induction motor fitted with a cage rotor

If this type of motor is started up by being switched directly across the supply, the starting current is about 4–7 times the full-load current, the actual value depending upon the size and design of the machine. Such a large current can cause a relatively large voltage drop in the cables and thereby produce an objectionable momentary dimming of the lamps in the vicinity. Consequently it is usual to start cage motors — except small machines — with a reduced voltage, using one of the following methods:

(1) Star–delta starter   The two ends of each phase of the stator winding are brought out to the starter which, when moved to the 'starting' position, connects the winding in star. After the motor has accelerated, the starter is quickly moved to the 'running' position, thereby changing the connections to delta. Hence the voltage per phase at starting is $1/\sqrt{3}$ of the supply voltage, and the starting torque, by expression (24.17), is one-third of that obtained if the motor were switched directly across the supply with its stator winding delta-connected. Also, the starting current per phase is $1/\sqrt{3}$ and that taken from the supply is one-third of the corresponding value with direct switching.

(2) Auto-transformer starter

In fig. 24.6, T represents a three-phase star-connected auto-transformer (see section 20.20) with a mid-point tapping on each phase so that the voltage applied to motor M is half the supply voltage. With such tappings, the supply current and the starting torque are only a quarter of the values when the full voltage is applied to the motor.

Fig. 24.6 Starting connections of an auto-transformer starter

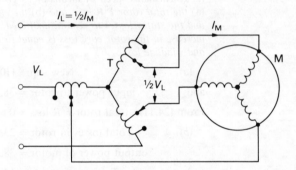

After the motor has accelerated, the starter is moved to the 'running' position, thereby connecting the motor directly across the supply and opening the star-connection of the auto-transformer.

Taking the general case where the output voltage per phase of the auto-transformer is n times the input voltage per phase, we have:

$$\frac{\text{starting torque with auto transformer}}{\text{starting torque with direct switching}} = \left(\frac{\text{output voltage}}{\text{input voltage}}\right)^2 = n^2.$$

Also, if Z_0 be the equivalent standstill impedance per phase of a star-connected motor, referred to the stator circuit, and if V_P be the star voltage of the supply, then,

with direct switching, starting current $= V_P/Z_0$.

With auto-transformer starting,

$$\text{starting current in motor} = nV_P/Z_0$$

and starting current from supply $= n \times$ motor current

$$= n^2 V_P/Z_0$$

∴ $$\frac{\text{starting current with auto-transformer}}{\text{starting current with direct switching}}$$

$$= (n^2 V_P/Z_0) \div (V_P/Z_0) = n^2.$$

The auto-transformer is usually arranged with two or three tappings per phase so that the most suitable ratio can be selected for a given motor, but an auto-transformer starter is more expensive than a star–delta starter.

24.12 Comparison of cage and slip-ring rotors

The cage rotor possesses the following advantages:

(*a*) cheaper and more robust;
(*b*) slightly higher efficiency and power factor;
(*c*) explosion proof, since the absence of slip-rings and brushes eliminates risk of sparking.

The advantages of the slip-ring rotor are:

(*a*) the starting torque is much higher and the starting current much lower;
(*b*) the speed can be varied by means of external rotor resistors (see section 24.10).

Summary of important formulae

$$\text{Synchronous speed} = f/p \tag{24.1}$$

$$\left.\begin{array}{l}\text{Fractional or} \\ \text{per unit slip, } s\end{array}\right\} = \frac{\text{synchronous speed} - \text{rotor speed}}{\text{synchronous speed}} \tag{24.1}$$

$$\text{Rotor frequency} = f_r = sf \tag{24.2}$$

$$\text{Rotor e.m.f. per phase} = E_r = sE_0 \tag{24.6}$$

$$\text{Rotor impedance per phase} = Z_r = \sqrt{\{R^2 + (sX_0)^2\}} \tag{24.7}$$

$$s = \frac{\text{total rotor } I^2R \text{ loss}}{\text{input power to rotor}} \tag{24.11}$$

$$\text{Torque on rotor} \propto \frac{\Phi^2 sR}{R^2 + (sX_0)^2} \tag{24.12}$$

$$\text{For small slip,} \quad \text{torque} \propto s/R \tag{24.14}$$

For large slip and low rotor resistance,

$$\text{torque} \propto R/S \tag{24.15}$$

For maximum torque,

$$R = sX_0 \tag{24.16}$$

$$\text{Starting torque} \propto (\text{stator applied voltage})^2 \tag{24.17}$$

EXERCISES 24

1. Explain how slip-frequency currents are set up in the rotor windings of a three-phase induction motor.
   A two-pole, three-phase, 50-Hz induction motor is running on load with a slip of 4 per cent. Calculate the actual speed and the synchronous speed of the machine.

Sketch the speed/load characteristic for this type of machine and state with which kind of d.c. motor it compares.

(U.E.I., O.2)

2. Explain why an induction motor cannot develop torque when running at synchronous speed. Define the slip speed of an induction motor and deduce how the frequency of rotor currents and the magnitude of the rotor e.m.f. are related to slip.

An induction motor has four poles and is energized from a 50-Hz supply. If the machine runs on full load at 2 per cent slip, determine the running speed and the frequency of the rotor currents. (N.C.T.E.C., O.2)

3. Give a clear explanation of the following effects in a three-phase induction motor: (*a*) the production of the rotating field; (*b*) the presence of an induced rotor current; (*c*) the development of the torque. (S.A.N.C., O.2)

4. If a six-pole induction motor supplied from a three-phase 50-Hz supply has a rotor frequency of 2.3 Hz, calculate (*a*) the percentage slip and (*b*) the speed of the rotor in revolutions per minute.

5. Show how a rotating magnetic field can be produced by three-phase currents.

A fourteen-pole, 50-Hz induction motor runs at 415 r/min. Deduce the frequency of the currents in the rotor winding and the slip. (App. El., L.U.)

6. Explain the principle of action of a three-phase induction motor and the meaning of the term *slip*. How does slip vary with the load?

A centre-zero d.c. galvanometer, suitably shunted, is connected in one lead of the rotor of a three-phase, six-pole, 50-Hz slip-ring induction motor and the pointer makes 85 complete oscillations per minute. What is the rotor speed?

(U.E.I.)

7. Describe, in general terms, the principle of operation of a three-phase induction motor.

The stator winding of a three-phase, eight-pole, 50-Hz induction motor has 720 conductors, accommodated in 72 slots. Calculate: (*a*) the distribution factor of the winding; (*b*) the flux per pole of the rotating field in the airgap of the motor, needed to generate 230 V in each phase of the stator winding.

(App. El., L.U.)

8. A three-phase induction motor, at standstill, has a rotor voltage of 100 V between the slip-rings when they are open-circuited. The rotor winding is star-connected and has a leakage reactance of 1 Ω/ph at standstill and a resistance of 0.2 Ω/ph. Calculate: (*a*) the rotor current when the slip is 4 per cent and the rings are short-circuited; (*b*) the slip and the rotor current when the rotor is developing maximum torque. Assume the flux to remain constant.

9. A three-phase, 50-Hz induction motor with its rotor star-connected gives 500 V (r.m.s.) at standstill between the slip-rings on open circuit. Calculate the current and power factor at standstill when the rotor winding is joined to a star-connected external circuit, each phase of which has a resistance of 10 Ω and an inductance of 0.04 H. The resistance per phase of the rotor winding is 0.2 Ω and its inductance is 0.04 H. Also calculate the current and power factor when the slip-rings are short-circuited and the motor is running with a slip of 5 per cent. Assume the flux to remain constant. (App. El., L.U.)

10. If the star-connected rotor winding of a three-phase induction motor has a resistance of 0.01 Ω/ph and a standstill leakage

reactance of 0.08 Ω/ph, what must be the value of the resistance per phase of a starter to give the maximum starting torque? What is the percentage slip when the starting resistance has been reduced to 0.02 Ω/ph, if the motor is still exerting its maximum torque?

11. A three-phase, 50-Hz, six-pole induction motor has a slip of 0.04 per unit when the output is 20 kW. The frictional loss is 250 W. Calculate: (*a*) the rotor speed, and (*b*) the rotor I^2R loss.

12. Sketch the usual form of the torque/speed curve for a polyphase induction motor and explain the factors which determine the shape of this curve.

    In a certain eight-pole, 50-Hz machine, the rotor resistance per phase is 0.04 Ω and the maximum torque occurs at a speed of 645 r/min. Assuming that the airgap flux is constant at all loads, determine the percentage of maximum torque: (*a*) at starting, and (*b*) when the slip is 3 per cent.   (App. El., L.U.)

13. Describe briefly the construction of the stator and slip-ring rotor of a three-phase induction motor. Explain the action of the motor and why the rotor is provided with slip-rings.

    A three-phase, 50-Hz induction motor has 4 poles and runs at a speed of 1440 r/min when the total torque developed by the rotor is 70 N · m. Calculate: (*a*) the total input (in kilowatts) to the rotor; (*b*) the rotor I^2R loss in watts.   (App. El., L.U.)

14. Explain how a rotating magnetic field may be produced by stationary coils carrying three-phase currents.

    Determine the efficiency and the output kilowatts of a three-phase, 400-V induction motor running on load with a fractional slip of 0.04 and taking a current of 50 A at a power factor of 0.86. When running light at 400 V, the motor has an input current of 15 A and the power taken is 2000 W, of which 650 W represent the friction, windage and rotor core loss. The resistance per phase of the stator winding (delta-connected) is 0.5 Ω.   (App. El., L.U.)

15. Calculate the relative values of (i) the starting torque and (ii) the starting current of a three-phase cage-rotor induction motor when started by: (*a*) direct switching; (*b*) a star–delta starter; and (*c*) an auto-transformer having 40 per cent tappings.

25 Power Electronics

25.1 Introductory

Although transistors can be used as a switch, generally their current-carrying capacity is small. There are many applications in which it would be advantageous to have a high-speed switch which could handle up to 1000 A: such a device is the thyristor, which also has the advantages of having no moving parts and no arcing.

25.2 Thyristor

The basic parts of the thyristor are its four layers of alternate p-type and n-type silicon semiconductors, forming three p–n junctions, A, B and C, as shown in fig. 25.1(a). The terminals connected to the n_1 and p_2 layers are the cathode and anode respectively. A contact welded to the p_1 layer is termed the *gate*. The British Standards graphical symbol for the thyristor is given in fig. 25.1(b). The direction of the arrowhead on the

Fig. 25.1 Thyristor arrangement and symbols

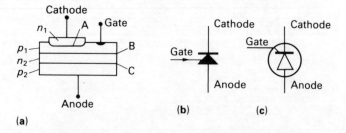

gate lead indicates that the gate contact is welded to a p-region and shows the direction of the gate current required to operate the device. If the gate contact is welded to an n-region, the arrowhead should point outwards from the rectifier. The graphical symbol shown in fig. 25.1(c) is, however, frequently used to represent a thyristor.

When the anode is positive with respect to the cathode, junctions A and C are forward-biased and therefore have a very low resistance, whereas junction B is reverse-biased and consequently presents a very high resistance, of the order of megohms, to the passage of a current. On the other hand, if the anode terminal be made *negative* with respect to the cathode terminal, junction B is forward-biased while A and C act as two reverse-biased junctions in series.

Let us now consider the effect of increasing the voltage applied across the thyristor, *with the anode positive relative to the cathode*. At first, the forward leakage current reaches saturation value due to the action of junction B. Ultimately, a breakover value is reached and the resistance of the thyristor instantly falls to a very low value, as shown in fig. 25.2. The

Fig. 25.2 Thyristor characteristic

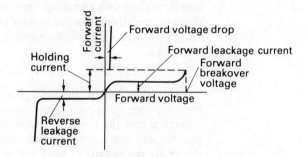

forward voltage drop is of the order of 1–2 volts and remains nearly constant over a wide variation of current. A resistor is necessary in series with the thyristor to limit the current to a safe value.

We shall now consider the effect upon the breakover voltage of applying a positive potential to the gate as in fig. 25.3. When switch S is closed, a bias current, I_B, flows via

Fig. 25.3 Gate control of thyristor breakover voltage

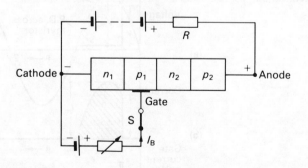

the gate contact and layers p_1 and n_1, and the value of the breakover voltage of the thyristor depends upon the magnitude of the bias current in the way shown in fig. 25.4. Thus, with $I_B = 0$, the breakover voltage is represented by OA, and remains practically constant at this value until the bias current is increased to OB. For values of bias current between OB and OD, the breakover voltage falls rapidly to nearly zero. An alternative method of representing this effect is shown in fig. 25.5.

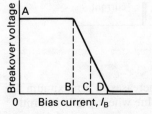

Fig. 25.4 Variation of breakover voltage with bias current

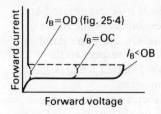

Fig. 25.5 Variation of breakover voltage with bias current, I_B

In the triggered condition, the thyristor approximates a single p–n diode, the anode current being limited only by the resistance R of the external circuit. Once the gate has triggered the device, *it loses control over the anode current*; and the only method of restoring the device to its high-resistance condition is to reduce the anode current below the *holding* value indicated in fig. 25.2. The value of the holding current is usually very small compared with the rated forward current; for instance, the holding current may be about 10 mA for a thyristor having a forward-current rating of 40 A.

If the thyristor is connected in series with a non-reactive load, of resistance R, across a supply voltage having a sinusoidal waveform and if it is triggered at an instant corresponding to an angle ϕ after the voltage has passed through zero from a negative to a positive value, as in fig. 25.6(a), the value of the applied voltage at that instant is given by:

$$v = V_m \sin \phi.$$

Up to that instant, the voltage across the thyristor has been growing from zero to v. When triggering occurs, the voltage across the thyristor instantly falls to about 1–2 volts and remains approximately constant while current flows, as shown by the slightly curved line in fig. 25.6(a). Also, at the instant of triggering, the current increases immediately from zero to i,

where $$i = \frac{v - \text{p.d. across thyristor}}{R}$$

$$= v/R \text{ when the p.d. across thyristor} \ll v.$$

Fig. 25.6 Phase-controlled half-wave rectification

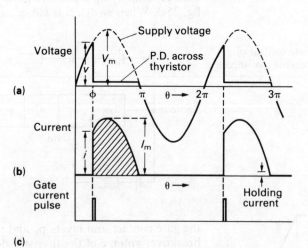

If ϕ is less than $\pi/2$, the current increases to a maximum I_m and then decreases to the holding value when it falls instantly to zero, as shown in fig. 25.6(b). The average value of the current over one cycle is the shaded area enclosed by the current wave divided by 2π.

If the p.d. across the thyristor, when conducting, is very small compared with the supply voltage, and if the holding current is negligible compared with I_m, the waveform of the current is practically sinusoidal for values of θ between ϕ and π,

$\therefore$ average value of current over one cycle

$$= I_{av} = \frac{1}{2\pi} \int_{\phi}^{\pi} I_m \sin \theta \cdot d\theta$$

$$= \frac{I_m}{2\pi} \left[-\cos \theta \right]_{\phi}^{\pi} = \frac{I_m}{2\pi} (1 + \cos \phi)$$

When $\phi = 0$, $I_{av} = I_m/\pi = 0.3185 I_m$.

When $\phi = \pi/2$, $I_{av} = 0.159 I_m$.

When $\phi = \pi$, $I_{av} = 0$.

Hence, by varying the instant at which the thyristor is triggered, it is possible to control the output current over a wide range. Triggering signals for this purpose are usually in the form of a positive pulse of short duration compared with the time of one cycle of the alternating voltage. A gate current of about 50 mA applied for about 5 μs is all that is required to trigger a device having current ratings between a few milliamperes and hundreds of amperes.

25.3 Some thyristor circuits

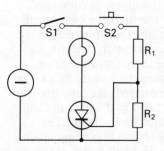

Fig. 25.7 Simple d.c. thyristor switch-on circuit

A simple switch-on arrangement incorporating a thyristor is shown in fig. 25.7. In this case, when the switch S1 is closed, the thyristor immediately blocks the passage of current through the lamp. When switch S2 is temporarily closed, some milliamperes of current flow into the gate and switch on the thyristor. Once the thyristor is switched on, it is no longer necessary to keep switch S2 closed.

In order to switch off the lamp in fig. 25.7, we could open switch S1, but this raises the question — is switch S1 not doing the same work as the thyristor and have we not duplicated our switching? Really we should be looking to S1 being replaced by the thyristor, and this can be achieved if we modify the circuit to that shown in fig. 25.8. The closure of switch S3 connects the capacitor so that it tries to discharge in opposition to the passage of current through the thyristor, with the result that for an instant there is no current passing through the thyristor. This short interruption is sufficient to stop the thyristor conducting and it therefore returns to the off condition.

When operating from an a.c. supply, the thyristor does not always receive sufficient gate current to switch it on. Instead we find that the forward conduction characteristic changes as

Fig. 25.8 Simple d.c.
thyristor on-off circuit

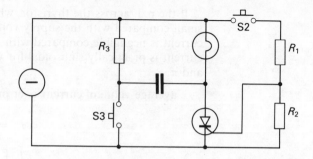

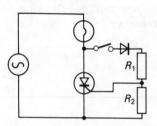

Fig. 25.9 Forward
conduction characteristic of
thyristor

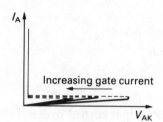

Fig. 25.10 Simple half-wave
a.c. switch circuit

the gate current is increased; thus if we have insufficient gate current to switch on the thyristor, it is nevertheless noticeable from the characteristic shown in fig. 25.9 that less anode–cathode voltage is required to raise the leakage current to a sufficient level that the avalanche breakdown takes place and the thyristor switches on of its own accord.

For the simple circuit shown in fig. 25.10, this form of switch-on action can be employed. When switch S1 is open, there is no gate current and the thyristor is able to block the applied alternating voltage even at its peak value. When S1 is closed, the applied voltage causes a current to pass through R_1 and R_2, causing volt drops across each. The voltage across R_1 causes some current to pass through the gate, thus reducing the breakdown voltage of the thyristor. Eventually a point is reached at which either there is sufficient gate current to directly switch on the thyristor or the gate current sufficiently reduces the breakdown voltage of the thyristor that it again switches on — usually we experience this latter form of action.

Once the thyristor commences to conduct, the volt drop across it falls almost to zero; thus the gate current almost disappears. When the half-cycle of the supply is completed, the anode current falls to zero and the thyristor switches off.

During the subsequent half-cycle, the thyristor cannot conduct; thus the lamp receives approximately a half-wave supply, depending on how soon the thyristor switches on after the commencement of the positive half-wave. By varying R_1, the instant of switch on can be delayed; thus the thyristor can be used to control the voltage developed across the lamp and consequently the power dissipated by the lamp. The diode in the circuit shown prevents reverse bias being applied to the thyristor gate during the negative half-cycles of the supply.

25.4 Limitations to thyristor operation

Due to the nature of the construction of a thyristor, there is some capacitance between the anode and the gate. If a sharply rising voltage is applied to the thyristor, then there is an inrush of charge corresponding to the relation

$i = C(\mathrm{d}v/\mathrm{d}t)$. This inrush current can switch on the thyristor, and it can arise in practice due to surges in the supply system, e.g. due to switching or due to lightning. Thus thyristors may be inadvertently switched on, and such occurrences can be avoided by providing C–R circuits in order to divert the surges from the thyristors.

The leakage current in any p–n junction depends on the temperature of the junction. It follows that if we were to raise the temperature of a thyristor, the leakage current would rise, and it approximately doubles for every 8°C rise in temperature. If the temperature is permitted to rise too much, again the leakage current could inadvertently switch on the thyristor; thus precautions must be taken in order to maintain the operating temperature of a thyristor at a reasonably low level. Alternatively, it is possible to make use of this observation and to use the thyristor as a switch which will complete a circuit should a predetermined temperature be exceeded.

25.5 The thyristor in practice

The thyristor has therefore been seen to be a switch, and you may wonder why we should use this form instead of the normal mechanical device, especially since we see from the simple circuits that we still must retain mechanical switches. The advantage of the thyristor is that it can operate without involving arcing; thus there are no parts being worn out either as a result of motion or of the burning of the arc. The switches used to turn the thyristor on and also off do not involve the interruption of large currents, so they are less prone to wear and tear.

Apart from the aspects of reliability and safety due to the lack of moving parts, the thyristor as a switch is much more definite in its action; thus we can determine, for instance, the instant at which it will commence to pass current during each cycle of an alternating current. This permits its use as a control device, thus making the thyristor available as a means of regulating the speed of a machine, regulating a voltage supply and regulating a host of other variable quantities necessary to the control engineer.

EXERCISES 25

1. Explain the function of the gate when the thyristor is switched into the conducting mode.
2. Explain the manner in which a thyristor is switched on without the injection of a gate current.
3. A thyristor is used to control the current to a lamp bulb which may be considered as a linear resistor. The resistance of the lamp

bulb is 500 Ω and the 50-Hz supply voltage is 240 V r.m.s. with a sinusoidal waveform.

Determine the average current in the lamp bulb over half a cycle if: (*a*) the thyristor conducts as soon as the voltage rises beyond zero; (*b*) the thyristor commences conduction 3.0 ms after the voltage rises beyond zero.

4. Determine the r.m.s. currents for Q. 3.
5. For the circuit given in Q. 3, determine the angle of switching in order that: (*a*) the average current is 0.25 A; (*b*) the r.m.s. current is 0.3 A.
6. Obtain the control circuit of a thyristor-controlled device and determine its manner of operation.

26.1 Electrical analogue indicating instruments

An indicating instrument is often fitted with a pointer which indicates on a scale the value of the quantity being measured. Such an instrument is said to be an *analogue* instrument; alternatively, some instruments have a display of numbers similar to a calculator and are said to be *digital* instruments. The moving system of an analogue instrument is usually carried by a spindle of hardened steel, having its ends tapered and highly polished* to form pivots which rest in hollow-ground bearings, usually of sapphire, set in steel screws. In some instruments, the moving system is attached to two thin ribbons of spring material such as beryllium–copper alloy, held taut by tension springs mounted on the frame of the movement. This arrangement eliminates pivot friction and the instrument is less susceptible to damage by shock or vibration.

Indicating analogue instruments possess three essential features:

(*a*) a *deflecting device* whereby a mechanical force is produced by the electric current, voltage or power;

(*b*) a *controlling device* whereby the value of the deflection is dependent upon the magnitude of the quantity being measured; and

(*c*) a *damping device* to prevent oscillation of the moving system and enable the latter to reach its final position quickly.

The action of the deflecting device depends upon the type of instrument, and the principle of operation of each of the instruments most commonly used in practice will be described in later sections.

---

* The necessity of handling measuring instruments with care can be realized from the fact that if a pivot has a circle of contact 0.05 mm in diameter and supports a mass of 3 grams (a normal mass for the moving system of an ammeter or voltmeter), the pressure is approximately 15 MN/m².

26.2 Controlling devices

There are two types of controlling devices, namely:

(*a*) spring control, and
(*b*) gravity control (not used in modern instruments).

The most common arrangement of spring control utilizes two spiral hairsprings, A and B (fig. 26.1), the inner ends of which are attached to the spindle S. The outer end of B is fixed, whereas that of A is attached to one end of a lever L, pivoted at P, thereby enabling zero adjustment to be easily effected. The hairsprings are of non-magnetic alloy such as phosphor–bronze or beryllium–copper.

The two springs, A and B, are wound in opposite directions so that when the moving system is deflected, one spring winds up while the other unwinds, and the controlling torque is due to the combined torsions of the springs. Since the torsional torque of a spiral spring is proportional to the angle of twist, the controlling torque is directly proportional to the angular deflection of the pointer.

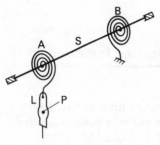

Fig. 26.1 Spring control

26.3 Damping devices

The combination of the inertia of the moving system and the controlling torque of the spiral springs or of gravity gives the moving system a natural frequency of oscillation (section 7.13). Consequently, if the current through an under-damped ammeter were increased suddenly from zero to OA (fig. 26.2), the pointer would oscillate about its mean position, as shown by curve B, before coming to rest. Similarly every fluctuation of current would cause the pointer to oscillate and it might be difficult to read the instrument accurately. It is therefore desirable to provide sufficient damping to enable the pointer to reach its steady position without oscillation, as indicated by curve C. Such an instrument is said to be *dead-beat*.

The two methods of damping commonly employed are:

(*a*) eddy-current damping, and
(*b*) air damping.

Fig. 26.2 Damping curves

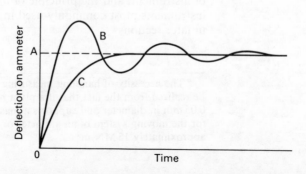

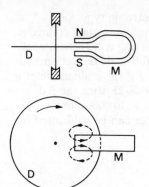

Fig. 26.3 Eddy-current damping

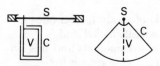

Fig. 26.4 Air damping

One form of eddy-current damping is shown in fig. 26.3, where a copper or aluminium disc D, carried by a spindle, can move between the poles of a permanent magnet M. If the disc moves clockwise, the e.m.f.s induced in the disc circulate eddy currents as shown dotted. It follows from Lenz's Law that these currents exert a force opposing the motion producing them, namely the clockwise movement of the disc.

Another arrangement, used in moving-coil instruments (section 26.5), is to wind the coil on an aluminium frame. When the latter moves across the magnetic field, eddy current is induced in the frame; and, by Lenz's Law, this current exerts a torque opposing the movement of the coil.

Air damping is usually obtained by means of a thin metal vane V attached to the spindle S, as shown in fig. 26.4. This vane moves in a sector-shaped box C, and any tendency of the moving system to oscillate is damped by the action of the air on the vane.

26.4 Types of ammeters, voltmeters and wattmeters

The principal types of electrical indicating instruments, together with the methods of control and damping, are summarized in the following table:

Type of instrument	Suitable for measuring	Method of control	Method of damping
Permanent-magnet moving-coil	Current and voltage, d.c. only	Hairsprings	Eddy current
Moving-iron	Current and voltage, d.c. and a.c.	Hairsprings	Air
Thermocouple	Current and voltage, d.c. and a.c.	As for moving coil	As for moving coil
Electrodynamic or dynamometer†	Current, voltage and power, d.c. and a.c.	Hairsprings	Air
Electrostatic	Voltage only, d.c. and a.c.	Hairsprings	Air or eddy current
Rectifier	Current and voltage, a.c. only	As for moving coil	As for moving coil
Electronic	Current, voltage and power, d.c. and a.c.	Not applicable	Not applicable

† 'Electrodynamic' is the term recommended by the British Standards Institution, but 'dynamometer' is the term more frequently used in practice.

Apart from the electrostatic and electronic types of voltmeter, all voltmeters are in effect milliammeters connected in series with a non-reactive resistor having a high resistance. For instance, if a milliammeter has a full-scale deflection with 10 mA and has a resistance of 10 Ω and if this milliammeter is connected in series with a resistor of 9990 Ω, then the p.d. required for full-scale deflection is 0.01 × 10 000, namely 100 V, and the scale of the milliammeter can be calibrated to give the p.d. directly in volts.

26.5 Permanent-magnet moving-coil ammeters and voltmeters

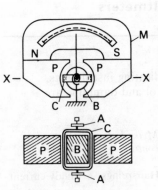

Fig. 26.5 Permanent-magnet moving-coil instrument

The high coercive force (section 3.11) of modern steel alloys, such as Alcomax (iron, aluminium, cobalt, nickel and copper), allows the use of relatively short magnets and has led to a variety of arrangements of the magnetic circuit for moving-coil instruments. The front elevation and sectional plan of one arrangement are shown in fig. 26.5, where M represents a permanent magnet and PP are soft-iron pole pieces. The hardness of permanent-magnet materials makes machining difficult, whereas soft iron can be easily machined to give exact airgap dimensions. In one form of construction, the anisotropic* magnet and the pole pieces are of the sintered type, i.e. powdered magnet alloy and powdered soft iron are compressed in a die to the required shape and heat-treated so that the magnet and the pole pieces become alloyed, thereby eliminating airgaps at the junctions of the materials.

An alternative arrangement is to cast the magnet and attach the soft-iron plate, in one piece, to the two surfaces of M which have been rendered flat by grinding, the joints being made by a resin-bonding technique. This construction enables the drilling of the cylindrical hole and the machining of the gaps to be done with precision.

The rectangular moving-coil C in fig. 26.5 consists of insulated copper wire wound on a light aluminium frame fitted with polished steel pivots resting on jewel bearings. Current is led into and out of the coil by spiral hairsprings AA, which also provide the controlling torque. The coil is free† to move in airgaps between the soft-iron pole pieces PP

---

* A magnet is made *anisotropic* by being cooled at a particular rate in a powerful directional magnetic field. As it solidifies, the magnetic domains (section 3.13) remain aligned in the same direction, thereby giving the magnet a high coercive force in that direction. An *isotropic* magnet material, on the other hand, has equal magnetic properties in all directions.

† Students often form the impression that the moving coil is wound on the soft-iron cylinder. It is important to realize that this cylinder is *fixed* and that the frame, on which the coil is wound, *does not touch* the cylindrical core.

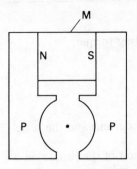

Fig. 26.6 An alternative
arrangement of the magnet
system

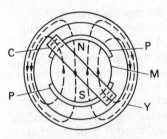

Fig. 26.7 Moving-coil
instrument with centre-core
magnet

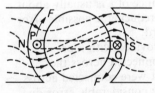

Fig. 26.8 Distribution of
resultant magnetic flux

and a soft-iron cylinder B supported by a brass plate (not
shown). The functions of core B are: (a) to intensify the
magnetic field by reducing the length of airgap across which
the magnetic flux has to pass, and (b) to give a radial
magnetic flux of uniform density, thereby enabling the scale
to be uniformly divided.

An alternative arrangement of the magnet system is shown
in fig. 26.6, where M represents the magnet and PP are soft-
iron pole pieces.

Fig. 26.7 shows an arrangement where the central
cylindrical core consists of an anisotropic magnet M, with
soft-iron pole pieces PP attached to M. The return path for
the magnetic flux is provided by a soft-iron ring or yoke Y,
concentric with the core and separated from it by an airgap
of uniform width, the distribution of the flux being as
indicated by the dotted lines in fig. 26.7. The moving coil C,
wound on an aluminium frame supported by jewel bearings
and controlled by spiral hairsprings, is free to move in the
airgap between the pole pieces and the yoke. This
arrangement gives a very compact movement and is
particularly suitable for small instruments.

The central-pole construction has the advantages:
(a) reduced cost; (b) reduction in the overall size of the
instrument; (c) better shielding from external magnetic fields.

The manner in which a torque is produced when the coil is
carrying a current may be understood more easily by
considering a single turn PQ, as in fig. 26.8. Suppose P to
carry current outwards from the paper; then Q is carrying
current towards the paper. Current in P tends to set up a
magnetic field in a counterclockwise direction around P and
thus strengthens the magnetic field on the lower side and
weakens it on the upper side. The current in Q, on the other
hand, strengthens the field on the upper side while weakening
it on the lower side. Hence, the effect is to distort the
magnetic flux as shown in fig. 26.8. Since the flux behaves as
if it was in tension and therefore tries to take the shortest
path between poles NS, it exerts forces FF on coil PQ,
tending to move it out of the magnetic field.

The deflecting torque $\propto$ current through coil

$$\times \text{ flux density in gap}$$

$$= kI \text{ for uniform flux density,}$$

where $\qquad k = $ a constant for a given instrument

and $\qquad I = $ current through coil.

The controlling torque of the spiral springs

$$\propto \text{ angular deflection}$$

$$= c\theta$$

where $\qquad c = $ a constant for given springs

and $\qquad \theta = $ angular deflection.

For a steady deflection,

controlling torque = deflecting torque,

hence

$$c\theta = kI$$

$$\therefore \qquad \theta = \frac{k}{c}I,$$

i.e. the deflection is proportional to the current and the scale is therefore uniformly divided.

A numerical example on the calculation of the torque on a moving coil is given in example 26.2, p. 566.

As already mentioned in section 26.3, damping is effected by eddy currents induced in the metal frame on which the coil is wound.

Owing to the delicate nature of the moving system, this type of instrument is only suitable for measuring currents up to about 50 milliamperes directly. When a larger current has to be measured, a *shunt* S (fig. 26.9), having a low resistance, is connected in parallel with the moving coil MC, and the instrument scale may be calibrated to read directly the total current I. Shunts are made of a material, such as manganin (copper, manganese and nickel), having negligible temperature coefficient of resistance. A 'swamping' resistor r, of material having negligible temperature coefficient of resistance, is connected in series with the moving coil. The latter is wound with copper wire and the function of r is to reduce the error due to the variation of resistance of the moving coil with variation of temperature. The resistance of r is usually about three times that of the coil, thereby reducing a possible error of, say, 4 per cent to about 1 per cent.

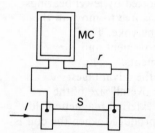

Fig. 26.9 Moving-coil instrument as an ammeter

The shunt shown in fig. 26.9 is provided with four terminals, the milliammeter being connected across the potential terminals. If the instrument were connected across the current terminals, there might be considerable error due to the contact resistance at these terminals being appreciable compared with the resistance of the shunt.

The moving-coil instrument can be made into a voltmeter by connecting a resistor R of manganin or other similar material in series, as in fig. 26.10. The scale may be calibrated to read directly the voltage applied to the terminals TT.

The main advantages of the moving-coil instrument are:

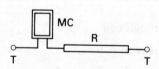

Fig. 26.10 Moving-coil instrument as a voltmeter

(i) high sensitivity;
(ii) uniform scale;
(iii) well shielded from any stray magnetic field.

Its main disadvantages are:

(i) more expensive than the moving-iron instrument;
(ii) only suitable for direct currents and voltages.

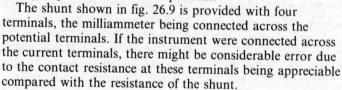

Example 26.1 *A moving coil instrument gives full-scale deflection with 15 mA and has a resistance of 5 Ω. Calculate the resistance required: (a) in parallel to enable the instrument to read up to 1 A, and (b) in series to enable it to read up to 10 V.*

(a) $\qquad$ Current through coil $=$ $\dfrac{\text{p.d. across coil}}{\text{resistance of coil}}$
(fig. 26.9)

$\therefore \qquad \dfrac{15}{1000} = \dfrac{\text{p.d. (in volts) across coil}}{5}$

so that $\qquad$ p.d. across coil $= 0.075\,\text{V}$.

For fig. 26.9,

current through S $=$ total current $-$ current through coil

$$= 1 - 0.015 = 0.985\,\text{A}.$$

Current through S $= \dfrac{\text{p.d. across S}}{\text{resistance of S}}$

$\therefore \qquad 0.985 = \dfrac{0.075}{\text{resistance of S (in ohms)}}$

and $\qquad$ resistance of S $= \dfrac{0.075}{0.985} = 0.076\,16\,\Omega$.

(b) For fig. 26.10,

current through coil $= \dfrac{\text{p.d. across TT}}{\text{resistance between TT}}$

$\therefore \qquad \dfrac{15}{1000} = \dfrac{10}{\text{resistance between TT}}$

so that $\qquad$ resistance between TT $= 666.7\,\Omega$.

Hence, resistance of resistor R required in series with coil

$$= \text{total resistance between TT} - \text{resistance of coil}$$

$$= 666.7 - 5 = 661.7\,\Omega.$$

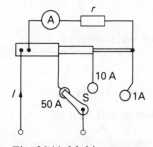

Fig. 26.11 Multi-range moving-coil ammeter

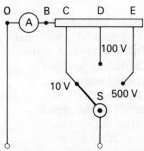

Fig. 26.12 Multi-range moving-coil voltmeter

The moving-coil instrument can be arranged as a multi-range ammeter by making the shunt of different sections as shown in fig. 26.11, where A represents a milliammeter in series with a 'swamping' resistor r of material having negligible temperature coefficient of resistance.

With the selector switch S on, say, the 50-A stud, a shunt having a very low resistance is connected across the instrument, the value of its resistance being such that full-scale deflection is produced when $I = 50\,\text{A}$. With S on the 10-A stud, the resistance of the two sections of the shunt is approximately five times that of the 50-A section, and full-scale deflection is obtained when $I = 10\,\text{A}$. Similarly, with S on the 1-A stud, the total resistance of the three sections is such that full-scale deflection is obtained with $I = 1\,\text{A}$. Such a multi-range instrument is provided with three scales so that the value of the current can be read directly.

A multi-range voltmeter is easily arranged by using a tapped resistor in series with a milliammeter A, as shown in fig. 26.12. For instance, with the data given in example 26.1, the resistance of section BC would be $661.7\,\Omega$ for the 10-V range. If D be the tapping for, say, 100 V, the total resistance between O and $D = 100/0.015 = 6666.7\,\Omega$, so that the resistance of section $CD = 6666.7 - 666.7 = 6000\,\Omega$.

Similarly, if E is to be 500-V tapping, section DE must absorb 400 V at full-scale deflection; hence the resistance of DE = 400/0.015 = 26 667 Ω. With the aid of selector switch S, the instrument can be used on three voltage ranges, and the scales can be calibrated to enable the value of the voltage to be read directly.

Example 26.2

The coil of a moving-coil instrument is wound with $42\frac{1}{2}$ turns. The mean width of the coil is 25 mm and the axial length of the magnetic field is 20 mm. If the flux density in the airgap is 0.2 T, calculate the torque, in newton metres, when the current is 15 mA.

Since the coil has $42\frac{1}{2}$ turns, one side has 42 wires and the other side has 43 wires.

From expression (2.1), force on the side having 42 wires

$$= 0.2\,[\text{T}] \times 0.02\,[\text{m}] \times 0.015\,[\text{A}] \times 42$$
$$= 2520 \times 10^{-6}\,\text{N}$$

∴ torque on that side of coil

$$= (2520 \times 10^{-6})\,[\text{N}] \times 0.0125\,[\text{m}]$$
$$= 31.5 \times 10^{-6}\,\text{N} \cdot \text{m}$$

Similarly, torque on side of coil having 43 wires

$$= 31.5 \times 10^{-6} \times 43/42$$
$$= 32.2 \times 10^{-6}\,\text{N} \cdot \text{m}$$

∴   total torque on coil $= (31.5 + 32.2) \times 10^{-6}\,\text{N} \cdot \text{m}$
$$= 63.7 \times 10^{-6}\,\text{N} \cdot \text{m}.$$

26.6 Moving-iron ammeters and voltmeters

Moving-iron instruments can be divided into two types:

(i) the *attraction* type, in which a sheet of soft iron is attracted towards a solenoid, and

(ii) the *repulsion* type, in which two parallel rods or strips of soft iron, magnetized inside a solenoid, are regarded as repelling each other.

These two types will now be described in greater detail.

Type (i). Fig. 26.13 shows an end elevation and a sectional front view (taken on XX) of the attracted-iron type. A soft-iron disc A is attached to a spindle S carried by jewelled centres J, and is so placed that it is attracted towards solenoid C when the latter is carrying a current.

Damping is provided by vane V attached to the spindle and moving in an air chamber, and control is provided by two spiral hairsprings SS, as shown in fig. 26.14.

Type (ii). This type is shown in fig. 26.14 where C represents the solenoid. A soft-iron rod or strip A is attached

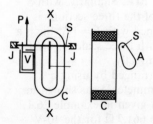

Fig. 26.13 Attraction-type moving-iron instrument

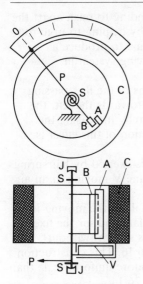

Fig. 26.14 Repulsion-type moving-iron instrument

Fig. 26.15 Distribution of magnetic flux in the repulsion-type moving-iron instrument

to the bobbin on which the coil is wound, and another soft-iron rod or strip B is carried by spindle D.

When a current flows through coil C, A and B are magnetized in the same direction and it is usual to say that B tries to move away from A because poles of the same polarity repel each other. Such a statement, however, gives no indication as to how the deflecting force is actually produced.

In section 2.3, it was shown that when two *permanent* magnets are placed side by side, with the N poles pointing in the same direction, as in fig. 2.5, the force of repulsion between the magnets is due to lateral pressure in the magnetic field occupying the space *between* the magnets. In the case of two parallel rods, A and B, situated inside a coil C carrying a current, as in fig. 26.15,* the distribution of the magnetic flux is roughly as shown dotted. There is practically no magnetic flux in the space between A and B, and therefore there cannot be any lateral pressure in that region.

All the magnetic flux must be linked with the whole or part of the coil, and fig. 26.15(*a*) shows the approximate distribution of the flux passing through the rods when they are close together. It was mentioned in section 2.3 that

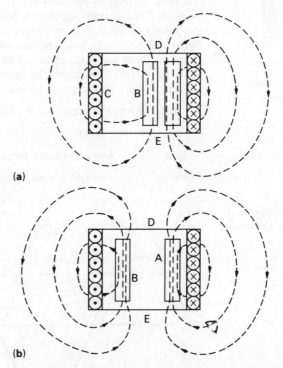

(a)

(b)

magnetic flux behaves as if it were in tension; hence the flux passing through rod B tends to shorten its paths by pulling B towards the left, as shown in fig. 26.15(*b*). (Rod A is assumed to be fixed.) In addition to this tension effect, there is also a

* The rods are shown much wider than is the case in an actual instrument. This is done to enable the dotted lines representing the flux passing through the rods to be shown clearly.

lateral pressure in regions D and E tending to push apart the fluxes passing through A and B, thereby helping to urge rod B towards the left. These effects combine to produce a clockwise deflecting torque on the moving system in fig. 26.14.

As a result of the shortened air paths of the flux passing through B when the latter has moved towards the left, the magnitude of the flux is increased, thereby increasing the inductance of the coil; but the implication of this effect is beyond the scope of this book.

In fig. 26.14 the controlling torque is exerted by hairsprings SS; and air damping is provided by vane V moving in an air chamber.

In commercial instruments, it is usual for the moving iron B to be in the form of a thin curved plate and for the fixed iron A to be a tapered curved sheet, as shown (without control and damping devices) in fig. 26.16. This construction can be arranged to give a longer and more uniform scale than is possible with the rods shown in fig. 26.14.

For both the attraction and the repulsion types it is found that for a given position of the moving system, the value of the deflecting torque is proportional to the square of the current, so long as the soft-iron is working below saturation. Hence, if the current waveform is as shown in fig. 26.17, the variation of the deflecting torque is represented by the dotted wave. If the supply frequency is, say, 50 Hz, the torque varies between zero and a maximum 100 times a second, so that the moving system — owing to its inertia — takes up a position corresponding to the mean torque, where

mean torque $\propto$ mean value of the square of the current

$$= kI^2$$

where k = a constant for a given instrument

and I = r.m.s. value of the current.

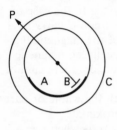

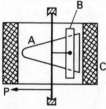

Fig. 26.16 Repulsion-type moving-iron instrument

Fig. 26.17 Deflecting torque in a moving-iron instrument

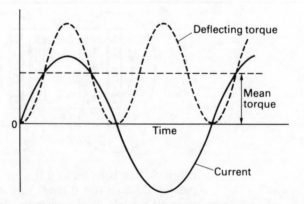

Hence the moving-iron instrument can be used to measure both direct current and alternating current, and in the latter case the instrument gives the r.m.s. value of the current. Owing to the deflecting torque being proportional to the square of the current, the scale divisions are not uniform, being cramped at the beginning and open at the upper end of the scale.

Since the strength of the magnetic field, and therefore the magnitude of the deflecting torque, depend upon the number of ampere-turns on the solenoid, it is possible to arrange different instruments to have different ranges by merely winding different numbers of turns on the solenoids. For example, suppose that full-scale deflection is obtained with 400 ampere-turns, then for

full-scale reading with 100 A, no. of turns = 400/100 = 4,

and full-scale reading with 5 A, no. of turns = 400/5 = 80.

A moving-iron voltmeter is a moving-iron milliammeter connected in series with a suitable non-reactive resistor.

Example 26.3   *A moving-iron instrument requires 250 ampere-turns to give full-scale deflection. Calculate: (a) the number of turns required if the instrument is to be used as an ammeter reading up to 50 A, and (b) the number of turns and the total resistance if the instrument is to be arranged as a voltmeter reading up to 300 V with a current of 20 mA.*

(*a*)   No. of turns = 250/50 = 5.

(*b*)   No. of turns = 250/0.02 = 12 500.

Total resistance = 300/0.02 = 15 000 Ω.

The advantages of moving-iron instruments are:

  (i) robust construction;
  (ii) relatively cheap;
  (iii) can be used to measure direct and alternating currents and voltages.

The disadvantages of moving-iron instruments are:

(i) Affected by stray magnetic fields. Error due to this cause is minimized by the use of a magnetic screen such as an iron casing (section 2.4).

(ii) Liable to hysteresis error when used in a d.c. circuit; i.e. for a given current, the instrument reads higher with decreasing than with increasing values of current. This error is reduced by making the iron strips of nickel–iron alloy such as Mumetal (section 3.11).

(iii) Owing to the inductance of the solenoid, the reading on moving-iron voltmeters may be appreciably affected by variation of frequency. This error is reduced by arranging for the resistance of the voltmeter to be large compared with the reactance of the solenoid.

(iv) Moving-iron voltmeters are liable to a temperature error owing to the solendoid being wound with copper wire. This error is minimized by connecting in series with the solenoid a resistor of a material, such as manganin, having a negligible temperature coefficient of resistance.

26.7 Thermocouple instruments

This type of instrument utilizes the thermoelectric effect observed by Seebeck in 1821, namely that in a closed circuit consisting of two different metals, an electric current flows when the two junctions are at different temperatures. Thus, if A and B in fig. 26.18 are junctions of copper and steel wires, each immersed in water, then if the vessel containing B is heated, it is found that an electric current flows from the steel to the copper at the cold junction and from the copper to the steel at the hot junction, as indicated by the arrowheads. A pair of metals arranged in this manner is termed a *thermocouple* and gives rise to a *thermo-e.m.f.* when the two junctions are at different temperatures.

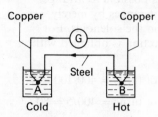

Fig. 26.18 A thermocouple

This *thermoelectric effect* may be utilized to measure temperature. Thus, if the reading on galvanometer G be noted for different temperatures of the water in which junction B is immersed, the temperature of junction A being maintained constant, it is possible to calibrate the galvanometer in terms of the difference of temperature between A and B. The materials used in practice depend upon the temperature range to be measured; thus, copper–constantan couples are suitable for temperatures up to about 400°C and steel–constantan couples up to about 900°C, constantan being an alloy of copper and nickel. For temperatures up to about 1400°C, a couple made of platinum and platinum–iridium alloy is suitable.

A thermocouple can be used to measure the r.m.s. value of an alternating current by arranging for one of the junctions of wires of dissimilar material, B and C (fig. 26.19), to be placed near or welded to a resistor H carrying the current *I* to be measured. The current due to the thermo-e.m.f. is measured by a permanent-magnet moving-coil microammeter A. The heater and the thermocouple can be enclosed in an evacuated glass bulb D, shown dotted in fig. 26.19, to shield them from draughts. Ammeter A may be calibrated by noting its reading for various values of direct current through H, and it can then be used to measure the r.m.s. value of alternating currents of frequencies up to several megahertz.

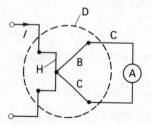

Fig. 26.19 A thermocouple ammeter

26.8 Electrodynamic or dynamometer instruments

The action of this type of instrument depends upon the electromagnetic force exerted between fixed and moving coils carrying current. The upper diagram in fig. 26.20 shows a sectional elevation through fixed coils FF and the lower diagram represents a sectional plan on XX. The moving coil M is carried by a spindle S and the controlling torque is

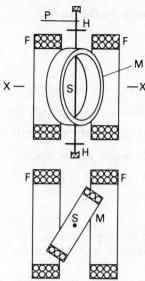

Fig. 26.20 Electrodynamic or dynamometer instrument

Fig. 26.21 Magnetic fields due to fixed and moving coils

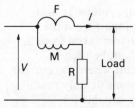

Fig. 26.22 Wattmeter connections

exerted by spiral hairsprings H, which also serve to lead the current into and out of M.

The deflecting torque is due to the interaction of the magnetic fields produced by currents in the fixed and moving coils; thus fig. 26.21(*a*) shows the magnetic field due to current flowing through F in the direction indicated by the dots and crosses, and fig. 26.21(*b*) shows that due to current in M. By combining these magnetic fields, it will be seen that when currents flow simultaneously through F and M, the resultant magnetic field is distorted as shown in fig. 26.21(*c*) and the effect is to exert a clockwise torque on M.

Since M is carrying current at right-angles to the magnetic field produced by F,

deflecting force on each side of M

$\propto$ (current in M)

$\times$ (density of magnetic field due to current in F)

$\propto$ current in M $\times$ current in F.

In dynamometer ammeters, the fixed and moving coils are connected in parallel, whereas in voltmeters they are in series

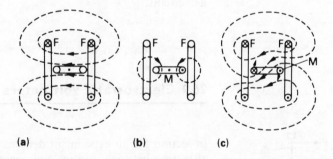

(a)   **(b)**   **(c)**

with each other and with the usual resistor. In each case, the deflecting force is proportional to the square of the current or the voltage. Hence, when the dynamometer instrument is used to measure an alternating current or voltage, the moving coil — due to its inertia — takes up a position where the average deflecting torque over one cycle is balanced by the restoring torque of the spiral springs. For that position, the deflecting torque is proportional to the mean value of the square of the current or voltage, and the instrument scale can therefore be calibrated to read the r.m.s. value.

Owing to the higher cost and lower sensitivity of dynamometer ammeters and voltmeters compared with moving-iron instruments, the former are seldom used commercially, but *electrodynamic* or *dynamometer wattmeters* are very important because they are commonly employed for measuring the power in a.c. circuits. The fixed coils F are connected in series with the load, as shown in fig. 26.22. The moving coil M is connected in series with a non-reactive resistor R across the supply, so that the current through M is proportional to and practically in phase with the supply voltage V; hence:

instantaneous force on each side of M

$\propto$ (instantaneous current through F)

$\times$ (instantaneous current through M)

$\propto$ (instantaneous current through load)

$\times$ (instantaneous p.d. across load)

$\propto$ instantaneous power taken by load

$\therefore$　average deflecting force on M

$\propto$ average value of the power over a complete number of cycles.

When the instrument is used in an a.c. circuit, the moving coil — due to its inertia — takes up a position where the average deflecting torque over one cycle is balanced by the restoring torque of the spiral springs; hence the instrument can be calibrated to read the mean value of the power in an a.c. circuit.

26.9 Electrostatic voltmeters

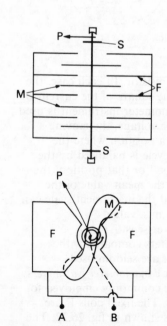

Fig. 26.23 Electrostatic voltmeter

In section 5.1 an experiment demonstrating the mutual attraction between positive and negative charges was described. This phenomenon is utilized in the electrostatic voltmeter. This instrument consists of fixed metal plates F, shaped as indicated in the lower part of fig. 26.23, and very light metal vanes M attached to a spindle controlled by spiral springs S and carrying a pointer P.

The voltage to be measured is applied across terminals A and B. In section 5.27 it was shown that the force of attraction between F and M is proportional to the square of the applied voltage; hence this instrument can be used to measure either direct or alternating voltage, and when used in an a.c. circuit it reads the r.m.s. value.

The main advantages of the electrostatic voltmeter are: (a) it takes no current from a d.c. circuit (apart from the small initial charging current) and the current taken from an a.c. circuit is usually negligible. Hence it can be used to measure the p.d. between points in a circuit where the current taken by other types of voltmeter might considerably modify the value of that p.d. (b) It is particularly suitable for measuring high voltages, since the electrostatic forces are then so large that its construction can be greatly simplified.

26.10 Rectifier ammeters and voltmeters

In this type of instrument a rectifier such as a copper-oxide rectifier is used to convert the alternating current into a unidirectional current, the mean value of which is measured on a permanent-magnet moving-coil instrument.

Rectifier ammeters usually consist of four rectifier elements arranged in the form of a bridge, as shown in fig. 26.24, where the apex of the black triangle indicates the direction in which the resistance is low, and A represents a moving-coil ammeter. During the half-cycles that the current is flowing from left to right in fig. 26.24, current flows through elements B and D, as shown by the full arrows. During the other half-cycles, the current flows through C and E, as shown by the dotted arrows. The waveform of the current through A is therefore as shown in fig. 26.25. Consequently, the deflection

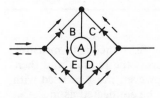

Fig. 26.24 Bridge circuit for full-wave rectification

Fig. 26.25 Waveform of current through moving-coil ammeter

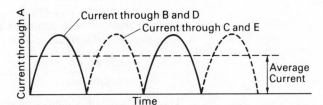

of A depends upon the average value of the current, and the scale of A can be calibrated to read the r.m.s. value of the current on the assumption that the waveform of the latter is sinusoidal with a form factor of 1.11.

In a rectifier voltmeter, A is a milliammeter and the bridge circuit of fig. 26.24 is connected in series with a suitable non-reactive resistor.

The main advantage of the rectifier voltmeter is that it is far more sensitive than other types of voltmeter suitable for measuring alternating voltages. Also, rectifiers can be incorporated in universal instruments, such as the Avometer, thereby enabling a moving-coil milliammeter to be used in combination with shunt and series resistors to measure various ranges of direct current and voltage, and in combination with a bridge rectifier and suitable resistors to measure various ranges of alternating current and voltage.

If a diode or a solid-state rectifier D is connected in series with a capacitor C across an a.c. supply, as shown in fig. 26.26, the capacitor is charged until its p.d. is practically equal to the peak value of the alternating voltage. The value of this voltage can be measured by connecting across the capacitor a microammeter A in series with a resistor R having a high resistance.

The capacitor C is charged during the very small fraction of each cycle that the applied voltage exceeds the p.d. across C. During the remainder of each cycle, the capacitor supplies the current flowing through A and R.

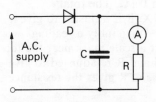

Fig. 26.26 A simple form of electronic voltmeter

The microammeter can be calibrated to read either the peak or the r.m.s. value of the voltage. In the latter case, it has to be assumed that the voltage is sinusoidal.

26.11 Measurement of resistance by the voltmeter–ammeter method

The most obvious way of measuring resistance is to measure the current through and the p.d. across the resistor and then apply Ohm's Law. This method, however, must be used with care; for instance if the instruments be connected as in fig. 26.27(*a*), and if the voltmeter V be other than the electrostatic type, the current taken by V passes through A and may be comparable with that through R if the resistance of the latter is fairly high. If the resistance of V is known, its current can be calculated and subtracted from the reading on A to give the current through R.

A better method is to connect the voltmeter across R and A as in fig. 26.27(*b*), then:

$$\frac{\text{reading on V}}{\text{reading on A}} = \text{resistance of (R + A)}.$$

The resistance of A can be easily calculated from the p.d. across A — obtained by connecting V as shown dotted — and the corresponding current through A.

Fig. 26.27 Measurement of resistance

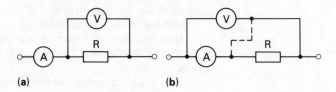

(a)   **(b)**

26.12 Measurement of resistance by substitution

(a) Series method

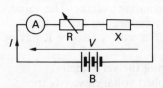

Fig. 26.28 Measurement of resistance

The unknown resistor X (fig. 26.28) is connected in series with a variable resistor R, such as a decade resistance box, which can be varied in steps of 1 or 0.1 Ω. The total resistance of R must exceed that of X. R is adjusted to give a convenient reading on ammeter A — preferably a reading such that the pointer is exactly over a scale mark near the top end of the scale. X is then removed and R re-adjusted to give the same reading on A. The increase of R gives the resistance of X.

A modification of this arrangement is frequently used in universal instruments. A moving-coil milliammeter A is connected in series with a variable resistor *r* and a cell B to

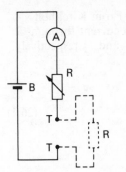

Fig. 26.29 An ohmmeter circuit

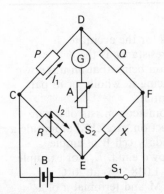

Fig. 26.30 Ammeter and ohmmeter scales

terminals TT, as in fig. 26.29. The procedure is to short-circuit terminals TT and adjust r to give full-scale deflection on A. The unknown resistor R is then connected across TT, and A can be calibrated to give the resistance of R directly, thereby making the instrument into an ohmmeter. Thus, suppose A to have full-scale deflection with 1 mA and B to have an e.m.f. of 1.5 V. For full-scale deflection with TT short-circuited, total resistance of r and B must be $1.5 \div 1/1000$, namely 1500 Ω. If a 1000-Ω resistor is connected across TT, the current is $1.5 \times 1000/2500$, namely 0.6 mA; i.e. a scale deflection of 0.6 mA corresponds to a resistance of 1000 Ω, as shown in fig. 26.30.

The resistance of r in fig. 26.29 is variable to allow for variation of the e.m.f. and internal resistance of cell B. It should be mentioned, however, that if the e.m.f. of B falls

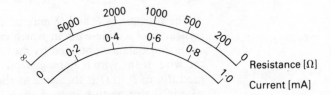

appreciably below its rated value, the resistance scale will only be approximately correct.

(b) Alternative circuit method

In fig. 26.31, X is the unknown resistor, R a known variable resistor and S a two-way switch. The reading on A is noted when S is on a. The switch is then moved over to b and R adjusted to give the same reading on A. The resistance of X is obviously the same as that of R.

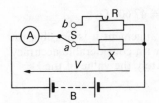

Fig. 26.31 Measurement of resistance

26.13 Measurement of resistance by the Wheatstone bridge

The four branches of the network CDFEC, in fig. 26.32, have two known resistances P and Q, a known variable resistance R and the unknown resistance X. A battery B is connected through a switch S_1 to junctions C and F; and a galvanometer G, a variable resistor A and a switch S_2 are in series across D and E. The function of A is merely to protect G against an excessive current should the system be seriously out of balance when S_2 is closed.

With S_1 and S_2 closed, R is adjusted until there is no deflection on G even with the resistance of A reduced to zero. Junctions D and E are then at the same potential, so that the p.d. between C and D is the same as that between C and E, and the p.d. between D and F is the same as that between E and F.

Suppose I_1 and I_2 to be the currents through P and R

Fig. 26.32 Wheatstone bridge

respectively when the bridge is balanced. From Kirchhoff's First Law it follows that since there is no current through G, the currents through Q and X are also I_1 and I_2 respectively.

But   p.d. across $P = PI_1$

and   p.d. across $R = RI_2$

$$\therefore \qquad\qquad PI_1 = RI_2 \qquad\qquad (26.1)$$

Also   p.d. across $Q = QI_1$

and   p.d. across $X = XI_2$

$$\therefore \qquad\qquad QI_1 = XI_2. \qquad\qquad (26.2)$$

Dividing (26.2) by (26.1), we have:

$$Q/P = X/R$$

and $$X = R \times Q/P. \qquad\qquad (26.3)$$

The resistances P and Q may take the form of the resistance of a slide-wire, in which case R may be a fixed value and balance obtained by moving a sliding contact along the wire. If the wire is homogeneous and of uniform section, the ratio of P to Q is the same as the ratio of the lengths of wire in the respective arms. A more convenient method, however, is to arrange P and Q so that each may be 10, 100 or 1000 Ω. For instance, if $P = 1000\,\Omega$ and $Q = 10\,\Omega$, and if R has to be 476 Ω to give a balance, then from (26.3):

$$X = 476 \times 10/1000 = 4.76\,\Omega.$$

On the other hand, if P and Q had been 10 and 1000 Ω respectively, then for the same value of R:

$$X = 476 \times 1000/10 = 47\,600\,\Omega.$$

Hence it is seen that with this arrangement it is possible to measure a wide range of resistance with considerable accuracy and to derive very easily and accurately the value of the resistance from that of R.

26.14 The potentiometer

One of the most useful instruments for the accurate measurement of p.d., current and resistance is the potentiometer, the principle of action being that an unknown e.m.f. or p.d. is measured by balancing it, wholly or in part, against a known difference of potential.

In its simplest form, the potentiometer consists of a wire MN (fig. 26.33) of uniform cross-section, stretched alongside a scale and connected across a secondary cell B of ample capacity. A standard cell SC of known e.m.f. E_1, for example a cadmium cell having an e.m.f. of 1.018 59 V at 20°C (section 1.7), is connected between M and terminal a of a two-way switch S, care being taken that the corresponding terminals of B and SC are connected to M.

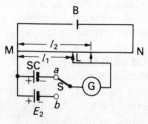

Fig. 26.33 A simple potentiometer

Slider L is then pressed momentarily against wire MN and its position adjusted until the galvanometer deflection is zero when L is making contact with MN. Let l_1 be the corresponding distance between M and L. The fall of potential over length l_1 of the wire is then the same as the e.m.f. E_1 of the standard cell.

Switch S is then moved over to b, thereby replacing the standard cell by another cell, such as a Leclanché cell, the e.m.f. E_2 of which is to be measured. Slider L is again adjusted to give zero deflection on G. If l_2 be the new distance between M and L, then:

$$E_1/E_2 = l_1/l_2$$

$$\therefore \qquad E_2 = E_1 \times l_2/l_1. \tag{26.4}$$

26.15 A commercial form of potentiometer

The simple arrangement described in the preceding section has two disadvantages: (i) the arithmetical calculation involved in expression (26.4) can introduce an error, and in any case takes an appreciable time; (ii) the accuracy is limited by the length of slide-wire that is practicable and by the difficulty of ensuring exact uniformity over a considerable length.

In the commercial type of potentiometer shown in fig. 26.34, these disadvantages are practically eliminated. R consists of 14 resistors in series, the resistance of each resistor being equal to that of the slide-wire S. The value of the current supplied by a lead–acid cell B is controlled by a slide-wire resistor W. A double-pole change-over switch T, closed

Fig. 26.34 A commercial potentiometer

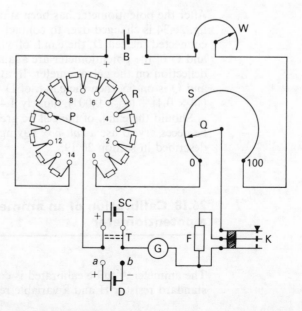

on the upper side as in fig. 26.34, connects the standard cell
SC between arm P and galvanometer G. A special key K,
when slightly depressed, inserts a resistor F having a high
resistance in series with G; but when K is further depressed,
F is short-circuited. The galvanometer is thereby protected
against an excessive current should the potentiometer be
appreciably out of balance when K is first depressed.

26.16 Standardization of the potentiometer

Suppose the standard cell SC to be of the cadmium type
having an e.m.f. of 1.018 59 V at 20°C. Arm P is placed on
stud 10 and Q on 18.59 — assuming the scale alongside S to
have 100 divisions. The value of W is then adjusted for zero
deflection on G when K is fully depressed. The p.d. between
P and Q is then exactly 1.018 59 V, so that the p.d. between
two adjacent studs of R is 0.1 V and that corresponding to
each division of S's scale is 0.001 V. Consequently, if P is
moved to, say, stud 4 and Q to 78.4 on the slide-wire scale,
the p.d. between P and Q = (4 × 0.1) + (78.4 × 0.001) =
0.4784 V. It is therefore a simple matter to read the p.d.
directly off the potentiometer.

Since most potentiometers have fourteen steps on R, it is
usually not possible to measure directly a p.d. exceeding
1.5 V.

26.17 Measurement of the e.m.f. of a cell

After the potentiometer has been standardized, switch T of
fig. 26.34 is changed over to contacts *a* and *b*, across which is
connected the cell D, the e.m.f. of which is required. Arms P
and Q of the potentiometer are again adjusted to give zero
deflection on the galvanometer. If, at balance, P is on stud 14
and Q is on 65, then the e.m.f. of D is
(14 × 0.1) + (65 × 0.001), namely 1.465 volts.

Should the e.m.f. of the cell be greater than 1.5 V, it would
be necessary to use a *volt-box* having a high resistance, as
described in section 26.19.

26.18 Calibration of an ammeter by means of
a potentiometer

The ammeter A to be calibrated is connected in series with a
standard resistor H and a variable resistor J across a cell L of

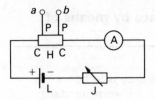

Fig. 26.35 Calibration of an ammeter

ample current capacity, as in fig. 26.35. The standard resistor H is usually provided with four terminals, namely two heavy current terminals CC and two potential terminals PP. The resistance between the potential terminals is known with a high degree of accuracy and its value must be such that with the maximum current through the ammeter, the p.d. between terminals PP does not exceed 1.5 V. For instance, suppose A to be a 10-A ammeter; then the resistance of H must not exceed 1.5/10, namely 0.15 Ω. Further, the resistance of H should preferably be a round figure, such as 0.1 Ω in this case, in order that the current may be quickly and accurately deduced from the potentiometer readings.

Terminals PP of the standard resistor are connected to terminals *ab* (fig. 26.34) of the potentiometer (cell D having been removed). After the potentiometer has been standardized, switch T is changed over to *ab*; and with the current adjusted to give a desired reading on the scale of ammeter A, arms P and Q are adjusted to give zero deflection on the galvanometer. For instance, suppose the current to be adjusted to give a reading of, say, 6 A on the ammeter scale, and suppose the readings on P and Q, when the potentiometer is balanced, to be 5 and 86.7 respectively, then the p.d. across terminals PP is 0.5867 V; and since the resistance between the potential terminals PP is assumed to be 0.1 Ω, the true value of the current through H is 0.5867/0.1, namely 5.867 A. Hence, the ammeter is reading high by 0.133 A.

26.19 Calibration of a voltmeter by means of a potentiometer

Suppose the voltmeter to be calibrated to have a range of 0–100 V. It is therefore necessary to use a *volt-box* to enable an accurately known fraction — not exceeding 1.5 V — to be obtained. The high-resistance volt-box *ae*, fig. 26.36, has tappings at accurately determined points. This arrangement enables voltmeters of various ranges to be calibrated; thus the 100-V voltmeter is connected across the 150-V tappings, the resistance between *ad* being 100 times that between *ab*. The 1.5-V tappings are connected to terminals *ab* of the potentiometer (fig. 26.34). Various voltages can be applied to the voltmeter by moving the slider along a resistor *f* connected across a suitable battery *g*.

Let us suppose that the voltmeter reading has been adjusted to 70 V and that the corresponding readings on P and Q to give a balance are 7 and 8.4 respectively. The p.d. across *ab* is 0.7084 V, and the true value of the p.d. across *ad* is therefore 70.84 V. Hence the voltmeter is reading low by 0.84 V.

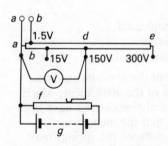

Fig. 26.36 Calibration of a voltmeter

26.20 Measurement of resistance by means of a potentiometer

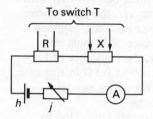

To switch T

Fig. 26.37 Measurement of resistance

The resistor X (fig. 26.37), whose resistance is to be determined, is connected in series with a known standard resistor R, an ammeter A and a variable resistor *j* across an accumulator *h*. The purpose of A is simply to check that the value of the current is not excessive.

Connections are taken from R to, say, the upper pair of terminals of switch T in fig. 26.34 (the standard cell having been removed), and those from X are taken to the lower pair of terminals of T. With a constant current through R and X, potentiometer readings are noted, first with switch T on the upper side to measure the p.d. across R, and then with T on the lower side to measure the p.d. across X. If these readings were 0.648 V and 0.1242 V respectively and if the resistance of R was 0.1 Ω, then for a current *I* amperes through R and X:

$$I \times 0.1 = 0.648 \quad \text{and} \quad IX = 0.1242,$$

$$\therefore \qquad X/0.1 = 0.1242/0.648$$

and

$$X = 0.019\,17\,\Omega.$$

This method is particularly suitable for the accurate measurement of low resistances, in which case it may be necessary to use knife-edge contacts, as indicated by the arrowheads in fig. 26.37, to give the precise points between which the resistance is being determined.

26.21 Calibration accuracy and errors

Much has been made throughout this chapter concerning the errors that may arise during the measurement of voltage, current and resistance. But why should there be error? Error occurs for three reasons:

    (i) the limitations of the instrument used;
    (ii) the operator is never infallible;
    (iii) the instrument may disturb the circuit.

(i) First there are the limitations of the instrument. These may arise from incorrect calibration of the instrument, which does not necessarily mean that the scale indications have been put in the wrong place. It could be that the meter has changed its deflection with age, for instance the springs may not be as stiff as when they were made. The friction of the bearings may have changed with time. So in a number of ways it is quite possible that the meter may not indicate exactly the quantity that it is supposed to measure. Because of this, every meter is permitted a margin of error which is stated as a percentage of the indication.

This is unusual because it is the general case in engineering that you state the desired measurement and the error permitted about that quantity. Thus you may wish to have a shaft made to the diameter of 100 mm with a tolerance of 1 mm, i.e. the error of manufacture is ±1 per cent about the desired diameter.

In meter measurement, however, it is usual to calculate errors on the incorrect basis of the indicated quantity and not the actual quantity. This procedure has no undue effect provided the error is less than about 2 per cent and makes the method of calculation very much easier. However, this method of error calculation must not be used when the error exceeds 3 per cent.

The reason that this change of basis is employed is that it would be difficult to set a supply to, say, 120 V and then read accurately the indicated voltage from the meter whereby the error could be ascertained. Instead we set the supply so that the meter indicates 120 V and then determine by potentiometer or similar means the correct supply voltage. Let this correct voltage be, say, 120.84 V; thus the difference between the indicated voltage and the correct voltage is 0.84 V, which is $0.84 \times 100/120 = 0.7$ per cent. However, the meter gave an indication that was lower than the correct voltage so the error is stated as -0.7 per cent.

Now let the supply voltage be varied so that the meter indicates 60 V. This time, we may find that the correct voltage as measured by the potentiometer is 59.94 V which means that the meter has overestimated the voltage by 0.06 V, which relative to 60 V means that the error is $+0.1$ per cent.

Thus we have the following rules:

(*a*) If the instrument error is positive then the indicated quantity is higher than the true quantity.

(*b*) If the instrument error is negative then the indicated quantity is lower than the true quantity.

It would be much more difficult if we were to calculate the errors relative to the correct quantities and the difference would scarcely be noticeable when the errors are so small.

Apart from the causes described, there are other sources of instrument error. Knowing that a range of error is acceptable (and inevitable) it follows that the manufacturer need not seek perfection during construction and can therefore use components which have a range of value, e.g. a multiplier for a voltmeter may acceptably have a resistance of, say, ±1 per cent about its stated value. Also the meter scales can be produced in number and therefore are not exactly matched to the individual instrument. However, even with such sources of possible error, it is relatively simple to make instruments of $\pm 2\frac{1}{2}$ per cent error and even these tend to be much better than the range suggests. A typical range of error is shown in fig. 26.38, and it will be seen that the error changes over the range of indication.

(ii) However, no matter how well a meter performs, it remains only as good as the operator who can be careless about reading the scale and thereby cause an error,

Fig. 26.38 Variation of error
with deflection

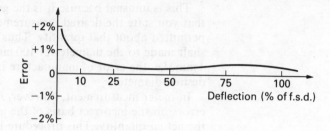

or there may be a limit as to how accurately the meter may
be read to the nearest division with certainty although usually
we try to divide up a division into ten parts so that we may
give a reading including a fraction of a division. This leads to
an unexpected error; let us look at the scale shown in
fig. 26.39.

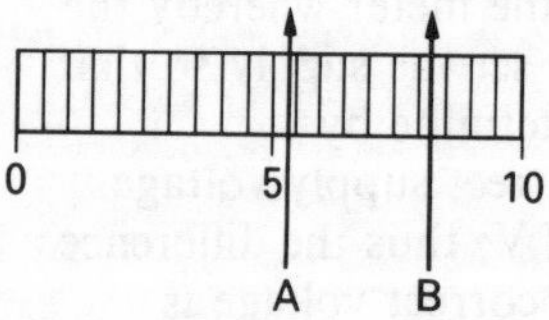

In position A, the pointer is indicating 5.4 which is what
many people would take it to be. However, without being
careless, some will read it as 5.5 whilst others will see it as
5.3. This is a defect in their judgement and for this reason, no
one should place too much reliance on such estimated figures.

Fig. 26.39 Scale indications

Position B illustrates another common reading error. The
pointer indicates approximately half way between 8.2 and 8.3.
This leads to some reading the indication as being 8.2 whilst
others take it as 8.3. It is the human desire to be helpful that
causes the error, and not the meter, and we must be aware
that this introduces such a source of error.

(iii) Finally, there is the error due to circuit disturbance.
Meters require a certain amount of power to cause operation.
Provided this power is small relative to the power in the
measured circuit, then little error will result. However, if the
meter power is comparable to the power in the circuit a
serious error will result. Before giving an example to illustrate
this, another point to bear in mind is that any meter will,
within the limitations of calibration, indicate the conditions at
its terminals correctly. Thus a voltmeter will indicate, within
the limitations of its normal accuracy, the terminal voltage
and similarly an ammeter will indicate the current passing
through it.

To illustrate these remarks consider the following example.

Example 26.4

*A voltage of 100 V is applied to a circuit comprising two 50-kΩ
resistors in series. A voltmeter, with an f.s.d. of 50 V and a figure of
merit 1 kΩ/V, is used to measure the voltage across one of the 50-kΩ
resistors. Calculate: (a) the voltage across the 50-kΩ resistor; (b) the
voltage measured by the voltmeter.*

Let V_1 be the voltage across a 50-kΩ resistor when the voltmeter
is not in circuit; thus

$$V_1 = \frac{R}{2R} \times V = \frac{50 \times 10^3}{100 \times 10^3} \times 100 = 50 \text{ V}.$$

Let R_V be the resistance of the voltmeter:

$$R_V = 50 \times 1000 = 50\,000 \ \Omega = 50 \text{ k}\Omega.$$

When the voltmeter is connected in circuit, it shunts the 50-kΩ resistor. If R_e is the resistance of the parallel networks, then for two parallel 50-kΩ resistances, $R_e = 25$ kΩ. The network is thus effectively changed into an equivalent resistance of 25 kΩ in series with a resistance of 50 kΩ. The voltage thus measured by the voltmeter is given by

$$\frac{R_e \cdot V}{R + R_e} = \frac{25 \times 100}{50 + 25} = 33.3\,\text{V}.$$

In example 26.4, clearly the voltage as indicated by the voltmeter is quite erroneous. The error has been caused by the effect of the voltmeter on the circuit. Because of the values chosen, the voltmeter takes the same current as the load whose voltage is being measured. The power taken by the voltmeter is equal to the power in the measured load.

Even if the resistance of the voltmeter had been ten times as great, the error would still have been almost 2 per cent. It can therefore be seen that the meter can affect the circuit to which it is being applied, i.e. the circuit has been disturbed.

Whilst the sources of error discussed are general, there are further sources of error that have been introduced but which are specific to the practice of alternating current, such as change of waveform causing error in certain a.c. meters. Such problems cause further errors due to limitations of the instrument in coping with extreme conditions.

26.22 Determination of error due to instrument errors

When using instruments to measure electrical quantities, you cannot avoid the introduction of error. It follows therefore that you ought to develop an appreciation of the quality of the measurements you make on the measurements you determine. For instance, if you have found the resistance of a resistor to be 100 Ω, you then have to decide if this is the correct value or if it is a value with a possible error of say ±5 per cent. In the second case, you know that the resistance has a value between 95 Ω and 105 Ω although probably it has a value near to 100 Ω.

Using other measurement techniques, we may determine the same resistance with a better (or a poorer) degree of accuracy. The accuracy you require depends on the application to which you intend to put the resulting information. In the case of a 1000-m underground cable which has developed a break in its insulation, you could determine the resistance or the capacitance from one end of the cable up to the point of breakdown. If this were done to an accuracy of 5 per cent, you would require to dig up 100 m of the cable to be sure of finding the fault! If your accuracy of determination were 0.5 per cent then the hole would shrink to 10 m long and if the accuracy were 0.1 per cent, the hole would only be 2 m long, which seems much more reasonable. In this case, therefore, we wish to ensure an accuracy of 0.1 per cent.

Ideally we would measure any electrical quantity directly by one instrument. For instance, we would wish to measure voltage directly by means of a voltmeter, an oscilloscope or a potentiometer. The voltmeter is the simplest to operate, so let us suppose that on a voltmeter operating with a full-scale deflection of 100 V we obtain an indication of 82.6 V. For the particular model of voltmeter, we note that the limit of error over the effective range expressed as a percentage of the scale range is 2.0 per cent; thus the greatest error on any indication is 2.0 per cent of 100 V which is 2.0 V. We may therefore conclude that for the voltmeter operating under the given conditions, the voltage across its terminals is 82.6 ± 2.0 volts. It follows that the error may be expressed as ±2.4 per cent of the indicated value.

As the indication relative to the full-scale deflection becomes lower, the specified limit of error becomes relatively more important; thus for the given voltmeter, the limit of error is still 2.0 V when the voltmeter indicates 50.0 V, in which case the percentage error limit is ±4.0 per cent of the indicated voltage. By the time the voltage indicated falls to 20.0 V, the percentage error limit rises to 10.0 per cent.

From these observations, we appreciate that a meter of apparently reasonable accuracy can in fact give indications with relatively large errors when operating under conditions which do not demand deflection of the meter approaching full-scale deflection. It is for this reason that you should always choose a meter scale that gives the greatest possible deflection within the scale of the meter.

By comparison, circuit components can be manufactured to have tolerances which remain the same at all settings. For instance, in a bridge network, we use decade resistors. If the decade resistor is manufactured to have a percentage error limit of say 1.0 per cent, then this is the percentage error limit at all settings.

Many experimental techniques involve the combination of two or more observations. The resistance of a circuit component is the ratio of the voltage across it to the current passing through it, and these quantities could be measured by a voltmeter and an ammeter. Each meter has an error, and this raises the question of the effect of combining the two errors. Similarly, a bridge determines the resistance, say, by comparing it with three other resistance values, each of which has an error. Again there is the question of the effect of combining the possible errors in each of the three values.

Suppose that

$$X = \frac{AB}{C}$$

then

$$\ln X = \ln A + \ln B - \ln C$$

and

$$\frac{dX}{X} = \frac{dA}{A} + \frac{dB}{B} - \frac{dC}{C}$$

Each of the small changes can be positive or negative; thus the righthand side of the relation is greatest when all the small terms produce errors of the same kind. Thus the maximum possible error occurs when

$$\frac{\mathrm{d}X}{X} = \pm\left(\frac{\mathrm{d}A}{A} + \frac{\mathrm{d}B}{B} + \frac{\mathrm{d}C}{C}\right) \tag{26.5}$$

This curious looking expression becomes more readily understood when we apply it to an example, as in example 26.5.

Example 26.5
The current in a circuit is measured as 235 µA and the accuracy of measurement is ±0.5 per cent. The current passes through a resistor of resistance 35 kΩ ± 0.2 per cent. Estimate the voltage across the resistor.

Estimate of voltages:

$$V = IR = 235 \times 10^{-6} \times 35 \times 10^3 = 8.225\,\text{V}.$$

Estimate of errors from relation (26.5): the maximum relative error is given by the sum of the relative errors; thus the maximum relative error = 0.5 + 0.2 = 0.7 per cent.

The voltage may therefore be expressed as 8.225 V ± 0.7 per cent but the basic value is given to a greater degree of accuracy than the error. The voltage should therefore be

$$8.23\,\text{V} \pm 0.7 \text{ per cent}.$$

It may be that we would wish to express the answer to this problem in volts, in which case we require to determine the maximum possible error. The error is ε and

$$\varepsilon = \frac{0.7}{100} \times 8.225 = 0.058\,\text{V}.$$

The voltage across the resistor may therefore be expressed as

$$8.225 \pm 0.058\,\text{V}.$$

Again it would be consistent to round off the figures and express the voltage as

$$8.23 \pm 0.06\,\text{V}.$$

In example 26.5, the relative error values were already determined. If we return to the instance of the voltmeter in which the limit of error was referred to the scale range, the determination of the final result becomes more complicated.

Example 26.6
The voltage across a resistor is measured by a voltmeter which gives an indication of 75.5 V when operating on a scale of range 0–100 V. The current in the resistor is measured by an ammeter which gives an indication of 3.45 A when operating on a scale of range 0.5 A. Both instruments have a limit of error of 1.0 per cent of the scale range. Determine the resistance of the resistor and express the value both in error form and in percentage error form.

Estimate of resistance:

$$R = \frac{V}{I} = \frac{75.5}{3.45} = 21.88\,\Omega.$$

Estimate of error: error of voltage is 1.0 per cent of 100 V, which is 1.0 V. Relative to the indication, the voltage error is

$$\varepsilon_V = \frac{1.0}{75.5} \times 100 = 1.32 \text{ per cent}.$$

Error of current is 1.0 per cent of 5 A, which is 0.05 A. Relative to the indication, the current error is

$$\varepsilon_I = \frac{0.05}{3.45} \times 100 = 1.45 \text{ per cent}.$$

The maximum relative error is therefore given by

$$\varepsilon = \varepsilon_V + \varepsilon_I = 1.32 + 1.45 = 2.77 \text{ per cent}.$$

The error in the determination of the resistance can therefore be as great as

$$\frac{2.77}{100} \times 21.88 = 0.61\,\Omega.$$

The resistance of the resistor is therefore

$$21.9 \pm 0.6\,\Omega$$

or $\qquad\qquad 21.9\,\Omega \pm 2.8 \text{ per cent}.$

When attempting to obtain the best possible accuracy from a number of related measurements, it is usual to determine the error characteristics of the instruments. A typical error characteristic relates percentage error to indication as shown in fig. 26.40. From such characteristics, we can relate the error appropriate to each indication and hence determine a corrected value. An instance of this operation is given in the following example.

Fig. 26.40 Error characteristic of an indicating instrument

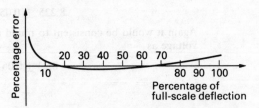

Example 26.7   *A voltmeter indicates that the voltage applied to a resistor is 21.8 V and an ammeter indicates that the current in the resistor is 0.68 A. The error characteristics for the instruments are shown in fig. 26.41. Determine the resistance of the resistor.*

Fig. 26.41

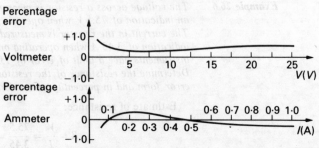

From the characteristics, the percentage error of the voltmeter is +0.6 when indicating 21.8 V and that of the ammeter −0.2 when indicating 0.68 A.

The estimated resistance is

$$R = \frac{V}{I} = \frac{21.8}{0.68} = 32.06\ \Omega.$$

The total error is given by

$$\varepsilon = \varepsilon_V - \varepsilon_I = +0.6 - (-0.2) = +0.8.$$

The resistance value is therefore 0.8 per cent high and the corrected value is therefore

$$32.06 \times \frac{100 - 0.8}{100} = 31.80\ \Omega.$$

To be consistent with the accuracy of the data supplied, the resistance should be given as 31.8 Ω.

One of the most common mistakes when dealing with instrument indications, and particularly when dealing with the results of calculations, is to attribute to them an accuracy that is not justified. For instance an instrument with an accuracy of 1.0 per cent should indicate to that accuracy and no more. Thus if it were a voltmeter of full-scale deflection, a reasonable indication might be given as 75.6 V but certainly not 75.64 V. Often the extra decimal figure is given as an indication on the part of the observer that the reading was between 75.6 and 75.7 and he does not intend to make out that the indication was quite so accurate — but that is not what is stated by the figure quoted.

This false accuracy arises from a failure to appreciate that the number 2.5 is not the same as 2.50. Whereas 2.5 indicates a value between 2.45 and 2.55, 2.50 indicates a number between 2.495 and 2.505. While this may sometimes be appreciated in part, it is surprising how often an experimental table of results shows such figures as 1.65, 1.75, 1.9, 2, 2.15, 2.25, etc. At one extreme the figures are given to a maximum error of 0.005 yet at the other extreme the maximum error can rise to 0.5. This has possibly arisen in part from the observer being too lazy to write 2.0 when he is satisfied with a mere 2 — but he has conveyed completely different information as a result.

Another difficulty arises when a figure such as 234 000 is given, suggesting that the number is correct to a maximum error of 0.5, i.e. the actual value lies between 233 999.5 and 234 000.5. More likely the intention is that the third significant figure is the last reliable figure, in which case the value should have been given as 234×10^3.

Finally, the result of a calculation cannot produce a value of greater accuracy than the accuracy of the information used to formulate the calculation. For instance, if a current of 11 A passes through a 23-Ω resistor, then each figure has been given to an accuracy of two significant places. The voltage across the resistor should therefore be determined as 250 V and not 253 V, which suggests three-figure accuracy. With the advent of calculators, such false accuracy can run riot, so that we find such extreme mistakes as a voltage of 65 V associated with a current of 3.2 A giving rise to a determined value of resistance of 20.3125 Ω. Alas, accuracy cannot be achieved so painlessly, but instead it must be won with great care in the choice of instrumentation and its application.

EXERCISES 26

1. Sketch and describe the construction of a moving-coil ammeter and give the principle of operation.

   A moving-coil instrument gives full-scale deflection with 15 mA and has a resistance of 5 Ω. Calculate the resistance of the necessary components in order that the instrument may be used as: (a) a 2-A ammeter; (b) a 100-V voltmeter.

   (App. El., L.U.)

2. Why is spring control to be preferred to gravity control in an electrical measuring instrument?

   The coil of a moving-coil meter has a resistance of 5 Ω and gives full-scale deflection when a current of 15 mA passes through it. What modification must be made to the instrument to convert it into: (a) an ammeter reading to 15 A; (b) a voltmeter reading to 15 V? (U.L.C.I., O.1)

3. If the shunt for Q. 2(a) is to be made of manganin strip having a resistivity of 0.5 $\mu\Omega \cdot$ m, a thickness of 0.6 mm and a length of 50 mm, calculate the width of the strip.

4. Draw a diagram to show the essential parts of a modern moving-coil instrument. Label each part and state its function.

   A moving-coil milliammeter has a coil of resistance 15 Ω and full-scale deflection is given by a current of 5 mA. This instrument is to be adapted to operate: (a) as a voltmeter with a full-scale deflection of 100 V; (b) as an ammeter with a full-scale deflection of 2 A. Sketch the circuit in each case, calculate the value of any components introduced and state any precautions regarding these components. (c) Explain how the moving-coil instrument can be adapted to read alternating voltage or current. (N.C.T.E.C., O.1)

5. (a) A moving-coil galvanometer, of resistance 5 Ω, gives a full-scale reading when a current of 15 mA passes through the instrument. Explain, with the aid of circuit diagrams, how its range could be altered so as to read up to: (i) 5 A, and (ii) 150 V. Calculate the values of the resistors required.

   (b) A uniform potentiometer wire, AB, is 4 m long and has resistance 8 Ω. End A is connected to the negative terminal of a 2-V cell of negligible internal resistance, and end B is connected to the positive terminal. An ammeter of resistance 5 Ω has its negative terminal connected to A and its positive terminal to a point on the wire 3 m from A. What current will the ammeter indicate? (S.A.N.C., O.1)

6. A moving-coil instrument, which gives full-scale deflection with 15 mA, has a copper coil having a resistance of 1.5 Ω at 15°C, and a temperature coefficient of 1/234.5 at 0°C, in series with a swamp resistor of 3.5 Ω having a negligible temperature coefficient. Determine: (a) the resistance of shunt required for a full-scale deflection of 20 A, and (b) the resistance required for a full-scale deflection of 250 V.

   If the instrument reads correctly at 15°C, determine the

percentage error in each case when the temperature is 25°C.

(App. El., L.U.)

7. Describe, with the aid of sketches, the effect on a current-carrying conductor lying in and at right-angles to a magnetic field.

   The coil of a moving-coil instrument is wound with $40\frac{1}{2}$ turns. The mean width of the coil is 4 cm and the axial length of the magnetic field is 5 cm. If the flux density in the gap is 0.1 T, calculate the torque in newton metres when the coil is carrying a current of 10 mA. (U.E.I., O.1)

8. Explain, with the aid of a circuit diagram, how a d.c. voltmeter may be calibrated by means of a potentiometer method.

   A moving-coil instrument, used as a voltmeter, has a coil of 150 turns with a width of 3 cm and an active length of 3 cm. The gap flux density is 0.15 T. If the full-scale reading is 150 V and the total resistance of the instrument is 100 000 Ω, find the torque exerted by the control springs at full scale.

(App. El., L.U.)

9. A rectangular moving coil of a milliammeter is wound with $30\frac{1}{2}$ turns. The effective axial length of the magnetic field is 20 mm and the effective radius of the coil is 8 mm. The flux density in the gap is 0.12 T and the controlling torque of the hairsprings is 0.5×10^{-6} newton metre per degree of deflection. Calculate the current to give a deflection of 60°.

10. A moving-iron ammeter is wound with 40 turns and gives full-scale deflection with 5 A. How many turns would be required on the same bobbin to give full-scale deflection with 20 A?

11. Describe with the aid of a diagram the construction of a repulsion-type moving-iron instrument with particular reference to the means used for: (a) deflection; (b) control; (c) damping.

    A moving-iron voltmeter, in which full-scale deflection is given by 100 V, has a coil of 10 000 turns and a resistance of 2000 Ω. Calculate the number of turns required on the coil if the instrument is converted for use as an ammeter reading 20 A full-scale deflection. (U.E.I., O.1)

12. Explain the principle of operation of *one* type of moving-iron instrument, showing how it is suitable for use on d.c. and a.c. systems.

    The total resistance of a moving-iron voltmeter is 1000 Ω and the coil has an inductance of 0.765 H. The instrument is calibrated with a full-scale deflection on 50 V, d.c. Calculate the percentage error when the instrument is used on (a) 25-Hz supply, (b) 250-Hz supply, the applied p.d. being 50 V in each case. (U.E.I., O.2)

13. If a rectifier-type voltmeter has been calibrated to read the r.m.s. value of a sinusoidal voltage, by what factor must the scale readings be multiplied when it is used to measure the r.m.s. value of: (a) a square-wave voltage; (b) a voltage having a form factor of 1.15?

14. A permanent-magnet moving-coil milliammeter, having a resistance of 15 Ω and giving full-scale deflection with 5 mA, is to be used with bridge-connected rectifiers (fig. 26.24), and a series resistor to measure sinusoidal alternating voltages. Assuming the 'forward' resistance of the rectifier units to be negligible and the 'reverse' resistance to be infinite, calculate the resistance of the series resistor if the instrument is to give full-scale deflection with 10 V (r.m.s.).

15. Give a summary of four different types of voltmeters commonly used in practice. State whether they can be used on a.c. or d.c. circuits. In *one* case, give a sketch showing the construction,

with the method of control and damping employed.

A d.c. voltmeter has a resistance of 28 600 Ω. When connected in series with an external resistor across a 480-V d.c. supply, the instrument reads 220 V. What is the value of the external resistance?   (U.E.I., O.1)

16. The resistance of a coil is measured by the ammeter–voltmeter method. With the voltmeter connected across the coil, the readings on the ammeter and voltmeter are 0.4 A and 3.2 V respectively. The resistance of the voltmeter is 500 Ω. Calculate (a) the true value of the resistance and (b) the percentage error in the value of the resistance if the voltmeter current were neglected.

17. A voltmeter is connected across a circuit consisting of a milliammeter in series with an unknown resistor R. If the readings on the instruments are 0.8 V and 12 mA respectively and if the resistance of the milliammeter is 6 Ω, calculate: (a) the true resistance of R, and (b) the percentage error had the resistance of the milliammeter been neglected.

18. A milliammeter, giving full-scale deflection with 2 mA, is used in the ohmmeter circuit of fig. 26.29. Battery B has an e.m.f. of 1.45 V. Resistor r is adjusted to give full-scale deflection when terminals TT are short-circuited. When an unknown resistor R is connected across TT, the milliammeter reads 0.8 mA. Calculate the resistance of R.

19. A high resistance was measured by the parallel substitution method (fig. 26.31). The resistance of R was 0.1 MΩ and that of galvanometer A was 1 kΩ. The galvanometer deflections were: (a) with resistor R, 65 divisions; (b) with the unknown resistor X, 28 divisions. Calculate the resistance of X.

20. Describe the principle of the Wheatstone bridge and derive the formula for balance conditions.

The ratio arms of a Wheatstone bridge are 1000 and 100 Ω respectively. An unknown resistor, believed to have a resistance near 800 Ω, is to be measured, using a resistor adjustable between 60 and 100 Ω. Sketch an appropriate circuit.

If the bridge is balanced when the adjustable resistor is set to 77.6 Ω, calculate the value of the unknown resistor.

State, with reasons, the direction in which the current in the detector branch would flow when the bridge is slightly off balance due to the adjustable resistor being set at too *low* a value.   (S.A.N.C., O.1)

21. Describe with the aid of a circuit diagram the principle of the Wheatstone bridge, and hence deduce the balance condition giving the unknown in terms of known values of resistance.

In a Wheatstone bridge ABCD, a galvanometer is connected between B and D, and a battery of e.m.f. 10 V and internal resistance 2 Ω is connected between A and C. A resistor of unknown value is connected between A and B. When the bridge is balanced, the resistance between B and C is 100 Ω, that between C and D is 10 Ω and that between D and A is 500 Ω. Calculate the value of the unknown resistance and the total current supplied by the battery.   (U.L.C.I., O.1)

22. Describe fully, with the aid of a circuit diagram, the Murray-loop test for the location of an earth fault on one core of a two-core cable, and derive an expression for the distance of the fault from the test end of the cable. How does a high fault resistance affect the practical application of the test?

A test was carried out on a 1000-m length of twin cable on which an earth fault had occurred and the balance point on a one-metre slide wire was found to be 45 cm from the end connected to the faulty core. Determine the distance of the fault from the test end of the cable. (W.J.E.C., O.2)

23. The arms of a Wheatstone bridge have the following resistances: AB, 10 Ω; BC, 20 Ω; CD, 30 Ω; and DA, 10 Ω. A 40-Ω galvanometer is connected between B and D and a 2-V cell, of negligible internal resistance, is connected across A and C, its positive end being connected to A. Calculate the current through the galvanometer and state its direction.

24. Describe, with the aid of a circuit diagram, how a simple potentiometer can be used to check the calibration of a d.c. ammeter.

    The current through an ammeter connected in series with a standard resistor of 0.1 Ω was adjusted to 8 A. A standard cell, of e.m.f. 1.018 V, gives a balance at 78 cm, while the potential difference across the standard resistor gives a balance at 60 cm when measured with a simple potentiometer. Calculate the percentage error of the ammeter. (U.E.I., O.1)

25. Explain, with the aid of appropriate circuit diagrams, the use of a direct-reading d.c. potentiometer to calibrate a voltmeter with a full-scale reading of 250 V.

    During such a calibration, the voltmeter is connected to a volt-box with a ratio of 100/1, and its reading is adjusted to 120 V. The potentiometer is balanced with stud and slide-wire settings of 11 and 85.4 respectively. Calculate the error in the voltmeter reading and state whether the instrument reads high or low at this point. (U.L.C.I., O.2)

26. Describe with the aid of a circuit diagram how a simple potentiometer can be used to measure the e.m.f. of a cell.

    A voltmeter, having a resistance of 100 Ω, registered 1.47 V when connected across the terminals of a dry cell. The simple potentiometer was used to measure the e.m.f. of the cell, and balance was obtained at 73 cm with the dry cell in circuit, and at 50 cm with a standard cell of 1.018 V. Calculate: (*a*) the e.m.f. of the dry cell, and (*b*) the internal resistance of the dry cell. (U.E.I., O.1)

27. Explain the theory of the simple slide-wire potentiometer.

    A simple slide-wire potentiometer is used to check the calibration of a 5-A full-scale-deflection ammeter by measuring the voltage drop across a 0.1-Ω standard resistor. When the current is adjusted so that the ammeter reads 5 A, balance is obtained with a slide-wire setting of 24 cm. If the calibration of the potentiometer is 0.02 V/cm, what is the percentage error of the ammeter at this reading? (E.M.E.U., O.1)

28. A simple slide-wire potentiometer and a potential divider (volt-box), of ratio 50/1, are used to check the calibration of a 50-V full-scale-deflection voltmeter. When the voltmeter reads 50 V, balance is obtained with a slide-wire setting of 49 cm. If the calibration of the potentiometer is 0.02 V/cm, what is the percentage error of the voltmeter at this reading? Draw a diagram of essential connections. (E.M.E.U., O.1)

29. The current through an ammeter connected in series with a standard shunt of 0.01 Ω, as in fig. 26.35, is adjusted to 40 A; and the readings on the studs and the slide-wire of the potentiometer, when balanced, are 4 and 12.3 respectively. Calculate the percentage error of the ammeter and state whether the instrument is reading high or low.

30. The reading on a voltmeter connected across the 0/300-V range of a volt-box (fig. 26.36) is adjusted to 250 V, and the readings on the studs and the slide-wire of the potentiometer, when balanced, are 12 and 34.7 respectively. Calculate the percentage

error of the voltmeter and state whether the instrument is reading high or low.

31. State Kirchhoff's Laws for d.c. circuits.

    A 2.2-V cell is connected through a rheostat R to a slide-wire XY of resistance 50 Ω and length 100 cm. A standard cell having an e.m.f. of 1.018 V, and an 85-Ω galvanometer in series with it, are connected between points X and Z, where Z is the tapping point on the slide-wire. This potentiometer is standardized by making XZ = 50.9 cm and adjusting R until the galvanometer shows no deflection. A voltage source, of negligible internal resistance, is then substituted for the standard cell, and balance is restored by changing XZ to 72 cm. Determine the galvanometer current if XZ is now reduced to 70 cm.

    (App. El., L.U.)

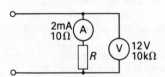

Fig. A

32. The resistance of a resistor is to be determined by measuring the voltage drop across it when passing a given current, as shown in fig. A. The instrument indications are 2 mA and 12.0 V, the ammeter having a resistance of 10 Ω and the voltmeter a resistance of 10 kΩ. Calculate: (*a*) the resistance of the resistor as indicated by the instrument readings; (*b*) the correct resistance of the resistor allowing for instrument error; (*c*) the percentage error of the uncorrected resistance with respect to the corrected resistance.

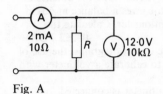

Fig. B

33. With reference to Q. 32, repeat the calculation assuming that the voltmeter and the ammeter had been connected as shown in fig. B yet had indicated the same readings as before.

34. A voltage of 80.0 V is applied to a circuit comprising two resistors of resistance 105 Ω and 55 Ω respectively. The voltage across the 55-Ω resistor is to be measured by a voltmeter of internal resistance 100 Ω/V. Given that the meter is set to a scale of 0–50 V, determine the voltage indicated. (Give your answer to the third significant figure.)

35. An industrial grade ammeter (0–50 mA) was compared with a precision grade ammeter and the following test data, after the indications of the precision-grade ammeter had been corrected, were obtained:

Industrial grade ammeter (mA)	0	10.0	20.0	30.0	40.0	50.0
Precision grade ammeter (mA)	0	9.6	18.4	28.9	39.2	49.5

    Determine whether or not the industrial grade ammeter is within the required ± 1.0 per cent error of full-scale range.

36. For the network shown in fig. C, determine the readings indicated by the ammeter and the wattmeter. The supply voltage may be assumed sinusoidal.

    A rectifier diode is connected into the circuit between the ammeter and the resistor; again determine the readings indicated by the ammeter and the wattmeter. The diode may be assumed ideal.

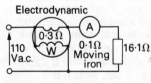

Fig. C

37. An ammeter is required to meet the following specification: a.c./d.c., 0–5 MHz, accuracy ± 2 per cent on a non-linear scale and responding to r.m.s. values.

    Given that the ammeter can also be easily overloaded, which of the following types of ammeter meets the specification: (*a*) moving-coil; (*b*) moving-iron; (*c*) moving-coil, rectifier; (*d*) thermocouple; (*e*) electrostatic; (*f*) electrodynamic; (*g*) electronic?

38. A voltmeter is required to meet the following specification: a.c./d.c., 25 Hz–1 MHz, accuracy ± 2 per cent on a linear scale and responding to peak values.

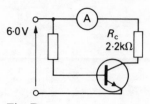

Fig. D

Which of the types of meter listed in Q. 37 meets the specification?

39. The voltage applied to the simple transistor network shown in fig. D is 6.0 V and the ammeter indicates exactly 1.7 mA. Calculate: (*a*) the apparent collector–emitter voltage; (*b*) the maximum and minimum values of the collector–emitter voltage, given that the accuracy of the ammeter is ± 2 per cent and the tolerance of the resistor R_c is ± 10 per cent; (*c*) the percentage error of the collector–emitter voltage in each case.

27.1 Introduction to valves

For the first fifty years of electronic systems, the principal amplifier component was the vacuum valve, which was a comparatively large device using a considerable amount of power. The development of the semiconductor range of devices, which require much less power and which are very much smaller, has caused a remarkable decline in the need for valves to the point that, old equipment apart, they are only to be found in a few very specialized applications.

However, the principles of the valves also apply to the operation of the cathode-ray tube found in television monitors and oscilloscopes. For these major applications, it is therefore still of interest to consider electronic vacuum devices.

27.2 The two-electrode vacuum valve or diode

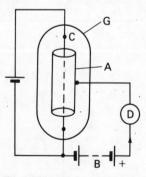

Fig. 27.1 A vacuum diode

If a metal cylinder A surrounds an incandescent filament C in an evacuated glass bulb G, as shown in fig. 27.1, and a battery B is connected in series with a milliammeter D between the cylinder and the negative end of the filament, it is found that an electric current flows through the milliammeter when the cylinder is made positive relative to the filament; but when the connections to battery B are reversed, so as to make A negative, there is no current through D.

Let us now consider the reason for this behaviour.

An electrical conductor contains a large number of mobile or free electrons that are not attached to any particular atom of the material, but move at random from one atom to another within the boundary of the conductor; and the higher the temperature of the conductor, the greater is the velocity attained by these electrons. In the case of an incandescent tungsten filament, for instance, some of these free electrons may acquire sufficient momentum to overcome the forces tending to hold them within the boundary of the filament. Consequently they escape outwards; but if there is no p.d. between the filament and the surrounding cylinder, the electrons emitted from the filament form a negatively-charged cloud or *space charge* around the wire, the latter being left positively charged. Hence the electrons near the surface experience a force urging them to re-enter the filament, and a condition of equilibrium is established in which electrons re-enter the surface at nearly the same rate as they are being emitted, the difference being the electrons which succeed in passing through to cylinder A and represent a current of the order of microamperes. This current is reduced to zero if the potential of the cylinder is made about 1 volt negative relative to the filament.

If cylinder A is made positive in relation to the filament, electrons are attracted outwards from the space charge, as indicated by the dotted radial lines in fig. 27.2, and fewer of the electrons emitted from the filament are repelled back into the latter. The number of electrons reaching the cylinder increases with increase in the positive potential on the cylinder, as shown in fig. 27.3, until ultimately all the electrons emitted from the filament travel to the cylinder. The corresponding rate of flow of electrons is referred to as the *saturation current*.

If the cylinder is made negative in relation to the filament, the electrons of the space charge are repelled towards the filament, so that none reaches the cylinder and the reading on milliammeter D (fig. 27.1) is zero.

Since the cylinder is normally positive in relation to the filament, the former is termed the *anode* and the latter the *cathode*; and since current can flow in one direction only, the arrangement is termed a *thermionic valve* or merely a *valve*. The liberation of electrons from an electrode by virtue of its temperature is referred to as *thermionic emission*. When a valve contains only an anode and a cathode, it is referred to as a *diode*. The p.d. between the anode and the cathode is termed the *anode voltage* and the rate of flow of electrons from cathode to anode constitutes the *anode current*. The *conventional* direction of the anode current is from the positive of battery B via the anode to the cathode, as indicated by the arrowhead in fig. 27.1; but it is important to realize that the anode current is actually a movement of electrons in the reverse direction.

The thermionic emission from a metallic surface is given approximately by Richardson's formula:

$$I = AT^2 e^{-b/T} \text{ amperes per square metre}$$

where T = thermodynamic temperature, in kelvins,

= 273.15 + temperature in °C,

e = Napierian base = 2.718

and A and b = constants for a given material.

The current represented by the above expression is the saturation current, namely the anode current when all the electrons emitted from the cathode are attracted to the anode. This condition is represented by range BC of the

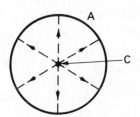

Fig. 27.2 Electron paths from cathode to anode

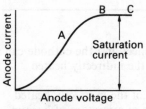

Fig. 27.3 Anode-current/anode-voltage characteristic of a vacuum diode

characteristic* in fig. 27.3; and the corresponding value of the anode current is referred to as the *temperature-limited current*, since, for a given cathode, it depends only upon the temperature of the latter.

Over range OB, the same number of electrons are being emitted from the cathode, but the number which reaches the anode depends upon the combined effect of the space charge and the anode voltage. The lower the anode voltage, the more effective is the opposition of the space charge to the movement of electrons from the cathode to the anode. Over range OA, the value of this *space-charge limited current* is represented approximately by the expression:

$$i_A = kv_A^{1.5}$$

where k = a constant for a given valve.

27.3 Construction of a diode

The anode is stamped out of nickel sheet and the cathode can be either of the directly heated or of the indirectly heated type.

Most diodes incorporate cathodes of the indirectly heated type, in which the cathode C consists of a mixture of barium and strontium oxides sprayed on a hollow nickel cylinder N, as in fig. 27.4. The cathode is heated by a tungsten filament H, known as the *heater*, embedded in an insulator I to prevent the heater making electrical contact with the cathode. Oxide emitters must be activated by special heat-treatment to produce a layer of metallic molecules of barium and strontium on the surface of the oxide.

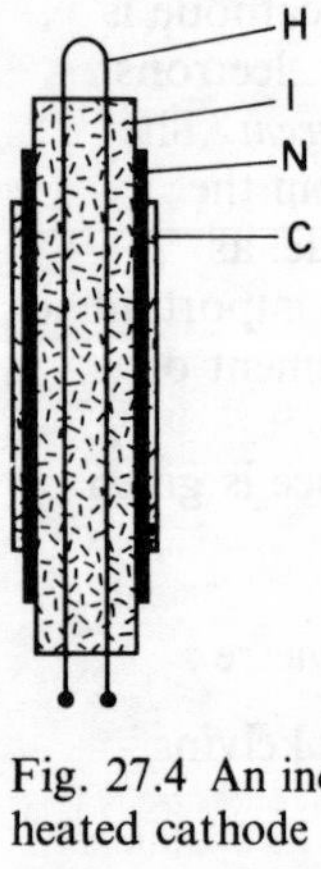

An indirectly heated cathode gives greater flexibility in the spacing of the electrodes and makes it possible to use an a.c. supply for heating the cathode. If a directly heated cathode were supplied from an a.c. source, the alternating p.d. across the filament would cause a corresponding variation in the value of the anode current. Also, in apparatus incorporating several valves, the indirectly heated cathode allows the heaters to be supplied from a common source.

The oxide-coated cathode has a higher efficiency and, in the absence of positive-ion bombardment, has a longer life than the other types of cathode and is almost universally used in small valves.

Fig. 27.4 An indirectly heated cathode

---

* The characteristic shown in fig. 27.3 is for a pure tungsten cathode. The anode current for an oxide-coated cathode continues to increase, but at a reduced rate, with increase of anode voltage.

27.4 The three-electrode vacuum valve or triode

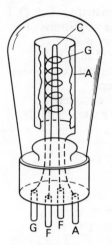

Fig. 27.5 A vacuum triode

Fig. 27.5 shows the general arrangement of a triode having a directly heated cathode C. The anode cylinder A is shown cut to depict more clearly the internal construction. Grid G is usually a wire helix attached to one or two supporting rods. The pitch of this helix and the distance between the helix and the cathode are the main factors that determine the characteristics of the triode.

In a triode, the potential of the anode A is always positive with respect to the filament, so that electrons tend to be attracted towards A from the space charge surrounding cathode C. The effect of making the grid G positive with respect to C is to attract more electrons from the space charge. Most of these electrons pass through the gaps between the grid wires, but some of them are caught by the grid as shown in fig. 27.6(*a*) and return to the cathode via the grid circuit. On the other hand, the effect of making the grid negative is to neutralize, partially or wholly, the effect of the positive potential of the anode. Consequently, fewer electrons reach the anode, the paths of these electrons being as shown in fig. 27.6(*b*). No electrons are now reaching the grid, i.e. there is no grid current when the grid is negative by more than about one volt with respect to the cathode. The paths of the electrons which are repelled back from the space charge into the cathode are not indicated in fig. 27.6.

Fig. 27.6 Influence of grid potential upon electron paths

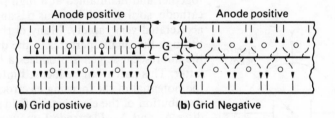

(a) Grid positive   **(b)** Grid Negative

It will be seen that the magnitude of the anode current can be controlled by varying the p.d. between the grid and the cathode; and since the grid is in close proximity to the space charge surrounding the cathode, a variation of, say, one volt in the grid potential produces a far greater change of anode current than that due to one volt variation of anode potential. The relationship between the anode current and the grid voltage for a given anode voltage is termed the *transfer* or *mutual characteristic* of the triode, and that between the anode current and the anode voltage for a given grid voltage is termed the *anode characteristic*.

27.5 Cathode-ray tube

The cathode-ray oscilloscope (usually abbreviated to 'C.R.O.') is almost universally employed to display the waveforms of alternating voltages and currents and has very many applications in electrical testing — especially at high frequencies. The cathode-ray tube is an important component of both the C.R.O. and the television receiver.

Fig. 27.7 shows the principal features of the modern cathode-ray tube. C represents an indirectly heated cathode

Fig. 27.7 A cathode-ray tube

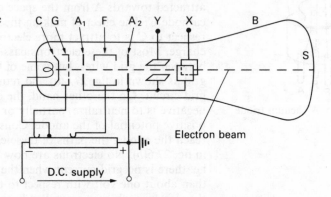

Electron beam

D.C. supply

and G is a control grid with a variable negative bias by means of which the electron emission of C can be controlled, thereby varying the brilliancy of the spot on the fluorescent screen S. The anode discs A_1 and A_2 are usually connected together and maintained at a high potential relative to the cathode, so that the electrons passing through G are accelerated very rapidly. Many of these electrons shoot through the small apertures in the discs and their impact on the fluorescent screen S produces a luminous patch on the latter. This patch can be focused into a bright spot by varying the potential of the focusing electrode F, thereby varying the distribution of the electrostatic field in the space between discs A_1 and A_2. Electrode F may consist of a metal cylinder or of two discs with relatively large apertures. The combination of A_1, A_2 and F may be regarded as an *electron lens* and the system of electrodes producing the electron beam is termed an *electron gun*. The glass bulb B is thoroughly evacuated to prevent any ionization.

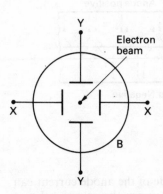

Fig. 27.8 Deflecting plates of a cathode-ray tube

27.6 Deflecting systems of a cathode-ray tube

(I) Electrostatic deflection

The electrons after emerging through the aperture in disc A_2 pass between two pairs of parallel plates, termed the X and Y plates and arranged as in fig. 27.8. One plate of each pair is

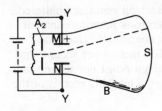

Fig. 27.9 Electrostatic deflection of an electron beam

usually connected to anode A_2 and to earth, as in fig. 27.7.

Suppose a d.c. supply to be applied across the Y-plates, as in fig. 27.9, then the electrons constituting the beam are attracted towards the positive plate M and the beam is deflected upwards. If an alternating voltage were applied to the Y-plates, the beam would be deflected alternately upwards and downwards and would therefore trace a *vertical* line on the screen. Similarly, an alternating voltage applied to the X-plates would cause the beam to trace a horizontal line. The method of calculating the deflection of an electron moving across an electric field is discussed in section 5.19.

(2) Electromagnetic deflection

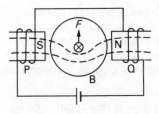

Fig. 27.10 Electromagnetic deflection of an electron beam (tube viewed from screen end)

If two coils, P and Q, were arranged outside the tube, with their axis perpendicular to the beam, and if a direct current were passed through the coil in the direction shown in fig. 27.10, the beam would be deflected upwards. This is due to the fact that the electron beam behaves as a flexible conductor and the *conventional* direction of the current in the beam is from the screen towards the cathode. Applying either the grip or the corkscrew rule, we find that the magnetic flux underneath the beam is strengthened and that above the beam is weakened, so that the resultant flux is distorted as shown in fig. 27.10. Consequently there is a force F urging the beam upwards. An alternating current through the coils would give a vertical line on the screen — similar to that obtained with an alternating voltage across the Y-plates.

27.7 Motion of an electron in a uniform transverse magnetic field

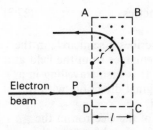

Fig. 27.11 Motion of an electron in a magnetic field when $l > r$

To find the value of the force on each electron

Suppose the area represented by the dotted rectangle ABCD in fig. 27.11 to represent the cross-section of a uniform magnetic field and suppose the direction of the magnetic flux to be outwards from the paper. Also, suppose a beam of electrons, travelling at v metres per second, to enter this field as shown in fig. 27.11, the direction of the beam being perpendicular to that of the magnetic field. As explained in section 27.6, there is a force urging the electrons in the direction shown in the diagram.

If N be the number per second passing any point P and if the negative charge on each electron is e coulombs, then the quantity of electricity passing P is Ne coulombs per second. Consequently the beam is equivalent to a conductor carrying a current of Ne amperes. When an electron beam is at right-angles to a magnetic field of density B teslas, there is a force of NeB newtons per metre acting on the beam (see section 2.8), the direction of the force being perpendicular to *both* the magnetic field and the beam.

Hence no energy is supplied to or taken from the electrons,

so that the kinetic energy of each electron remains unaltered. This means that the *speed* of the electrons remains constant though the direction of the beam is continually changing.

Since the speed of the electrons is v metres per second and the number of electrons passing a given point per second is N, it follows that N electrons occupy a length v metres of the beam;

$$\therefore \quad \text{no. of electrons per metre length of beam} = N/v.$$

Hence force on each electron traversing the magnetic field

$$= \frac{\text{force per metre length of beam}}{\text{no. of electrons per metre length of beam}}$$

$$= \frac{NeB}{N/v} = Bev \text{ newtons} \tag{27.1}$$

To find the locus of the electron path in the magnetic field

From Mechanics, it is known that when a constant force is acting on a body at right-angles to its direction of motion, the body moves in a circle of radius r and has an acceleration v^2/r, where v is the constant speed of the body. Hence the electron follows a circular path while it is moving through the uniform magnetic field.

Force, in newtons, on electron to maintain this circular path

$$= \text{mass of electron in kilograms}$$

$$\times \text{acceleration in metres per second}^2$$

$$= \mathbf{m}v^2/r \text{ newtons} \tag{27.2}$$

Equating expressions (27.1) and (27.2), we have:

$$Bev = \mathbf{m}v^2/r$$

$$\therefore \qquad r = \frac{\mathbf{m}}{e} \cdot \frac{v}{B} \text{ metres} \tag{27.3}$$

If r is less than l, the electron describes a semicircle in the magnetic field, as in fig. 27.11, and emerges from the field at a distance $2r$ from the point of entry. It is then travelling in a direction directly opposite to its original direction, with the magnitude of its velocity unchanged.

If $l \ll r$, as in fig. 27.12, the path of the electron in the magnetic field is an arc EF of a circle of radius r and the

Fig. 27.12 Deflection of an electron beam in a magnetic field when $l \ll r$

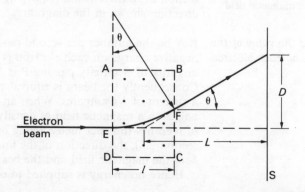

beam is deflected through an angle θ,

where

$$\theta = \frac{\text{arc EF}}{r} \simeq \frac{l}{r} \text{ radian} \tag{27.4}$$

Substituting for r from expression (27.3), we have:

$$\theta = \frac{lB}{v} \cdot \frac{\mathbf{e}}{\mathbf{m}} \text{ radian} \tag{27.5}$$

Example 27.1 *An electron has a velocity of 10^7 m/s when it enters a magnetic field perpendicularly to the direction of the flux. If the flux density is uniform at 0.5 mT and the axial length of the magnetic field is 2 cm, calculate: (a) the radius of curvature of the electron path in the magnetic field, and (b) the angle through which the electron is deflected. Assume the ratio $\mathbf{e}/\mathbf{m}$ to be 1.76×10^{11} C/kg.*

(a) From expression (27.3),

$$r = \frac{10^{-11}}{1.76} \times \frac{10^7}{0.5 \times 10^{-3}} = 0.1136 \text{ m}$$

$$= 11.36 \text{ cm}.$$

(b) Since $l \ll r$, we have from expression (27.4):

$$\theta \simeq l/r = 2/11.36 = 0.176 \text{ radian}$$

$$= 0.176 \times 57.3 = 10.1 \text{ degrees}.$$

27.8 Electrostatic and magnetic deflections on the screen of a cathode-ray tube

(a) Electrostatic deflection

It was shown in section 5.19 that the deflection of an electron passing through an electric field is given by:

$$x = \frac{1}{2} \cdot \frac{\mathbf{e}}{\mathbf{m}} \cdot \frac{V}{d} \cdot \left(\frac{l}{v}\right)^2 \text{ metres} \tag{27.6}$$

where e = charge on electron, in coulombs,

m = mass of electron, in kilograms,

V = p.d. between deflecting plates, in volts,

d = distance between deflecting plates, in metres,

l = axial length of deflecting plates, in metres

and v = initial velocity of electron, in metres per second.

When the electrons leave the electric field, they continue in a straight line to screen S, as indicated in fig. 27.13. If this straight line is produced backwards, it can be shown from the geometry of the parabolic path of the electrons in the electric field, that this line meets the original axis of the electron movement at a point A midway along the electric field.

If L = distance from A to the screen, in metres,

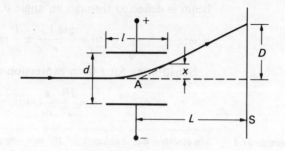

Fig. 27.13 Electrostatic
deflection of electrons in a
cathode-ray tube

and D = deflection on the screen, in metres,

then $\dfrac{x}{D} = \dfrac{l/2}{L}$

$$\therefore \quad D = \frac{2L}{l} \times \frac{1}{2} \times \frac{e}{m} \times \frac{V}{d} \times \left(\frac{l}{v}\right)^2 = \frac{e}{m} \times \frac{V}{d} \times \frac{Ll}{v^2} \qquad (27.7)$$

If V_A is the accelerating voltage, i.e. the p.d. between the cathode and the final anode, then, from expression (5.21),

$$\tfrac{1}{2}mv^2 = eV_A.$$

Substituting for $e/(mv^2)$ in expression (27.7), we have:

$$D = \frac{Ll}{2d} \times \frac{V}{V_A} \qquad (27.8)$$

(b) Magnetic deflection In a C.R.T., the length l of the magnetic field in the initial direction of the electron beam is very small compared with the radius of curvature of that beam; hence the deflection of the electron beam is given by expression (27.5),

i.e. $$\theta = \frac{lB}{v} \cdot \frac{e}{m}.$$

After emerging from the magnetic field, the electrons travel along the tangent to the arc at F, as shown in fig. 27.12. This tangent cuts the original axis of the beam at a point roughly midway along the magnetic field. Hence,

if D = deflection on screen, in metres

and L = distance of screen from mid-point of magnetic field, in metres,

$$\frac{D}{L} = \tan\theta \simeq \theta = \frac{lB}{v} \cdot \frac{e}{m}$$

$$\therefore \quad D = \frac{LlB}{v} \cdot \frac{e}{m} \qquad (27.9)$$

From expression (5.21), $v = \sqrt{(2V_A e/m)}$, where V_A is the p.d. between the cathode and the final anode.

Substituting for v in expression (27.9), we have:

$$D = BLl\left(\sqrt{\frac{e}{m}}\right) \times \frac{1}{\sqrt{(2V_A)}} \qquad (27.10)$$

Comparison of expressions (27.8) and (27.9) shows that the electrostatic deflection is inversely proportional to the accelerating voltage V_A, whereas the magnetic deflection is inversely proportional to the square root of this voltage. Also, the magnetic deflection is a function of e/m, whereas the electrostatic deflection is independent of this ratio.

27.9 Cathode-ray oscilloscope

Most oscilloscopes are general-purpose instruments and the basic form of their operation is illustrated in fig. 27.14. For simplicity, we shall restrict our interest to displaying one signal, although most oscilloscopes are capable of displaying two.

Fig. 27.14 Basic schematic diagram of a cathode-ray oscilloscope

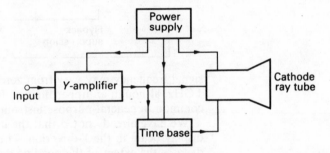

The input signal is amplified by the Y-amplifier, so called because it causes the beam to be driven up and down the screen of the cathode-ray tube in the direction described as the Y-direction by mathematicians.

The time base serves to move the beam across the screen of the tube. When the beam moves across the screen, it is said to move in the X-direction. It would not be appropriate if the movements in the X- and Y-directions were not coordinated; hence the time base may be controlled by the output of the Y-amplifier. This interrelationship is quite complex and therefore requires further explanation.

However, before proceeding, we require another major component which is the power supply. This serves to energize the grid and anode systems of the cathode-ray tube, as well as to energize the brilliance, focus and astigmatism controls of the beam. The power supply also energizes the amplifiers for the control of the beam.

Assuming that you are already familiar with the operation of the cathode-ray tube, it remains to consider those parts of the overall instrument which give rise to controls that we must operate in order to use the oscilloscope. A more complete schematic diagram therefore is shown in fig. 27.15.

The circuitry of an oscilloscope has to be capable of handling a very wide range of input signals varying from a few millivolts to possibly a few hundred volts, whilst the input

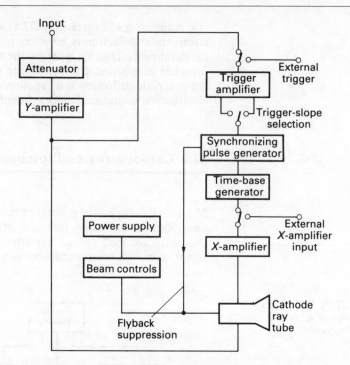

Fig. 27.15 Schematic diagram of a cathode-ray oscilloscope

signal frequency may vary from zero (d.c.) up to possibly 1 GHz, although an upper limit of 10 to 50 MHz is more common in general-purpose instruments.

We have already noted that the input signal drives the display beam in the Y-direction. The height of the screen dictates the extent of the possible deflection. The output of the Y-amplifier therefore has to be of a sufficient magnitude to drive the beam up and down the screen in order to give as large a display as possible without the display disappearing off the edge of the screen. Nevertheless, the Y-amplifier operates with a fixed gain and it is necessary to adjust the magnitude of the input signal to the amplifier, this being done with an attenuator. An attenuator is a network of resistors and capacitors, and its function is to reduce the input signal.

This may seem to be a peculiar function but the hardest task required of the amplifier is to increase the voltage of a small signal in order to obtain a display across the greatest extent of the screen. This determines the amplifier gain, but, unless something is done about it, greater input signals would cause the display to extend beyond the screen. These greater signals, however, can readily be cut down to size by an attenuator, thus leaving the Y-amplifier to continue to operate with its gain fixed to suit the smallest signal.

The attenuator has a number of switched steps, the lowest normally being 0.1 V/cm and the highest being 50 V/cm. For instance, if we set the control to 10 V/cm and apply an input signal of 50 V peak-to-peak, it follows that the height of the display on the screen is 50/10 = 5 cm. As many screens give a display 8 cm high, this is the best possible scale; and such a display would disappear at the top and bottom of the screen.

Associated with the vertical scale control, we also have a

Y-shift control which permits us to centralize the display vertically on the screen.

In oscilloscopes, the Y-amplifier may consist of a single stage in a very basic model, but generally a number of stages are incorporated, especially in those oscilloscopes used for measurements as opposed to simple waveform displays.

The X-amplifier is normally identical to the Y-amplifier and has an associated X-shift control comparable to the Y-shift control. In each case, the shift can be achieved by adjusting the bias voltage to the amplifier, thus causing a shift of the mean output voltage.

The most usual mode of oscilloscope operation has the X-amplifier fed from a time-base generator, the function of which is to drive the beam at a steady speed across the screen and, when it reaches the righthand side of the screen, the beam is then made to fly back to the lefthand side and start out again across the screen. To produce such an output, the input signal to the X-amplifier must take the form of a sawtooth waveform, which is illustrated in fig. 27.16. As the signal steadily increases, the beam is moved across the screen. When the signal reaches its peak, ideally it should drop to zero, thus instantaneously returning the beam to the beginning of its travel. In practice, there are two (possibly three) differences between the ideal waveform and that actually experienced.

The ramp of the waveform is not linear but is derived from an R–C circuit transient. Provided that the time constant of the R–C circuit is very much greater than that of the time required for the beam to scan across the screen, the ramp is almost linear. The difference has been exaggerated in fig. 27.16(b) for clarity.

When the waveform reaches its peak, it is not possible, for reasons which we have observed in our studies of transients, for the signal to suddenly return to zero; thus there is a short period during which the voltage decays. This is called the flyback time, because it is the period during which the beam flies back to the start of its travel.

The flyback time is made as short as possible partly to save the display time lost and partly to reduce the trace of the beam returning across the screen. To help eliminate this unwanted display, the beam current is reduced during the flyback time by means of a flyback suppression pulse.

There may be a short interval between the end of the flyback period and the following scanning period. This is necessary when the time base is controlled from an external trigger source, which is described later. The short delay ensures that the scan starts at the same point in the display waveform, thus causing the display to appear stationary on the screen. In most applications, the time base is controlled by a pulse generator synchronized to the signal from the Y-amplifier, and there is no need to have a delay between the end of the flyback period and the beginning of the scanning period.

The signal to the X-amplifier comes from the time-base generator which may operate in any of the following modes:

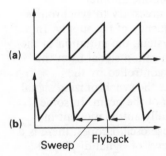

(a)

(b)

Sweep   Flyback

Fig. 27.16 X-amplifier sawtooth waveforms:
(a) idealized form;
(b) practical form

(*a*) self-oscillating;

(*b*) self-oscillating and synchronized;

(*c*) externally triggered.

In the purely self-oscillating arrangement, the time-base voltage rises to a preset value, at which instant the beam has reached the righthand extremity of its travel. When the preset value is reached, the flyback is automatically initiated and, as soon as the original value at the beginning of the scan is obtained, the generator starts generating the next sweep across the screen. The problem with the self-oscillating arrangement is that it works independently of the input signal; thus in the first sweep it may start when the input signal is at a positive maximum value yet the next sweep starts at a negative maximum, thus giving a completely different trace. This sort of variation at best gives rise to an apparently moving display and at worst to two or three displays which are superimposed on one another.

To overcome this problem, it is necessary to synchronize the time-base generator to the frequency of the supply. In a self-oscillating and synchronized system, the initiation of the flyback is controlled by a synchronization signal from the Y-amplifier. Because the flyback is controlled by the synchronizing signal, it follows that the synchronizing signal also controls the start of the sweep of the beam. In some oscilloscopes, this arrangement is fully automatic, but in many of the cheaper general-purpose oscilloscopes, there is a stabilizing control which sets the level of signal display at which the flyback is initiated.

It is necessary to appreciate the reason why this arrangement operates from the finish of the display and not from the beginning. The time-base generator causes the beam to sweep across the display at regular intervals. Let us assume that this is taking place with a frequency 50 Hz and also let us assume that the frequency of the signal to be displayed is 150 Hz. During the sweep time, the input signal undergoes three cycles; thus we would hope to see these three cycles being displayed. In practice, a bit of one cycle would be lost because not all of the time is available for display, the remaining time being taken up by the flyback period. However, the main problems are to commence the trace at the same point in the input signal each time. Let us assume that we wish to start when the input signal is positive and rising. Ideally this would coincide with the instant at which the sweep was due to commence; thus three cycles (almost) would be displayed, followed by the flyback, and everything would be ready for the next sweep to commence displaying the following three input signal waves.

However, what happens if we just miss the start of an input signal wave? If we wait for the next instant of the signal being positive and increasing, then we have to wait for almost a complete cycle, which would be lost to the display. And, even more awkward, what happens if the frequency of the signal to be displayed is 152 Hz? After all, as the signal frequency increases we expect to see more than three cycles, so that if the frequency is 200 Hz for example, we expect the display of four cycles.

The answer is not to wait for the chosen instant but rather to get on with the display up to the time of the chosen instant. In this way, we do not miss anything by waiting (although we shall miss that short period of display during the flyback) but, having reached the chosen instant, the beam is caused to fly back and to recommence the sweep with the minimum delay. It now starts no matter what is happening, and continues again up to the chosen point at which the flyback is again initiated. In this way, we can display any number of cycles or fractions of a cycle in excess of one cycle.

The stabilizing control has to be adjusted appropriately to synchronize the flyback of the time-base generator to the output of the Y-amplifier. In most cases, this can be readily achieved, but sometimes the quality of the signal to be displayed is not sufficiently reliable, in which case the time-base generator must be controlled from an external source, which provides a suitable trigger. In this case, the trigger initiates each individual time-base sweep and the flyback then follows automatically when the time-base signal has reached a preset value. In this case, the time-base generator remains inactive until the trigger releases another sweep. This means that possibly a significant part of the display can be omitted. For this reason, some oscilloscopes are provided with gain controls to the X-amplifier whereby the display can be expanded and we can examine the display in greater detail.

If we wish to make time or frequency measurements, the time-base control must have a calibration setting at which the display time coincides with the control markings. For instance, if the time-base control is set to $10\,\mu s/cm$, then the X-amplifier control is set to the calibration mark and we know that each centimetre of the display in the X-direction represents $10\,\mu s$ of time. A typical range of time-base control settings is $0.5\,s/cm$ to $1\,\mu s/cm$.

There are several applications of the oscilloscope in which we do not require the time-base generation at all but instead we drive the X-amplifier from another signal source in a similar manner to the operation of the Y-amplifier. For this reason, many oscilloscopes afford direct access to the X-amplifier and we shall look at such an application in section 27.10.

The input impedance of most general-purpose oscilloscopes is $1\,M\Omega$ shunted by a capacitance of 20 to 50 pF according to the model used. The effect of the capacitance becomes significantly effective only at high frequencies, i.e. in excess of $1\,GHz$. Such a high input impedance makes the oscilloscope suitable for many measurement techniques, since the oscilloscope scarcely modifies the network into which it has been introduced.

Mention has already been made of the calibration of oscilloscopes, and many have built-in calibration circuits. Generally these give a square or trapezoidal waveform of known peak-to-peak magnitude and cycle duration. This calibration signal is fed into the oscilloscope and the gain of the Y-amplifier is adjusted to give the appropriate vertical display. Similarly, the gain of the X-amplifier is adjusted to give a signal display of appropriate length. These adjustments are usually made by potentiometers with a screw adjustment operated by a screwdriver. In this way, calibration adjustment cannot be confused with the other controls of the oscilloscope.

This brief description of the operation of the principal components has indicated the main controls that we require to use when displaying and measuring waveforms and phase differences by means of the oscilloscopes.

27.10 Use of the cathode-ray oscilloscope in waveform measurement

A discussion of the use of the oscilloscope falls naturally into two parts: the use of the instrument itself, and the methods of connecting the instrument to the circuits in which the measurements are to be made. For ease of introduction, let us assume that the signals applied to the oscilloscope are suitable.

Once the connections between the source of the signal and the oscilloscope have been made, the oscilloscope should be switched on and given time to warm up. Generally a trace will appear on the screen, but should this not occur, some useful points to check are that the vertical and horizontal shift controls are centralized, that the brilliance control is centralized, that the trigger is set to the automatic position (where appropriate) and that the stabilizing control is varied to ensure that the display time base is operated. Normally these checks ensure that the display appears, but if these do not work, then you have to check out the full procedure in accordance with the manufacturer's operating manual.

Once the display has been established, adjust the brilliance to obtain an acceptable trace which is not brighter than necessary. Too bright a trace, especially if permitted to remain in the one position for a considerable period of time, can damage the fluorescent material on the screen, hence the reason for minimizing the brilliance. It also does no harm to check that the beam is focused and that there is minimum astigmatism. These do not vary much with operation but sometimes the controls are adjusted incorrectly.

The display is next centralized vertically and the scale control adjusted to give the highest possible display that can be contained within the screen.

Having the display clearly in view, it may be that the trace is stationary but it could also be slipping slowly in a horizontal direction. In the latter case, the stabilizing control requires to be adjusted until the trace is locked in position and remains stationary.

Unless you have some unusual observations to make on

the waveform, the X-gain amplifier should be set to the calibration position and the X-shift control readjusted to centralize the display horizontally. This display may contain only part of the waveform or a great many waveforms; this is changed by adjusting the time-base control until the desired number of waveforms are displayed.

This again is a brief description of the setting-up procedure of the oscilloscope and it only serves to highlight the common form of operation. Different models of oscilloscope vary in detail, but the procedure is essentially that indicated. However, words cannot substitute for the practical experience of operating oscilloscopes and you will readily obtain a better appreciation of the oscilloscope from a few minutes' experimentation with one in a laboratory.

To aid observation of the display on an oscilloscope, a set of squares is marked on the transparent screen cover. This marking is termed a graticule and is illustrated in fig. 27.17.

Graticules are marked out with a 1-cm grid and are presently 10 cm across by 8 cm high. Older models had graticules 8 cm by 8 cm or sometimes 10 cm by 10 cm. To avoid parallax error, you should always observe the trace directly through the graticule and not from the side.

Let us now consider the interpretation of the basic forms of display, which are sine waves, square waves and pulses. A typical sine waveform display as seen through a graticule is shown in fig. 27.18. To obtain this display, let us assume that the vertical control is set to 2 V/cm and the time-base control to 500 μs/cm.

The peak-to-peak height of the display is 4.8 cm; hence the peak-to-peak voltage is $4.8 \times 2 = 9.6$ V. This may be a direct measurement of a voltage or the indirect measurement of, say, a current. In the latter case, if a current is passed through a resistor of known resistance, then the current value is obtained by dividing the voltage by the resistance.

You will note that the voltage measured is the peak-to-peak value. If the signal is sinusoidal, then the r.m.s. value is obtained by dividing the peak-to-peak value by $2\sqrt{2}$, which in this instance gives an r.m.s. value of 3.4 V.

The oscilloscope therefore can be seen to suffer the disadvantage when compared with electronic voltmeters that it is more complex to operate and to interpret. However, we are immediately able to determine whether we are dealing with sinusoidal quantities, which is not possible with other meters and which is essential to interpreting the accuracy of the measurement of alternating quantities. The oscilloscope is therefore an instrument whereby we observe the waveform in detail and measurements of magnitude of the signal are essentially those of peak-to-peak values.

Returning to the display shown in fig. 27.18, the length of one cycle of the display is 8.0 cm; hence the period of the waveform is

$$8.0 \times 500 \times 10^{-6} = 4.0 \times 10^{-3}\,\text{s} = 4.0\,\text{ms}$$

It follows that the frequency of the signal is

$$1/(4.0 \times 10^{-3}) = 250\,\text{Hz}$$

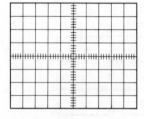

Fig. 27.17 Cathode-ray oscilloscope graticule

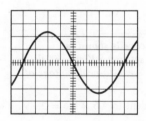

Fig. 27.18 Sine waveform display on an oscilloscope

In each case, the accuracy of measurement is not particularly good. At best, we cannot claim an accuracy of measurement on the graticule that is better than to the nearest millimetre; thus the accuracy at best is about 2 per cent.

If we wish to determine the values of a waveform, such as the average and r.m.s. values of non-sinusoidal waveforms or the mark-to-space ratio of a pulse waveform, it is better to take a photograph of the trace. This is easily done as most oscilloscopes have camera attachments which take photographs of the type that are developed within a minute. Such photographs can be examined at leisure, whereas maintaining a trace for such a length of time on the cathode-ray tube could damage the screen, a point that has already been mentioned.

Typical traces for square waveforms and pulses are shown in fig. 27.19.

Fig. 27.19 Cathode-ray oscilloscope displays for square waves and pulses: (*a*) square wave; (*b*) pulse

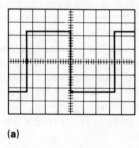

(a)

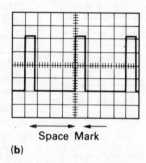

Space Mark

(b)

Example 27.2

The trace displayed by a cathode-ray oscilloscope is shown in fig. 27.19(a). The signal amplitude control is set to 0.5 V/cm and the time-base control to 100 μs/cm. Determine the peak-to-peak voltage of the signal and its frequency.

Height of display is 4.6 cm. This is equivalent to $4.6 \times 0.5 = 2.3$ V. The peak-to-peak voltage is therefore 2.3 V.

The width of the display of one cycle is 7.0 cm. This is equivalent to a period of $7.0 \times 100 \times 10^{-6} = 700 \times 10^{-6}$ s. It follows that the frequency is given by $1/(700 \times 10^{-6}) = 1430$ Hz.

Example 27.3

An oscilloscope has a display shown in fig. 27.19(b). The signal amplitude control is set to 0.2 V/cm and the time-base control to 10 μs/cm. Determine the mark-to-space ratio of the pulse waveform and the pulse frequency. Also determine the magnitude of the pulse voltage.

The width of the pulse display is 0.8 cm and the width of the space between pulses is 3.2 cm. The mark-to-space ratio is therefore $0.8/3.2 = 0.25$.

The width of the display from the commencement of one pulse to the next is 4.0 cm. This is equivalent to $4.0 \times 10 \times 10^{-6} = 40 \times 10^{-6}$ s, being the period of a pulse waveform. The pulse frequency is therefore given by $1/(40 \times 10^{-6}) = 25\,000$ Hz $= 25$ kHz.

The magnitude of the pulse voltage is determined from the pulse height on the display, this being 4.2 cm. The pulse voltage is therefore $4.2 \times 0.2 = 0.84$ V.

At the start of this section, we had to assume that the signals applied to the oscilloscope were suitable. Now we must determine in what way a signal may be thought of as suitable for an oscilloscope.

Most oscilloscopes operate with the body or chassis of the instrument at earth potential. Also most oscilloscopes are connected to the signal source by means of a coaxial cable, the outer conductor of which is connected to the body of the oscilloscope and is therefore at earth potential. It follows that one of the connections from the oscilloscope will connect one terminal of the signal source to earth.

The effect of this observation can be illustrated by considering the test arrangement shown in fig. 27.20. A resistor and a capacitor are connected in series and supplied from a signal generator. It is usual that one terminal of the signal generator is also at earth potential; thus the connection diagram shown in fig. 27.20 is suitable for the circuit and for the oscilloscope. It is suitable because the earth point of the oscilloscope is connected to the earth point of the generator and they are therefore at the same potential.

Fig. 27.20 An experiment involving an oscilloscope

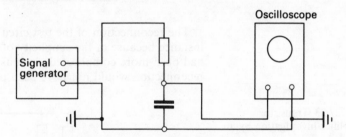

The oscilloscope displays the waveform of the voltage across the capacitor. However, what if we wished to display the waveform of the voltage across the resistor? We appear to have two choices; either we reconnect the oscilloscope across the resistor, or we reconnect the test circuit to the signal generator. Let us reconnect the oscilloscope as shown in fig. 27.21.

Fig. 27.21 Unsuitable reconnection of the oscilloscope

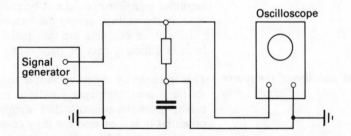

Although the signal into the oscilloscope is now that of the voltage across the resistor, the test circuit has been seriously changed. The earth connection from the oscilloscope to the junction between the resistor and the capacitor causes the capacitor to be short-circuited, i.e. the current from the signal generator passes through the resistor, into the earth connection of the oscilloscope and back to the generator through the earth connection of the generator.

In this particular test, this problem can easily be overcome by interchanging the components of the tests circuit as shown in fig. 27.22. You should also notice that the connection from the earth terminal of the oscilloscope to the test circuit is not actually required, since it duplicates the connection available through earth. As the connection also provides a screen for the coaxial cable, thereby minimizing interference from other sources, it is good practice to retain the second connection.

Fig. 27.22 Suitable reconnection of the oscilloscope

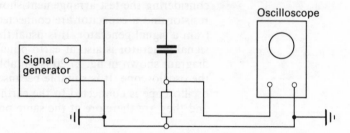

The reconnection of the test circuit was possible in this instance because of the simplicity of the circuit. If the circuit had been more complicated, such as that shown in fig. 27.23, reconnection would not have been possible. For instance, the

Fig. 27.23 Transistor amplifier investigation by an oscilloscope

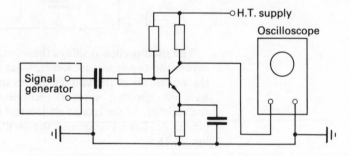

amplifier transistor could not be reconnected in order to observe the voltage across the base–collector junction.

In such cases, there are four possible methods whereby this form of difficulty may be overcome.

(a) Isolation of the source from earth

This is generally the most simple solution to achieve. If the source is energized from a battery, no connection is made to earth; hence the network that is supplied can take up any potential it wishes, and the only connection to earth is that of the oscilloscope. In such an instance, the battery source is isolated from earth and there is no return path for current trying to leak away from the network.

Many signal sources are energized from the mains supply, in which case they are connected to earth for safety by the third wire of the supply flex. They are also indirectly connected to earth by the neutral wire, since the neutral wire of any mains supply is connected to earth back at the supply substation. However, when such instruments are used in a laboratory, special supply arrangements may be provided whereby the frame of the source is not earthed and the 240-V supply is isolated from earth by means of an isolating transformer. In effect this has the same result as that of the source, which was battery operated, there being no return path to the source for any current trying to leak away from the network.

Such isolation procedures make for good safety practice in laboratories and workshops. If there is a return path through earth and two pieces of equipment in close proximity during testing are at different potentials, there is a risk of shock to a person touching both. The discontinuity of the earth connection minimizes such a risk in these operating conditions.

For the arrangement shown in fig. 27.23, if the signal generator were not connected to earth, it would be quite possible to connect the oscilloscope across the base–collector junction of the transistor without affecting the operation of the circuit.

(b) Isolation of the load from the source

This can be achieved by an isolating transformer, but the use of a transformer limits the applications to those of alternating current. An example of an experiment involving an isolating transformer is shown in fig. 27.24. The transformer windings

Fig. 27.24 Experiment involving the isolation of the load from the source

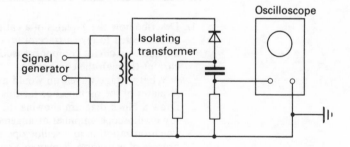

each have the same number of turns, thus the input and output of the transformer are essentially the same. However, the secondary winding and the load which it supplies are free of connection to earth; therefore the introduction of the earth-connected oscilloscope does not interfere with the operation of the network.

(c) Double-beam oscilloscope with difference-of-signals facility

This facility is not available in most general-purpose oscilloscopes, although most oscilloscopes are double-beam instruments. A double-beam oscilloscope has two beams and therefore can display two traces at the one time. Thus, going back to the experiment shown in fig. 27.20, one beam could have displayed the voltage across the capacitor whilst the other displayed the supply voltage. It is the difference of these two displays which gives the voltage across the resistor, and better oscilloscopes have this facility built into them.

It follows that for the test arrangement in fig. 27.23, the base–collector voltage could be obtained by connecting one input to receive the collector-to-earth voltage and the other input to receive the base-to-earth voltage. The controls could then be set to show the difference of these signals, which would be the base–collector voltage.

(d) Isolation of the oscilloscope from earth

This arrangement is not generally favoured, partly because of the high voltages involved within an oscilloscope, but can be employed when none of the other methods is suitable. For instance, if we wish to observe the voltage across a component of a circuit which cannot be isolated readily from earth, e.g. part of a power circuit, and which is at a considerable potential to earth at both terminals, it is better to isolate the oscilloscope and to set it up prior to the application of power to the principal circuit. Under such conditions, it is not possible to adjust the oscilloscope during the test as it then takes up the potential of the power circuit. It is possible to take records of its display by camera, this being remotely controlled. This form of operation is an advanced part of engineering technology and should only be carried out under the supervision of an experienced electrical engineer.

EXERCISES 27

1. Describe how the *Y*-plates of a cathode-ray oscilloscope would be connected to give a trace of (*a*) the alternating voltage applied to a circuit component (such as a coil) and (*b*) the current through the component.

    What voltage waveform would normally be applied to the *X*-plates of the oscilloscope for this purpose?

2. Draw a block diagram showing the principal parts of a cathode-ray oscilloscope amplifier arrangement. What is the purpose of synchronization in an oscilloscope and why is it essential to the process of waveform display in a cathode-ray oscilloscope?

3. Explain with the aid of a circuit diagram showing the connections made to the cathode-ray oscilloscope, how it may be used to determine the r.m.s. value of an a.c. signal which is: (*a*) sinusoidal; (*b*) non-sinusoidal.

    An oscilloscope is used to display a sinusoidal alternating voltage. The display shows a sine wave of amplitude 3.3 cm and the *Y*-amplifier sensitivity is set to 5 V/cm. Determine the r.m.s. value of the voltage.

4. Draw a block schematic diagram of a cathode-ray oscilloscope. Include in your diagram a switch for the selection of internal/external time base, and describe an application of the oscilloscope in which the external time base would be used.

    Explain the limitation of the *Y*-amplifier with regard to the accurate observation of rectangular pulses.

5. The display given by a cathode-ray oscilloscope is shown in fig. A. Given that the sensitivity of the *Y*-amplifier is set to 10 V/cm and the time base control to 5 ms/cm, determine: (*a*) the peak-to-peak value of the voltage; (*b*) the period of the signal; (*c*) the frequency.

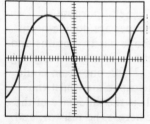

Fig. A

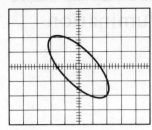

Fig. B

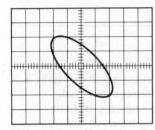

Fig. C

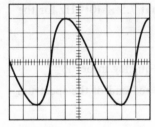

Fig. D

6. The display of a cathode-ray oscilloscope is given in fig. B. The figure is produced by the application of two signals of equal amplitude to the X- and Y-plates. Determine the phase angle between the signals.

7. The display of a cathode-ray oscilloscope is shown in fig. C. The signal applied to the X-amplifier is a sinusoidal signal of frequency 600 Hz and voltage 20 V peak-to-peak. The sensitivities of the X- and Y-amplifiers are set to the same values. Determine: (a) the frequency of the voltage applied to the Y-amplifier; (b) its peak-to-peak value; (c) the phase angle between the signals.

8. The display of a cathode-ray oscilloscope is shown in fig. D. Describe the harmonic content of the signal.

9. An electron beam, after being accelerated by a p.d. of 1500 V, travels through a uniform magnetic field of density 1.5 mT, the direction of the magnetic field being normal to the initial direction of the beam. If the width of the magnetic field traversed by the electron beam is 15 mm, calculate: (a) the radius of curvature of the beam while it is travelling through the magnetic field, and (b) the angle through which the beam is deflected. Assume e/m to be 1.76×10^{11} C/kg.

10. It is found that an electron beam is deflected 8 degrees when it traverses a uniform magnetic field, 3 cm wide, having a density of 0.6 mT. Calculate: (a) the speed of the electrons, and (b) the force on each electron. The direction of the beam is normal to that of the flux.

11. The Y-deflecting plates of a cathode-ray tube have an axial length of 20 mm and are spaced 8 mm apart. Their centre (point A in fig. 27.13) is 150 mm from the screen. The final anode potential is 4 kV. Calculate: (a) the beam velocity as it enters the electric field of the Y-plates, and (b) the p.d. between the Y-plates to give a deflection of 30 mm on the screen.

12. A cathode-ray tube, with magnetic deflection, has its screen 20 cm from the centre of the magnetic field. The width of the uniform magnetic field is 3 cm and the final anode potential is 6 kV. Calculate the density of the magnetic field to produce a deflection of 4 cm on the screen. Assume $e/m = 1.76 \times 10^{11}$ C/kg.

28 Electric Lamps and Illumination

28.1 The spectrum

Newton discovered that when white light, such as that given by the sun, is passed through a glass prism, a band of light is obtained having colours in the following sequence: red, orange, yellow, green, blue, violet. It is known that light is a form of electromagnetic radiation having a velocity of 3×10^8 m/s and that the various colours have different frequencies ranging from about 4×10^{14} Hz for the extreme visible red end of the spectrum to about 7.5×10^{14} Hz for the extreme visible violet end.

Since velocity in metres per second

$$= \text{wavelength in metres} \times \text{frequency in hertz}$$

$\therefore$ wavelength of the extreme visible red end of the spectrum

$$= \frac{3 \times 10^8}{4 \times 10^{14}} = 0.75 \times 10^{-6} \text{ m}$$

$$= 750 \text{ nanometres} = 750 \text{ nm}$$

where 1 nanometre $= 10^{-9}$ metre. Similarly,

wavelength of the extreme visible violet end of the spectrum

$$= \frac{3 \times 10^8}{7.5 \times 10^{14}} = 0.4 \times 10^{-6} \text{ m}$$

$$= 400 \text{ nm}.$$

Electromagnetic radiations that are immediately beyond the red and the violet ends of the spectrum are termed *infra-red* and *ultraviolet* respectively. In the case of light emitted by incandescent solid bodies, the spectrum is continuous, but the relative intensity of the different colours depends upon the temperature of the body — the higher the temperature, the more pronounced is the violet end of the spectrum compared with the red end. When light is obtained from a gaseous discharge, the spectrum is discontinuous, i.e. it consists of one or more coloured lines. Thus in the case of the sodium lamp, the spectrum consists mainly of two yellow lines very close together with wavelengths of 589.0 and 589.6 nm. These two wavelengths are so close to each other that the light from a sodium lamp is said to be 'monochromatic', namely a light having only one wavelength.

Fig. 28.1 Relative luminous
sensitivity of the human eye

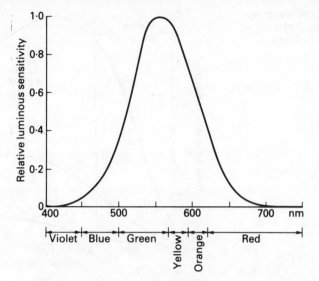

Fig. 28.1 shows how the sensitivity or the brightness
sensation of the average human eye varies at different
wavelengths for ordinary levels of illumination, the power
radiated as light being assumed constant for the different
wavelengths. It will be seen that the eye is most sensitive to
light having a wavelength of about 555 nm in the green
portion of the spectrum. The ratio of the visual effect of light
of a given wavelength to that at a wavelength of 555 nm is
termed the *relative luminous sensitivity* of the human eye; thus
the relative luminous sensitivity of the eye to light having a
wavelength of either 510 nm or 610 nm is about 0.5.

The most common and useful source of light is the sun,
and the graph in fig. 28.2 shows the relative values of the
power radiated at different wavelengths for an average
summer midday sunlight. It will be seen that the maximum
power is radiated at about 500 nm, which is approximately
the wavelength at which the human eye is most sensitive.

Fig. 28.2 Spectral power
distribution curve for
sunlight

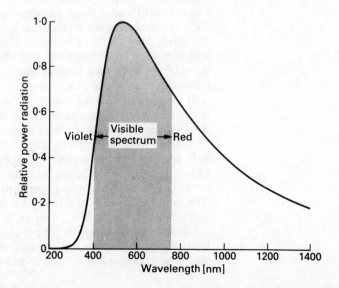

Fig. 28.3 Spectral power
distribution curves for an
incandescent filament

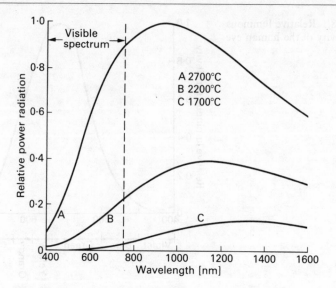

Fig. 28.3 gives the relative values of the power radiated
from an incandescent filament at different temperatures. These
curves show that the lower the temperature, the lower is the
amount of energy radiated in the visible range and emphasize
the necessity of operating incandescent lamps at the highest
practicable temperature.

28.2 Units of luminous intensity and illuminance

(a) The standard originally used in photometry was a wax
candle, but, owing to its unreliability, it was replaced by a
lamp burning vaporized pentane. The luminous intensity of
this lamp was equal to about ten of the original candles. Even
this standard was difficult to reproduce accurately, and in
1909 the incandescent filament lamp was adopted as the
standard. The luminous intensity of a number of such lamps
was measured by comparison with the pentane lamp and
these lamps were then used as standards for determining the
luminous intensity of other lamps.

It is essential that a primary standard should be capable of
being accurately reproduced from a specification; and in 1948
it was decided to base the unit of luminous intensity upon the
luminance (or objective brightness) of a small aperture due to
light emitted from a radiator maintained at the temperature
of solidification of platinum, namely 1773°C. The
construction of this primary standard is shown in fig. 28.4.
The radiator consists of a tube of fused thoria (thorium
oxide), about 45 mm long, with an internal diameter of about
2.5 mm, the bottom of the tube being packed with powdered
fused thoria. The tube is supported vertically in pure
platinum contained in a fused thoria crucible. The latter has a

Fig. 28.4 Primary standard
of light

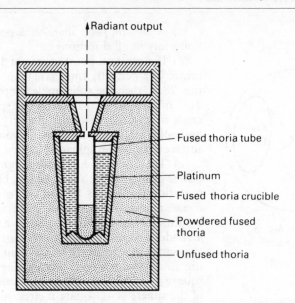

↑Radiant output

— Fused thoria tube

— Platinum

— Fused thoria crucible

— Powdered fused
thoria

— Unfused thoria

lid with a small hole in the centre, about 1.5 mm in diameter, and is almost embedded in powdered fused thoria in a larger refractory container having a funnel-shaped opening. The reason for the use of pure fused thoria is that this material is unaffected at the temperature of melting platinum and does not contaminate the latter. The presence of any impurity in the platinum would alter the melting-point temperature and therefore affect the luminance of the aperture.

The platinum is first melted by eddy currents induced in it by a high-frequency current in a coil surrounding the outer container and is then allowed to cool very slowly. While the platinum is changing from the liquid to the solid state, the temperature remains constant sufficiently long for measurements to be made of the luminous intensity of the light beam passing through the aperture of known diameter. The luminance of this new primary standard was found to be 589 000 international candles per square metre, the international candle being the unit of luminous intensity previously used. In 1948 the International Conference of Weights and Measures decided to adopt 600 000 units per square metre as the luminance of the platinum primary standard. This new unit of luminous intensity, termed the *candela*,* is therefore defined as *the luminous intensity, in the perpendicular direction, of a surface of* 1/600 000 *square metres of a black body at the temperature of freezing platinum under standard atmospheric pressure.*

The *luminous intensity* of a lamp is defined as the light-radiating capacity of a source in a given direction, expressed in candelas (symbol: cd).

(*b*) A *point source* of light is a source which, for photometric purposes, can with sufficient accuracy be considered as concentrated at a point.

---

* Pronounced with the first syllable slightly accentuated, as in 'candle'.

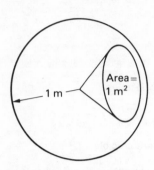

Fig. 28.5 Relationship between the candela and the lumen

(c) A *uniform point source* is a point source emitting light uniformly in all directions.

(d) If a uniform point source of 1 candela is placed at the centre of a perfectly transparent sphere of, say, 1 m radius (fig. 28.5), then the solid angle subtended at the centre by 1 m² of area on the surface of the sphere is termed a *unit solid angle* or *steradian*; and the quantity of light emitted through a unit solid angle and therefore passing through 1 m² of the surface area of the sphere is termed a *lumen* (symbol: lm).

(e) The *luminous flux* from a light source is the radiant power evaluated according to its ability to produce visual sensation (section 28.1). The unit of luminous flux is the *lumen*, namely the luminous flux emitted in unit solid angle by a uniform point source having a luminous intensity of 1 candela. Thus for a uniform point source of 1 candela at the centre of a perfectly transparent sphere of 1 m radius (fig. 28.5), the luminous flux passing outwards through each square metre of the surface is 1 lumen. Since the surface area of the sphere is 4π square metres, it follows that the total luminous flux from a uniform point source of 1 candela is 4π lumens. Hence, for a point source having a luminous intensity of I candelas,

total luminous flux emitted in solid angle $d\omega$

$$= d\Phi = I \cdot d\omega \text{ lumens}$$

or $\qquad\qquad I = d\Phi/d\omega \text{ candelas},$

i.e. the luminous intensity, in candelas, of a source in a given direction is the luminous flux, in lumens, emitted in a very narrow cone containing the direction, divided by the solid angle of the cone. In the case of an actual lamp, it is necessary to estimate the optical centre of the lamp and to regard the light as being emitted from that point.

(f) The *mean spherical luminous intensity* of a luminous source is the average value of the luminous intensity in all directions. Hence if a luminous source has a mean spherical luminous intensity of I candelas, the total luminous flux emitted by that source is $4\pi I$ lumens. It is not correct to say that I candelas are equal to $4\pi I$ lumens, since the 'candela' is the unit of 'intensity', whereas the 'lumen' is the unit of 'flux'.

(g) The *illuminance* at a point of a surface is the luminous flux per unit area at that point. The unit of illuminance is the *lumen per square metre* and termed the *lux* (symbol: lx). With a uniform point source of 1 candela at the centre of a hollow sphere of 1 metre internal radius, the illuminance on the surface* is 1 lx and the rays of light reach the surface normally.

If the internal radius of the sphere is increased from 1 metre to r metres, the surface area is increased from 4π to $4\pi r^2$ square metres. With a uniform point source of 1 candela

---

* It is assumed that no light is reflected by the internal surface. This condition can be fulfilled by using a blackened surface.

at the centre,

$$\left.\begin{array}{c}\text{number of lumens per square metre}\\ \text{on a sphere of radius } r \text{ metres}\end{array}\right\} = \frac{4\pi}{4\pi r^2} = \frac{1}{r^2}.$$

Hence the illuminance of a surface is inversely proportional to the square of its distance from the source. It follows that if the luminous intensity of a source S in direction SA (fig. 28.6) is I candelas and if the distance between S and A is d metres, then for a surface at right-angles to the incident ray SA,

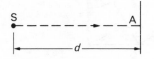

Fig. 28.6 Illuminance on a surface normal to the incident rays

$$\text{illuminance at A} = I/d^2 \text{ lux}.$$

The *inverse square law of illuminance* strictly applies only to a point source of light in a completely black room. When the size of the source is small compared with the distance of the surface from the source, the above expression can be used to calculate the illuminance on a plane normal (or perpendicular) to the incident rays.

(*h*) *The cosine law of illuminance.* Suppose that the surface (ABCD in fig. 28.7) to be illuminated is so placed that the angle between the incident ray and the normal to the surface is θ and that the incident rays are parallel. The luminous flux falling on ABCD is exactly the same as that which would fall on a surface EFGH, normal to the rays. It follows that:

$$\frac{\text{illuminance on ABCD}}{\text{illuminance on EFGH}} = \frac{\text{area of EFGH}}{\text{area of ABCD}} = \cos\theta$$

$\therefore$ illuminance on ABCD = illuminance on EFGH $\times \cos\theta$

Hence, if the surface illuminated from a point source (fig. 28.6) is tilted about A so that the angle of incidence at A is θ, as in fig. 28.8,

$$\text{illuminance at A} = \frac{I}{d^2} \cdot \cos\theta \qquad (28.1)$$

(*i*) The *luminance** of a source in a given direction is the luminous intensity in that direction per unit of projected area. It is usual to express the luminance in candelas per metre^2 of projected area. Approximate values of the luminance of

---

* The term 'brightness' by itself is ambiguous, since it is necessary to distinguish between 'objective' and 'subjective' brightness. The brightness of a source, as judged by the eye, depends upon a number of factors such as the brightness of the surrounding surface; for instance, a light source appears brighter if the surroundings are dark. Such interpretation of brightness is referred to as *subjective brightness* or *luminosity*. If the brightness of a source is measured photometrically in terms of the candelas per unit of projected area, the value is a definite quantity for a given source and is independent of such factors as the surrounding surfaces. Hence, this interpretation of brightness is referred to as *objective brightness* or *luminance*. Luminosity and luminance are not proportional to each other; for instance, a surface having twice the luminance of another surface may not *look* twice as bright.

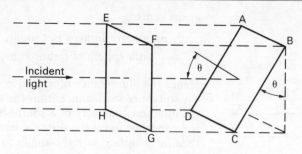

Fig. 28.7 Cosine law of illuminance

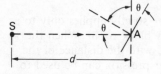

Fig. 28.8 Cosine law of illuminance

different sources are given in the following table:

Source	Luminance (in cd/m²)
Zenith sun	16×10^8
Crater of carbon arc (25 A)	2×10^8
Tungsten, gas-filled, clear (100 W)	6.5×10^6
Tungsten, gas-filled, pearl (100 W)	8×10^4
Mercury, high-pressure, clear (400 W)	120×10^4
Mercury, low-pressure, fluorescent (80 W)	0.9×10^4
Sodium, low-pressure, clear (140 W)	8×10^4
Clear blue sky	0.4×10^4

It is extremely important in practice to avoid considerable contrast of brightness in the line of vision of the eye, since this causes glare and an intense eyestrain. From the above table it will be seen that the luminance of a pearl lamp is only about $\frac{1}{80}$th of that of the corresponding lamp with a clear bulb; consequently there is far less risk of discomfort or glare with the former than with the latter. The very low luminance of the fluorescent lamp constitutes one of its main advantages.

28.3 Illuminance on a surface

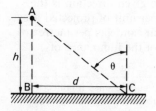

Fig. 28.9 Illuminance on a surface

Suppose A in fig. 28.9 to represent a luminous source suspended h metres above the ground; then:

$$\text{illuminance at B} = \frac{\text{luminous intensity in direction AB}}{h^2}$$

and

$$\text{illuminance at C} = \frac{\text{luminous intensity in direction AC} \times \cos \theta}{AC^2}$$

$$= \frac{(\text{luminous intensity in direction AC}) \times h}{(h^2 + d^2)^{3/2}}.$$

If the luminous intensity of the lamp is uniform in the lower hemisphere, then:

$$\text{illuminance at C} = \text{illuminance at B} \times \cos^3 \theta$$

$$= \frac{\text{illuminance at B}}{\{1 + (d/h)^2\}^{3/2}}.$$

Fig. 28.10 Illuminance due
to a source of uniform
luminous intensity

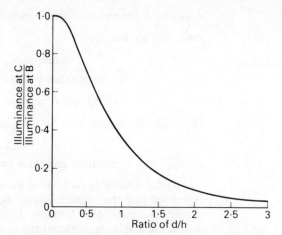

The above relationship is represented graphically in
fig. 28.10, assuming the luminous intensity of the lamp to be
uniform in the lower hemisphere. It will be seen that the
illuminance on the ground falls off very rapidly with increase
of distance from the point directly underneath the lamp. This
effect can be reduced by using a luminaire* to distribute most
of the light at an angle of about 60–75° to the vertical.

By using a number of lamps, suitably spaced, it is possible
to obtain fairly even illuminance over the working surface.
With such an arrangement, the following method may be
used to calculate the illuminance. Suppose the surface to be
illuminated to have an area of A square metres and the
average illuminance on that surface to be E lux. The useful
luminous flux is therefore EA lumens. But the lamps must
emit a larger number of lumens, since some of the light from
the lamps is absorbed by the ceiling, the walls and the
luminaires. The ratio of the useful number of lumens to the
total lumens emitted by the lamps is termed the *utilization
factor*.

$$\text{Hence, total lumens from lamps} = \frac{EA}{\text{utilization factor}}.$$

Tables of the values of the utilization factor for rooms of
various shapes, different kinds of luminaires, different
spacing/height ratios, and various colours of ceilings and
walls are available. With open reflectors, the utilization factor
lies between 0.4 and 0.8; whereas with pendant fittings giving
indirect lighting, the light has to be reflected from the ceiling
and walls, and the utilization factor may be as low as 0.1.

Finally, it is necessary to allow for the depreciation in the
value of the useful luminous flux due partly to the
accumulation of dust on the lamp bulbs and on the
luminaires and partly to the fall in the light output of each
lamp during its life. This effect is taken into account by

* A *luminaire* is a term used internationally for a lamp fitting
which distributes, filters or transforms the light given by a lamp or
lamps and which includes all the items necessary for fixing and
protecting these lamps and for connecting them to the supply circuit.

applying a *maintenance factor*, where

maintenance factor

$$= \frac{\text{illuminance at any given time}}{\text{illuminance with lamps new and fittings clean}}$$

$= 0.85$ to 0.6, depending upon circumstances.

Hence, total lumens from lamps when new

$$= \frac{EA}{(\text{utilization factor}) \times (\text{maintenance factor})} \quad (28.2)$$

The most suitable number of luminaires depends upon the type used and their height above the working plane which is usually assumed to be horizontal and 0.85 m above the floor. Too large a spacing between the luminaires may result in excessive variation of the illuminance over the working plane. The minimum illuminance at any point on the working plane should not be less than 70 per cent of the maximum value. In general, the spacing between the luminaires should be about 1.0 to 1.5 times their height above the working plane.

The illuminance of the working interior of a factory or an office should not be less than 400 lx; and where fine work such as fine assembly is involved, the illuminance should be about 1 to 2 klx. An investigation in a factory producing leather handbags and similar articles showed that when the illuminance was increased from 300 to 1000 lx, the output increased by 11 per cent. In this case an increase of only 1 per cent in output would have made the increased illuminance a commercial proposition.

In addition to providing adequate illuminance on the working surface, a lighting system should also provide *shadow* — not the kind of shadow that makes for danger but that which gives an object its three dimensions and enables people to recognize shapes quickly and easily by ensuring that the shadow is a familiar one.

Another requirement of a good lighting scheme is that it makes the principal object the brightest thing in the field of vision, the immediate background a little less bright and the general surroundings (walls, etc.) still less bright. An illuminance ratio of 10:3:1 has been suggested as the best for task/background/surroundings. This ratio gives the conditions under which the eye works best and has the added advantage of emphasizing the importance of the workpiece and avoiding distraction from things around.

Example 28.1

A room, 10 m × 5 m × 4 m, is to have an average illuminance of about 200 lx on a working plane 0.85 m above the floor. Assume the utilization factor to be 0.6 and allow for a maintenance factor of 0.8. Calculate the number of 200-W lamps to be installed and indicate on a diagram a suitable arrangement. Assume each lamp, when new, to emit 2700 lm and a spacing/height ratio of about 1.0.

Area of working plane $= 10 \times 5 = 50 \, \text{m}^2$

$\therefore$ luminous flux on working plane $= 200 \times 50 = 10\,000 \, \text{lm}$.

$$\text{Luminous flux from lamps} = \frac{10\,000}{0.6 \times 0.8} = 20\,800 \,\text{lm}$$

and number of lamps $= 20\,800/2700 = 8$.

Let us assume that the luminaires are installed about 0.5 m below the ceiling, then:

height of luminaires above working surface

$$= 4 - (0.85 + 0.5) = 2.65 \,\text{m}.$$

With a spacing/height ratio of about 1.0, it follows that the 8 luminaires can be installed in two rows, spaced as in fig. 28.11.

Fig. 28.11 Spacing of luminaires in example 28.1

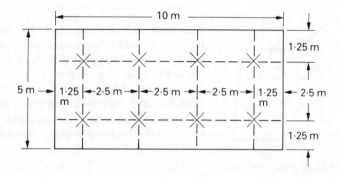

28.4 Measurement of luminous intensity

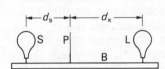

Fig. 28.12 Measurement of luminous intensity

It is assumed that a standard lamp S is available, the luminous intensity of which in a given direction is known. One method of determining the luminous intensity of another lamp L is to mount the two lamps as in fig. 28.12 on an optical bench B in a dark room, with a photometer head P so arranged that its position can be varied until its two sides are equally bright. Care must be taken to ensure the correct alignment of the lamps with the photometer head. If the distances of S and L from P are d_s and d_x respectively when balance is obtained, and if I_s and I_x represent the luminous intensities of S and L respectively in the direction of the photometer head, then:

$$\frac{I_s}{d_s^2} = \frac{I_x}{d_x^2}$$

$$\therefore \qquad I_x = I_s \times (d_x/d_s)^2.$$

Alternatively, an illumination meter is placed at a convenient known distance from the standard lamp S, and the reading on the meter is noted. Test lamp L is then substituted for S and the reading on the meter is again noted. Then:

$$I_x = I_s \times \frac{\text{meter reading with test lamp L}}{\text{meter reading with standard lamp S}}.$$

A variant of the latter method is to adjust the respective distances d_s and d_x of the standard and test lamps from the illumination meter to give the same deflection; then:

$$I_x = I_s \times (d_x/d_s)^2.$$

This method has the advantage of eliminating error due to any non-linearity of the photo-cell of the illumination meter.

28.5 Polar curves of light distribution

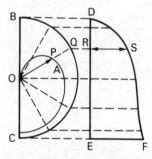

Fig. 28.13 A polar curve of light distribution and the corresponding Rousseau diagram

It is often of importance to know the luminous intensity of a lamp in different directions and particularly the effect of a shade or reflector upon the distribution of the luminous intensity. This effect is most conveniently represented by polar co-ordinates. Thus, curve A in fig. 28.13 represents the relationship between the luminous intensity in a vertical plane through the lamp and the angle made with the vertical axis for a pearl-type lamp without a shade.

28.6 Measurement of the luminous flux and the mean spherical luminous intensity of a lamp

The number of lumens emitted by a lamp and its mean spherical luminous intensity can be determined from the polar curve of light distribution, but *not* by merely taking the mean value of the polar coordinates. The construction usually employed is that known as the *Rousseau diagram* and is shown in fig. 28.13. Suppose curve A to represent the polar curve of light distribution for the lamp under consideration. With centre O, a semi-circle of any convenient radius OB is drawn and a number of radii are inserted. From the ends of these radii, horizontal lines are drawn to cut a vertical line DE. These lines are extended to the right of DE to represent to scale the luminous intensities, in candelas, in the direction of the respective radii; thus, for radius OQ, RS is made equal to the luminous intensity OP in direction OQ. The points on the horizontal lines are then joined by a smooth curve DSF. It can be shown that the average width between the vertical line DE and the curve DSF gives the mean spherical luminous intensity of the lamp.

Once the mean spherical luminous intensity of a lamp has been determined as described above, that lamp can then be used as a standard for measuring the mean spherical luminous intensity and the lumen output of other lamps by means of an *integrating photometer* (or *photometric integrator*). In its most accurate form the integrating photometer consists of a hollow sphere S (fig. 28.14), with matt white surface on

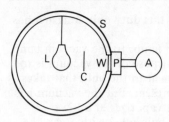

Fig. 28.14 An integrating
photometer

the inside, so that the light from lamp L is thoroughly
diffused and the internal surface of the sphere thereby
uniformly illuminated. A milk-glass window W is shielded by
a screen C from the direct rays of the lamp. Consequently the
illumination of W is entirely due to diffused light, and this is
proportional to the total luminous flux of the lamp.

The relative luminance of the milk-glass window W with
different lamps in the sphere can be determined by placing a
photocell P, corrected for colour, facing towards W, as shown
in fig. 28.14. The reading on microammeter A, connected to
the photocell, is first noted with a standard lamp of known
mean spherical luminous intensity (m.s.l.i.) in the sphere, and
then with the standard lamp replaced by the lamp the m.s.l.i.
of which is required. Then:

$$\frac{\text{m.s.l.i. of test lamp}}{\text{m.s.l.i. of standard lamp}} = \frac{\text{reading with test lamp}}{\text{reading with standard lamp}}.$$

The number of lumens emitted by the test lamp is 4π times
its mean spherical luminous intensity.

28.7 Measurement of illuminance and luminance

An illumination meter (or lightmeter) utilizes a photovoltaic
cell such as that described in section 28.14. With a suitable
resistor in series in the external circuit between ring R and
plate P in fig. 28.22, the current through the microammeter is
practically proportional to the illuminance directly in lux.

The sensitivity of the photovoltaic cell to light of different
wavelengths is not exactly the same as that of the human eye,
the cell being relatively more sensitive at each end of the
visible spectrum. Hence most modern lightmeters have a filter
which matches cell response to eye response. If the meter has
an uncorrected photocell, a correction factor has to be
applied when it is used to measure illuminance due to light of
a colour other than that used for its calibration.

The *luminance* of a surface in a given direction is obtained
by dividing the luminous intensity of the light in that
direction by the projected area of the surface viewed from
that direction. It is often convenient to place an aperture of
accurately-known area in front of the surface, particularly if
there is a variation in the luminance of the surface.

28.8 Incandescent electric lamp

In this type of lamp the filament must be capable of being
operated for long periods at a high temperature without

appreciable deterioration; and for this duty only tungsten has proved satisfactory.

There are two types of tungsten lamp: (i) the vacuum lamp, and (ii) the gas-filled lamp. The purpose of the vacuum is to prevent loss of heat from the filament to the bulb that takes place by convection when gas is present. But the vacuum has the disadvantage that the filament vaporizes at a lower temperature than it does with the bulb filled with a gas, the effect being very similar to the variation in the boiling point of water with surface pressure.

The vaporization of the filament not only reduces the sectional area of the filament, thereby increasing its resistance and reducing the temperature and the luminous intensity of the lamp, but it also allows tungsten to condense on the internal surface of the bulb, blackening the latter and reducing the intensity still further. Consequently the highest temperature at which it is practicable to operate the filament in a vacuum lamp is limited to about 2100°C.

The introduction of an inert gas, e.g. nitrogen or argon, enables the filament temperature to be raised to about 2500°C before blackening takes place at an excessive rate. If no other change were made except to introduce a gas, it would be found that the heat lost by convection would be so great that the power required to maintain the filament at 2500°C would have increased more in proportion than the light given out by the lamp. Consequently the efficacy of the gas-filled lamp would be lower than that of the vacuum lamp. This difficulty is overcome by winding the filament as a very close helix, as shown in fig. 28.15. Langmuir discovered that a thin layer of the gas adheres to the filament and that the convection current of gas merely glides over this fixed layer. The effect is similar to that observed on a river; the main stream may run swiftly, but the water alongside the banks may be practically stationary. It follows that if the clearance between adjacent turns of the filament is less than the thickness of two of these layers, the latter prevent any gas passing between the turns. Hence the surface with which the gas can come into contact is practically the same as that of a rod of diameter d and length l (fig. 28.15); and since this area is far less than the surface area of the filament, the loss of heat by convection is very considerably reduced.

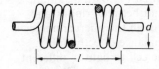

Fig. 28.15 A helical filament

In 40-W to 100-W lamps, the coiled filament is wound into a coarser helix, thereby reducing still further the effective filament surface exposed to the gas. The coiled-coil lamp has an efficacy of about 10–20 per cent higher than that of the corresponding lamp with the filament wound as a single helix.

The latest development in this type of lamp is the tungsten-halogen* lamp used for projectors, headlights, etc. The most common halogen used is iodine. At temperatures between 250°C and 750°C, iodine vapour and tungsten combine to form tungsten iodide: but at temperatures above 1250°C, tungsten iodide splits into tungsten and iodine.

---

* *Halogens* is the name given to a group of elements consisting of fluorine, chlorine, bromine and iodine.

The bulb wall temperature is designed to be between 250°C and 750°C. Hence the iodine vapour combines with any evaporated tungsten to form tungsten iodide vapour before the tungsten reaches the bulb wall. The iodide vapour is carried over the hot filament by gas convection where it splits into tungsten and iodine. The tungsten is deposited back on the filament. Consequently the bulb never blackens. The vapour pressure in the bulb is about 5 atmospheres (compared with about 1 atmosphere in the glass-bulb lamp), and the bulb is made of quartz to withstand the higher temperature and pressure.

The volume of a tungsten-halogen lamp is only about 1 per cent of that of its conventional counterpart. This results in increased efficacy† (about 20 lm/W compared with about 12 lm/W for the ordinary 100-W gas-filled lamp), and the permanently clean interior of the bulb gives it a longer useful life.

28.9 Electric discharge lamps

When the p.d. between two electrodes in a tube containing gas exceeds a critical value, ionization occurs and the current has to be limited by connecting an external resistor or inductor in series. The ionized gas emits radiation which may be in the visible, ultraviolet or infra-red regions of the spectrum. All discharge lamps must therefore have:

(*a*) an envelope containing gas which can be ionized by the voltage available and which is capable of emitting visible radiation, directly or indirectly;

(*b*) an external inductor or resistor to limit the current;

(*c*) an arrangement for initiating the discharge.

28.10 High-pressure* mercury–vapour lamp

Fig. 28.16 shows one arrangement of this type of lamp. For simplicity, some of the constructional features, such as lamp cap, have been omitted. The lamp consists of an inner bulb B, generally made of fused silica, which can withstand higher temperature than glass. This bulb contains a small quantity of

---

† The term *luminous efficiency*, applied to *lumens per watt*, is incorrect, since efficiency is the ratio of the output power to the input power when both are expressed in the *same* unit.

* A high-pressure mercury–vapour lamp operates at a vapour pressure of about 1 bar ($= 10^5 \, \text{N/m}^2 \simeq 1$ atmosphere). In extra-high-pressure mercury–vapour lamps, the vapour pressure reaches the order of 10 bars.

Fig. 28.16 High-pressure
mercury-vapour lamp

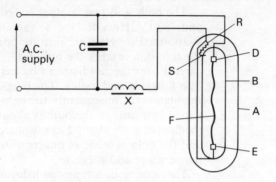

mercury and argon, and is protected from draughts by a glass
bulb A, which may be cylindrical, as in fig. 28.16, or elliptical
(or egg-shaped). The latter shape is being more commonly
adopted as it enables the temperature of the middle of the
envelope to be limited to about 270°C, compared with about
500°C with the cylindrical envelope. The space between bulbs
A and B contains nitrogen at a pressure of about half an
atmosphere.

The discharge tube B has three electrodes, namely the main
electrodes D and E, and a starting electrode S. The latter is
connected through a resistor of about 10 to 30 kΩ to the
main electrode E situated at the other end of the tube.
Electrodes D and E consist of tungsten-wire helices filled with
electron-emissive materials; usually barium and strontium
carbonates mixed with thoria.

When the lamp is switched on, the high voltage gradient
between electrodes D and S ionizes the argon and enables the
arc F to strike between the main electrodes D and E. The
p.d. between D and E gradually increases from about 20 V up
to about 100–120 V, and during this time the colour changes
and increases in intensity.

The function of inductor X is to limit the current to a safe
value; but since the presence of X produces a lagging power
factor, a capacitor C is connected to improve the power
factor to about 0.9 lagging.

If the lamp is switched off, it cannot be restarted
immediately. The vapour pressure is too high to allow the
supply voltage to start ionization, and the lamp must cool
sufficiently for the vapour pressure to fall to the value at
which ionization between electrodes D and S can take place.

If a mercury lamp were operated at a low vapour pressure,
most of the output would be radiated at a wavelength of
253.7 nm, which is in the ultraviolet range and is well below
the visible range. The light from such a lamp is greenish-blue
and therefore produces colour distortion. When the vapour
pressure is increased, much of the ultraviolet energy emitted
by the arc is absorbed by the surrounding mercury vapour
and is re-radiated at wavelengths that are in the visual range.

The adoption of quartz instead of glass for the inner tube
has led to the use of operating pressures of about
10 atmospheres, resulting in a higher proportion of emission
at longer wavelengths.

The colour quality of the light emitted from a high-pressure mercury lamp is further improved (*a*) by the addition of a mixture of metallic iodides inside the discharge tube to give multi-line spectrum in the visual range and (*b*) by coating the inner surface of the outer glass bulb with fluorescent phosphors such as yttrium vanadate; these phosphors absorb ultraviolet radiation emitted by the mercury discharge and radiate energy in the red region of the spectrum.

The luminous efficacy of a 400-W mercury lamp having a coating of yttrium vanadate is about 50 lm/W, and that of a 1-kW mercury-iodide lamp is about 85 lm/W.

28.11 Low-pressure and high-pressure sodium–vapour lamps

One type of sodium–vapour lamp consists of a U-tube A (fig. 28.17) containing sodium together with neon at a pressure of about $1 \, \text{kN/m}^2$ and about 1 per cent of argon, the function of the argon being to reduce the initial ionizing voltage. The oxide-coated tungsten electrodes are connected

Fig. 28.17 Low-pressure sodium-vapour lamp

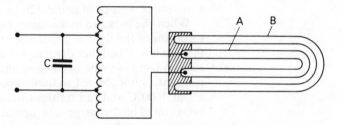

to a step-up auto-transformer T, designed to have a relatively large leakage reactance. When this transformer is on open circuit, its output voltage is about 500 V, and this is sufficient to initiate a discharge through the gas. This discharge has a reddish colour, but as the temperature rises and the sodium is vaporized, the discharge changes to monochromatic light consisting of two yellow lines having wavelengths of 589.0 and 589.6 nm.

An alternative starting method for the low-pressure sodium lamp is the use of a semiconductor circuit in conjunction with a series current-limiting inductor. The semiconductor circuit produces a high voltage which ionizes the gas. This effect is repeated in each half cycle until the sodium metal is vaporized. As the sodium vapour pressure increases, the power from the semiconductor is reduced until, when the lamp attains its normal working condition, the semiconductor circuit ceases to play any part in the operation of the lamp. This method dispenses with the high leakage reactance transformer, thereby reducing the power absorbed and requiring a smaller capacitor to raise the power factor to a given value. For example, the power absorbed by a 35-W l.p. sodium lamp with a semiconductor starter and a choke circuit is only about 46 W, compared with about 64 W taken by the same lamp fed from a leaky-reactance transformer, and a p.f. of 0.85 is attainable with a 7-μF capacitor compared with a 15-μF capacitor.

The efficacy of the lamp decreases rapidly as the current density is increased above a certain value. Consequently, the lamp has to be operated at a low current density, and this necessitates a large surface area of tube compared with the power dissipated. Also, for maximum efficacy, the temperature of the tube has to be about 220°C; and in order to maintain this temperature with low power per unit length of tube, it is necessary to insulate the tube thermally. One method of doing this is to enclose the discharge tube in a double-walled vacuum jacket B — somewhat similar to that of the domestic vacuum flask — as shown in fig. 28.17.

Hot sodium vapour is very active chemically and the U-tube A (fig. 28.17) has therefore to be made of ply glass, the inner layer being a special sodium-resisting glass and the outer ordinary soda glass. Since the sodium solidifies when the tube cools, it is necessary to ensure that the sodium is deposited reasonably uniformly along the whole of the tube and not concentrated at one end. Consequently, the lamp must be used horizontally. Low-pressure sodium lamps have a luminous efficacy of about 120 to 160 lm/W.

When the pressure in a sodium lamp is increased, the spectrum of the light is broadened to include colours other than yellow. The inner discharge tube of a high-pressure sodium lamp is made of sintered aluminium oxide which is resistant to hot ionized sodium vapour up to a temperature of about 1600°C and can transmit over 90 per cent of visible radiation. The discharge tube operates at a pressure of about half an atmosphere and is enclosed in an evacuated glass envelope to maintain the tube at the correct temperature. The lamp gives a rich golden light which enables colours to be easily distinguished.

The discharge tube of a high-pressure sodium lamp contains sodium and mercury, with argon or xenon at a low pressure added for starting purposes. A voltage pulse of about 2.5 kV is required to initiate the discharge. A 400-W high-pressure sodium lamp has a luminous efficacy of about 100 lm/W.

28.12 Low-pressure fluorescent mercury–vapour tubes

The fluorescent lamp usually consists of a long glass tube T (fig. 28.18), internally coated with a fluorescent powder. The tube contains a small amount of argon together with a little mercury. At each end of the tube there is an electrode E

Fig. 28.18 A fluorescent
mercury-vapour tube with
glow starter

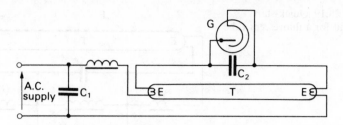

consisting of a coiled tungsten filament coated with a mixture
of barium and strontium oxides. Each electrode has attached
to it two small metal plates — one at each end of the filament.
These plates act as nodes for withstanding bombardment by
electrons during the half-cycles when the electrode is positive.
During the other half-cycles, the adjacent hot filament acts as
the cathode, emitting electrons.

Circuits for the control of fluorescent tubes can be divided
into two main groups, namely *switch-start* circuits and *quick-
start* circuits.

The switch-start circuit in general use is a voltage-operated
device referred to as a *glow* switch (the *thermal* switch has
been almost entirely superseded by the glow switch since the
latter is relatively cheap, simple and reliable). Fig. 28.18
shows a fluorescent tube fitted with a glow switch G. The
latter consists of two bimetallic strip electrodes in a glass bulb
filled with a mixture of helium, hydrogen and argon at low
pressure. The contacts are normally open, but the application
of the supply voltage starts a glow discharge between the
electrodes of G and the heat is sufficient to bend the
bimetallic strips until they make contact, thereby closing the
circuit between electrodes EE. A relatively large current flows
through these electrodes, raising them to incandescence and
the gas in their immediate vicinity is ionized. After a second
or two, the bimetallic strips cool sufficiently to break contact,
and the sudden reduction of current induces in choking coil X
an e.m.f. of the order of 800–1000 V. This surge is sufficient to
ionize the argon in the space between electrodes EE. The heat
generated in the tube vaporizes the mercury and the p.d.
across the tube falls to about 100–110 V. This p.d. is not
sufficient to restart the glow in the starter switch G. The final
vapour pressure in the tube is about 1 N/m^2 for an ambient
temperature of 20°C. A $0.02\text{-}\mu\text{F}$ capacitor C_2 is connected
across the glow switch to suppress radio interference. Without
C_2, high-frequency voltage oscillations may occur across the
starter contacts.

The power factor of the lamp circuit, including choke X, is
about 0.5, and a capacitor C_1 is introduced to raise the
power factor to about 0.9 lagging.

Quick-start circuits are referred to in various ways,
e.g. switchless, instant start, etc., and one arrangement is
shown in fig. 28.19. Electrode pre-heating is provided by a
small auto-transformer, the primary winding D being
connected across the lamp electrodes EE. A narrow metal or
metallized strip F is attached to the outer surface of the tube
and connected to earth through the lamp caps.

Fig. 28.19 Quick-start
circuit for a fluorescent tube

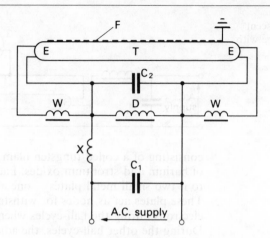

Immediately the lamp is switched across, say, a 240-V
supply, almost the whole of this voltage is applied across the
tube electrodes EE, and the p.d. applied from windings WW
to the filaments raises the latter to incandescence. The
combination of this pre-heating and of the relatively high
potential gradient between each electrode and the earth strip
F is sufficient to start ionization in the vicinity of the
electrodes, and this ionization then spreads to the whole tube.
Immediately this occurs, the p.d. across the primary winding
D falls to about 100–110 V, so that the voltage across
filaments EE is practically halved, the temperature of the
filaments being maintained partly by the reduced current and
partly by the heat of the discharge.

It was mentioned in section 28.10 that most of the energy
of a low-pressure mercury–vapour lamp is radiated at a
wavelength of 253.7 nm, which is in the ultraviolet range. The
fluorescent coating absorbs this ultraviolet energy and
converts it into visible radiation. Different fluorescent powers
re-radiate the absorbed energy at different wavelengths; for
instance, magnesium tungstate gives pale blue radiation, zinc
silicate green, zinc–beryllium silicate gives yellow to orange,
cadmium borate and yttrium vanadate each give red
radiation. By using an appropriate combination of various
fluorescent powders it is possible to obtain any desired
colour. The average efficacy of the fluorescent lamp (warm-
white type) is about 60 lm/W and its luminance is about
9000 cd/m².

The ultimate failure of a fluorescent lamp is caused by the
exhaustion of the oxides with which the filaments are coated,
this material being gradually disintegrated during the life of
the lamp, particularly each time the lamp is switched on. The
average life of the fluorescent tube is about 5000 hours
compared with about 1000 hours for the tungsten-filament
lamp.

The light output of all discharge lamps fluctuates at twice
the supply frequency, but with fluorescent lamps, this
fluctuation is considerably reduced by the persistence of glow
of the fluorescent powder. Nevertheless, this flicker can
produce undesirable stroboscopic effects with rotating

machinery. This trouble can be practically eliminated by:
(*a*) using groups of three lamps distributed between the three
phases of a three-phase supply; (*b*) using twin lamps on a
single-phase supply, one being connected in series with an
inductor and the other in series with an inductor and a
capacitor of such capacitance that the current leads the
supply voltage by about 60°, so that the currents in the two
lamps differ in phase by about 120° and the power factor of
the combined circuits is practically unity; (*c*) introducing a
full-wave rectifier between the inductor and the lamp so that
the current through the latter is unidirectional and
approximately constant.

28.13 Phototransistor

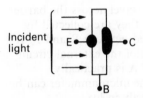

Fig. 28.20 Phototransistor

An ordinary transistor can be enclosed in a metal or other
opaque container. If light waves are allowed to fall on a
germanium or silicon crystal, the energy supplied by the
photons is sufficient to create electron-hole pairs, exactly as
created by an increase of temperature.

The phototransistor is similar to a normal transistor except
that a window in the container allows light to fall on the
base, as indicated in fig. 28.20, so as to penetrate as near as
possible to the collector junction, thereby creating electron-
hole pairs in the base.

Fig. 28.21 shows how an n–p–n phototransistor P can be
arranged to energize the coil of a relay D. The two parallel
arrows outside the envelope are the symbol indicating that
the device is operated by light or other radiation. The
sensitivity of the phototransistor is greatly increased by
connecting the base to the emitter through a resistor R of the
order of 5 kΩ. The function of rectifier A is to allow the
current in D to flow through A immediately after the incident
light on P is cut off, thereby allowing the current in D to
decay relatively slowly and thus preventing a high e.m.f. being
induced in D.

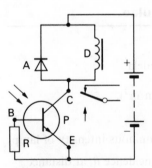

Fig. 28.21 Relay operated
by a phototransistor

The maximum sensitivity of germanium occurs at a
wavelength of about 1550 nm, which is well in the infra-red
region. That of silicon occurs at about 850 nm (see fig. 28.1).

Phototransistors are used to operate burglar alarms, smoke
detectors, counters, etc.

28.14 Photovoltaic cells

The sensitive element in a photovoltaic cell is a
semiconductor of such material that light flux falling on it
displaces electrons from some of the atoms. The two

Fig. 28.22 Selenium
photovoltaic cell

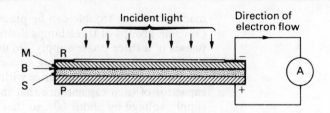

semiconductors found most suitable for photovoltaic cells are
selenium and cuprous oxide, and fig. 28.22 shows the
arrangement of a cell having selenium as the active material.
A steel plate P is coated with a thin layer S of selenium at
about 200°C and annealed at about 80°C to produce the
crystalline form. This selenium layer is covered with a very
thin transparent film M of metal, and a collecting ring R of
metal is sprayed around the edge of the film. Between the
selenium S and the film M, there appears to be a 'barrier
layer' B. When the light falls on the cell, it passes through the
transparent film M and causes electrons to be released from
the metallic selenium. These electrons travel across the barrier
layer to the metal film M, from which they are collected by
ring R. A microammeter A is connected between R and P. It
is found that with a suitable resistance of the external circuit
between R and P, the current through A is practically
proportional to the illuminance and the microammeter can be
calibrated to read the illuminance directly in lux
(section 28.2).

Summary of important formulae

Unit of luminous intensity is the candela.
Unit of luminous flux is the lumen.
Unit of illuminance is the lux ($= 1\,\mathrm{lm/m^2}$).

No. of lumens emitted by a lamp

$$= 4\pi \times \text{mean spherical luminous intensity of lamp}.$$

For a surface having an angle of incidence θ, at distance
d metres from a source having luminous intensity of
I candelas,

$$\text{illuminance} = \frac{I}{d^2} \cdot \cos\theta \text{ lux} \tag{28.1}$$

To produce an average illuminance of E lux on a surface of
A square metres,

total lumens from lamps when new

$$= \frac{EA}{(\text{utilization factor}) \times (\text{maintenance factor})} \tag{28.2}$$

where utilization factor

$$= \frac{\text{no. of lumens on working plane}}{\text{no. of lumens emitted by lamps}}$$

and maintenance factor

$$= \frac{\text{illuminance at any given time}}{\text{illuminance with lamps new and fittings clean}}$$

EXERCISES 28

1. A metal-filament gas-filled lamp takes 0.42 A from a 230-V supply and emits 1120 lm. Calculate: (*a*) the number of lumens per watt; (*b*) the mean spherical luminous intensity (or m.s.l.i.) of the lamp; and (*c*) the m.s.l.i. per watt.

2. Six lamps are used to illuminate a certain room. If the luminous efficacy of each lamp is 11 lm/W and the lamps have to emit a total of 10 000 lumens, calculate: (*a*) the m.s.l.i./lamp, and (*b*) the cost of the energy consumed in 4 h if the charge for electrical energy is 5.2 p/kWh.

3. The effective area of the filament of a certain pearl-type lamp, when viewed from below, was 20 cm², and the luminous intensity in the downward direction was 153 cd. Calculate the luminance of the lamp when viewed from that direction.

4. A fluorescent tube in a dark room is shielded by a large metal sheet alongside the tube. The screen has a circular hole, 12 mm diameter, the centre of the hole being midway along and directly opposite the axis of the tube. The illuminance indicated on a lightmeter having its surface normal to the light beam and situated 200 mm from the screen is 17.3 lx. Calculate the luminance of the lamp. Assume the shielding to be such that the reading on the lightmeter is zero when the aperture is covered.

5. A lamp having a mean spherical luminous intensity of 80 cd has 70 per cent of the light reflected uniformly on to a circular screen 3 m in diameter. Find the illuminance on the screen.

6. In a test on a photometric bench, a filament lamp was placed at a distance of 50 cm from the photometer head, the corresponding distance of a standard 60-cd lamp being 35 cm. Find the luminous intensity of the test lamp in the direction of the head.

7. In a photometric bench test, balance is obtained when a standard lamp of 25 cd in the horizontal direction is 1 m and the lamp being tested is 1.25 m from the photometer screen. What is the luminous intensity of the test lamp in the direction of the screen?
    If the light from the test lamp is reduced by 15 per cent, what will be the respective distances of the lamps from the photometer screen? In this case, the lamps are fixed 2.5 m apart, and the photometer screen moves between them.

8. Two lamps of 16 cd and 24 cd respectively are 2 m apart. A screen is placed between them 0.8 m from the 16-cd lamp. Calculate the illuminance on each side of the screen. Where must the screen be placed in order to be equally illuminated on both sides?

9. A lamp giving 200 cd in all directions below the horizontal is suspended 2 m above the centre of a square table of 1 m side.

Calculate the maximum and minimum illuminances on the surface of the table.

10. Explain what is meant by *luminous flux* and define the unit in which it is expressed. A circular area of radius 6 m is to be illuminated by a single lamp vertically above the circumference of the circle. The minimum illuminance is to be 6 lx, and the maximum illuminance 20 lx. Find the mounting height in metres and the mean spherical luminous intensity of the lamp. Assume the luminous intensity to be uniform in all directions.

(E.M.E.U.)

11. Four lamps are suspended 8 m above the ground at the corners of a square of 4 m side. Each lamp gives 250 cd uniformly below the horizontal plane. Calculate the illuminance: (*a*) on the ground directly under each lamp; (*b*) at the centre of the square.

12. Two identical lamps, each having a uniform intensity in all directions below the horizontal, are mounted at a height of 4 m. Determine what the spacing of the supports must be in order that the illuminance of the ground midway between the supports shall be one-half of the illuminance directly beneath a lamp.

(W.J.E.C., O.2)

Note. The simplest way of solving this problem is by trial and error, starting with the assumption that the illuminance directly under one lamp due to the other lamp is negligible.

13. A filament lamp is enclosed in a fitting that absorbs 25 per cent of the light flux from the lamp and distributes the remainder uniformly over the lower hemisphere. The lamp may be assumed to have a luminous efficacy of 12 lm/W. The fitting is suspended 2 m above the centre of a horizontal surface 3 m square. Determine the necessary rating (in watts) of the lamp if no point on the horizontal surface is to have an illuminance less than 18 lux. Reflected light from other surfaces may be ignored.

14. Two lamps, each of the same rating and equipped with an industrial-type reflector giving a distribution curve as shown below, are suspended 4 m apart and 2 m above a horizontal working plane. Calculate the illuminance on this plane: (*a*) at the point A, vertically below one of the lamps; (*b*) at the point B, 1 m from A along the line joining A with the point vertically below the other lamps. The polar curve for a lamp and reflector is as follows:

Angle (in degrees measured from horizontal axis)	15	30	45	60	75	90
Luminous intensity (candelas)	50	125	240	190	155	140

(App. El., L.U.)

15. A 200-cd lamp emits light uniformly in all directions and is suspended 5 m above the centre of a working plane which is 7 m square. Calculate the illuminance, in lux, immediately below the lamp and also at each corner of the square. If the lamp is fitted with a reflector which distributes 60 per cent of the light emitted uniformly over a circular area 5 m in diameter, calculate the illuminance over this area.

16. A room measuring 15 m by 20 m is to be provided with an illuminance of 200 lx over the horizontal plane using fluorescent tubes, each 2 m long. Each tube gives an output of 3200 lm, 65 per cent of which is effective over the working plane. If a maintenance factor of 0.8 is to be allowed for, find the number of tubes required and sketch a plan view of a suitable arrangement for them.

17. A room, 10 m by 20 m, is to be illuminated by 8 lamps and the average illuminance is to be 300 lx. If the utilization factor is

0.48 and the maintenance factor is 0.8, calculate the mean spherical luminous intensity per lamp.

18. A room, 40 m by 15 m, is to be illuminated by 80-W fluorescent tubes mounted 3 m above the working plane on which an average illuminance of 180 lx is required. Using a maintenance factor of 0.8 and a utilization factor of 0.5, design and sketch a suitable layout. The 80-W fluorescent tube has an output of 4500 lm. (W.J.E.C., O.2)

19. Describe an experiment to determine the constants α and β in the relationship luminous intensity $= \alpha V^{\beta}$ for a metal-filament lamp.

If for a given lamp the value of β is 4, find the percentage change in the luminous intensity of the lamp for a change of ± 5 per cent in the voltage across the lamp.

20. Two 110-V lamps, one of 60 W and the other of 75 W, are connected in series across a 220-V supply. Calculate the current taken by and the p.d. across each lamp, neglecting any variation in resistance. Assuming the luminous intensity to be proportional to the fourth power of the voltage, calculate the percentage value for each lamp compared with normal operation on a 110-V circuit. (U.E.I.)

21. A certain lamp gave the following distribution of luminous intensity, in candelas, in a vertical plane, the angle being measured from the axis through the centre of the lamp cap:

Angle, in degrees	0	15	30	45	60	90	120	150	165	180
Candelas	0	84	154	200	224	254	266	276	283	288

Find: (*a*) the mean spherical luminous intensity, and (*b*) the luminous flux of the lamp.

22. Describe briefly *three* applications of photoelectric cells and state the type of cell used for each.

The light from a certain lamp falls perpendicularly upon a photoelectric cell at a distance of 3 m from the lamp, the lamp emitting 60 cd uniformly in all directions. The photoelectric cell characteristic has a slope of 30 μA/lm and the cathode area is 10 cm^2. If the cell is connected in series with a resistor of 5 MΩ across a d.c. supply, calculate the difference in the p.d. across this resistor when the lamp is switched on and off. (U.L.C.I.)

Answers to Exercises

Exercises 1. Page 28

1. $0.4 \, \text{m/s}^2$.   2. 2 kN.
3. 14.715 kN; 220.7 MJ, 61.3 kWh; 368 kW, 26 kWh.
4. 16 370 N m.   5. 212 N m.
6. 94.7 A, 641.2 p.   7. 83.8 per cent, 146 kJ; 67.2 Ω.
8. 12.82 kW, 26.7 A, 102.56 kWh.
9. 59.5 Ω, 0.38 p.   10. 13.93 kW, £2.23.
11. 1.98 g.   12. 241 min.
13. 4.035 V, 780 J.   14. 26.57 V, 13.8 W.   15. 5 Ω.
16. A: 2.4 A, 14.4 V, 9.6 V; B: 1.6 A, 16 V, 8 V; 96 W.
17. 10.96 V; 121.7 A, 78.3 A; 1.33 kW, 0.86 kW.
18. 11.33 Ω, 208 W.   19. 40 Ω, 302.5 W.
20. 0.569 mm.   21. 70°C.
22. 30 Ω.   23. 0.0173 $\mu\Omega$ m, 0.398 Ω.
24. 91.8°C.   25. 4.035 Ω.   26. 99.6°C.
27. 40 Ω; 6 V, 0.9 W; 0.8 W.   28. 1 A, 12 A, 13 A; 104 V.
29. 1.069 A, discharge; 0.1145 A, discharge; 94.66 V.
30. 5; if A is positive with respect to B, load current is from D to C.
31. 72.8 mA, 34.27 mA, 1.927 V.   32. 11.55 A, 3.86 A.
33. I_A, 0.183 A, charge; I_B, 5.726 A, discharge; I_C, 5.543 A, discharge; 108.55 V.
34. 0.32 A.   35. B, 2 volts above A.
36. 2.84 Ω, 1.45 Ω.   37. 20.6 mA from B to E.
38. 0.047 A.   39. 0.192 A.
40. 2.295 A, 0.53 A, 1.765 A, both batteries discharging; 4.875 A.
41. AB, 183.3 Ω; BC, 550 Ω; CA, 275 Ω.
42. A, 4.615 Ω; B, 12.31 Ω; C, 18.46 Ω.
43. $R_{AB} = 247.8 \, \Omega$, $R_{BC} = 318.6 \, \Omega$, $R_{CA} = 223 \, \Omega$; 80 Ω.
44. 244.4 V, 240 V, 238.8 V; 2.68 kW; 21.44 kWh.
45. 475 mm^2, 86 p.   46. 261.5 V, 247.6 V, 1.502 kW.
47. AB, 100 A; BC, 20 A; AC, 80 A; V_B, 241 V; V_C, 240.4 V.
48. 106.25 A, 46.25 A, 33.75 A, 63.75 A; 231.5 V, 229.65 V, 232.35 V.
49. 243.43 V, 239.93 V.

Exercises 2. Page 46

1. 60 N.   2. 50 A.   3. 0.333 T.   4. 0.452 N.
5. 173 N · m, 12.7 kW.   6. 0.667 T.
7. 10 m/s, 0.8 N, 0.48 J.   8. 0.375 T, 0.1406 N.

9. 12 N, 1.2 V, 120 W.   10. 2.33 mV.   11. 0.333 V.
12. 4.42 μV.   13. 37.5 V.   14. 0.32 V.
15. −33.3 V, 0, 100 V.
16. 2700 V, direction same as that of current.
17. 24 V, 224 V.   18. 5.6 V.   19. 1400 r/min.

Exercises 3. Page 84

1. 476 A, 1.19 × 10⁶ A/Wb, 1120.
2. 720 A/m, 663; 2400 A/m, 398.   3. 553 A, 147 μWb.
4. 1.5 × 10⁶ A/Wb, 144.
5. 4.5 A; 0.4375 × 10⁶ A/Wb, 306 A.
6. 121 A, 582.   7. 0.0825 A.   8. 1100 A.
9. 1220 A.   10. 5.65 A.   11. 1520 A.
12. 0.64 T, 358 A/m.
13. 308 A/m, 0.593 T; 616 A/m, 1.04 T; 924 A/m, 1.267 T; 1232 A/m,
    1.41 T; 1540 A/m, 1.49 T.
14. 1.22 T, 1620.   15. 1.23 × 10⁻⁹ C/division.
16. 1800 J, 1.4 T, 11.55 W at 50 Hz.
17. 2.63 W.   18. 0.333 W.
19. 1.11 J.   20. 337 W.
21. 0.37 T; −26 200 A/m.   22. 11.85 cm × 5 cm².
23. 7.95 cm × 0.38 cm².   24. 4.2 cm × 4.4 cm².
25. 9.55 cm × 3.12 cm².   26. 0.218 T.
27. 0.236 T.   28. 578 μV.   29. 0.007 64 J.
30. 1.37 A.   31. 13.13 A, 29.85 N.
32. 955 A.   33. 995 A/m, 1.25 mT.
34. 6000 A; 500 000 A/Wb; 2650. 35. 125 N/m.
36. A, 4780 A/m, 6 mT; B, 6370 A/m, 8 mT; C, 2390 A/m, 3 mT;
    A, O; B, 3185 A/m, 4 mT; C, 2390 A/m, 3 mT.

Exercises 4. Page 115

1. 0.375 H.   2. 0.15 H.   3. 160 A/s.   4. 3.5 V.
5. 1.25 mH, 0.1 V.
6. 530 A/m, 354 A/m, assuming circular cross-section; 2.66 A;
   18.8 μH.
7. 0.24 H, 0.06 H.   8. 1492, 47.1 mH, 1.884 V.
9. 157 μH, 94.2 mH.
10. 15 H; 300 V, assuming zero residual flux.
11. 0.96 H, 182.4 V.   12. 0.15 H, 750 V, 0.1 H.
13. 2.68 A, 13.32 mH, 7.14 V, assuming zero residual flux.
14. 0.09 H, 180 V.   15. 26.67 μH.
16. 100 μH, 6 mV, 0.527.   17. 2.23 mm, 0.125 H.
19. 25 V to 125 V during first second; 100 V during next second;
    87.5 V to −12.5 V during last 2 seconds.
20. 0.0833 H, 0.041 65 s, 1.667 V.   21. 20 A/s, 1.55, 0.23 s.
22. 20 A/s, 0.1 s, 1.5 A.
23. 0.316 A, 632 Ω, 1.896 H, 0.0946 J.
24. (a) 1 A, (b) 2.5 A, (c) 0, (d) 200 V; (a) 2.5 A, (b) 2.5 A;
    (c) 700 V, (d) 500 V.

25. 0.316 A, 155.5 V. 26. 40 A/s, 10 J.
27. 12 V. 28. 0.0625 H.
29. − 10 V. 30. 15.06 μH. 31. 1.35 mH.
32. 150 μC. 33. 740 μH.
34. 12.8 × 10^{-10} C, 10.65 μH. 35. 1.612 mH, 0.417.
36. 0.0275 J, 0.0175 J; 0.3535.
37. 32 mH, 48 mH, 160 mJ, 96 mJ.
38. 290 μH, 440 μH. 39. 0.6.

Exercises 5. Page 154

1. 150 V, 3 mC. 2. 45 μF, 4.615 μF.
3. 3.6 μF, 72 μC, 288 μJ. 4. 1200 μC; 120 V, 80 V; 6 μF.
5. 500 μC; 250 V, 166.7 V, 83.3 V; 0.0208 J.
6. 15 μF in series. 7. 2.77 μF, 18.46 μC.
8. 4.57 μF, 3.56 μF. 9. 3.6 μF, 5.4 μF.
10. 200 V; 1.2 mC, 2 mC, 3.2 mC on A, B and C respectively.
11. 360 V, 240 V; 40 μF, 0.8 J. 12. 267 V; 0.32 J, 0.213 J.
13. 150 V, 100 V, 600 μC; 120 V, 480 μC, 720 μC.
14. 62.5 V. 15. 120 V, 0.06 J, 0.036 J.
16. 590 pF, 0.354 μC, 200 kV/m, 8.85 μC/m^2.
17. 1593 pF, 0.796 μC, 167 kV/m, 8.85 μC/m^2.
18. 664 pF, 0.2656 μC, 100 kV/m, 4.43 μC/m^2.
19. 177 pF, 1062 pF, 33.3 V, 0.0354 μC.
20. 619.5 pF, 31 ms, 0.062 μC. 21. 2212 pF.
22. 1.416 mm.
23. 1290 pF; 1.82 kV/mm (paper), 0.453 kV/mm; 0.0161 J.
24. 0.245 m^2, 2.3 MV/m in airgap.
25. 8.58 kV/cm in airgap.
26. 20 μC, 15 μC; 1000 μJ, 2250 μJ; 35 μC; 140 V; 2450 μJ.
27. 8.8 × 10^{-12} F/m; 100 kV/m, 0.88 μC/m^2.
28. 2.81; 30 kV/m, 0.7425 μC/m^2; 0.4455 μJ.
29. 397 pF, 8.82 × 10^{-12} F/m. 30. 0.5 μC, 3.12 × 10^{12} electrons.
31. 3.744 × 10^{21} electrons/s. 32. 11.2 mA.
33. 115.2 J, 7.2 × 10^{20} electronvolts; 8.39 × 10^6 m/s.
34. 18.72 × 10^{15} electrons/s; 6.75 × 10^{20} electronvolts, 108 J;
    6.5 × 10^6 m/s.
35. 1.28 × 10^{-14} N, 6.41 × 10^{-17} J, 8.43 × 10^{-10} s.
36. 2.04 mm.
37. 16.4 kV/m, 1.155 × 10^7 m/s, 16.1°.
38. 5.93 × 10^6 m/s, 6.74 × 10^{-9} s, 10^{16} electrons/s.
39. 4 × 10^{-15} N.
41. 0.8 s, 125 V/s, 12.5 mA, 0.01 C, 0.5 J.
43. 18.4 μA, 4000 μJ. 44. 100 V, 0.1056 ms.
45. 50 mA, 500 mA, 1.25 J. 46. 1342 s.
47. 100 MΩ.
48. 39.7 m (approx.), if wound spirally.
49. 4.75 kV, 1 mJ, 0.4 N. 50. 3.75 μF, 1500 μC.
51. 300 V; 4 μJ, 16 μJ. 52. 2510 μJ/m^3.
53. 20.8 N/m^2, 20.8 J/m^3.
54. 15 930 pF, 3000 kV/m, 159.3 μC/m^2, 71.7 N.

Exercises 6. Page 175

1. 125.6 V, 33.3 Hz.
2. 20 Hz, 35.5 V, 38.5 V.
3. 28.27 sin 157t volts; 8.74 V.
4. 900 r/min.
5. 16 poles.
6. 60 Hz.
7. 52.2 V, average; 57.3 V, r.m.s.
8. 14.2 A, 16.4 A, 1.155.
9. 81.5 V, 1.087.
10. 5 V, 5.77 V, 1.154.
11. 30 A, r.m.s.; 20 A, average.
12. 10 V, 11.55 V, 1.155.
13. 0.816, none.
14. 0.637 A, 1.0 A.
15. 0.825 A, moving-coil ammeter; 1.296 A, moving-iron ammeter.
16. 2.61 kV, 3 kV.
17. 7.77 mA, 11 mA.
18. 5.06 A; 1.08, assuming ammeter to read r.m.s. value of sinusoidal current.
19. 500 sin (314t + 0.93) V, 10 sin (314t + 0.412) A, −300 V, −9.15 A, 29.5°, current lagging.
20. 150 Hz, 6.67 ms, 0.68 ms, 7.35 ms, 0.5 kWh.
21. 35 A.
22. 68 V, 48.1 V.
23. 55.9 A, 1.2° lead.
24. 23.1 sin (314t + 0.376) A, 16.3 A, 50 Hz.
25. 262 sin (314t + 0.41) V, 185.5 V, 50 Hz.
26. 121 V, 9.5° leading; 101.8 V, 101° leading.
27. 72 sin (ωt + 0.443).
28. 208 sin (ωt − 0.202), 11° 34′ lagging, 18° 26′ leading, 76 sin (ωt + 0.528).
29. 10.75 A, 8.5° lagging.

Exercises 7. Page 213

1. 125.6 V (max.), 1.256 A (max.), 6.03 N m (max.).
2. 500 W.
3. Zero.
4. 2.65 A, 0.159 A.
5. 6.23 A, 196 V, 156 V, 51.5°.
6. 21.65 Ω, 0.0398 H, 2165 W.
7. 25.12 Ω, 29.3 Ω, 8.19 A, 0.512, 1008 W.
8. 2.64 A, 69.7 W, 0.1147.
9. 0.0435 H, 16 Ω, 58.6°.
10. 170 V, 46°; 3.07 Ω, 0.06 H.
11. 254 Ω, 1.168 mH.
12. 5.34 A, 0.857.
13. 77.5 W, 5.5 Ω.
14. 41.4 A, 4.5°; 41.4 sin (314t − 0.078) A; 4.825 Ω, 4.81 Ω, 0.38 Ω.
15. 61.1 Ω, 0.229 H, 34° 18′.
16. 78.6 W, 0.0376 H, 0.716.
17. 174 V, 75.6 V; 109.5 W, 182.5 W; 0.47.
18. 0.129 H, 0.0818 H, 0.706 A or 1.72 A.
19. 49.6 mA, 0.0794; 189 mA, 0.303.
20. 9.95 A.
21. 133 sin (ωt − 0.202) V, 6.65 sin (ωt + 0.845) A.
22. 25.8 μF, 0.478 leading, 145.5 V.
23. 1.54 A, 1.884 A, 2.435 A, 50° 42′, 308 W, 0.633 leading.
24. 11.67 Ω, 144 μF; 14.55 A, 0.884 leading.
25. 10.3 A, 1060 W, 0.447 lagging, 103 V, 515 V, 309 V.
26. 7.27 A, 5.56 A, 468 V, 0.795 lagging.
27. 7.07 A, 0.707 leading, 5 A, 5 A, 500 var.
28. 10 A, 13.42 A, 5.62 A; 20.75 A, 0.95 lagging.
29. 19.75 W, 504 μF.
30. 10.28 A, 10.2 A, 13.75 A, 0.96 leading, 3160 W.
31. 2 A, 4 A, 2 A, 4.56 A, 15° 15′ lagging.
32. 127.4 μF, 56 A, 11.2 kW.
33. 8.5 A, 0.794, 13.53 Ω, 10.74 Ω, 8.25 Ω.
34. 6.05 A, 15.2° lagging.

35. 159 pF, 5 mA; 145 pF, 177 pF.
36. 920 Hz, 173 V.   37. 15.8.
38. 400 V, 401 V, 10 V, 0.5 A, 15 W.
39. 50 Hz, 12.95 Ω, 0.154 H, 207.5 W; 21.85 sin $(314t + \pi/4)$ A.
40. 31.5 V, 60 Ω, 0.167 H, 4220 pF.
41. 84.5 μH, 30 Ω, 17.7.   42. 57.2 Hz, 167 Ω.
43. 125 kΩ.   44. 76.5 pF, 0.87 MHz.
45. 28.7 μF.   46. 5.02 kVA.
47. 10.3 kVA, 0.97 leading; 20.6 kVA, 0.97 lagging.
48. 324 μF.
49. 17.35 A, 4140 W, 390 var, 0.996 lagging.
50. 59 kW, 75.5 kVA, 0.78 lagging; 47 kvar, in parallel.
51. 37.1 A, 0.371, 2750 W; 17.9 A, 0.895, 3200 W; 4000 W.
52. 106.7 W, 0.1433 H.   53. 1.592 A, 11.4 Ω.
54. 21.2 μF, 0.952 A.   55. 0.0126 rad.
56. 1.74 MΩ, 0.061 μF.

Exercises 8. Page 233

1. $10 + j0$, $10 \angle 0°$; $2.5 - j4.33$, $5 \angle -60°$; $34.64 + j20$, $40 \angle 30°$.
2. $20 + j0$, $0 + j40$, $26 - j15$, $5 - j8.66$, $51 + j16.34$; 53.5, $17° 46'$ lead.
3. $11.2 \angle 26° 34'$, $8.54 \angle -69° 27'$.
4. $10 + j17.32$, $28.28 - j28.28$.
5. $13 - j3$, $13.35 \angle -13°$.   6. $7 + j13$, $14.77 \angle 61° 42'$.
7. $38.28 - j10.96$, $39.8 \angle -16°$.
8. $-18.28 + j45.6$, $49.1 \angle 111° 48'$.
9. 50 Ω, 0.0956 H; 30 Ω, 63.7 μF; 76.6 Ω, 0.2045 H; 20 Ω, 92 μF.
10. $0.0308 - j0.0462$ S, $0.0555 \angle -56° 18'$ S; $0.04 + j0.02$ S, $0.0447 \angle 26° 34'$ S; $0.0188 - j0.006\,84$ S, $0.02 \angle -20°$ S; $0.0342 + j0.094$ S, $0.1 \angle 70°$ S.
11. $0.69 - j1.725$ Ω, $1.86 \angle -68° 12'$ Ω; $10.82 + j6.25$ Ω, $12.5 \angle 30°$ Ω.
12. 4 Ω, 1910 μF; 20 Ω, 31.85 mH; 1.44 Ω, 1274 μF; 3.11 Ω, 8.3 mH.
13. $91.8 + j53$ V, $2.53 - j1.32$ A; $106 \angle 30°$ V, $2.85 \angle 27° 30'$ A.
14. $0.104 + j2.82$ A; $2.82 \angle 87° 53'$ A.
15. 25 Ω, 50 Ω.
16. 24 Ω, 57.3 mH; 12 Ω, 145 μF; $0.0482 \angle 18° 10'$ S, $18° 10'$, current leading.
17. 7.64 A, 8° 36' lagging.
18. $13.7 - j3.2$ Ω, capacitive; 2250 W; 13° 9'.
19. $1.192 - j0.538$ A.
20. 20 Ω, 47.8 mH; 10 Ω, 53 μF; $0.0347 - j0.0078$ S, 12° 40' lagging.
21. 2.83 A, 30° lagging.
22. 2.08 A, lagging 34° 18', from supply; 4.64 A, lagging 59°, through coil; 2.88 A, leading 103° 30', through capacitor.
23. 18.75 A, 0.839 lagging.   24. 21.33 Ω, 93.3 mH.
25. 0.68 μF.   26. 174 Ω, 80.1 mH.
27. 940 W; 550 var, lagging.   28. 1060 W; 250 var, leading.
29. 4 kVA, 3.46 kW; 8 kVA, 8 kW; 11.64 kVA, 11.46 kW; $44.64 + j37.32$ A.

Exercises 9. Page 256

2. 25, 25, 625.
3. 488, 130 000, 63 400 000, 156 000.　　　4. 470.
5. 750, 2000, 8000, 250 Ω, 40 mA.
6. 20.6, 38.　　　　7. 0.63 V.　　　　8. 20 dB.
9. 0.775 V, 2.45 V, 0.245 V.
10. 54 dB, 115 dB, 64.5 dB, 9 W.
11. 2.4 V, 75.8 dB.　　　　12. 1.39 kΩ, 1610.
13. 4.50 µV, 72 µV, 33.7 dB.　　　14. −0.013, 26.1.
15. 41.5.　　　　16. 36.　　　　17. 97, −0.0916.
18. 108.8, 28.26, 35.86, 40.7 dB, 29 dB, 31.1 dB.
19. −0.0071, −30.1 dB.　　　20. 4.87, 13.75 dB.
21. 97.8, 101.9.　　　　22. 51.3, 20 000, 5.3 per cent.
23. −0.053, 37.3 mV.

Exercises 10. Page 282

11. 0 mA, 20 mA, 220 mA.　　　12. 0 mA, 7 mA, 63 mA.
13. 800 Hz.　　　　14. 8.1 A, 1.7 Ω.
15. 354 mA, 112 mA, 177 mA, 31.3 W.
16. 26 Ω, 833 V, 0.78.　　　18. 250 Ω, 1.54 W.
20. 0.35 A, 0.59 A.　　　21. 1.6 Ω, 0.72.
22. 2.6 Ω, 13.8 A.　　　23. 40 Ω, 0.07.
24. 22.2 Ω, 100 Ω, about 20–70 mA, 11–17 V, 0.17.

Exercises 11. Page 311

10. 74 µA; 32.0 dB; 33.6 dB.　　11. 2.06 mV; 1290; 27.5.
12. 25 Ω; 10 kΩ.　　　13. 4.9 mA; 8.6 V; 3.3 V; 21.
14. 4.6 V; 1.9 mA; 0.9 mA; 45.　　15. 113 kΩ; 1.1 V; 1.2 mW.
16. 30; 30; 900.
17. 36.9; 26.7; 178; 123; 7200; 3280.
18. 63.6; 42.4; 2700.　　　19. 50; 250; 12 500; 10 mW.
20. 41.7; 20.9; 870.　　　21. 1.046 V; 5960; 37.8 dB.
22. 0.33 V.　　　23. 2.07 mA; 3.74 V; 1390; 31.4 dB.
24. 476; 1455; 693 000.　　　25. 0.32 mA; 0.23 mA; 0.46 V; 8.9.
26. 0.95 V; 0°; 60.　　　27. −4.26 dB.
28. 1.33 V; 44.2 dB.　　　29. 80; 320; 44.1 dB.
30. 101; 40.5; 4100.　　　31. 81; 48; 2300.
34. 220 kΩ; 2.5 mA/V.
35. 50 kΩ; 2.4 mA/V; 120; for G, 8 mA; 10 V; −0.49 V.
36. 95; 57.　　　37. 1.19 V; 2.85 V.
38. 5 M; 800.　　　39. 2 mA/V.
40. 3.66 nF; 11.　　　41. 389 mV.
42. 200 Ω; 36.4 V.　　　43. 24.2 V.

Exercises 12. Page 331

1. 3750; 20 000.
2. 1200; 288 000.
3. 33.3; 33.3; 1110; 1 230 000.
4. 150; 4.
5. 100 mV; 1 V; 0.16 μW.
6. 0.125 mV; 7.94 mV; 26 pW.
7. 60 dB; 58.7 dB.
9. 9 V.
11. 1 MΩ.
12. 2.8 V.
13. 33.3 kΩ.
14. 0.5 sin ωt; 2.12 sin ($\omega t + \pi/4$).
15. 6.44; 7.44.
16. -34.2 V.
17. 0.9 mV.

Exercises 13. Page 347

1. 1; A + B; A + B.
2. A + B; 0; A.
3. A + B; A . B; A + B . $\bar{C}$.
4. 0; 1; 1.
6. A . B . C.
7. $\bar{A}$. $\bar{B}$ + C.
8. A.
9. $\bar{B}$; $\bar{A}$. B + $\bar{B}$. C; $\bar{B}$. $\bar{C}$ + $\bar{A}$. B . C; no simplification; B . $\bar{D}$ + $\bar{B}$. D.
12. B . (A . $\bar{C}$ + $\bar{A}$. C).
14.

A	B	C	F
0	0	0	0
0	0	1	0
0	1	0	1
0	1	1	1
1	0	0	0
1	0	1	1
1	1	0	0
1	1	1	1.

15. A . (B + C); $\bar{A}$. B . C; A . $\bar{B}$ + $\bar{A}$. B.
16. A . C + $\bar{A}$. $\bar{C}$ + B.
17. $\bar{A}$. B . $\bar{C}$.

Exercises 14. Page 370

Many of the exercises permit alternative programs, hence only a few are illustrated below.

```
 4. LDA  0260
    ADD  0261
    STA  0262
    BRK
14. LDA  0266
    STA  0267
    SHL  0267
    SHL  0267
    LDA  0267
    ADD  0266
    STA  0268
    BRK
```

```
16. LDA  #1E
    STA  0290
    ORA  0293
    STA  0292
    BRK
```

Exercises 15. Page 392

1. 240 A, 498 V; 720 A, 166 V; 119.5 kW.
2. 120 A, 301 V, 36.12 kW; 240 A, 150.5 V, 36.12 kW.
3. 432 V.   4. 0.02 Wb.   5. 946 r/min.   6. 0.0274 Wb.
7. 1065 r/min.   8. 140 V.   9. 0.0192 Wb.   10. 0.0333 Wb.
11. 200 At, 3400 At.   14. 4440 At, 555 At; 0.4625 A.
15. 8.   16. 3150 At.   17. 0.443 Ω.

Exercises 16. Page 410

1. 589 r/min.   2. 714 r/min.   3. 190 V.
4. 468 V, 1075 r/min, 333 N·m.   5. 140 A.
6. 208 V, 41 A.   7. 439 r/min; 400 N·m.
8. 730 r/min, 226 N·m.   9. 13.85 mWb, 114.7 N·m.
10. 36.2 N·m.   11. 35 Ω, 962 r/min.   12. 5.8 Ω, 60 V.
13. 1 A, 2005 r/min; 2 A, 1150 r/min; 3 A, 958 r/min, etc.
14. 391 r/min.
15. 5 A, 445 rad/s; 15 A, 148 rad/s; 25 A, 101 rad/s, etc.
16. 5.5 Ω, 23.5 Ω.   17. 356 N·m.
18. 210.3 V, 0.0127 Wb, 114.3 N·m.
19. 595 r/min.   20. 40 A, 695 r/min.   21. 333 N·m.
22. 5 A, 2090 r/min; 10 A, 1035 r/min; 15 A, 738 r/min, etc.
23. 1.91 Ω, 80 V.   24. 2.767 Ω, 76.7 V.
25. 18.6 rad/s (or 178 r/min).   26. 917 r/min, 2560 N·m.
27. 40 Ω.   28. 0.94 Ω, 932 r/min.
29. 6.53 Ω.   30. 3.05 Ω, 5.67 Ω, 9.2 Ω.

Exercises 17. Page 421

1. 0.0219 Wb.   2. 8, 0.0321 Wb.   3. 0.123 Ω, 534.5 V.
4. 282.66 V.   5. 467 V, 373 V, 0, 240 Ω.
6. 400 V, 240 Ω.   7. 71 Ω, 83 V.
8. 258 V, 100 A (approx).   9. 6, 0.0296 Ω.

Exercises 18. Page 429

1. 800 W.   2. 1.87 A, 766 r/min, 10.3 A.
3. 6.49 kW, 0.843 p.u.   4. 4.61 kW, 0.835 p.u.
5. 0.8925 p.u.
6. 6.9 kW, 0.857 p.u., 0.871 p.u., 9.24 kW.

Exercises 19. Page 448

1. 4240 V, 70.7 A.   2. 80.6 A.
3. 6.18 A, 5.05 A, 9.87 A, 325 V.   4. 246 V; 346 V, 200 V, 200 V.
5. 11.56 kW, 34.7 kW.   6. 96.2 A, 55.6 A.
7. 8.8 A, 416 V, 15.25 A, 5810 W.
8. 4.26 Ω, 0.0665 H, 18.75 A, 32.45 A.
9. 1.525 A; 2.64 A, 2.64 A, 4.57 A; 1.32 A, 1.32 A, 0; 210 V.
10. 61.7 A, 0.858, 38 kW.   11. 90.8 A, 240 V.
12. 72.3 μF/ph.
13. 3.81 kvar, 70.5 μF, 23.5 μF, 0.848 lagging.
14. 9.24 A, 5120 W, 0.8 lagging, 14.3 μF, 7.78 A.
15. 115 Ω, 0.274 H, 1.667 A, 956 W; 1.445 A, 1.445 A, 0.
16. 462 V, 267 V, 16.7 kvar, 5.87 Ω, 8.92 Ω, 75 A, 33 kW.
17. 43.3 A, 26 A, 17.3 A, 22.9 A.   18. 17.3 A, 31.2 kW.
19. 4 A, 6.67 A, 3.08 A; 6.84 A in R, 10.33 A in Y, 5.79 A in B.
20. 21.6 A in R, 49.6 A in Y, 43.5 A in B.
21. 1.1 kW, 0.715.   22. 8 kW.
23. 0.815.   24. 13.1 kW, 1.71 kW.
25. 21.8 A, 36° 35′ lagging; 7 kW; 13 kW.
26. 0.655, 0.359.   27. 15 kW, 0.327.
29. 11.7 kW, 5.33 kW.
30. Line amperes × line volts × sin ϕ = 8570 var.

Exercises 20. Page 480

1. 462, 28, 42 800 mm².   2. 500, 690 V, 4450 mm².
3. 475, 18.   4. 1200 V, 0.675 T.
5. 146.7 V; 22.73 A, 341 A; 0.033 Wb.
6. 60.6 A, 833 A, 0.0135 Wb, 1100 turns.
7. 1.57 kVA, 64 turns.   8. 14, 22, 3500, 805 (primary).
9. 25.9 A.   10. 7 A, 0.735 lagging.
11. 4.43 A, 1.0.
12. 1016 V; 32.9 A, secondary; 8.225 A, 14.25 A, primary.
13. 10. 87.   14. 317.5 V, 952.5 V.
15. Primary and secondary phase currents, 65.8 A; 38.06 A in
    resistors; 114 A from supply; 65.8 A to load; 87 kW; 29 kW.
16. 5.18 mWb, 345 W, 4.77 A (r.m.s.).
17. 300, 1.44 mm, 88°.   18. 2.3 A, 0.164.
19. 176.5 V.   20. 900 V.
21. 377.6 V.   22. 4 per cent.
23. 2.13 per cent, 225.1 V; 4.22 per cent, 220.3 V; −0.81 per cent,
    231.86 V; 0.960 p.u., 0.9585 p.u.; 56.25 kVA.
24. 2.54 per cent.   25. 0.892 p.u.   26. 0.981 p.u.
27. 0.9626 p.u.   28. 0.905 p.u.
29. 0.9598 p.u., 8.65 kVA, 0.9601 p.u. at unity power factor.
30. 0.9868 p.u., 0.9808 p.u.
31. 0.961 p.u., 0.9628 p.u., 23.3 kW.
32. A, 93.77, 96.24, 96.97 per cent; B, 96.42, 97.34, 96.97 per cent;
    A, 40 kVA; B, 23.1 kVA.
33. A, 0.9553 p.u.; B, 0.9682 p.u.; £3.15.
34. 0.9225 p.u.   35. 44 W, 10 W.
36. 1600 W, 800 W.   37. 362.8 W.

Exercises 21. Page 494

1. $0.55\ \mu\text{N} \cdot \text{m/degree}$.
2. $33.1°$.
3. $1.91\,\text{H}$, $1.01\,\text{H}$.
4. $1.23\,\text{H}$, $295\,\text{N} \cdot \text{m}$, $18.5\,\text{kW}$.
5. $1.11\,\text{H}$, $239\,\text{V}$.
6. $0.6\,\text{A}$, $181.5\,\text{V}$, $2605\,\text{r/min}$, $109\,\text{W}$, 0.883 lagging, 85.8 per cent.

Exercises 22. Page 514

1. $3600\,\text{r/min}$, $0.0227\,\text{T}$.
4. 1.195.
5. 1.24.
6. $4.19\,\text{N} \cdot \text{m}$.
7. 0.958.
8. $2560\,\text{V}$.
9. 0.960, $503\,\text{V}$.
10. 8 poles, $32.3\,\text{mWb}$.
11. 8.
12. $3.54\,\text{kV}$.
13. $818\,\text{V}$.
14. 0.831, $1600\,\text{V}$.
15. 0.966.
16. $20.5\,\text{V}$.
17. $0.0862\,\text{Wb}$, $525\,\text{A}$.
18. $6.4\,\text{kW}$.
19. $500\,\text{r/min}$.
20. 4 poles.
21. $150\,\text{Hz}$.

Exercises 23. Page 533

1. 4.4 per cent, 14.0 per cent, -7.3 per cent.
2. $2412\,\text{V}$.
3. $1585\,\text{V}$.
4. 12.5 per cent increase.
5. $39.9\,\text{A}$, $3290\,\text{V}$, $131.3\,\text{kW}$.
6. $111\,\text{A}$, $1845\,\text{V}$, $615\,\text{kW}$.
7. $71.1\,\text{A}$, $1920\,\text{V/ph}$.
8. $2212\,\text{V}$, $2040\,\text{V}$, $2380\,\text{V}$.
9. $234\,\text{V}$, $198.5\,\text{V}$, $268.5\,\text{V}$.
10. $206\,\text{V}$, $14°\,2'$; $9.8\,\text{A}$, 0.913 lagging.
11. $500\,\text{r/min}$, $1165\,\text{N} \cdot \text{m}$.
12. $1200\,\text{kW}$, 0.3875 leading.
13. $289\,\text{kVA}$, 0.692 leading.
14. $74.8\,\text{A}$, 0.995 lagging.
15. $0.225\,\text{H}$, $223\,\text{N} \cdot \text{m}$.
16. $16.55\,\text{A}$.

Exercises 24. Page 549

1. $2880\,\text{r/min}$, $3000\,\text{r/min}$.
2. $1470\,\text{r/min}$, $1\,\text{Hz}$.
4. 4.6 per cent, $954\,\text{r/min}$.
5. $1.585\,\text{Hz}$, 3.17 per cent.
6. $971.7\,\text{r/min}$.
7. 0.960, $9\,\text{mWb}$.
8. $11.33\,\text{A}$; 0.2, $40.8\,\text{A}$.
9. $10.67\,\text{A}$, 0.376; $21.95\,\text{A}$, 0.303.
10. $0.07\ \Omega$, 37.5 per cent.
11. $960\,\text{r/min}$, $855\,\text{W}$.
12. 27.5 per cent, 41 per cent.
13. $11\,\text{kW}$, $440\,\text{W}$.
14. 0.857 p.u., $25.57\,\text{kW}$.
15. (i) $1:0.333:0.16$; (ii) $1:0.333:0.16$.

Exercises 25. Page 557

3. 0.432 A; 0.344 A.   4. 0.480 A; 0.418 A.

5. 81°; 96.6°.

Exercises 26. Page 588

1. 0.0378 Ω; 6662 Ω.   2. 5005 μΩ; 995 Ω.
3. 8.325 mm.   4. 19.985 kΩ; 0.0378 Ω.
5. 0.015 04 Ω; 9.995 kΩ; 0.231 A.
6. 0.003 753 Ω, 16.662 kΩ; −1.2 per cent, −0.000 36 per cent.
7. 810 mN m.   8. 30.4 μN m.
9. 25.6 mA.   10. 10.
11. 25.   12. 0.72 per cent; 36 per cent.
13. 0.9; 1.036.   14. 1785 Ω.
15. 33.8 kΩ.   16. 8.13 Ω; 1.6 per cent low.
17. 60.67 Ω; 9.9 per cent high.   18. 1087.5 Ω.
19. 233.5 kΩ.   20. 776 Ω.
21. 5000 Ω; 21.45 mA.   22. 90 m.
23. 3.08 mA from D to B.   24. 2.17 per cent high.
25. 1.46 V high.   26. 1.486 V; 1.09 Ω.
27. 4.17 per cent high.   28. 2.04 per cent high.
29. 2.98 per cent low.   30. 1.24 per cent high.
31. 0.41 mA from Z to X.   32. 6 kΩ, 15 kΩ, 60 per cent.
33. 6 kΩ, 5.99 kΩ, 1.69 per cent.   34. 27.3 V.
35. Outwith limit.   36. 6.67 A; 734 W; 4.72 A; 519 W.
37. (d).   38. (g).
39. 2.26 V; 1.81 V; 2.71 V; 19.9 per cent; 19.9 per cent.

Exercises 27. Page 614

3. 23.3 V.   5. 60 V; 38 ms; 26.3 Hz.
6. 137°.   7. 600 Hz; 20 V; 134°.
8. Fundamental plus even harmonics.
9. 87 mm; 9.9°.
10. 22.7×10^6 m/s; 2.18×10^{-15} N.
11. 37.5×10^6 m/s; 640 V.   12. 1.74 mT.

Exercises 28. Page 637

1. 11.6 lm/W; 89.2 cd; 0.923 cd/W.
2. 132.6 cd; 18.92 p.   3. 76 500 cd/m².
4. 6130 cd/m².   5. 99.5 lx.
6. 122.5 cd.
7. 39.06 cd; standard lamp, 1.16 m; test lamp, 1.34 m.
8. 25 lx; 16.67 lx; 0.9 m from 16-cd lamp.

9. 50 lx; 41.9 lx.
10. 10.82 m; 2340 cd.
11. 11.6 lx; 13.1 lx.
12. 9.4 m.
13. 155.5 W; 55.7 lx.
14. 37.3 lx at A; about 38.7 lx at B.
15. 8 lx; 2.87 lx; 76.8 lx.
16. 36.
17. 1550 cd.
18. 60 lamps.
19. 21.5 per cent; -18.5 per cent.
20. 0.606 A, 122.2 V, 97.8 V; 152.3 per cent; 62.5 per cent.
21. 236 cd, 2960 lm.
22. 1.0 V.

Index